D0163108

Introduction to the Mathematics of Operations Research with Mathematica®

Second Edition

PURE AND APPLIED MATHEMATICS

A Program of Monographs, Textbooks, and Lecture Notes

MONOGRAPHS AND TEXTBOOKS IN PURE AND APPLIED MATHEMATICS

Recent Titles

F. W. Steutel and K. van Harn, Infinite Divisibility of Probability Distributions on the Real Line (2004)

G. S. Ladde and M. Sambandham, Stochastic versus Deterministic Systems of Differential Equations (2004)

B. J. Gardner and R. Wiegandt, Radical Theory of Rings (2004)

J. Haluska, The Mathematical Theory of Tone Systems (2004)

C. Menini and F. Van Oystaeyen, Abstract Algebra: A Comprehensive Treatment (2004)

E. Hansen and G. W. Walster, Global Optimization Using Interval Analysis, Second Edition, Revised and Expanded (2004)

M. M. Rao, Measure Theory and Integration, Second Edition, Revised and Expanded (2004)

W. J. Wickless, A First Graduate Course in Abstract Algebra (2004)

R. P. Agarwal, M. Bohner, and W-T Li, Nonoscillation and Oscillation Theory for Functional Differential Equations (2004)

J. Galambos and I. Simonelli, Products of Random Variables: Applications to Problems of Physics and to Arithmetical Functions (2004)

Walter Ferrer and Alvaro Rittatore, Actions and Invariants of Algebraic Groups (2005)

Christof Eck, Jiri Jarusek, and Miroslav Krbec, Unilateral Contact Problems: Variational Methods and Existence Theorems (2005)

M. M. Rao, Conditional Measures and Applications, Second Edition (2005)

A. B. Kharazishvili, Strange Functions in Real Analysis, Second Edition (2006)

Vincenzo Ancona and Bernard Gaveau, Differential Forms on Singular Varieties: De Rham and Hodge Theory Simplified (2005)

Santiago Alves Tavares, Generation of Multivariate Hermite Interpolating Polynomials (2005)

Sergio Macías, Topics on Continua (2005)

Mircea Sofonea, Weimin Han, and Meir Shillor, Analysis and Approximation of Contact Problems with Adhesion or Damage (2006)

Marwan Moubachir and Jean-Paul Zolésio, Moving Shape Analysis and Control: Applications to Fluid Structure Interactions (2006)

Alfred Geroldinger and Franz Halter-Koch, Non-Unique Factorizations: Algebraic, Combinatorial and Analytic Theory (2006)

Kevin J. Hastings, Introduction to the Mathematics of Operations Research with *Mathematica*®, Second Edition (2006)

Introduction to the Mathematics of Operations Research with Mathematica®

Second Edition

Kevin J. Hastings

Knox College
Galesburg, Illinois, U.S.A.

Chapman & Hall/CRC
Taylor & Francis Group
Boca Raton   London   New York

Chapman & Hall/CRC is an imprint of the
Taylor & Francis Group, an informa business

Published in 2006 by
CRC Press
Taylor & Francis Group
6000 Broken Sound Parkway NW, Suite 300
Boca Raton, FL 33487-2742

© 2006 by Taylor & Francis Group, LLC
CRC Press is an imprint of Taylor & Francis Group

No claim to original U.S. Government works
Printed in the United States of America on acid-free paper
10 9 8 7 6 5 4 3 2 1

International Standard Book Number-10: 1-57444-612-6 (Hardcover)
International Standard Book Number-13: 978-1-57444-612-8 (Hardcover)

Taylor & Francis Group
is the Academic Division of Informa plc.

Visit the Taylor & Francis Web site at
http://www.taylorandfrancis.com

and the CRC Press Web site at
http://www.crcpress.com

To my wife Gay Lynn, without whose patience and advice during the course of a never-ending stream of books, I would never be able to live as a complete human being.

PREFACE

In the time that has elapsed since the first edition of the book, titled *Introduction to the Mathematics of Operations Research*, was published in 1989, changes have occurred in the discipline of Operations Research. The field is in the midst of a crisis, partly a result of unnecessarily poor image and partly because of real problems. The result is that its members question the future of operations research. Meanwhile, better, faster, and more widely available technology has made its way into the workplace of the O.R. professional and into the mathematics curriculum, and lively discussion has taken place about pedagogy, especially revolving around the passivity of many students and the need to get them more actively involved with their courses. I saw that the time had certainly come to revisit the first edition of this book in an attempt to attune the book to current circumstances.

The first edition sprang from the following observations (paraphrased from its preface). In industry, problems involving such areas as telecommunications, scheduling, inventory, production, transportation, and finance abound. Besides the inherent interest of these problems, there is also aesthetic beauty in the mathematics. Operations Research is both an assemblage of descriptive and analytical techniques to facilitate decision making in business and industry, and a way of approaching problems. There are concrete questions such as: what is the best way to schedule servers at a service facility, what is the best mix of several kinds of products using scarce raw materials, and how does one best maintain a machine that is deteriorating with time? But looked at as a problem-solving approach, O.R. involves defining and modeling the problem precisely, with enough detail to capture its essence without making the problem intractable; deciding on objectives; coming up with a solution, often an algorithm to improve a current configuration; implementing that algorithm; and finally observing the consequences of the answer. Much as in computer programming, the solution process is often a cycle in which the researcher goes back to the beginning to refine the model, the objectives, or the algorithm one or more times. The point of view taken by the first edition of this book was that the vast assortment of apparently unrelated questions in the field of O.R. is unified by the common features of the mathematical models used to describe them, and the way of going about solving problems. So the text was designed to show the mathematics that underlies the applied problems, and subsequently to show the "real-world" problems as examples of the application of the mathematical and algorithmic thinking that will live on indefinitely as the passage of time changes the kinds of problems that capture the attention of practitioners.

Also, it is as true now as it was fifteen years ago that there is a general shortage of faculty experienced in O.R, especially at the small university and private college level.   The breadth of Operations Research and the corresponding voluminous nature of most sources add to the difficulty of course design for the non-specialist.   I wanted a concise book whose focus was on the mathematics of Operations Research, which would be a more suitable introduction to the subject in a mathematical sciences department with limited resources than other texts might be.

There is much in the first edition that remains meaningful, and which validates the approach, given the criticism that has been leveled from inside and outside of the field of Operations Research. One hears that Operations Research groups are being phased out in many organizations because the groups are not worth the investment. Criticisms usually include: our O.R. people are trained to execute a few algorithms under stringent assumptions but when it comes to an actual messy problem that does not fit a stereotype they are lost; or, O.R. people prefer to do their esoteric research on some little corner of the field about which only a few people really care. But my first book took the point of view that a student of Operations Research cannot and should not simply step through every method for every problem in every application area without a feel for the core of the field or an understanding of the complete problem-solving process. I believe that the subject is still vital, useful, and an excellent part of an undergraduate mathematics major because it gives deeper perspective on mathematics and its use, it exposes students to mathematical modeling in situations grounded in reality, and, done correctly, it greatly enhances their general reasoning and problem-solving ability. Even if the phrase "Operations Research" dies out and even if O.R. departments disappear, these kinds of skills will always be valuable to organizations in the private and public sectors. And at least a few specific topics will always occupy an important position in applied mathematics: representations of problems using graphs, optimization of linear functions subject to linear constraints, modeling and prediction of random events occurring through time, and the optimal control of such random events. This is, and will remain, the governing structure of the book: Graph Theory, Linear Programming, Stochastic Processes, and Dynamic Programming.

The challenge in producing a new edition was to retain the character of the book, yet take into account new developments in the spheres of mathematics pedagogy and the field of O.R. In keeping with the comments above, the following are the main areas in which the second edition differs from the first:

1. The book is more interactive. Self-check questions, and suggestions to investigate the material further are interspersed in the development.

2. Technology is smoothly integrated into the development in such a way as to expose new issues and possibilities, enhance students' desire to experiment, and drastically reduce computational burden.

3. The problem sets emphasize problem solving even more. Longer projects are included that do not fit into existing molds, for which the students must develop their own techniques.

4. A few new topics are included for more breadth: the traveling salesman problem and other famous graph theory problems are introduced briefly in a new section of Chapter 1, simulation has been integrated into Chapters 4 and 5, and a treatment of Brownian motion has been appended to Chapter 5, which permits examples of problems in the growing field of mathematical finance to be presented.

5. The review of topics from probability has been moved to an appendix, so as not to interrupt the flow unnecessarily. Students taking this course ought to have a course in probability as a prerequisite anyway.

6. Not the least important, answers to selected exercises are in another appendix. Publication timing problems in the first edition prevented them from being included there.

The integration of technology requires special discussion. At the time I wrote the first edition there were lots of programs to execute the simplex algorithm for linear programming, and a few others for other kinds of special problems, but there was no common environment for doing operations research, from pictorial representation, to symbolic derivation, to computation, to technical typesetting of reports. Since then there have arisen such environments. In fact, it has become possible to have an electronic, fully executable version of the printed text with which the students can interact directly; in short, a living textbook. While there are several possible symbolic algebra-graphical packages that can suffice, and countless other very powerful and very specialized professional programs, I prefer the one that I think will be left standing after intense competition: *Mathematica*. This package is extremely general, and more importantly, programmable, and with the advent of its most recent versions (3.0 and higher) it provides the ability for students to create professionally typeset mathematical documents with text integrated with computation. *Mathematica* already has facilities to support much of the material in the book, and what it does not have directly is easily programmable. I have found that it helps to teach the meaning of the simplex algorithm very well, and greatly simplifies the burdensome computations in graph theory and dynamic programming. Its simulation capability is quite good because it provides simple tools that students can adapt, and in the process learn more about model building and better understand the system they are trying to simulate. I have also found students doing significantly higher quality work when asked to turn in typeset *Mathematica* notebooks than they do by hand. Perhaps the professional appearance of their product gives them more of a sense of pride in it, which induces them to do even better work next time. In fact, the program is such an integral part of this second edition that the title has been modified to: *Mathematics of Operations Research with Mathematica.* This is a completely self-contained printed text, accompanied by an electronic version, together with a package of useful commands that I have written. The electronic version is in the form of *Mathematica* notebooks, one per section, and all *Mathematica* input cells will be live, so that the students can reexecute commands, edit them, devise new ones, etc. In this way, the student can direct his or her own study, which increases greatly the level of involvement, and one hopes, the level of comprehension and problem-solving.

Here are a few of the ways in which *Mathematica* has significant impact on the book:

1. A *Mathematica* tool for drawing labeled graphs allows students to redraw graphs in graph algorithms conveniently.

2. Students can experiment with large powers of adjacency matrices of large graphs to verify the theorem about path counting in Chapter 1, and to check regularity of Markov chains in Chapter 5.

3. Students are asked to implement some algorithms in *Mathematica*, which forces more thorough understanding.

4. Students can make good use of *Mathematica*'s equation-solving tools to construct feasible regions of linear programming problems in Chapters 2 and 3, and to use the "dictionary" method to solve them without headaches, and yet with understanding of how the method proceeds from step to step.

5. In Chapters 4 and 5, students can write simulators in *Mathematica* for processes such as Markov chains, Poisson processes, and Brownian motions, not only to observe their properties, but also to aid their understanding of the defining conditions of those processes.

6. Naturally recursive problems such as first passage times and absorption probabilities can be solved recursively in *Mathematica*.

7. Theoretically simple but tedious probabilistic computations regarding Poisson processes and queues in Chapter 6 are made easier to carry out using *Mathematica*'s distribution tools.

8. *Mathematica*'s symbolic algebra ability can be used to greatly simplify the task of solving dynamic programming problems, permitting longer time horizons and larger state and action spaces to be used, and focusing attention back on the modeling aspect of such problems where it belongs.

9. In general, the shift in emphasis from hand to computer computations facilitates examination of sensitivity of solutions to parameter changes.

It remains true that Operations Research is an endless source of interesting problems, which has never failed in my experience to stimulate the students of mathematics who I have taught, and to open their eyes to ideas and applications that they never before imagined. My wish continues to be that students take this book as a jumping-off point to further work in Operations Research or related areas such as Statistics, Management, Applied Mathematics, or Finance, as many of my students have done.

Finally, I would like to think the staff at Taylor & Francis publishing, including Kevin Sequeira and Fred Coppersmith, for all their help in bringing this project to fruition.

Kevin J. Hastings
Knox College
August 31, 2005

Note on *Mathematica* Packages and Electronic Book

This is a book that exists not only in the print medium but also electronically. The CD that accompanies the print version contains *Mathematica* notebooks, one per section, which together contain all of the material in the book and which should run quite well in *Mathematica* versions 5.0 or later, and perhaps (with no guarantees) in earlier versions. It also contains special packages that I have written with commands to support the book. To use them, simply make a new folder called KnoxOR in the AddOns/ExtraPackages directory of your *Mathematica* folder, and copy into it from the CD the files Graphs.m, LinearProgramming.m, StochasticProcesses.m, and DynamicProgramming.m.

When you boot up *Mathematica* and open one of the notebook files, you will notice that the output cells are not included; but if you select the Kernel menu command to execute all initialization cells, then the output that is contained in the printed text should be regenerated automatically. Some graphics in GraphicsArray cells will need to be resized to look well, and in general graphics would need to be sized and centered in order to look precisely like those in the printed text. The manufacturer of *Mathematica*, Wolfram Research, has made some changes since I first started this edition of the book and wrote the packages, including relocating some of its commands that my packages call on into different packages, and they may do so again in the future. So far, these path problems have not affected the notebooks so badly that any commands would not run, although warning messages are generated. In particular, the notebooks that use the StochasticProcesses.m package produce shadowing warnings relative to the names Type, Distribution, Absolute, and Relative. I decided to leave things as they were so that the notebooks would run on earlier versions of *Mathematica*, but if problems develop, you are encouraged to look using a text editor at the four ".m" packages near the top of the file to see what *Mathematica* packages are being loaded in, and correct the names of those packages as the warning messages indicate.

In its most recent versions, *Mathematica* has come up with a more refined ShowGraph command in its DiscreteMath`Combinatorica` package, which probably outshines the DisplayGraph command in my KnoxOR`Graphs` package. This change also took place as I was writing. But instead of rewriting the whole text I decided to stay with my own version, which is somewhat more attuned to what I wanted to use it for anyway. You might want to experiment with ShowGraph yourself.

Finally, bear in mind that the usual copyright privileges apply to the electronic version; you should no sooner share the notebook files with others than allow others to duplicate the printed text.

Contents

Graph Theory and Network Analysis

Introduction

In this chapter we are concerned with problems of optimization on a network of points connected by weighted edges. To illustrate one such problem, suppose that in Figure 1.1 the points represent stations among which communication is to be maintained. The weight of a line segment, or *edge*, is the cost of direct communication between the two stations connected by the edge. It might happen that it is impossible for a pair of stations to *directly* communicate, so that there may not be an edge between every pair. It is not even desirable for all stations to be linked directly to all others, as long as each station can reach each other station through one or more intermediaries. The problem is to find a set of edges of minimum cost that does not break communication between any pair of stations in the network.

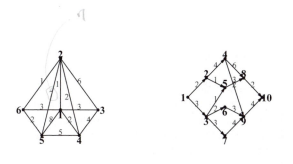

Figure 1.1 – Finding a sparse network   **Figure 1.2** – Project completion

A second type of problem involving networks is that of finding paths of maximal weight. In Figure 1.2, suppose that the edges represent tasks and the weight of an edge is the time required to finish the task. Some tasks may require the completion of a previous task before they can begin. For this reason we give a direction to edges; if task x points to a node on the graph, and task y points away from the same node, then x must be completed before y. For instance, both of the tasks represented by (2,5) and (3,5) must be done before the task represented by (5,8). Task (1,2) requires two time units, (1,3) requires three, etc. The problem is to find a path from node 1 to node 10 with maximum weight. This sequence of tasks will be such that if there is an unexpected delay at any stage, then the entire project will be delayed.

Another problem of network theory is exemplified by the assignment of jobs to workers in such a way as to maximize the total effectiveness of all workers at the jobs assigned to them. In the graph of Figure 1.3, the nodes on the left are workers, those on the right are jobs, and the weights on the edges are measures of effectiveness. For instance, worker 1 has a rating of 4 at job 4. We will find a way of matching workers uniquely to jobs to solve the maximization problem mentioned above, as a specific instance of the more general class of *matching problems* for graphs.

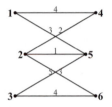

Figure 1.3 – Matching problem

Section 1.1 introduces the basic notions relevant to such graphs as are depicted in Figures 1.1, 1.2, and 1.3. In Section 1.2, we discuss *spanning trees*, which are the sparsest possible connected subgraphs of a graph. Section 1.3 contains algorithms for the solution of minimal cost network problems of the sort illustrated by Figure 1.1. The algorithms to solve the critical path problem of Figure 1.2, and other problems of maximal flow through a network, are given in Sections 1.4 and 1.5. The matching problem is solved in Section 1.6. Several other important graph theory problems, including the so-called *traveling salesman problem*, are discussed in the concluding Section 1.7. Along the way we will use *Mathematica* to characterize and display graphs, to make computations, and to implement algorithmic solutions to graph theoretic problems. In the electronic version of the text, you may want to open up the closed cells preceding each of the figures above to see the *Mathematica* code that generated the graphics. We will learn shortly how to produce such code.

1.1 Definitions and Examples

The intuitive meanings of "graph" and "directed graph" should be clear from the preceding discussion. A graph is a collection of vertices (or nodes) and edges (or arcs) connecting those vertices. The graph is directed if there is a notion of direction for its edges. A more precise set–theoretic definition is as follows.

DEFINITION 1. A *graph* G is a pair (V, E) where $V = \{v_1, v_2, ..., v_n\}$ is a finite set of elements called *vertices* and $E = \{\{v_i, v_j\}\}$ is a set of two-element subsets of V. Each member of E is called an *edge*. A *directed graph* is similar, except that edges are ordered pairs (v_i, v_j).

We allow the possibility of an empty graph ($n = 0$), but henceforth we usually dismiss it as a trivial case without special mention.

Figure 1.1 is a graph with $V = \{1, 2, 3, 4, 5, 6\}$ and edge set

$$E = \{ \{1,2\}, \{1,3\}, \{1,4\}, \{1,5\}, \{1,6\}, \{2,3\}, \{2,4\}, \{2,5\},$$
$$\{2,6\}, \{3,4\}, \{4,5\}, \{5,6\} \}$$

Figure 1.2 is a directed graph, with the vertex set $V = \{1, 2, 3, 4, 5, 6, 7, 8, 9, 10\}$, and edge set

$$E = \{(1, 2), (1, 3), (2, 4), (2, 5), (3, 5), (3, 6), (3, 7), (4, 8), (4, 9),$$
$$(5, 8), (6, 9), (7, 9), (8, 10), (9, 10)\}$$

Activity 1 – Write the formal description of the graph in Figure 1.3. Try to describe carefully what special geometry this graph has.

In the following definitions we use the word "graph" generically to mean either a directed or an undirected graph, when the item being defined makes sense in both cases. When necessary we include the adjectives "directed" and "undirected."

DEFINITION 2. A *subgraph* of a graph $G = (V, E)$ is another graph $G' = (V', E')$ such that $V' \subseteq V$ and $E' \subseteq E$. A graph $G = (V, E)$ is *weighted* by a *weight function* w if $w : E \rightarrow \mathbb{R}$.

The problems discussed in the introduction involve the location of an optimal subgraph of a weighted graph, satisfying certain constraints.

A graph may be characterized by a matrix of 0's and 1's. There is a row and a column for each vertex in the graph, and the matrix has 1 in component (i, j) if and only if there is an edge from vertex i to vertex j in the graph. We have spoken in the context of directed graphs, but note that an undirected edge $\{v_i, v_j\}$ may be viewed as two directed edges (v_i, v_j) and (v_j, v_i). Thus, undirected graphs are just special cases of directed graphs in which $(v_j, v_i) \in E$ whenever $(v_i, v_j) \in E$.

DEFINITION 3. Vertex v_j is *adjacent* to v_i if $(v_i, v_j) \in E$. The *adjacency matrix* of a graph G with n vertices is an $n \times n$ matrix $A = (a_{ij})$ with components

$$a_{ij} = \begin{cases} 1 & \text{if } (v_i, v_j) \in E \\ 0 & \text{otherwise} \end{cases}$$

The *weight matrix* of a weighted graph G with n vertices and weight function $w(v_i, v_j)$ is the $n \times n$ matrix $W = (w_{ij})$ with components

$$w_{ij} = \begin{cases} w(v_i, v_j) & \text{if } (v_i, v_j) \in E \\ 0 & \text{otherwise} \end{cases}$$

For example, the graph of Figure 1.3 has adjacency matrix

$$A = \begin{pmatrix} 0 & 0 & 0 & 1 & 1 & 0 \\ 0 & 0 & 0 & 1 & 1 & 1 \\ 0 & 0 & 0 & 0 & 1 & 1 \\ 1 & 1 & 0 & 0 & 0 & 0 \\ 1 & 1 & 1 & 0 & 0 & 0 \\ 0 & 1 & 1 & 0 & 0 & 0 \end{pmatrix}$$

By the remark prior to the definition, the adjacency matrix of an undirected graph is symmetric. We have not forbidden vertices to be adjacent to themselves; indeed, in Chapter 4 we will study transition diagrams of Markov chains that have this property. But in this chapter we will usually have no self-loops, that is, edges from a vertex into itself, and hence the adjacency matrices will usually be zero along their diagonals.

Activity 2 – Write the adjacency matrix for the graph of Figure 1.2

The KnoxOR`Graphs` *Mathematica* package that is available with the electronic version of the text contains a function to display graphs. You should have installed the KnoxOR suite of packages in your ExtraPackages directory, and to access the commands for graph theory you must load the package as in the first line of input below. The syntax of the DisplayGraph function follows. Note that it assumes the graph is represented as an adjacency matrix or weight matrix.

```
Needs["KnoxOR`Graphs`"]
```

```
(****    DisplayGraph[graph,options]    ****)
```

```
Options[DisplayGraph]
```

```
{GraphType → Undirected, VertexLabels → Automatic,
 VertexPositions → Automatic,
 VertexLabelPositions → Automatic,
 EdgeLabels → Automatic,
 EdgeLabelPositions → Automatic,
 EdgeStyle → Thickness[0.005], EdgeSeparation → 0.01,
 DisplayFunction → (Display[$Display, #1] &),
 AspectRatio → 1,
 LoopPositions → Automatic, LoopSize → 0.05}
```

There are several important options to control the appearance of the graph. You can set GraphType to Directed to sketch a graph with arrows for directed edges. VertexLabels can be set to a list of names, one per vertex, in order that vertices can be displayed with names other than the standard integer names $\{1, 2, 3, \ldots, n\}$. In other functions to come, however, when individual vertices are referred to it is by their number not their name. The option VertexPositions can be set to a list of pairs $\{\{x_1, y_1\}, \{x_2, y_2\}, \ldots\}$, which are the coordinates in the plane of the vertices 1, 2, ... Sometimes graph features can overlap, and so to enhance visibility, the user has some control over where labels appear. For example, the VertexLabelPositions option can be set to a list such as {Above, Below, ToLeft, ToRight,...} to indicate where the vertex labels should appear relative to the vertex dots. Take care with this and other options to make sure that sizes are consistent; for example, the value set for VertexLabelPositions should be a list of the same length as the number of vertices in the graph. EdgeLabels can be set to a matrix, usually the weight matrix, of labels to place on the edges. Like VertexLabelPositions, the option EdgeLabelPositions can be set to a matrix of the same size as the adjacency matrix whose entries are the words Above, Below, ToLeft, or ToRight to indicate where, relative to the edge midpoint, the edge labels should appear. The option EdgeStyle can be used to apply a style to the edges, such as coloration, dashing, or boldfacing. EdgeSeparation controls the space between arrows in a double edge. For graphs that have self-loops, the options LoopSize and LoopPositions can be used. LoopSize controls the size of the loops by setting the fraction of the overall picture size to be used as the loop radius. And LoopPositions, like VertexLabelPositions, can be set to a list of entries that can be Above, Below, etc., to control where the loop appears relative to the vertex. By default, the loop positions will be above the vertices. The length of the LoopPositions list should be the same as the number of vertices with self-loops. Standard options DisplayFunction and AspectRatio, just as in *Mathematica*'s Plot command, are also accepted. DisplayFunction→Identity suppresses printing, which is handy when you want to produce but not print graphs to combine later into one picture with the Show command (resetting DisplayFunction to be $DisplayFunction as a Show option). AspectRatio controls the shape of the picture, which is sometimes helpful in making a graph more aesthetically appealing.

EXAMPLE 1. (a) Figure 1.4 is the graph whose adjacency matrix is

$$A = \begin{pmatrix} 0 & 1 & 1 & 1 \\ 1 & 0 & 0 & 1 \\ 1 & 0 & 0 & 0 \\ 1 & 1 & 0 & 0 \end{pmatrix}$$

This would be entered into *Mathematica* as follows, as a list of rows of the matrix.

```
adjmatrix = {{0, 1, 1, 1},
    {1, 0, 0, 1}, {1, 0, 0, 0}, {1, 1, 0, 0}};
```

To draw this graph in *Mathematica*, we first plan out the coordinates of the vertices against the backdrop of an unseen coordinate system. In this case, to produce the square shape of the graph in Figure 1.4, we use the VertexPositions option to place the four vertices at $(0, 1)$, $(1, 1)$, $(0, 0)$, and $(1, 0)$ respectively. For variety we give the vertices labels of v_1 through v_4 using the VertexLabels option, and for visibility we use the VertexLabelPositions option to put the first two labels above the vertices and the second two below the vertices. Because the graph is undirected and unweighted, we do not need the GraphType, EdgeLabels, or EdgeLabelPositions options. This plan leads to the DisplayGraph command below.

```
DisplayGraph[adjmatrix, VertexPositions →
    {{0, 1}, {1, 1}, {0, 0}, {1, 0}},
    VertexLabels → {"v₁", "v₂", "v₃", "v₄"},
    VertexLabelPositions →
    {Above, Above, Below, Below}, AspectRatio → 1];
```

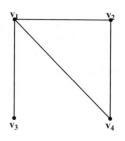

Figure 1.4 – An undirected graph

You can compute that

$$A^2 = \begin{pmatrix} 3 & 1 & 0 & 1 \\ 1 & 2 & 1 & 1 \\ 0 & 1 & 1 & 1 \\ 1 & 1 & 1 & 2 \end{pmatrix}$$

and it is easy to see from the graph that for each pair of vertices (i, j), $A^2(i, j)$ is the number of paths of length 2 from i to j. We will return to this idea later.

(b) The weight matrix of the graph of Figure 1.1 is below:

$$W = \begin{pmatrix} 0 & 1 & 3 & 2 & 8 & 3 \\ 1 & 0 & 6 & 2 & 7 & 1 \\ 3 & 6 & 0 & 4 & 0 & 0 \\ 2 & 2 & 4 & 0 & 5 & 0 \\ 8 & 7 & 0 & 5 & 0 & 2 \\ 3 & 1 & 0 & 0 & 2 & 0 \end{pmatrix}$$

Recall that when an edge does not exist in the graph, we adopt the convention that the corresponding weight in the matrix is zero. The total weight of the path 1, 2, 3, 4, for example, is $1 + 6 + 4 = 11$. If weights are viewed as costs, then to move from 1 to 4 it is cheaper to use edge {1, 4} than this path. However it is not cheaper to move directly from 1 to 5 (weight 8) than it is to use the indirect path 1, 6, 5 (weight $3 + 2 = 5$). The path 1, 2, 5, 1 is an example of a path that begins and ends at the same node, and will be called a *cycle* (see the definition below). ■

The following definitions should be self-explanatory.

DEFINITION 4. For $m \geq 1$, a *path* of length m from vertex v to vertex u is a sequence of vertices $v = v_0, v_1, v_2, ..., v_m = u$ such that each successive pair (v_i, v_{i+1}) is in the edge set E. A *path of length 0* from v_0 to itself is the singleton v_0. If there is a path from v to u, we call v an *ancestor* or *predecessor* of u, and we call u a *descendant* of v. If the graph is weighted by a weight function w, then the *weight of the path* is

$$\sum_{i=0}^{m-1} w(v_i, v_{i+1})$$

A path $v_0, v_1, ..., v_m$ is *simple* if for all $i \neq j$, $v_i \neq v_j$. A path is a *cycle* if it has length at least 3 in the undirected case, and 2 in the directed case, and $v_0 = v_m$ but no other pair of its vertices is equal.

EXAMPLE 2. Referring to the graph of Figure 1.4, v_1, v_2, v_4 is a simple path, and v_2, v_4, v_1, v_2 is a cycle. The length of this cycle is 3, hence it does qualify under the definition. The path v_1, v_2, v_1 is not a cycle because of this length requirement. Without the stipulation that $m \geq 3$, every edge of an undirected graph would give rise to a cycle. This is not true in a directed graph such as the one in Figure 1.5. In this graph, 2, 3, 2 is a cycle, as is 1, 2, 3, 4, 5, 1. The total weight of the latter path is $2 + 1 + 3 + 4 + 3 = 13$. Pay attention again to the syntax of the DisplayGraph command. We have defined the weight matrix as W, placed the five vertices in a roughly pentagonal shape with appropriate coordinates using VertexPositions, produced a directed graph with GraphType, and used the entries of the weight matrix itself as the EdgeLabels. In addition, we use EdgeLabelPositions to place the edge labels conveniently. ∎

```
W = {{0, 2, 0, 0, 0}, {0, 0, 1, 0, 0}, {0, 1, 0, 3, 5},
    {0, 0, 0, 0, 4}, {3, 0, 0, 7, 0}};
MatrixForm[W]
DisplayGraph[W, VertexPositions →
    {{0, 1}, {2, 2}, {4, 1}, {3, 0}, {1, 0}},
   EdgeLabels → W, GraphType → Directed,
   VertexLabelPositions →
   {ToLeft, Above, ToRight, Below, Below},
   EdgeLabelPositions → {{0, Above, 0, 0, 0}, {0, 0,
       Below, 0, 0}, {0, Above, 0, ToRight, ToLeft},
      {0, 0, 0, 0, Above}, {ToLeft, 0, 0, Below, 0}},
   EdgeSeparation → .012];
```

$$\begin{pmatrix} 0 & 2 & 0 & 0 & 0 \\ 0 & 0 & 1 & 0 & 0 \\ 0 & 1 & 0 & 3 & 5 \\ 0 & 0 & 0 & 0 & 4 \\ 3 & 0 & 0 & 7 & 0 \end{pmatrix}$$

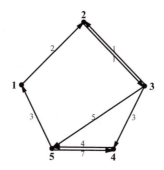

Figure 1.5 – A directed graph

Activity 3 – Try to find the longest path from vertex 1 to vertex 10 in the graph of Figure 1.2.

The proof of the following is easy, and is left as an exercise for the reader (see Exercise 2).

THEOREM 1. If there is a path from vertex v to vertex w, then there is also a simple path from v to w. If the number of vertices in G is n, then this simple path must have length less than or equal to $n - 1$. ∎

The last of the elementary ideas that we wish to introduce is the *degree* of a vertex. We use this concept only in studying undirected graphs; in particular it comes into play when we look at properties of trees in Section 1.2.

DEFINITION 5. Let v be a vertex of an undirected graph. Then the *degree* of v, denoted $d(v)$, is the number of edges $\{v, w\}$ to which v belongs.

In Figure 1.4, for example, v_1 has degree 3 and v_2 has degree 2.

Consider the total of the degrees of all the vertices in a graph $G = (V, E)$. Each edge $\{v, w\}$ contributes 1 to $d(v)$ and 1 to $d(w)$. Thus, adding the degrees will double count the edges, and the following relation is clear:

$$\sum_{v \in V} d(v) = 2 \text{ (number of edges in } E) \tag{1}$$

Connectivity

Next we study some ideas pertaining to the connectivity properties of graphs.

DEFINITION 6. A graph is *connected* if there is a path from every vertex to every other vertex. A directed graph is called *quasi-connected* if, for every pair of vertices u and v, there is a vertex w such that paths (of length 0 or more) exist from w to u and from w to v. That is, u and v have a common ancestor w (which might be one of u or v themselves).

EXAMPLE 3. In the case of undirected graphs, the above definition coincides well with our intuition about the meaning of the word "connected." The graphs in Figures 1.1 and 1.4 are connected graphs, for example, because every vertex can be reached from every other vertex. But for directed graphs, the definition that we have given, usually referred to as *strong connectivity*, may be stronger than it would appear to be at first glance. To see this consider Figure 1.6, in which an undirected graph and a similar directed graph are displayed. The graph of Figure 1.6(a) is not connected, since there is no path from 2 to 5, for instance. Both of the two subgraphs with vertex sets $\{1, 2, 3\}$ and $\{4, 5, 6, 7\}$ are connected, however. These two vertex sets will be called the *connected components* of the overall graph. The directed graph in Figure 1.6(b) is not connected, nor are the aforementioned subgraphs. For example, there is no path from vertex 2 to vertex 1, and no path from 7 to 6. But the vertex set $\{4, 5, 7\}$, together with the corresponding edge set, does form a connected graph, since paths exist from every vertex to every other vertex in this subset.

Quasi-connectivity is a weaker condition than connectivity. It is easy to check in Figure 1.6(b) that $\{4, 5, 6, 7\}$ forms a quasi-connected graph, since each pair of vertices has a common ancestor (namely vertex 6). ■

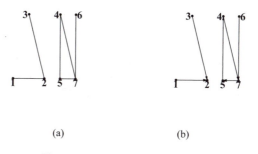

(a)   (b)

Figure 1.6 – Two disconnected graphs

Next we investigate the relation between connectivity and the adjacency matrix.

THEOREM 2. Let A be the adjacency matrix of a graph G of n vertices. Then $A^m(i, j)$ is the number of paths of length m from vertex i to vertex j. Thus, G is (strongly) connected if and only if for every pair of vertices (i, j), $A^k(i, j) \geq 1$ for at least one $k = 1, 2, ..., n$.

Proof. The proof of the first statement is by induction on m. When $m = 1$, the statement is true by the definition of the adjacency matrix. Now assume that for $k = 1, 2, . .., m$, $A^k(i, j)$ is the number of paths from i to j of length k. We have by the definition of matrix multiplication:

$$A^{m+1}(i, j) = \sum_{v \in V} A^m(i, v) A(v, j) = \sum_{v|(v,j) \text{ is an edge}} A^m(i, v)$$

The set of paths of length $m + 1$ from i to j can be expressed as the disjoint union, over the set of vertices v such that $(v, j) \in E$, of the set of paths for which v precedes j:

$$B_v = \{(v_0, v_1, ..., v_m, v_{m+1}) : v_0 = i, v_m = v, v_{m+1} = j\}$$

The number of paths of length $m + 1$ from i to j is therefore

$$\sum_{v|(v,j) \text{ is an edge}} n(B_v)$$

where $n(B_v)$ is the size of B_v. But B_v is in one-to-one correspondence with the paths of length m from i to v, hence by induction $n(B_v) = A^m(i, v)$, from which the first statement of the theorem follows. The second statement is true because G is connected iff there is a path of some length k between every pair of vertices. The longest possible length of such a path is n, in the case that in order to move from a vertex v back to itself, one must go through all n vertices and finish at v. ∎

EXAMPLE 4. The graph of Figure 1.6(a) has the adjacency matrix defined below:

```
grapha = {{0, 1, 0, 0, 0, 0, 0},
    {1, 0, 1, 0, 0, 0, 0}, {0, 1, 0, 0, 0, 0, 0},
    {0, 0, 0, 0, 1, 0, 1}, {0, 0, 0, 1, 0, 0, 1},
    {0, 0, 0, 0, 0, 0, 1}, {0, 0, 0, 1, 1, 1, 0}};
MatrixForm[grapha]
```

$$\begin{pmatrix} 0 & 1 & 0 & 0 & 0 & 0 & 0 \\ 1 & 0 & 1 & 0 & 0 & 0 & 0 \\ 0 & 1 & 0 & 0 & 0 & 0 & 0 \\ 0 & 0 & 0 & 0 & 1 & 0 & 1 \\ 0 & 0 & 0 & 1 & 0 & 0 & 1 \\ 0 & 0 & 0 & 0 & 0 & 0 & 1 \\ 0 & 0 & 0 & 1 & 1 & 1 & 0 \end{pmatrix}$$

The second, third, and fourth powers of this adjacency matrix are computed in *Mathematica* as:

```
{MatrixForm[MatrixPower[grapha, 2]],
 MatrixForm[MatrixPower[grapha, 3]],
 MatrixForm[MatrixPower[grapha, 4]]}
```

$$\left\{ \begin{pmatrix} 1 & 0 & 1 & 0 & 0 & 0 & 0 \\ 0 & 2 & 0 & 0 & 0 & 0 & 0 \\ 1 & 0 & 1 & 0 & 0 & 0 & 0 \\ 0 & 0 & 0 & 2 & 1 & 1 & 1 \\ 0 & 0 & 0 & 1 & 2 & 1 & 1 \\ 0 & 0 & 0 & 1 & 1 & 1 & 0 \\ 0 & 0 & 0 & 1 & 1 & 0 & 3 \end{pmatrix} \right. ,$$

$$\begin{pmatrix} 0 & 2 & 0 & 0 & 0 & 0 & 0 \\ 2 & 0 & 2 & 0 & 0 & 0 & 0 \\ 0 & 2 & 0 & 0 & 0 & 0 & 0 \\ 0 & 0 & 0 & 2 & 3 & 1 & 4 \\ 0 & 0 & 0 & 3 & 2 & 1 & 4 \\ 0 & 0 & 0 & 1 & 1 & 0 & 3 \\ 0 & 0 & 0 & 4 & 4 & 3 & 2 \end{pmatrix} , \begin{pmatrix} 2 & 0 & 2 & 0 & 0 & 0 & 0 \\ 0 & 4 & 0 & 0 & 0 & 0 & 0 \\ 2 & 0 & 2 & 0 & 0 & 0 & 0 \\ 0 & 0 & 0 & 7 & 6 & 4 & 6 \\ 0 & 0 & 0 & 6 & 7 & 4 & 6 \\ 0 & 0 & 0 & 4 & 4 & 3 & 2 \\ 0 & 0 & 0 & 6 & 6 & 2 & 11 \end{pmatrix} \right\}$$

Notice that the structure of these powers remains block diagonal, with a non-zero block corresponding to the first three vertices, another corresponding to the last four vertices, and zero blocks off the diagonal. This comes from the fact that vertices 1, 2, and 3 do not communicate with vertices 4, 5, 6, and 7. Inspecting the second power, for example, you see that there are two paths of length 2 from vertex 4 to itself (namely 4,5,4 and 4,7,4), and there is just one path of length 2 from vertex 5 to 7 (namely 5,4,7). The third power tells us, for example, that there are three paths of length 3 from vertex 5 to vertex 4 (try to list them), and two paths of length 3 from vertex 1 to vertex 2. Another observation that we can make from the fourth power matrix is that there are numerous paths of length 4 from every vertex in $\{4, 5, 6, 7\}$ to every other vertex in that set, hence that subgraph is connected. And, inspecting the $\{1, 2, 3\}$ blocks of the third and fourth power matrices, we see that there exist paths of length either 3 or 4 from every vertex in $\{1, 2, 3\}$ to every other vertex in that set, hence the subgraph including vertices 1, 2, 3, and the corresponding edges is connected. Exercise 8 asks you to argue from the matrix powers that the graph as a whole cannot be connected. ∎

In the second part of the statement of Theorem 2, the path length k was allowed to depend on the vertices being connected. An even stronger sort of connectivity postulates a path length that is uniformly good for all pairs of vertices.

> **DEFINITION 7.** A graph is *regular* if there is $m > 0$ such that $A^m(i, j) > 0$ (i.e., there is a path of length m from i to j) for all pairs of vertices (i, j).

EXAMPLE 5. The directed graph in Figure 1.7 is clearly regular, since by inspection the reader can see that there is a path of length 2 from each vertex to each other vertex. But the graph in Figure 1.8, in which edges are oriented in a clockwise direction, is not regular, due to the cyclic nature of the edges. The closed cell above the figures contains the definitions of the adjacency matrices for the two graphs. You should try raising the adjacency matrix for Figure 1.7 to the second power to see that all entries are non-zero, and you should see what happens to the adjacency matrix for Figure 1.8 as you raise it to higher and higher powers.

One is tempted to guess that if there are n vertices, then to check for regularity it suffices to check the powers of the adjacency matrix only up to n. To see that this is not quite the case, examine the graph of Figure 1.9. There is no path of length 2 from vertex 1 to vertex 2; there is no path of length 3 from vertex 1 to itself; but there is a path of length 4 between every pair of vertices. Hence $m = 4$ suffices, but no smaller m. (See Exercise 13.) ∎

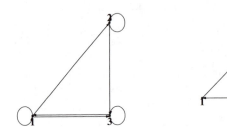

Figure 1.7 – A regular graph   **Figure 1.8** – A graph that is not regular

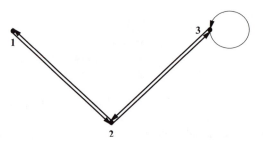

Figure 1.9 – Paths of length 4 exist between each pair of vertices

---

Activity 4 – Use *Mathematica* to check whether the graph of Figure 1.1 is regular.

Regularity is a property that will be important to the study of limiting distributions of Markov chains in Chapter 4. In that chapter as well, it will be helpful to group vertices into subsets, such that all vertices can communicate with all other vertices in the subset. We are led to the following definition.

DEFINITION 8. A *connected component* G' of a graph G is a maximal connected subgraph of G.

The word "maximal" in this definition means that there is no strictly larger connected subgraph containing G'. In Figure 1.6(a), for example, the set {4, 5, 7} certainly forms a connected set, but since vertex 6 is connected to these vertices, the subgraph is not maximal. It is clear that the undirected graph in Figure 1.6(a) can be partitioned into two connected components: {1, 2, 3} and their related edges, and {4, 5, 6, 7} and their edges. Such a partition is not possible for the directed graph of Figure 1.6(b). There, the set of vertices {4, 5, 7} and their edges form the only connected component.

Next we give an algorithm to find the connected component containing a given vertex of an undirected graph. The algorithm itself shows the existence of such a component, and, in so doing, implies that undirected graphs may be partitioned into connected components. The reason for this is that we can begin with any vertex, find the connected component containing it using the algorithm, proceed to another vertex that has not already been used, find its connected component, and so on, until the supply of vertices is exhausted. The idea of the algorithm is to begin with the given vertex v, and include all vertices connected by an edge to v into the current component of v. At each step, we scan all edges of the graph that have an endpoint at one of the vertices in the current component, and all of the vertices on the other end of these new edges. Any new vertices are added to the current component. If no new vertices are found, then the entire connected component has been located and the process ends. The reader is asked to prove that the algorithm works in Exercise 15.

CONNECTED COMPONENTS ALGORITHM
    1. Initialize vertex set $V_0 = \{v\}$.
    2. Let $m = 0$.
    3. Repeat a – b until $V_m = V_{m-1}$:
        a. Let $m = m + 1$
        b. Let
$V_m = V_{m-1} \cup \{w \in V : \text{there is an edge } \{v, w\} \text{ for some } v \in V_{m-1}\}$

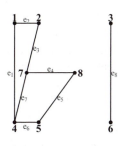

Figure1.10 – A graph with two connected components

EXAMPLE 6. Let us develop a *Mathematica* command to implement the connected components algorithm, and execute it on the graph shown in

Figure 1.10, with initial vertex 1. (In the closed cell above the figure, the *Mathematica* name g10 has been given to the adjacency matrix.)

It is an important fact that the practitioner of operations research must not only understand solution algorithms at a very deep level, but also be able to convert them to actual programs. This is because many problem situations require new solutions that existing software is not built to handle. And the philosophy of this book is that the process of forming such programs brings you to the deeper level of understanding required. An added benefit is to be confident that the algorithm will work in the cases that it is meant to be used for, which only happens after detailed understanding of the algorithm and its implementation.

When one looks carefully at the algorithm, one notices some space and time inefficiencies that can be remedied in the implementation. For example the algorithm suggests to keep, for each value m takes on as the loop in step 3 is executed, a set of vertices V_m currently in the component. We do not really need all of these vertex sets, which have many members in common. All we really need is the cumulative set of vertices that have been added to the connected component so far, the most recent set of vertices found on the previous pass through the loop, and the new candidate vertices that are the neighbors of the most recent set of vertices, some of which may not yet have been included into the component, and so should be added in the current pass. That way, instead of having to reexamine neighbors of vertices added in past steps, we search only among the most recent vertices, whose neighbors are possible new members of the component. When the set of new candidate vertices has no vertices that have not already been found, we know we are done.

The development of a working program requires attention to many things: the form of the input; assumptions about the input under which the program should be guaranteed to function properly; what is to be output, printed, or returned by the program; what local variables, and of what structure, should be introduced to do the job; and whether it will be helpful to build supporting programs to do some of the tasks the main program needs, so that the main program may be made more simple.

For our connected components problem, the program will need the adjacency matrix of an undirected graph, which we assume is a square, symmetric matrix of 0's and 1's in the *Mathematica* form of a list of row lists. The program will also need the number of the vertex whose component we are to find. We will assume that the vertices are numbered successively 1, 2, 3, ..., n, and that the vertex number is within the range specified by the size of the square adjacency matrix. Our goal is to have the program return the complete list of vertex numbers that make up the connected component. Because we would like to see something of the sequence of operations, we will have our program print out the current vertex set and the set of newly added vertices at every pass through the loop. Local set variables, in the form of *Mathematica* lists, will be needed for all vertices in the component so far, the most recently added vertices, and the new vertices to be added. A utility to return a list of all neighbors of a given vertex, given the adjacency matrix of the graph, will also help. Fortunately, such a function already exists in the KnoxOR`Graphs` package:

```
(****  FindNeighbors[adjmatrix,vertex]  ****)
```

We will initialize the component, and the most recently added set of vertices, to be the given vertex. Then at each pass through the main loop, we find the complete set of neighbors of recent vertices, and determine, by complementation, which have not yet been added. This becomes the new set of vertices, which we union with the current component and then reset the current set of vertices to be the new set in preparation for the next pass through the loop. The preceding discussion and the comments within the program below should make the operation of our function clear.

```
Components[thegraph_, vertex_] :=
 Module[{component,
    currentvertices, newvertices, neighbors},
   (* begin by adding in the starting vertex *)
   component = {vertex};
   currentvertices = {vertex}; newvertices = {};
   Print[component, "    ", newvertices];
   While[currentvertices ≠ {},
    neighbors = {};
     (* union together the neighbor
       sets of all current vertices *)
    Do[neighbors = Union[neighbors, FindNeighbors[
        thegraph, currentvertices[[i]]]],
      {i, 1, Length[currentvertices]}];
     (* let newvertices be neighbors that
       have not been added in yet *)
    newvertices = Complement[
      neighbors, component];
     (* add them in *)
    component = Union[component, newvertices];
     (* display results of pass *)
    Print[component, "    ", newvertices];
     (* reset lists for next pass *)
    currentvertices = newvertices;
    newvertices = {}];
   (* at loops end the entire component
     has been found, now return it *)
   component]
```

```
Components[g10, 1];
```

```
{1}    {}

{1, 2, 4}    {2, 4}

{1, 2, 4, 5, 7}    {5, 7}

{1, 2, 4, 5, 7, 8}    {8}

{1, 2, 4, 5, 7, 8}    {}
```

As the output shows, for the graph of Figure 1.10, vertices 2 and 4 are found first, then 5 and 7, and finally 8. A last unsuccessful search produces no new vertices, so that the connected component containing vertex 1 has vertex set $\{1, 2, 4, 5, 7, 8\}$ and edge set $\{e_1, ..., e_7\}$. Below we execute the command starting from vertex 3, and find the connected component of 3 to be vertices $\{3, 6\}$ and edge e_8. In the electronic version of the text you should try issuing similar commands to find the connected component starting with vertices 2, and then 6. Try also finding the connected components of each of vertices 1, 2, 3, and 4 in the graph of Figure 1.4. (You may need to reenter the definition of adjmatrix.) ■

```
Components[g10, 3];
```

```
{3}     {}

{3, 6}     {6}

{3, 6}     {}
```

Activity 5 – Think about how the FindNeighbors command might be implemented.

EXAMPLE 7. The algorithm does not quite work in the directed case, as shown by Figure 1.11. Beginning with vertex 1, we find vertices 2, 3, and 4. On the next pass, we obtain vertices 5 and 6, then on the third pass, 7 and 8, and finally we stop after seeing that vertices 7 and 8 have no new neighbors. But the set of vertices $\{1, 2, 3, 4, 5, 6, 7, 8\}$ does not form a connected component because, for instance, vertex 7 cannot reach vertex 1. The algorithm does return some useful information, however: namely, the set of vertices reachable from vertex 1, i..e. the *descendants* of 1. (See Exercise 9.) This set is *closed* in the sense that no vertex outside the set can be reached from a vertex in the set. (See Exercise 16.) ■

Figure 1.11 – Connected components algorithm fails for directed graphs

Exercises 1.1

1. (*Mathematica*) For the graph below, write the adjacency matrix A, compute A^3, and verify that for each i and j, $A^3(i, j)$ is the number of paths from i to j of length 3 by listing those paths.

Exercise 1

2. Prove Theorem 1.

3. Show that the graph whose adjacency matrix is below has no cycles.

$$\begin{pmatrix} 0 & 0 & 0 & 0 & 0 & 0 & 0 & 1 & 0 \\ 0 & 0 & 0 & 0 & 0 & 0 & 0 & 1 & 0 \\ 0 & 0 & 0 & 0 & 0 & 0 & 0 & 1 & 0 \\ 0 & 0 & 0 & 0 & 1 & 0 & 0 & 0 & 1 \\ 0 & 0 & 0 & 1 & 0 & 1 & 0 & 0 & 0 \\ 0 & 0 & 0 & 0 & 1 & 0 & 0 & 0 & 0 \\ 0 & 0 & 0 & 0 & 0 & 0 & 0 & 1 & 0 \\ 1 & 1 & 1 & 0 & 0 & 0 & 1 & 0 & 1 \\ 0 & 0 & 0 & 1 & 0 & 0 & 0 & 1 & 0 \end{pmatrix}$$

4. Show that there is no four-vertex undirected graph with degrees $d(v_1) = 3$, $d(v_2) = 2$, $d(v_3) = 2$, and $d(v_4) = 2$.

5. (*Mathematica*) Write a *Mathematica* command that takes the adjacency matrix of a graph and a vertex, and returns the degree of that vertex. Test it on all the vertices of the graph of Figure 1.4.

6. Let A be the adjacency matrix of an undirected graph G. Show that $A^2(i, i) = d(i)$.

7. Two graphs $G_1 = (V_1, E_1)$ and $G_2 = (V_2, E_2)$ are called *isomorphic* if there is a one-to-one onto function $f : V_1 \to V_2$ such that for all $v, w \in V_1$ edge $(v, w) \in E_1$ if and only if edge $(f(v), f(w)) \in E_2$. Show that the two directed graphs below cannot be isomorphic.

Exercise 7

8. Argue, using the adjacency matrix only, that the graph in Figure 1.6(a) is not connected.

9. (*Mathematica*) There is a function in the KnoxOR`Graphs` package called

```
(**** FindChildren[adjmatrix,parentlist] ****)
```

This command returns a list of all children of vertices in the given list of parents, where adjmatrix is the adjacency matrix of a directed graph and a vertex v is a child of a vertex u iff there is an edge (u, v) in the graph. Revise the Components function of Section 1.1 using FindChildren to produce a function called Descendants[adjmatrix, vertex] that returns the set of all vertices reachable by some path from the given vertex in the directed graph characterized by the given adjacency matrix. For each vertex in the graph of Figure 1.11, use the Descendants function to find the set of all descendants of that vertex.

10. Prove that if the vertex set of a directed graph can be partitioned into three subsets V_1, V_2, and V_3 such that edges only exist from V_1 into V_2, or from V_2 into V_3, or from V_3 into V_1, then the graph is not regular. Give an example of such a graph with eight vertices.

11. Decide whether the following graph is (a) connected or (b) quasi-connected.

```
ex12 =
   {{0, 1, 1, 0, 0}, {0, 0, 1, 1, 0}, {1, 0, 0, 0, 0},
    {0, 0, 0, 0, 1}, {0, 1, 1, 1, 0}};
DisplayGraph[ex12, GraphType → Directed,
   VertexLabelPositions →
      {ToLeft, Above, Below, Above, ToRight},
   EdgeSeparation → .02, AspectRatio → .7];
```

Exercise 11

12. Show that a connected directed graph is quasi-connected. Show that an undirected graph is quasi-connected if and only if it is connected.

13. (*Mathematica*) For the graph of Figure 1.9, verify that A^m is not entirely non-zero for any $m < 4$.

14. (*Mathematica*) Find all connected components of the graph below.

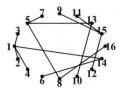

Exercise 14

15. Argue that for undirected graphs, the connected components algorithm does find the connected component of the given initial vertex.

16. Prove that for directed graphs, the connected components algorithm finds the set of vertices that can be reached from a given initial vertex v. Prove that this set is a closed set (see Example 7), and that if in addition every vertex u in the set can reach the initial vertex v, then this set is a connected component.

17. For an undirected graph with the adjacency matrix below, find the connected components.

$$\begin{pmatrix} 0 & 1 & 0 & 0 & 0 & 0 & 0 & 0 & 0 & 1 \\ 1 & 0 & 0 & 0 & 0 & 0 & 0 & 0 & 0 & 0 \\ 0 & 0 & 0 & 1 & 0 & 1 & 0 & 0 & 0 & 0 \\ 0 & 0 & 1 & 0 & 0 & 0 & 0 & 0 & 0 & 0 \\ 0 & 0 & 0 & 0 & 0 & 0 & 0 & 1 & 0 & 0 \\ 0 & 0 & 1 & 0 & 0 & 0 & 1 & 0 & 0 & 0 \\ 0 & 0 & 0 & 0 & 0 & 1 & 0 & 0 & 0 & 0 \\ 0 & 0 & 0 & 0 & 1 & 0 & 0 & 0 & 1 & 0 \\ 0 & 0 & 0 & 0 & 0 & 0 & 0 & 1 & 0 & 0 \\ 1 & 0 & 0 & 0 & 0 & 0 & 0 & 0 & 0 & 0 \end{pmatrix}$$

18. (*Mathematica*) Write a *Mathematica* command that takes a weight matrix of a graph, and a list of vertices forming a path in the graph, and returns the weight of the path.

19. Adjacency matrices are not the only way of representing graphs. An *adjacency list* representation of a graph is a list, vertex-by-vertex, of the vertices that are adjacent to that vertex. For example, for the graph of Figure 1.4 one would have the adjacency list

$$v_1 : v_2, v_3, v_4$$
$$v_2 : v_1, v_4$$
$$v_3 : v_1$$

For graphs with many vertices and not very many edges, this representation can result in a substantial savings in the amount of information recorded. Write adjacency lists for the graphs of (a) Figure 1.3; (b) Figure 1.5; (c) Figure 1.10.

20. (*Mathematica*) Devise a way of implementing in *Mathematica* an adjacency list representation of a graph. (See Exercise 19.) Write a function that converts an adjacency list to an adjacency matrix.

1.2 Spanning Trees

In many of the network problems that we will consider, the networks are special types of graphs called *trees*. These are subgraphs that retain the connectivity of the original graph, but are sparse in the sense of having as few edges as possible. In the definition below, the *underlying graph* of a directed graph is the undirected graph with the same vertex set, and with undirected edges $\{v, w\}$ for every directed edge (v, w) in the directed graph. Loosely speaking, we erase the arrows.

DEFINITION 1. An undirected graph is called a *tree* if it is connected and has no cycles. A directed graph is a *tree* with *root* (or *source*) v_1 if its underlying graph is a tree and if v_1 is an ancestor of every vertex $v \in V$.

We will occasionally use the word "root" in the sense of this definition even when the graph is not a tree.

EXAMPLE 1. Figure 1.12 shows an undirected tree. The fact that our definition requires vertices to have a common ancestor implies that the graph in Figure 1.13(a) is not a directed tree. The problem is the edge (v_5, v_3). Figure 1.13(b) illustrates a directed tree with root v_1, in which the direction of this edge has been reversed to (v_3, v_5). ∎

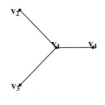

Figure 1.12 – An undirected tree

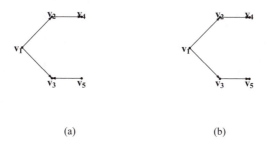

(a)   (b)

Figure 1.13 – (a) Directed graph; not a tree; (b) Directed tree

---

Activity 1 – Take a moment to draw for yourself some other examples of directed and undirected trees.

We will usually be considering trees as sparse subgraphs of other larger graphs, but since we desire connectivity of all vertices in the larger graph, we are led to make the following definition.

DEFINITION 2. A *spanning tree* of a graph G is a subgraph G' of G that is a tree containing all vertices of G.

Undirected Spanning Trees

First, consider undirected spanning trees. For example for the graph of Figure 1.4 in the previous section, the subgraph

$$G' = (\{v_1, v_2, v_3, v_4\}, \{\{v_1, v_3\}, \{v_1, v_2\}, \{v_2, v_4\}\})$$

is a spanning tree of the graph G. Recall also the graph of Figure 1.1, where the vertices represent communication stations, and the weight of an edge is the cost of setting up direct communication between the stations connected by the edge. The spanning tree formed by edges $\{1, 2\}$, $\{1, 3\}$, $\{1, 4\}$, $\{1, 5\}$, and $\{1, 6\}$ has total cost 17. Another spanning tree has edges $\{6, 2\}$, $\{2, 3\}$, $\{3, 4\}$, $\{4, 5\}$, and $\{1, 2\}$ and also has total cost 17. [See Figures 1.14(a) and (b).]

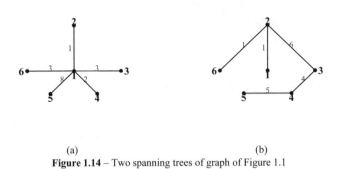

(a)   (b)

Figure 1.14 – Two spanning trees of graph of Figure 1.1

The interesting questions are: (1) how do we find a spanning tree and (2) how do we find a spanning tree of minimal cost, if there are edge weights representing costs? The latter is addressed in the next section. First, we will develop an algorithm to answer question (1). Theorem 1, to be stated shortly, supplies a termination condition for the spanning tree algorithm for undirected graphs. The proof of Theorem 1 requires three lemmas that are of interest in themselves (see [18], Theorems 3.17, 3.18, 3.19).

LEMMA 1. If G is a finite, undirected graph whose vertices each have degree ≥ 2, then G has a cycle.

Proof. We give the idea of the proof and leave the details to the reader. Begin with any vertex and move to a neighboring vertex. Since this new vertex has degree ≥ 2, we may move along an edge that has not already been used to another vertex. Continuing in this way, since the graph is finite, we must eventually return to a vertex already encountered, thus forming a cycle. ∎

LEMMA 2. A tree has at least one vertex of degree 1.

Proof. Since a tree has no cycles, this is an immediate consequence of Lemma 1. ∎

LEMMA 3. If G is connected and has strictly fewer edges than vertices, then G has a vertex of degree 1.

Proof. If, on the contrary, all vertices have degree at least 2, then we have from formula (1) of Section 1.1,

(number of edges) $= (1/2) \sum_{v \in V} d(v) \geq (1/2) \sum_{v \in V} 2 =$ (number of vertices) which is a contradiction of the hypothesis of the lemma. ■

We are ready to prove the main theorem characterizing trees ([18], Theorem 3.20).

THEOREM 1. Suppose G is connected graph of n vertices. Then G is a tree if and only if G has $n - 1$ edges.

Proof. It helps to restate the theorem slightly. The assertion is that a connected graph is a tree if and only if the number of vertices equals the number of edges + 1.

We show the forward implication by induction on the number of edges m of G. Assume G is a tree. When $m = 1$, by connectivity, G must have only two vertices. (For undirected graphs, we forbid loops from a single vertex to itself, so there must be more than one vertex.) Suppose now that the theorem is true when $m = k$, i.e., a tree of k edges has $k + 1$ vertices. Let G be a tree of $k + 1$ edges. By Lemma 2, G has a vertex v of degree 1. Delete v and the edge that is incident to it from G, and we still have a connected graph without cycles, i.e., a tree. The new tree has k edges, hence by the inductive hypothesis, it has $k + 1$ vertices. Thus, since v was the only vertex deleted, G has $k + 2$ vertices and the induction is complete.

We leave it to the reader (Exercise 7) to show the converse by a similar argument based on Lemma 3. ■

We will not be able to find a spanning tree if the given graph is not connected. Such a graph would be improper input into the algorithm described below. The next result shows that for a finite graph, this is the only kind of improper input.

THEOREM 2. If G is a finite, connected graph, then G has a spanning tree.

Proof. If G has no cycles, then G is already a tree. If there is a cycle, delete any edge in that cycle and the graph will still be connected (see Figure 1.15). Continue to delete edges in this way until there are no remaining cycles. ■

Activity 2 – If, in Theorem 2, the graph G is a directed graph, do you think the conclusion of the theorem is still true? What sort of connectivity should the hypothesis of the theorem refer to?

Figure 1.15 – Deleting an edge from a cycle does not disconnect the graph

Figure 1.16 – Inserting an edge between different components does not create a cycle

The algorithm for finding spanning trees is of the "greedy" variety; that is, we examine one edge of the graph at a time and include it into the edge set of the tree if we can. We must therefore have a way of telling whether the inclusion of an edge would produce a cycle. A useful device is to keep track of the connected component containing each vertex. If the vertices of the edge being examined are in different components (each of which had no cycles), then there will still be no cycles after the edge is included. Figure 1.16 shows this; you are asked for a proof in Exercise 8. If the vertices on the edge being considered are already in the same component, then adding the edge would create a cycle, so we should skip the edge. Using this criterion for including edges, we stop when we reach $n-1$ included edges. By construction, our spanning tree candidate T is a graph with no cycles and $n-1$ edges. By Theorem 1, it remains only to check that all of the vertices of G are in a single connected component of T.

The algorithm has at its disposal the number of vertices n of a connected, undirected graph G and the edge set of G, labeled $\{E(1), \ldots, E(M)\}$. The vertices have been labeled $1, \ldots, n$, and we will let $C(i)$ be the current component to which vertex i belongs. The letter T will stand for the set of edges currently in the tree candidate, and K will be the number of edges in the set T (see [18], Algorithm 3.22 and ensuing discussion).

UNDIRECTED SPANNING TREE ALGORITHM
1. Initialize: $T = \phi$, $K = 0$, $C(i) = i$ for each i, current edge $E(1)$.
2. If $K = n - 1$, stop; else repeat 3-4 for the current edge until $K = n - 1$.
   3. Examine the vertices i and j of the current edge. If
    $C(i) \neq C(j)$, do a-c.
   a. Include the current edge into T.
   b. Add 1 to K.
   c. If $C(i) < C(j)$, change $C(j)$, and all component
   numbers $C(k)$ matching $C(j)$, to $C(i)$; else change $C(i)$,
   and all component numbers $C(k)$ matching $C(i)$, to $C(j)$.
   4. Consider the next edge.

Proof that the algorithm yields a spanning tree. We will show that at the end of execution of the algorithm, $v(T) = n$, where $v(T)$ denotes the number of vertices in the spanning tree candidate, and we will show that the number L of connected components of T is 1. We do this by finding expressions relating n, K, L, and $v(T)$. To simplify the presentation, we treat the algorithm as if component numbers are not assigned until a vertex is examined. Then the addition of an edge has three cases, with the changes to $v(T)$, K, and L shown in Figure 1.17. In all cases one new edge is added, so K increases by 1. If two new vertices that have not yet been included into T are used, then $v(T)$ increases by 2; if the added edge links a vertex in T with a vertex not in T, then $v(T)$ increases by 1; and if the added edge connects two old vertices that happened to be in different components, then $v(T)$ does not change. You should check the L column for yourself. Let n_1, n_2, and n_3, respectively, be the numbers of times that edge addition was of type 1, 2, and 3 in the execution of the algorithm. Then

$$v(T) = 2\,n_1 + n_2$$
$$K = n_1 + n_2 + n_3$$
$$L = n_1 - n_3$$

Certainly $n \geq v(T)$, hence

$$n - v(T) = K + 1 - v(T) = n_3 - n_1 + 1 = -L + 1 \geq 0$$

Since $L > 0$, it must be that $L = 1$. Therefore all vertices are in the same connected component at the end. But then the above computation implies that $n = v(T)$. Hence the spanning tree candidate spans all vertices, is connected, and was built in a way that forbade cycles. It is therefore a spanning tree. ∎

Case	$v(T)$	K	L
1. new vertex $--$ new vertex	$+2$	$+1$	$+1$
2. old vertex $--$ new vertex	$+1$	$+1$	0
3. old vertex $--$ old vertex	0	$+1$	-1

Figure 1.17 – Adding a new edge in the spanning tree algorithm

As we discussed in Section 1.1, it is advantageous both from the pedagogical standpoint and the practical standpoint to be able to implement an algorithm like this one, and to do so one must consider assumptions, input, output, structuring of data, local variables, and supporting functions. The description of the algorithm here suggests that instead of characterizing the graph as an adjacency matrix, we could instead characterize it by two variables: the number n of vertices and the list of undirected edges $\{\{u_1, v_1\}, \{u_2, v_2\}, \ldots, \{u_k, v_k\}\}$. We assume that our input constitutes a connected, undirected graph, and that the vertex numbers referred to in the edge list are within the range 1, 2, ..., n. If we would like to use Display-Graph to show the graph, however, we are forced to produce a utility function that accepts such a list of edges, and the number of vertices, and returns the adjacency matrix of the graph. I have already included a function in the KnoxOR`Graphs` package that does this, as follows:

```
(**   ConvertToAdjMatrix[
        edgelist,numberofvertices,opts] **)
```

```
Options[ConvertToAdjMatrix]
```

```
{GraphType → Undirected, Weighted → False}
```

If the option GraphType is left at its default value of Undirected, then each edge $\{u, v\}$ in the list will be assumed to be an undirected edge, which produces two entries in the adjacency matrix in both the u, v and v, u positions. Otherwise if it is set to Directed only the u, v entry is set. I built the function so that if each entry in the list of edges has a third component signifying the weight of the edge, then the weight matrix is returned. This is the role of the option Weighted in the command; when set to True the input is expected to be a list of triples and the output will be the weight matrix. In Exercise 10 you are asked to write a simpler version of ConvertToAdjMatrix without the options.

Another useful supporting function would have the responsibility of updating component numbers when a new edge $\{u, v\}$ is inserted. The main algorithm will need to keep track of these component numbers at each step, and therefore we will need as a local variable a list of component numbers, $\{c_1, c_2, \ldots, c_n\}$, one for each vertex. I have written another function, contained in KnoxOR`Graphs`, to do this. For your information, the complete code for AdjustComponents is shown below. You should try running it on a few sample component lists and vertex numbers. Notice from the code that after finding the number of vertices n and making a copy of the input component list, the smaller and larger of the component numbers of the two vertices u and v are computed. Then a Do loop is used to examine each member of the component list, changing all that equal the larger component number to the smaller.

```
(****  AdjustComponents[u,v,components]  ****)
```

```
AdjustComponents[u_ , v_ , components_] :=
 Module[{newcomponents, smallcomp, bigcomp, n},
  n = Length[components];
  newcomponents = components;
  smallcomp =
   Min[components[[u]], components[[v]]];
  bigcomp = Max[components[[u]],
     components[[v]]];
  (* these are the connected component
     numbers of u and v *)
  (* now cruise through the list making
     all elements with the larger
component number have the smaller
     component number *)
  Do[If[newcomponents[[i]] == bigcomp,
       newcomponents[[i]] = smallcomp],
    {i, 1, n}];
  newcomponents]
```

The following example call to the function shows that when each of six vertices has its own component number, and edge $\{1, 3\}$ is to be added, the component of vertex 3 is adjusted down to that of vertex 1.

```
AdjustComponents[1, 3, {1, 2, 3, 4, 5, 6}]
```

```
{1, 2, 1, 4, 5, 6}
```

You are asked to produce a full program for the undirected spannning tree problem in Exercise 11. But the KnoxOR`Graphs` package contains a function called SpanningTreeOneStep that performs one step of the loop in the algorithm, which we can use sequentially to solve problems.

```
(* SpanningTreeOneStep[treelist,edgelist,
   edgenumber,componentlist, opts] *)
```

```
Options[SpanningTreeOneStep]
```

```
{ShowTree → True, Weighted → True,
 GraphType → Undirected, VertexLabels → Automatic,
 VertexPositions → Automatic,
 VertexLabelPositions → Automatic,
 EdgeLabels → Automatic,
 EdgeLabelPositions → Automatic,
 EdgeStyle → Thickness[0.005], EdgeSeparation → 0.01,
 DisplayFunction → (Display[$Display, #1] &),
 AspectRatio → 1,
 LoopPositions → Automatic, LoopSize → 0.05}
```

The full version of SpanningTreeOneStep takes the list of edges currently in the tree, the complete list of edges in the graph, the number of the edge in the graph edge list that is currently being considered for addition to the tree, and the current list of components of the vertices. It returns a pair {newtreelist, newcomponentlist}, which are the revised lists of tree edges and vertex components after the current edge has been added (if possible). SpanningTreeOneStep has options as shown above, which are just the options of DisplayGraph together with two new ones: ShowTree can be left at True or set to False, respectively, according to whether you do or do not want to see the new tree. The option Weighted is True by default to indicate a weighted graph, in which the edgelist and treelist entries each have a third component for the weight, or Weighted can be set to False in the case of this section where there are no weights on the edges. I do not want to load you down with the details of implementing options in *Mathematica*, and so in the cell below we look at just the lines that are directly relevant to the algorithm rather than the complete function. If you are curious, you can use a text processor to open up the graphs.m file in the KnoxOR subdirectory of the *Mathematica* ExtraPackages directory to see the full code. But read the stripped-down code below, paying attention to the comments, and relate it to the algorithm.

```
(* Important parts of SpanningTreeOneStep *)
numvertices = Length[componentlist];
newcomplist = componentlist;
newtreelist = treelist;
(* the new edge to be considered is v1,v2 *)
v1 = edgelist[[edgenumber]][[1]];
v2 = edgelist[[edgenumber]][[2]];
(* its current component
   numbers are comp1 and comp2 *)
comp1 = componentlist[[v1]];
comp2 = componentlist[[v2]];
If[comp1 != comp2, (* add the new
     edge and adjust component numbers *)
   AppendTo[newtreelist, edgelist[[edgenumber]]];
   newcomplist =
   AdjustComponents[v1, v2, componentlist]];
(* prepare the revised tree for display *)
A = ConvertToAdjMatrix[newtreelist, numvertices];
DisplayGraph[A];
(* return the new tree and component lists *)
{newtreelist, newcomplist}
```

To solve a problem using SpanningTreeOneStep, we would do the initializations of the algorithm, that is, define the edge list, set the tree list to be empty, and set the component list to {1, 2, ..., *n*}. Then successively set {treelist, componentlist} to be the output of SpanningTreeOneStep, stepping along the index of the edge to be tested, until the full tree is built. We illustrate with the following example.

EXAMPLE 2. Applying the algorithm to the graph of Figure 1.1, with the edges labeled in lexicographical order, results in the following sequence of operations. Here are the initializations:

```
edgelist =
  {{1, 2}, {1, 3}, {1, 4}, {1, 5}, {1, 6}, {2, 3},
    {2, 4}, {2, 5}, {2, 6}, {3, 4}, {4, 5}, {5, 6}};
treelist = {};
componentlist = {1, 2, 3, 4, 5, 6};
vposex2 = {{0, 0}, {0, 1},
    {2, 0}, {1, -.5}, {-1, -.5}, {-2, 0}};
vlabelposex2 = {Below, Above, ToRight,
    Below, Below, ToLeft};
```

```
{treelist, componentlist} = SpanningTreeOneStep[
    treelist, edgelist, 1, componentlist,
    Weighted → False, VertexPositions → vposex2,
    VertexLabelPositions → vlabelposex2,
    AspectRatio → .7]
```

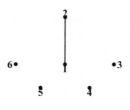

```
{{{1, 2}}, {1, 1, 3, 4, 5, 6}}
```

Figure 1.18 – Adding the first edge to a spanning tree

The first edge {1, 2} has been successfully added, and the component number of vertex 2 was changed to 1. Now we make a similar function call using the revised tree and component lists and edge 2.

```
{treelist, componentlist} = SpanningTreeOneStep[
   treelist, edgelist, 2, componentlist,
   Weighted → False, VertexPositions → vposex2,
   VertexLabelPositions → vlabelposex2,
   AspectRatio → .7]
```

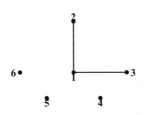

$$\{\{\{1, 2\}, \{1, 3\}\}, \{1, 1, 1, 4, 5, 6\}\}$$

Figure 1.19 – Adding the second edge to a spanning tree

Edge {1, 3} has been added. Note that vertices 1, 2, and 3 all belong to connected component 1, and vertices 4, 5, and 6 belong to their own components. The next edges in our list are {1, 4}, {1, 5}, and {1, 6}, and clearly they can each be added without producing a cycle. Here are the last three steps with the graphs suppressed:

```
{treelist, componentlist} = SpanningTreeOneStep[
   treelist, edgelist, 3, componentlist,
   Weighted → False, ShowTree → False]
{treelist, componentlist} = SpanningTreeOneStep[
   treelist, edgelist, 4, componentlist,
   Weighted → False, ShowTree → False]
{treelist, componentlist} = SpanningTreeOneStep[
   treelist, edgelist, 5, componentlist,
   Weighted → False, ShowTree → False]
```

$$\{\{\{1, 2\}, \{1, 3\}, \{1, 4\}\}, \{1, 1, 1, 1, 5, 6\}\}$$

$$\{\{\{1, 2\}, \{1, 3\}, \{1, 4\}, \{1, 5\}\}, \{1, 1, 1, 1, 1, 6\}\}$$

```
{{{1, 2}, {1, 3}, {1, 4}, {1, 5}, {1, 6}},
 {1, 1, 1, 1, 1, 1}}
```

The algorithm therefore finds the spanning tree of Figure 1.14(a). In the electronic version of the text, the closed cell below contains code to produce all the intermediate graphs, which can be selected and animated. You will observe that the spanning tree that is returned by the algorithm is highly dependent on the order in which the edges of the graph appear in the edgelist. (See Exercise 2.) In the next section we take advantage of this fact to produce an easy adaptation of our algorithm to the problem of finding minimal weight spanning trees. ∎

Directed Spanning Trees

The condition characterizing existence of directed spanning trees is a bit different. As Figure 1.13(a) shows, connectivity of the underlying graph is not strong enough to guarantee the existence of a directed spanning tree. Strong connectivity is sufficient, but not the weakest possible sufficient condition, as Figure 1.13(b) shows. It turns out that quasi-connectivity is the correct condition. If a directed graph has a spanning tree, then the root of the tree is a common ancestor of every pair of vertices, and so the graph is quasi-connected. This is half of Theorem 3 below. The converse will be proved constructively by devising an algorithm to find a spanning tree, assuming that the graph is quasi-connected ([47], Theorem 5.5.3).

THEOREM 3. A directed graph has a directed spanning tree if and only if it is quasi-connected. ∎

We would like to construct an algorithm to produce a directed spanning tree of a given directed graph, given its root. We should first pay some attention, however, to the problem of locating the root of a quasi-connected graph. This is not at all an obvious thing to do, especially for large graphs such as the one in Figure 1.20.

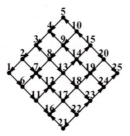

Figure 1.20 – Where is the root?

We will not give a program to find the root, which is likely to be a very time-consuming one. However, if you did Exercise 9 of Section 1.1, you saw how to use a function called FindChildren[adjmatrix, parentlist] contained in KnoxOR`Graphs` to produce a function called Descendants[adjmatrix, vertex], which returns the set of all descendants of the given vertex in the directed graph characterized by the given adjacency matrix. This Descendants function is very much like the Components function of the last section in the sense that it begins with the vertex and step-by-step fans out to child vertices of those vertices most recently examined, labeling those as new descendants. The code for my version of this function is in the closed cell below this paragraph. If you have a guess at which vertex is the root, you can then apply Descendants to that vertex, and if the set of descendants is the entire set of vertices in the graph, then this vertex is a root. (There may be more than one root.) So this function does not quite do the complete job of finding the root, but it does provide a useful check.

```
(****     Descendants[adjmatrix,vertex]     ****)
```

The adjacency matrix for the graph of Figure 1.20 was defined in the closed cell above the figure, and was given the name graph120. I designed that graph to have a root at vertex 14, and the function call below shows that it is indeed a root because all vertices are descendants of vertex 14.

```
Descendants[graph120, 14]
```

{1, 2, 3, 4, 5, 6, 7, 8, 9, 10, 11, 12, 13,
  14, 15, 16, 17, 18, 19, 20, 21, 22, 23, 24, 25}

Activity 3 – In the electronic version of the text, try to find out what other vertices are also roots of the graph of Figure 1.20.

Next, to construct a directed spanning tree, we search for vertices, beginning at the root and fanning out to children of the root, then their children, etc., much like the connected components algorithm. The directed spanning tree algorithm requires a quasi-connected, directed graph G of n vertices and its root w. The set-valued variables KT and ET indicate, respectively, the current set of vertices in the tree and the current set of edges in the tree. The variables NEW1 and NEW2, respectively, represent the most recent set of vertices that has been added and the next set of vertices to be added. At each pass through the loop in step 2, each vertex in the NEW1 set is examined, and if its children have not been included into the vertex set of the tree yet, then those edges are included into the tree and the children are included into the vertex set and also marked as newly added vertices for the next pass.

DIRECTED SPANNING TREE ALGORITHM
   1. Initialize KT $= \{w\}$, ET $= \phi$, NEW1 $= \{w\}$, NEW2 $= \phi$.
   2. Do a – b while NEW1 $\neq \phi$.
       a. For each $v \in$ NEW1:
           For each edge (v, u) such that $u \notin$ KT, include (v, u)
           into ET and u into NEW2 and KT.
       b. Let NEW1 $=$ NEW2, NEW2 $= \phi$.
   3. Return ET.

The algorithm makes sure that at every stage there are no edges pointing back into the current set of vertices, hence there are no cycles. The underlying graph is connected at every stage, since all new vertices are added with the edge connecting them to their parent, who is already connected to the rest of the graph. Note in particular that there is always a path in the tree from the root to every new vertex that is added. To see that a directed spanning tree results at the end of the execution of the algorithm, suppose instead that there is some vertex $v \in G$ that is never included into KT. Since G is quasi-connected, there is a path $w = v_0, v_1, ..., v_m = v$ in G from the root w to v. Let k be the smallest integer $\leq m$ such that v_k is not in KT. Such a k exists since $v_m = v \notin$ KT, and $k > 0$ since $w = v_0 \in$ KT. Then $v_0, ..., v_{k-1} \in$ KT but $v_k \notin$ KT. Then v_{k-1} was in NEW1 for exactly one execution of step 2. But in that case v_k would have been included into KT at step 2a, which is a contradiction. Therefore, under the assumption of quasi-connectivity, the algorithm generates a directed spanning tree. We have also established Theorem 3.

Activity 4 – If a directed graph is strongly connected, can the algorithm above be applied starting at any vertex w? Why, or why not?

EXAMPLE 3. To illustrate the application of the spanning tree algorithm for directed graphs, consider the graph of Figure 1.21(a). You can check that 5 is a root of this graph. The algorithm first includes edges (5, 2) and (5, 4); then edges (2, 1) and (4, 3) while ignoring edge (2, 4); then it checks and ignores edges (1, 3), (1, 4), (3, 1), and (3, 4). We see from this example that to improve the efficiency of the algorithm, we might add a check on the size of KT; if it is n, then stop execution. The spanning tree we obtain is shown in part (b) of the figure. ∎

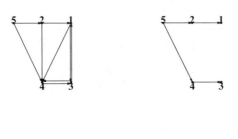

(a)   (b)
Figure 1.21 – Finding a directed spanning tree

Let us close this section by developing a *Mathematica* function for the directed spanning tree problem. Then we will test it out on the more complicated graph of Figure 1.20.

Because the "fanning out" method, called by the computer scientists *breadth-first search*, is used, we will give our function the name BreadthFirst-Tree. The function will receive the adjacency matrix in the usual form, and the vertex number w of the root as its input parameters. Our version will return the edge list of the directed tree. Recall that the ConvertToAdjMatrix function can be used to change the edge list form of the tree graph to adjacency matrix form, in order to prepare it for display if we like. We assume that the vertex number of the root is in the appropriate range determined by the dimension of the adjacency matrix, and that the graph is quasi-connected and the vertex is indeed a root. As local variables we will keep KT, the list of vertices in the tree so far, ET, the list of edges in the tree so far, and the lists NEW1 and NEW2 of most recently added and newly added vertices. The algorithm shows us exactly how to initialize and update these lists. As a supporting function we can use FindChildren[adjmatrix, parentlist], with one parent vertex from NEW1 at a time, to find all children of that vertex. Then we use complementation to pick out those children not already in KT, in order to add them into KT, add the edge to ET, and mark them as newly labeled. This entails the creation of other local variables called children and newchildren to hold onto the lists of all children of the current vertex, and those that are new. By the end of the loop, ET has the complete list of edges. Here is the function. Inspect the code carefully to be sure that you know what each line does and how it relates to the algorithm.

```
BreadthFirstTree[adjmatrix_, w_] :=
 Module[{numvertices, KT, ET,
   NEW1, NEW2, children, newchildren},
   (* initializations *)
   numvertices = Length[adjmatrix];
   KT = {w}; ET = {}; NEW1 = {w}; NEW2 = {};
   While[NEW1 ≠ {} && Length[KT] < numvertices ,
    (* for each vertex in NEW1,
      scan children for new vertices *)
    Do[children = FindChildren[
       adjmatrix, {NEW1[[vertex]]}];
     newchildren = Complement[children, KT];
     (* include the new children
       into the tree and the NEW2 set *)
     NEW2 = Join[NEW2, newchildren];
     KT = Join[KT, newchildren];
     (* put edges into ET *)
     Do[AppendTo[ET,
       {NEW1[[vertex]], newchildren[[i]]}],
      {i, 1, Length[newchildren]}],
     {vertex, 1, Length[NEW1]}];
    (* set up for the next pass *)
    NEW1 = NEW2;
    NEW2 = {}];
   ET]
```

EXAMPLE 4. Here is the spanning tree of the graph of Figure 1.21(a). Of course the edges are exactly the ones in Figure 1.21(b).

```
BreadthFirstTree[graph121a, 5]
```

{{5, 2}, {5, 4}, {2, 1}, {4, 3}}

Now we apply the function to the graph of Figure 1.20, using vertex 14 as the root. The tree is displayed in Figure 1.22 using the ConvertToAdjMatrix function. Notice how the order of edges in the edgelist tells the order in which the edges were added in; starting from 14 we get {14, 9} and {14, 15}, then from vertex 9 we get {9, 4} and {9, 8}, etc. Trace a few more of these edges on the tree shown in Figure 1.22 to see how the algorithm is fanning out through the graph breadthwise. ■

```
edgelist = BreadthFirstTree[graph120, 14]
```

{{14, 9}, {14, 15}, {9, 4}, {9, 8},
 {15, 10}, {15, 20}, {4, 3}, {8, 7}, {8, 13},
 {10, 5}, {20, 19}, {20, 25}, {3, 2}, {13, 12},
 {13, 18}, {19, 24}, {2, 1}, {18, 17}, {18, 23},
 {1, 6}, {17, 16}, {17, 22}, {6, 11}, {22, 21}}

```
vposits = {{-2, 0}, {-1.5, .5},
    {-1, 1}, {-.5, 1.5}, {0, 2},
    {-1.5, -.5}, {-1, 0}, {-.5, .5},
    {0, 1}, {.5, 1.5},
    {-1, -1}, {-.5, -.5}, {0, 0}, {.5, .5}, {1, 1},
    {-.5, -1.5}, {0, -1},
    {.5, -.5}, {1, 0}, {1.5, .5},
    {0, -2}, {.5, -1.5}, {1, -1},
    {1.5, -.5}, {2, 0}};
amatrix = ConvertToAdjMatrix[edgelist,
    25, GraphType → Directed];
DisplayGraph[amatrix, GraphType → Directed,
    VertexPositions → vposits];
```

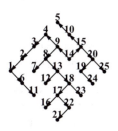

Figure 1.22 – A directed spanning tree for the graph of Figure 1.20

Exercises 1.2

1. Find a spanning tree of the graph below using the undirected spanning tree algorithm. Work by hand on this problem rather than using *Mathematica*. Assume that the order of the edges is:

$\{1, 2\}, \{2, 5\}, \{2, 3\}, \{4, 7\}, \{2, 6\}, \{1, 4\}, \{2, 4\}, \{3, 4\}, \{6, 7\}, \{5, 6\}, \{4, 5\}$

Exercise 1

2. (*Mathematica*) Referring to Example 2 of this section, what spanning tree does the SpanningTreeOneStep function find when the order of edges is:

(a) $\{\{5, 6\}, \{4, 5\}, \{3, 4\}, \{2, 6\}, \{2, 5\},$
    $\{2, 4\}, \{2, 3\}, \{1, 6\}, \{1, 5\}, \{1, 4\}, \{1, 3\}, \{1, 2\}\}$
(b) Determined by increasing order of cost. (See Figure 1.1)

Compute the total edge cost for each of the two spanning trees in (a) and (b).

3. Suppose that $G = (V, E)$ is a connected graph and $\{u, v\} \in E$ is an edge in some cycle. Show that the graph $G' = (V, E) - \{u, v\}$ is connected. (This fact was used in the proof of Theorem 2.)

4. Prove that a connected, undirected graph G is a tree if and only if for each edge $\{u, v\} \in G$, $G - \{u, v\}$ is not connected.

5. (*Mathematica*) The graph below shows computer links between an official vote-tallying center at vertex 1 and several precincts. For the sake of secrecy, links can be made secure, but since this is an expensive process, it is desired to secure the minimum possible number of links and let the transmissions occur only on those links. How should this be done? (Use the SpanningTree-OneStep function to display intermediate graphs and component lists.)

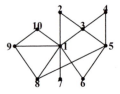

Exercise 5

6. Prove that if G is an undirected tree with more than one vertex, then G contains at least two vertices of degree 1.

7. Finish the proof of Theorem 1, that is, if G is a connected graph with n vertices and $n - 1$ edges, then G is a tree.

8. Consider two connected components of an undirected graph G, and suppose each has no cycles. Let G' be a new graph whose vertex set is the union of the vertex sets of the two components and whose edge set is the union of the two edge sets, together with a single edge $\{u, v\}$, where u is in one component and v is in the other. Show that G' has no cycles.

9. Is it possible to construct an undirected tree whose eight vertices have degrees 1, 2, 3, 3, 1, 1, 3, and 2, respectively? Why, or why not?

10. (*Mathematica*) Write your own version of the ConvertToAdj-Matrix[edgelist, n] command without options, which takes a list of edges of an undirected graph and the number of vertices in the graph, and returns the adjacency matrix.

11. (*Mathematica*) Using the work already done in creating the SpanningTree-OneStep function, write a full, simplified version of the complete undirected spanning tree algorithm, without the options, which takes the list of edges and the number of vertices in the graph, and returns the list of edges in a spanning tree.

12. A *complete* undirected graph is a graph such that edges exist between every pair of vertices. Find an upper bound for the number of spanning trees a complete graph can have.

13. Find a directed spanning tree of the following graph if one exists. Is the tree unique? Do this by hand and not in *Mathematica*.

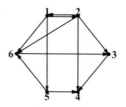

Exercise 13

14. (*Mathematica*) Vertices 8 and 18 are also roots in the graph of Figure 1.20. Check this using the Descendants function, and find directed spanning trees using each of these roots.

15. A forced-air heat distribution system in a building must get heat from the central furnace at vertex 1 in the figure to each of the rooms located at the other vertices. It is possible to mount ductwork, assumed unidirectional, along each of the edges shown on the graph. Find a sparsest possible system of ductwork to heat the building.

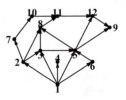

Exercise 15

16. (*Mathematica*) A directed graph has the adjacency matrix below. Use the BreadthFirstTree function to find a directed spanning tree. Determine the root using the Descendants function.

$$A = \begin{pmatrix} 0 & 1 & 1 & 1 & 0 & 0 & 0 & 0 & 0 \\ 0 & 0 & 0 & 1 & 1 & 0 & 0 & 0 & 0 \\ 0 & 0 & 0 & 1 & 0 & 0 & 0 & 0 & 0 \\ 0 & 0 & 0 & 0 & 1 & 1 & 1 & 0 & 0 \\ 0 & 0 & 0 & 0 & 0 & 0 & 0 & 1 & 0 \\ 0 & 0 & 0 & 0 & 0 & 0 & 0 & 1 & 1 \\ 0 & 0 & 0 & 0 & 0 & 0 & 0 & 0 & 1 \\ 0 & 0 & 0 & 0 & 0 & 0 & 0 & 0 & 1 \\ 0 & 0 & 0 & 0 & 0 & 0 & 0 & 0 & 0 \end{pmatrix}$$

Exercise 16

$$B = \begin{pmatrix} 0 & 0 & 0 & 0 & 1 & 0 & 0 & 0 & 0 & 0 & 0 & 0 & 0 & 0 & 0 & 0 \\ 1 & 0 & 0 & 0 & 0 & 0 & 0 & 0 & 0 & 0 & 0 & 0 & 0 & 0 & 0 & 0 \\ 0 & 1 & 0 & 1 & 0 & 0 & 0 & 0 & 0 & 0 & 0 & 0 & 0 & 0 & 0 & 0 \\ 0 & 0 & 0 & 0 & 0 & 0 & 0 & 1 & 0 & 0 & 0 & 0 & 0 & 0 & 0 & 0 \\ 0 & 0 & 0 & 0 & 0 & 0 & 0 & 0 & 1 & 0 & 0 & 0 & 0 & 0 & 0 & 0 \\ 0 & 1 & 0 & 0 & 1 & 0 & 0 & 0 & 0 & 1 & 0 & 0 & 0 & 0 & 0 & 0 \\ 0 & 0 & 1 & 0 & 0 & 1 & 0 & 1 & 0 & 0 & 1 & 0 & 0 & 0 & 0 & 0 \\ 0 & 0 & 0 & 0 & 0 & 0 & 0 & 0 & 0 & 0 & 0 & 1 & 0 & 0 & 0 & 0 \\ 0 & 0 & 0 & 0 & 0 & 0 & 0 & 0 & 0 & 0 & 0 & 0 & 0 & 0 & 0 & 0 \\ 0 & 0 & 0 & 0 & 0 & 0 & 0 & 0 & 1 & 0 & 0 & 0 & 0 & 1 & 0 & 0 \\ 0 & 0 & 0 & 0 & 0 & 0 & 0 & 0 & 0 & 1 & 0 & 1 & 0 & 0 & 1 & 0 \\ 0 & 0 & 0 & 0 & 0 & 0 & 0 & 0 & 0 & 0 & 0 & 0 & 0 & 0 & 0 & 1 \\ 0 & 0 & 0 & 0 & 0 & 0 & 0 & 0 & 1 & 0 & 0 & 0 & 0 & 0 & 0 & 0 \\ 0 & 0 & 0 & 0 & 0 & 0 & 0 & 0 & 0 & 0 & 0 & 0 & 1 & 0 & 0 & 0 \\ 0 & 0 & 0 & 0 & 0 & 0 & 0 & 0 & 0 & 0 & 0 & 0 & 0 & 1 & 0 & 0 \\ 0 & 0 & 0 & 0 & 0 & 0 & 0 & 0 & 0 & 0 & 0 & 0 & 0 & 0 & 1 & 0 \end{pmatrix}$$

Exercise 17

17. (*Mathematica*) Repeat Exercise 16 for the graph whose adjacency matrix is above. The root vertex is 7.

18. Prove or disprove. A directed graph is a tree if and only if it is connected and has no directed cycles.

19. Prove or disprove. A directed graph is a tree if and only if it is quasi-connected and has no directed cycles.

20. Would the directed spanning tree algorithm also find an undirected spanning tree if the given graph was connected and undirected? Explain.

1.3 Minimal Cost Networks

Undirected Graphs

We now show how to find a minimal cost spanning tree, as defined below, for a weighted, undirected graph such as that of Figure 1.1. Let G be such a graph.

> **DEFINITION 1**. A *minimal spanning tree* for G is a spanning tree whose total weight is less than or equal to that of any other spanning tree.

The algorithm to find minimal spanning trees is a variation of the undirected spanning tree algorithm from the last section called *Kruskal's algorithm*. The idea is that before executing the spanning tree algorithm, we sort the edges in order of increasing weight. Then, we greedily pick edges of small weight, without creating a cycle. The input to the algorithm is: the number n of vertices, the set of edges $\{E(1), ..., E(M)\}$, and the costs $c(E(i))$, for $i = 1, ..., M$.

> **KRUSKAL'S ALGORITHM FOR MINIMAL**
> **UNDIRECTED SPANNING TREES**
> 1. Sort the edges in increasing order of cost, i.e., rename edges such that
> $$c(E(1)) \le c(E(2)) \le ... \le c(E(M))$$
> 2. Execute the spanning tree algorithm to find a spanning tree T.
> 3. Add $c(E(i))$ for those edges $E(i) \in T$ to find the minimal cost.

Proof that Kruskal's algorithm yields a minimal undirected spanning tree. Denote by T^* the spanning tree found by Kruskal's algorithm. Since there are only finitely many spanning trees, there must exist at least one of minimal cost. Let T be a minimal cost spanning tree with the following property: the intersection of the edge set of T^* with that of T has at least as many elements as the intersection of the edge set of T^* with any other minimal cost spanning tree. That is, T is a "best-fitting" minimal cost spanning tree for T^*. We prove the optimality of T^* by obtaining a contradiction of the assumption that $T \neq T^*$. If the latter is true, then there is an edge e in $T^* - T$ such that $c(e) \leq c(e')$ for all edges e' in $T^* - T$. But we will find an edge $e' \in T^* - T$ whose cost is strictly less than the cost of e. This contradiction suffices to prove the result.

By Theorem 1 of Section 1.2, if edge e is included into T, a cycle must form. Within this cycle, there must be an edge $f \neq e$ that did not belong to T^*, else T^* would have had a cycle. If we form a new graph $T' = T - \{f\} \cup \{e\}$, then T' is still a spanning tree and the total tree costs $C(T)$ and $C(T')$ satisfy

$$C(T') - C(T) = c(e) - c(f) \geq 0$$

by the optimality of T. But if equality held in the above, then T' would be optimal and would share one more edge with T^* than T does, contradicting our choice of T. Thus, $c(e) > c(f)$. Again by Theorem 1, if edge f is included into T^*, a cycle forms. Then, just as above, there is an edge $e' \neq f$ in this cycle such that $e' \in T^* - T$. We must have $c(e') \leq c(f)$, else Kruskal's algorithm would have added f to T^* before considering e'. But this means that

$$c(e') \leq c(f) < c(e)$$

which is the contradiction of the choice of e for which we sought. ∎

Activity 1 – Devise some examples of real problems in which the goal is to find a minimal spanning tree.

We can again solve problems step-by-step using the SpanningTreeOne-Step function. Remember that if we leave the option Weighted at its default value of True, then the command accepts edge lists and tree lists, which are lists of triples, the third element of which is the weight of the edge determined by the first two elements. In the initialization phase, we would define the edge list, sort the edges, set the tree list to be empty, and set the component list to $\{1, 2, \ldots, n\}$. Then as before, successively set {treelist, componentlist} to be the output of SpanningTreeOneStep, incrementing the index of the edge to be tested, until there are $n - 1$ edges.

You can either enter the edges manually in increasing order of cost, or use the following utility in the KnoxOR`Graphs` package.

```
Needs["KnoxOR`Graphs`"]
```

```
(********    SortEdges[edgelist]    ********)
```

SortEdges takes the list of triples and returns a list of the same structure, with entries sorted in increasing order of the third component.

EXAMPLE 1. Let us apply Kruskal's algorithm to the communications network of Figure 1.1, redisplayed below for your convenience.

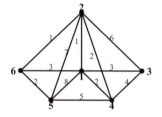

Figure 1.1 (reprise)

We start with the edges labeled in lexicographical order as in the last section, then apply SortEdges, and continue.

```
unsortededgelist =
  {{1, 2, 1}, {1, 3, 3}, {1, 4, 2}, {1, 5, 8},
    {1, 6, 3}, {2, 3, 6}, {2, 4, 2}, {2, 5, 7},
    {2, 6, 1}, {3, 4, 4}, {4, 5, 5}, {5, 6, 2}};
edgelist = SortEdges[unsortededgelist]
treelist = {};
componentlist = {1, 2, 3, 4, 5, 6};
vposits = {{0, 0}, {0, 1},
    {2, 0}, {1, -.5}, {-1, -.5}, {-2, 0}};
vlabelposits = {Below, Above, ToRight,
    Below, Below, ToLeft};
```

```
{{1, 2, 1}, {2, 6, 1}, {1, 4, 2}, {2, 4, 2},
 {5, 6, 2}, {1, 3, 3}, {1, 6, 3}, {3, 4, 4},
 {4, 5, 5}, {2, 3, 6}, {2, 5, 7}, {1, 5, 8}}
```

```
{treelist, componentlist} =
  SpanningTreeOneStep[treelist, edgelist, 1,
    componentlist, VertexPositions → vposits,
    VertexLabelPositions → vlabelposits,
    ShowTree → False]
```

```
{{{1, 2, 1}}, {1, 1, 3, 4, 5, 6}}
```

```
{treelist, componentlist} =
  SpanningTreeOneStep[treelist, edgelist, 2,
    componentlist, VertexPositions → vposits,
    VertexLabelPositions → vlabelposits,
    ShowTree → False]
```

```
{{{1, 2, 1}, {2, 6, 1}}, {1, 1, 3, 4, 5, 1}}
```

```
{treelist, componentlist} =
 SpanningTreeOneStep[treelist, edgelist, 3,
  componentlist, VertexPositions → vposits,
  VertexLabelPositions → vlabelposits,
  ShowTree → False]
```

{{{1, 2, 1}, {2, 6, 1}, {1, 4, 2}}, {1, 1, 3, 1, 5, 1}}

```
{treelist, componentlist} =
 SpanningTreeOneStep[treelist, edgelist, 4,
  componentlist, VertexPositions → vposits,
  VertexLabelPositions → vlabelposits,
  ShowTree → False]
```

{{{1, 2, 1}, {2, 6, 1}, {1, 4, 2}}, {1, 1, 3, 1, 5, 1}}

```
{treelist, componentlist} =
 SpanningTreeOneStep[treelist, edgelist, 5,
  componentlist, VertexPositions → vposits,
  VertexLabelPositions → vlabelposits,
  ShowTree → False]
```

{{{1, 2, 1}, {2, 6, 1}, {1, 4, 2}, {5, 6, 2}},
 {1, 1, 3, 1, 1, 1}}

```
{treelist, componentlist} =
 SpanningTreeOneStep[treelist, edgelist, 6,
  componentlist, VertexPositions → vposits,
  VertexLabelPositions → vlabelposits,
  EdgeLabelPositions → elabelpos,
  AspectRatio → .7]
```

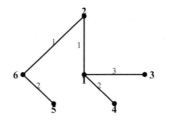

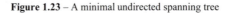

$\{\{\{1, 2, 1\}, \{2, 6, 1\}, \{1, 4, 2\}, \{5, 6, 2\}, \{1, 3, 3\}\},$
$\{1, 1, 1, 1, 1, 1\}\}$

Figure 1.23 – A minimal undirected spanning tree

In order, we add edges {1, 2}, {2, 6}, and {1, 4}, then skip edge {2, 4}, which would form a cycle. Edges {5, 6} and {1, 3} complete the tree. The total cost of this minimal spanning tree is $1 + 1 + 2 + 2 + 3 = 9$. ∎

There is a command in the KnoxOR`Graphs` package called Kruskal. The details follow

```
(*****     Kruskal[edgelist,n,opts]     *****)
```

```
Options[Kruskal]
```

```
{ShowTree → True, Weighted → True,
 GraphType → Undirected, VertexLabels → Automatic,
 VertexPositions → Automatic,
 VertexLabelPositions → Automatic,
 EdgeLabels → Automatic,
 EdgeLabelPositions → Automatic,
 EdgeStyle → Thickness[0.005], EdgeSeparation → 0.01,
 DisplayFunction → (Display[$Display, #1] &),
 AspectRatio → 1,
 LoopPositions → Automatic, LoopSize → 0.05}
```

Kruskal performs the full algorithm described above for minimal undirected spanning trees, given the weighted edge list and the number of vertices. It accepts the ShowTree and Weighted options as in SpanningTreeOneStep, and the options of DisplayGraph. Apart from the complications involved in implementing the options, Kruskal is a straightforward program that does the initializations with which we are familiar, sorts the edge list, and repeatedly examines edges, adding them to the tree and adjusting component numbers when the edge does not form a cycle. We illustrate its use in the next example.

Activity 2 – Open the closed cell below this in the electronic text to study the code for the Kruskal command.

EXAMPLE 2. A manufacturing plant is installing a computer network in the facility so that employees can enter quality control data directly at their stations. The costs of connecting various neighboring stations are given in the graph below.

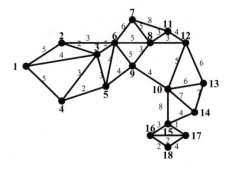

Figure 1.24 – Finding an optimal computer network

We must produce the initial list of weighted edges, in any order, to use as the first argument in Kruskal. We use lexicograhic ordering to define edges24 below. The second argument in Kruskal is the number of vertices, which is 18. (In the closed cell that generated the graph in Figure 1.24, the VertexPositions option has been set to a list called vertpos, the VertexLabelPositions option is vertlabelpos, the EdgeLabels are in a matrix called graph24, and the EdgeLabelPositions option has been set to elabelpos. These graph options will be passed to Kruskal below.)

```
edges24 = {{1, 2, 5}, {1, 3, 4},
   {1, 4, 5}, {2, 3, 2}, {2, 6, 3}, {3, 4, 3},
   {3, 5, 3}, {3, 6, 2}, {4, 5, 2}, {5, 6, 4},
   {5, 9, 4}, {6, 7, 6}, {6, 8, 5}, {6, 9, 5},
   {7, 8, 5}, {7, 11, 8}, {8, 9, 3}, {8, 11, 3},
   {8, 12, 3}, {9, 10, 4}, {10, 12, 5},
   {10, 13, 6}, {10, 14, 7}, {10, 15, 8},
   {11, 12, 4}, {12, 13, 6}, {13, 14, 7},
   {14, 15, 4}, {15, 16, 3}, {15, 17, 1},
   {16, 17, 2}, {16, 18, 2}, {17, 18, 4}};
Kruskal[edges24, 18, VertexPositions → vertpos,
   VertexLabelPositions → vertlabelpos,
   EdgeLabels → graph24, EdgeLabelPositions →
   elabelpos, AspectRatio → .7];
```

```
Edges in minimal spanning tree:
 {{15, 17}, {2, 3}, {3, 6}, {4, 5}, {16, 17}, {16, 18},
  {3, 4}, {8, 9}, {8, 11}, {8, 12}, {1, 3}, {5, 9},
  {9, 10}, {14, 15}, {7, 8}, {10, 13}, {10, 14}}

Total weight of spanning tree: 57
```

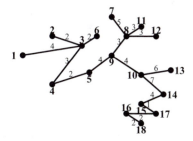

Figure 1.25 – Minimal spanning tree for computer network

The Kruskal function finds the minimal spanning tree shown in Figure 1.25. If you inspect the list of edges, you see that Kruskal first picks the only edge of cost 1, namely {15, 17}, then puts in all of the cost 2 edges, {2, 3}, {3, 6}, {4, 5}, {16, 17}, and {16, 18}, then moves on to whichever cost 3 edges it can include without creating a cycle, then passes to the cost 4 edges, etc. Exercise 3 asks you to compare the cost of this minimal tree, 57, to those of two other spanning trees. In situations where the cost is actually in units of thousands of dollars or greater, the savings achieved by designing a network using a minimal spanning tree can be substantial. ∎

Directed Graphs

A second type of minimal cost network problem involves weighted directed graphs. Suppose that vertex 1 in Figure 1.26 represents a production center, and the directed edges are possible routes to locations 2, 3, etc. that are to receive items from the production center, and then act as distributors to other locations to which they are connected. The weights, as usual, are transportation costs. All vertices except the root vertex 1 must be serviced, but it is superfluous for a location to be serviced by more than one previous distribution location. Hence the problem is to find a directed tree rooted at 1 that supplies every station and minimizes the transportation cost to every station.

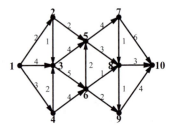

Figure 1.26 – A directed distribution system

DEFINITION 2. Let G be a quasi-connected, weighted, directed graph of n vertices rooted at v_1. A *minimal directed spanning tree T* is a directed tree of n vertices rooted at v_1 such that for all vertices v_k, $k = 2, ..., n$, the weight of the path in T from v_1 to v_k is smaller than or equal to the weight of any other path in G from v_1 to v_k.

Observe that if T is a directed spanning tree rooted at v_1, then for any vertex v in G there is a unique path in T, denoted by $P(v)$, from v_1 to v. We can clearly characterize such a tree by its paths to each vertex. If we denote the cost of a path from v_1 to v by $C(P(v))$, then we can restate the problem in the following form. Find a directed spanning tree T^* rooted at v_1 with paths $\{P^*(v)\}$, such that for each $v \in \{v_2, \ldots, v_n\}$,

$$C(P^*(v)) = \inf \{ C(P(v)) : P(v) \text{ is a path in } G \text{ from } v_1 \text{ to } v\} \qquad (1)$$

An algorithm will be given in which we begin with an arbitrary directed spanning tree, and then improve it step-by-step until optimality is reached. To see the idea behind this successive improvement strategy, we need the following theorem. It gives an example of an equation called a *dynamic programming* equation. The algorithm based on the theorem illustrates *policy improvement* for dynamic programming problems. The idea behind the DP equation is that the optimal cost to a position should be the smallest possible sum of the optimal cost to a predecessor position, plus the one-step cost from the predecessor to the current position. We will see much more of this idea in Chapter 6.

THEOREM 1. A directed spanning tree T^* rooted at v_1 is minimal if and only if for all vertices $v \neq v_1$ the paths $P^*(v)$ in the tree from the root to v satisfy

$$C(P^*(v)) = \inf_{(u,v) \in E} \{C(P^*(u)) + c(u, v)\} \qquad (2)$$

where $c(u, v)$ denotes the cost of edge (u, v). (As usual, E is the edge set of the graph G.)

Proof. Let T^* be a minimal tree, and suppose that there is an edge (u, v) such that

$$C(P^*(v)) > C(P^*(u)) + c(u, v) \qquad (3)$$

But then the path $P^*(u),v$ from the root v_1 to v has smaller total cost than $P^*(v)$, contradicting the optimality of T^*. Thus,

$$C(P^*(v)) \leq C(P^*(u)) + c(u, v) \qquad (4)$$

for all edges (u, v). Equality occurs for those edges (u, v) in T^*. This proves equation (2).

Conversely, suppose equation (2) holds for a directed spanning tree T^*, but suppose by contradiction that T^* is not optimal. Then there is a path, say,

$$P(v) : v_1 = w_1, w_2, \ldots, w_m = v$$

such that $C(P(v)) < C(P^*(v))$. There is a smallest integer k such that $C(P(w_k)) < C(P^*(w_k))$, since the set of all such k is a non-empty bounded set of positive integers. Then,

$$
\begin{aligned}
C(P^*(w_k)) \; & > \; C(P(w_k)) \\
& = \; C(P(w_{k-1})) + c(w_{k-1}, \, w_k) \\
& \geq \; C(P^*(w_{k-1})) + c(w_{k-1}, \, w_k)
\end{aligned}
$$

which contradicts equation (2). Therefore, the tree T^* is optimal. ∎

Equation (2) suggests an algorithm to find minimal cost directed spanning trees. Begin with an arbitrary directed spanning tree, say T, rooted at v_1. Breadth-first search will locate such a tree if one exists. If for all edges $(u, v) \in G$

$$C(P(v)) \leq C(P(u)) + c(u, v) \tag{5}$$

then by Theorem 1 the tree T is already optimal. Otherwise, there is at least one edge such that the *slack* of the edge is positive, where the slack $S(u, v)$ of the edge is the difference between the path cost to v and the sum of the path cost to u and the cost of edge (u, v), i.e.,

$$S(u, v) \equiv C(P(v)) - [C(P(u)) + c(u, v)] > 0 \tag{6}$$

Since the edge set is finite, there is an edge (u, v) of maximum slack. This edge is not already in the edge set of the tree T because the slack of all edges in T is clearly 0 (since the path to v is just the path to u with edge (u, v) adjoined). So, adjoin the edge (u, v) to T, while deleting the edge (u_0, v) in T that had pointed to v (see Figure 1.27). We prove shortly that this results in a strict improvement of cost. This is important because it implies that the policy improvement cannot cycle, hence the finiteness of the graph forces convergence to an optimal tree T^*. Repeat the process of edge substitution until equation (5) is true (equivalently $S(u, v) \leq 0$) for all edges (u, v).

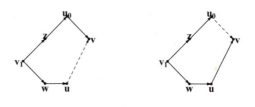

Figure 1.27 – Edge substitution in the minimal directed spanning tree algorithm

---

Activity 3 – When edge substitution occurs as shown in Figure 1.27, which vertices can and which cannot have their path costs changed? Bear in mind that there may be more to the graph than the small piece shown.

The input to the following algorithm is (1) the vertex set $\{v_1, \ldots, v_n\}$ of G, where v_1 is the root; (2) the edge sets of G and the initial tree T, denoted by EG and ET, respectively; and (3) the costs $c(u, v)$ of all edges in G. The quantities $P(v)$ and $C(P(v))$ are as described above.

MINIMAL DIRECTED SPANNING TREE ALGORITHM
   1. Repeat steps a–d until the maximum slack in the current tree is less than or equal to 0.
      a. For all vertices $v \neq v_1$, find the path $P(v)$ in T and calculate $C(P(v))$.
      b. Find an edge $(u, v) \in$ EG $-$ ET of maximum slack M.
      c. If $M > 0$, then do step d.
      d. Find the edge $(u_0, v) \in$ ET, and let
         ET $=$ ET $- \{(u_0, v)\} \bigcup \{(u, v)\}$.

Proof that algorithm yields a minimal directed spanning tree. Recall that we begin with a quasi-connected graph, so that an initial directed spanning tree can be found. We show that the loop a–d can only be executed finitely many times.

When edge (u_0, v) is replaced by edge (u, v) in step d, the only path costs that change are the ones whose terminal vertices are descendants of v (including v itself). Let w be one such vertex. Denote by $P_{\text{new}}(w)$ the new path to w, as usual let $P(w)$ denote the old path to w, and denote by $P(v, w)$

the path from v to w, which does not change during the substitution of (u, v) for (u_0, v). Then,

$$
\begin{aligned}
C(P_{\text{new}}(w)) &= C(P(u)) + c(u, v) + C(P(v, w)) \\
&< C(P(v)) + C(P(v, w)) \\
&= C(P(w))
\end{aligned}
$$

by the fact that edge (u, v) has positive slack. Thus, descendants of v receive strictly smaller costs as a result of the change in edges, and all other vertices maintain the same path cost. But there are only finitely many possible path costs to each vertex. This implies that loop a–d of the algorithm is only executed finitely many times. Since the only means of exiting the loop is by achieving the condition $M \leq 0$, we have at the end of execution $S(u, v) \leq 0$ for all edges (u, v). By Theorem 1, this implies that the final tree is minimal. ∎

Since the computations involved in carrying out the minimal directed spanning tree algorithm are simple, but very tedious, I have included a function in KnoxOR`Graphs` that performs one step of the process and graphs the new tree.

```
(**   DirectedSpanningTreeOneStep[
        theGraph,theTree,root,newedge,opts]   **)
```

```
Options[DirectedSpanningTreeOneStep]
```

```
{ShowTree → True, GraphType → Undirected,
 VertexLabels → Automatic,
 VertexPositions → Automatic,
 VertexLabelPositions → Automatic,
 EdgeLabels → Automatic,
 EdgeLabelPositions → Automatic,
 EdgeStyle → Thickness[0.005], EdgeSeparation → 0.01,
 DisplayFunction → (Display[$Display, #1] &),
 AspectRatio → 1,
 LoopPositions → Automatic, LoopSize → 0.05}
```

The arguments of DirectedSpanningTreeOneStep are the adjacency matrix of the full directed graph, the adjacency matrix of the current spanning tree, the number of the vertex that is the root, and the edge $\{u, v\}$ that is to be inserted to form the next tree. It returns the next tree in adjacency matrix form, and if the option ShowTree is left at its default value of True, the tree will be also be displayed with unused edges shown dashed, and new slack values computed and shown next to the unused edges. The command accepts the options of DisplayGraph to control the design. To go from step to step in the policy improvement algorithm, we need to inspect the graph of the previous tree to find an edge with the highest positive slack, if any, and let that edge be the fourth argument in the next call to DirectedSpanningTree-OneStep.

Since it is also tedious to set up the first graph and compute and display the slack values upon which the first edge substitution decision is made, there is a similar command, DirectedSpanningTreeFirstStep shown below, to display the first directed spanning tree. It takes the adjacency matrices of the complete graph and initial directed spanning tree as its first two arguments, the root of the tree as the third argument, and it has the same options as DirectedSpanningTreeOneStep. You will get a display of the initial tree, with edges that have been omitted from the full graph shown dashed and annotated by their slack values. For both of these functions, each vertex will be also be annotated with the total path cost to that vertex from the root.

```
(**  DirectedSpanningTreeFirstStep[
      theGraph,initTree,root,opts]  **)
```

We illustrate with the next example.

EXAMPLE 3. Let us use the policy improvement algorithm to find an optimal distribution network for the graph of Figure 1.26. Begin with the tree in Figure 1.28, generated by the breadth-first search spanning tree algorithm. The dashed edges are those *not* in ET and the solid edges are in ET. For example, the slack of the omitted edge (2, 3) is 1, because the current path to 3 has length 4, and the cost of the current path to 2 plus the edge cost of edge (2, 3) is $2 + 1 = 3$, hence the difference is positive 1. You should check all of the other slack values yourself. The adjacency matrix of the full graph was given the name adjmatrix24 in the closed cell that generated Figure 1.26, and the VertexPositions, VertexLabelPositions, and EdgeLabelPositions options were given the values vertpos, vlabelpos, and elabelpos, which you see in the options to our DirectedSpanningTree functions.

```
initialTree = {{0, 2, 4, 3, 0, 0, 0, 0, 0, 0},
    {0, 0, 0, 0, 2, 0, 0, 0, 0, 0},
    {0, 0, 0, 0, 0, 5, 0, 0, 0, 0},
    {0, 0, 0, 0, 0, 0, 0, 0, 0, 0},
    {0, 0, 0, 0, 0, 0, 4, 3, 0, 0},
    {0, 0, 0, 0, 0, 0, 0, 0, 2, 0},
    {0, 0, 0, 0, 0, 0, 0, 0, 0, 6},
    {0, 0, 0, 0, 0, 0, 0, 0, 0, 0},
    {0, 0, 0, 0, 0, 0, 0, 0, 0, 0},
    {0, 0, 0, 0, 0, 0, 0, 0, 0, 0}};
DirectedSpanningTreeFirstStep[adjmatrix24,
    initialTree, 1, VertexPositions → vertpos,
    VertexLabelPositions → vlabelpos,
    EdgeLabelPositions → elabelpos,
    AspectRatio → .8];
```

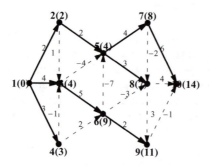

Figure 1.28 – Initial spanning tree for graph of Figure 1.26

In the first step, we include edge (8, 10) and discard edge (7, 10) because among the unused edges, (8, 10) has the largest positive slack of 4. In preparation for the next step, we let currTree receive the value returned by DirectedSpanningTreeOneStep, that is, the new tree adjacency matrix.

```
currTree = DirectedSpanningTreeOneStep[
    adjmatrix24, initialTree, 1,
    {8, 10}, VertexPositions → vertpos,
    VertexLabelPositions → vlabelpos,
    EdgeLabelPositions → elabelpos,
    AspectRatio → .8];
```

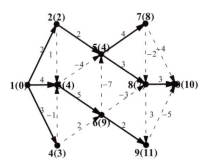

Figure 1.29 – Second spanning tree for graph of Figure 1.26

Next, since the largest positive slack on an unused edge is 3 on edge (8, 9), we discard edge (6, 9) in favor of (8, 9). The new tree is in Figure 1.30.

```
currTree =
  DirectedSpanningTreeOneStep[adjmatrix24,
    currTree, 1, {8, 9}, VertexPositions → vertpos,
    VertexLabelPositions → vlabelpos,
    EdgeLabelPositions → elabelpos,
    AspectRatio → .8];
```

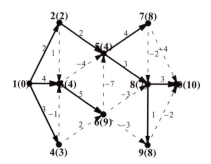

Figure 1.30 – Third spanning tree for graph of Figure 1.26

Two more steps of the same sort suffice: first add edge (4, 6) then add (2, 3). The final tree is in Figure 1.32. Observe that all slacks are now negative, which by Theorem 1 implies that the tree is optimal. Shortest paths in the original network are found by following the tree edges, and the figure also reports the lengths of those paths. ∎

```
currTree =
  DirectedSpanningTreeOneStep[adjmatrix24,
    currTree, 1, {4, 6}, VertexPositions → vertpos,
    VertexLabelPositions → vlabelpos,
    EdgeLabelPositions → elabelpos,
    AspectRatio → .8];
```

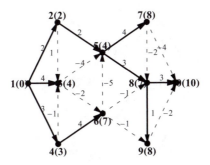

Figure 1.31 – Fourth spanning tree for graph of Figure 1.26

```
currTree =
  DirectedSpanningTreeOneStep[adjmatrix24,
    currTree, 1, {2, 3}, VertexPositions → vertpos,
    VertexLabelPositions → vlabelpos,
    EdgeLabelPositions → elabelpos,
    AspectRatio → .8];
```

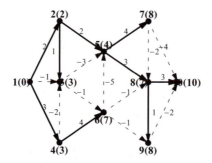

Figure 1.32 – Final spanning tree for graph of Figure 1.26

The whole process may be automated, of course, but by doing so you do not get the educational benefit of choosing the edge substitutions yourself. Nevertheless I have also placed in the KnoxOR`Graphs` package the command DirectedSpanningTree, whose full documentation appears after the query below. It takes the same arguments as DirectedSpanningTreeFirst-Step. I encourage you not to use it to do problems, but to read the code in the package to see how it was put together. You may also use it as a check on your work, or to experiment easily with what happens when the algorithm starts with different initial spanning trees.

```
? DirectedSpanningTree
```

DirectedSpanningTree[theGraph, initialTree,
    theRoot, opts] takes a given initial spanning
    tree of a given directed, quasi-connected graph,
    both in adjacency matrix form, and the vertex
    number of the root of the tree. It performs the
    full minimal directed spanning tree algorithm,
    displaying all intermediate graphs unless the
    option ShowTree is set to False, and returns   the
    minimal spanning tree in adjacency matrix form. The
    display options of DisplayGraph may be passed in.

Activity 4 – What is the implication of having an edge with slack 0 in the final minimal directed spanning tree?

Exercises 1.3

1. Use Kruskal's algorithm to find a minimal cost spanning tree for a graph whose vertices are labeled $\{1, 2,..., 8\}$ and whose edges have the costs below:

edge	cost	edge	cost
{1, 2}	2	{3, 7}	4
{1, 3}	2	{4, 5}	5
{1, 4}	1	{4, 6}	3
{2, 3}	2	{5, 6}	6
{3, 4}	3	{6, 7}	1
{3, 6}	2	{7, 8}	2

$$\begin{pmatrix}
- & 2 & - & 4 & - & - & - & - & - \\
2 & - & 6 & 3 & 5 & 8 & - & - & - \\
- & 6 & - & - & - & 1 & - & - & - \\
4 & 3 & - & - & 4 & - & 3 & - & - \\
- & 5 & - & 4 & - & 3 & 2 & 1 & 4 \\
- & 8 & 1 & - & 3 & - & - & - & 6 \\
- & - & - & 3 & 2 & - & - & 2 & - \\
- & - & - & - & 1 & - & 2 & - & 5 \\
- & - & - & - & 4 & 6 & - & 5 & -
\end{pmatrix}$$

Exercise 1   **Exercise 2**

2. Cables are to connect several components of a sound system. The vertices in the graph below represent the components, and the edges are possible connections. The matrix above gives the lengths of cable required to connect each pair of components. Find the system of connections that requires the least total amount of cable.

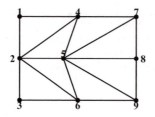

Exercise 3

3. In Example 2 compute the cost of the spanning trees formed by (a) breadth-first search; (b) ordering edges lexicographically. By how much do these costs differ from the total cost of the minimal spanning tree?

4. (*Mathematica*) Suppose that the distances between fifteen cities are as in the table below, An airline wishes to institute service among these cities. Assuming that flight cost is directly proportional to distance, find an optimal routing system that provides service to all cities.

City	1	2	3	4	5	6	7	8	9	10	11	12	13	14	15
1		24	20	22	15	60	20	24	31	34	36	27	28	29	34
2			32	45	12	21	14	50	41	34	40	18	20	25	37
3				14	20	35	80	46	35	28	49	38	57	35	41
4					18	20	26	40	35	72	26	47	22	36	46
5						31	19	30	46	38	35	27	21	50	43
6							20	40	34	58	25	32	18	33	60
7								35	31	27	36	22	47	61	35
8									25	17	32	43	71	54	43
9										25	30	32	40	45	21
10											15	26	29	33	36
11												27	29	31	40
12													24	16	19
13														31	26
14															23

5. Show that if a weighted, undirected graph G is connected and no two of its edges have the same cost, then there is a unique minimal spanning tree.

6. Explain why every vertex has component number 1 at the end of execution of Kruskal's algorithm.

7. Prove or disprove. Let T be a minimal spanning tree of an undirected graph G and fix a vertex v_0. Then for each vertex $u \neq v_0$, the cost of the path in T from v_0 to u is minimal among all paths in G from v_0 to u.

8. (*Mathematica*) An amusement park wishes to run a tram line among several of its rides. The rides are nodes in the graph below, and the weights of the edges are distances between the nodes. Design a connecting system such that the least possible length of track will be used.

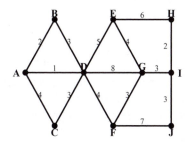

Exercise 8

9. (*Mathematica*) An alternative algorithm for finding a minimal undirected spanning tree of a graph of n vertices is called *Prim's algorithm*. Begin with a single vertex. At any stage, check edges not in the spanning tree that have one vertex in the current incomplete tree (and one not in it). Add the edge of smallest cost of this kind, and add the new vertex of that edge to the vertex set. Continue until the candidate has $n - 1$ edges. Write a command in *Mathematica* to implement Prim's algorithm, and use your command to find a minimal spanning tree in the graph of Figure 1.1.

10. Prove that Prim's algorithm of Exercise 9 yields a minimal spanning tree if the graph is connected. (Hint: Prove by induction that at each step the subgraph created by Prim's algorithm is connected and has no cycles, hence it is a tree, and moreover the Prim tree is contained in some minimal spanning tree.)

11. (*Mathematica*) Information is to flow from a source v_0 to each of seven other locations labeled $v_1,\ldots, v_7$ in the diagram below. Find a least costly way

of doing this if the edge weights represent the costs of direct communication between nodes.

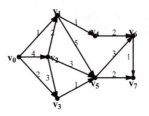

Exercise 11

12. (*Mathematica*) The matrix below gives the weights of directed edges connecting certain pairs of vertices in a directed graph. List shortest possible paths from vertex 1 to each other vertex in the network.

$$
\begin{pmatrix}
- & 4 & 3 & 4 & 2 & - & - & - & - & - & - \\
- & - & - & - & - & 3 & - & - & - & - & - \\
- & - & - & - & - & 6 & 2 & 2 & - & - & - \\
- & - & - & - & - & - & - & 1 & - & - & - \\
- & - & - & - & - & - & - & 5 & 2 & - & - \\
- & - & - & - & - & - & - & - & - & 4 & \\
- & - & - & - & - & - & - & - & - & 2 & \\
- & - & - & - & - & - & - & - & - & 3 & \\
- & - & - & - & - & - & - & - & - & 4 & \\
- & - & - & - & - & - & - & - & - & - & -
\end{pmatrix}
$$

13. The vertices in the graph below are grain elevators, some of which can be connected by chutes to neighboring elevators, for the purpose of shifting grain from one location to another. The edges are directed because the chutes are inclined, to allow passage of grain by gravity in only one direction. Find a chute system that allows each elevator to be reached from the main elevator at vertex 1 with the shortest possible path. Is the solution unique? Do this problem by hand, rather than with *Mathematica*.

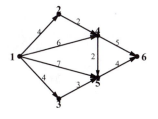

Exercise 13

14. (*Mathematica*) An alternative algorithm for finding the shortest path from the root v_0 to each vertex v in a directed graph, called *Dijkstra's algorithm*, is as follows. Initialize the cost $C(v)$ of a path to vertex v to be $c(v_0, v)$ if there is such an edge, and $+\infty$ otherwise. Initialize a "special" set of vertices S to be $\{v_0\}$, and initialize the predecessor of v to be $P(v) = v_0$ for all $v \neq v_0$. Then do the following loop $n - 1$ times, i.e., until S contains all vertices:

    1. Select a vertex $w \notin S$ such that $C(w)$ is minimized.
    2. Put w into S.
    3. Revise the costs $C(v)$ for $v \notin S$ (to reflect any cost reduction) by
        $C(v) = \min\{C(v), C(w) + c(w, v)\}$
    if there is an edge (w, v).
    4. For each $v \notin S$, if the minimum in step 3 is $C(w) + c(w, v)$ (i.e., cost was reduced), then label the predecessor of v as $P(v) = w$.

Then, to find the minimal path to each v, trace the path from v_0 to v by backtracking through predecessors: $P(v)$, $P(P(v))$, ..., v_0. Write a *Mathematica* command to implement Dijkstra's algorithm. (Hint: To break down this rather difficult program into manageable pieces, consider writing some supporting functions to do smaller tasks that the main Dijkstra program needs, such as revising the costs and predecessors in steps 3 and 4, and producing the path to each vertex given the predecessor list, etc.)

15. Prove that if a quasi-connected, directed graph with root v_0 and positive costs is input to Dijkstra's algorithm (see Exercise 14), then for each $v \neq v_0$, a shortest path from v_0 to v is returned. (Hint: Show inductively on the number of vertices added to S that the path found by Dijkstra to each vertex in S is the shortest possible.

16. Use Dijkstra's algorithm (see Exercise 14) to list shortest paths to all vertices $v_1,...,v_7$ in the graph of Exercise 11. Do this by hand, rather than with *Mathematica*.

17. (*Mathematica*) A reservoir at vertex 1 in the diagram below is to supply water to several pumping stations. The edge weights are costs of laying pipe from one station to another. How should the pipe be laid so that all stations are served, but cost is minimized? Use (a) the policy improvement algorithm and (b) Dijkstra's algorithm (see Exercise 14).

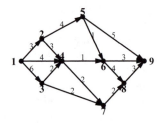

Exercise 17

1.4 Critical Path Algorithm

We now turn to the maximum weight path problem that was discussed in the introduction in the context of the completion time of a multiple stage project. Recall that in graphs such as Figure 1.2, displayed again below for your convenience, we think of edges as tasks. Vertices are present primarily to link edges, showing when one task must be preceded by another. Edge weights are task completion times. The problem is to find sequences of tasks that are the most time-consuming.

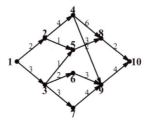

Figure 1.2 – Project completion

To avoid certain problems involving nonexistence of a solution, and to specialize to directed graphs of the form of Figure 1.2, we define the following special type of directed graph.

DEFINITION 1. A *directed network* with *root* (or *source*) r and *terminus* (or *sink*) t is a quasi-connected, directed graph with no directed cycles such that r is an ancestor of each vertex and t is a descendant of each vertex.

To say that a directed network has no *directed* cycles is not the same as to say that it is a directed tree in the earlier sense. As Figure 1.2 shows, the underlying graph is permitted to have *undirected* cycles. But if, as in Figure 1.33, we permit directed cycles, then paths of arbitrarily large cost could be created and logically difficulties would result. Note that task {2, 3} precedes task {3, 4}, which precedes task {4, 2}. By using the cycle 2, 3, 4, 2 arbitrarily often, paths of arbitrarily large cost from vertex 1 to vertex 5 can result, but even more troubling is that one can never do task {2, 3} because it requires {4, 2} and hence {3, 4} to be done, but {3, 4} cannot be done until {2, 3} is done. Physically, we do not wish the completion of several tasks to depend on one another in a cyclic way, else the project could not be completed at all.

Figure 1.33 – Cyclic task dependence

In the rest of this section, G is a directed network with root $r = v_1$, terminus $t = v_n$, and other vertices v_2, ..., v_{n-1}. The cost of an edge is denoted by $c(u, v)$. The path $P(v)$ from the root r to vertex v and the total cost $C(P(v))$ of the path are as defined in the last section. $E\,G$ and $E\,T$ will denote the edge sets of the full graph and a spanning tree, respectively.

DEFINITION 2. A *critical path* of G is a path $P^*(t)$ from the root r to the terminus t that has the longest possible total path cost among all paths from r to t.

Note that in the case where an edge cost represents time to completion of a task, such a path $P^*(t)$ gives a sequence of tasks of maximal total completion time. Therefore, if any of the tasks in $P^*(t)$ are delayed, the entire project will be delayed. It is possible to have more than one critical path in a directed network, as discussed below.

To find critical paths, we will use an algorithm similar to the minimal directed tree algorithm of Section 1.3. The quasi-connectivity of the graph implies as before that a directed spanning tree exists. The algorithm will find a *maximal directed tree* just as the earlier algorithm found a minimal tree. It is fairly clear what we mean by "maximal directed tree"; in Definition 2 of Section 1.3, simply replace the words "smaller than" by "greater than." Then the unique path from r to t in the maximal directed tree is a critical path. If, however, there is another maximal tree, then there may be a different critical path, and hence we are obliged to investigate the problem of non-uniqueness.

First, we have a counterpart to the dynamic programming equation of Section 1.3.

THEOREM 1. A directed spanning tree T^* with paths $\{P^*(v)\}$ is maximal if and only if for all vertices $v \neq r$,

$$C(P^*(v)) = \sup_{(u,v)\in EG} \{C(P^*(u)) + c(u, v)\} \; \blacksquare \qquad (1)$$

Activity 1 – Try to construct a proof of this theorem similar to that of Theorem 1 of the last section for minimal directed spanning trees. The idea, once again, is that for the tree to be maximal, the cost of the path to v must be the largest possible sum of the maximal cost to a predecessor vertex u, plus the cost from u to v.

The policy improvement algorithm is also similar to the one for the minimal tree problem. The only change is that an edge of minimal (negative) slack is sought for the purpose of substitution. After the maximal tree is found, we need only trace the path from the root to the terminus. The algorithm assumes that an initial directed spanning tree T has been found. The notation is as usual, and recall that the slack of edge (u, v) is defined as before, namely, the difference between the cost of the current path to v and the sum of the current path cost to u and the cost of edge (u, v):

$$S(u, v) = C(P^*(v)) - [C(P^*(u)) + c(u, v)] \qquad (2)$$

CRITICAL PATH ALGORITHM
   1. Repeat steps a–d until the minimum slack in the current tree is greater than or equal to 0.
      a. For all vertices $v \neq r$, find the path $P(v)$ in T and calculate $C(P(v))$.
      b. Find an edge $(u, v) \in EG - ET$ of minimum slack M.
      c. If $M < 0$, then do step d.
      d. Substitute (u, v) for the edge (u_0, v) currently pointing to v in T.
   2. Find and return the path from r to t.

Activity 2 – Explain why the critical path algorithm looks for edges of minimal negative slack to use in edge substitution.

For the problem of finding the time of completion of a project represented by a directed network, it is useful to know the final slack values:

$$S(u, v) = C(P^*(v)) - [C(P^*(u)) + c(u, v)]$$

for the edges not in the maximal tree. This is because $S(u, v)$ is the difference between the critical time to v and the critical time to u plus the completion time of task (u, v). Task (u, v) may then be delayed by this amount without delaying the project.

In the minimization problem, uniqueness of the optimal directed spanning tree is not very important. To find one best network is enough. But here we are really interested in finding all delay-causing paths, so that the issue of uniqueness deserves closer study. The key to an understanding of non-uniqueness lies in the unused slack zero edges in the final tree, as the following theorem shows.

THEOREM 2. (a) If $P^*(t)$ is a critical path, then there is a maximal directed tree containing it.
(b) Suppose T^* is a maximal directed tree. Let E_0 be the set of edges (u, v) not in T^* such that $S(u, v) = 0$. Then, given any maximal directed tree T^{**}, there exists a subset E_1 of E_0 such that

$$T^{**} = T^* \bigcup E_1 - \{(u_0, v) \in T^* : \exists\, (u, v) \in E_1\}$$

In other words, all maximal trees, and therefore all critical paths, may be found by beginning with one maximal tree and performing all possible combined substitutions of slack zero edges for edges in the current maximal tree.

Proof. (a) Let $P^*(t)$ be a maximal path from r to t that traverses the vertices $u_0 = r$, u_1, ..., $u_m = t$. We claim first that the subpath $P^*(u_i)$ is maximal for each $i = 1$, ..., m. Assuming the contrary, there is $i \in [1, m - 1]$ such that $P^*(u_i)$ is not maximal. Therefore, there is a path $P'(u_i)$ such that $C(P'(u_i)) > C(P^*(u_j))$. Then the path $P''(t)$ formed by following P' from r to u_i, then following P^* from u_i to t, has higher cost than $P^*(t)$, a contradiction of the choice of P^*. Hence, P^* must be maximal to each vertex on its path.

It is easy to see from the breadth-first search algorithm (see the Directed Spanning Tree Algorithm in Section 1.2) that if we first initialize KT and NEW1 to the set of vertices in the path $P^*(t)$ and initialize ET to the set of edges in $P^*(t)$, then a spanning tree containing $P^*(t)$ will result. That is, there is a spanning tree T containing this path. We claim furthermore that the Critical Path Algorithm applied to the initial tree T must preserve all of the edges in the path $P^*(t)$, resulting in a maximal spanning tree containing $P^*(t)$, as desired. To show this, we need only show that at any stage of execution of loop a–d such that $P^*(t)$ is still within the current tree, $P^*(t)$ will remain within the tree after the choice of a new edge. Suppose the current tree has paths $\{P(v)\}$, and consider an edge (w, u_i) pointing to a vertex u_i in the path $P^*(t)$. We have

$$
\begin{aligned}
S(w, u_i) &= C(P(u_i)) - [C(P(w)) + c(w, u_i)] \\
&= C(P^*(u_i)) - [C(P(w)) + c(w, u_i)] \\
&\geq 0
\end{aligned}
$$

since $P^*(u_i)$ is a maximal path to u_i. Since only edges of negative slack are selected for substitution, (w, u_i) will not be substituted for (u_{i-1}, u_i) in step d of the algorithm. This proves that $P^*(t)$ will be contained in the final tree returned by the Critical Path Algorithm, which finishes the proof of part (a).

(b) Let a maximal directed spanning tree T^* with paths $\{P^*(v)\}$ be given. To prove part (b), it suffices to show that for an arbitrary edge (u, v) not in T^*,

1. If $S(u, v) > 0$, then (u, v) cannot be contained in any maximal tree; and
2. If $S(u, v) = 0$, then the spanning tree created by substituting (u, v) for the edge (u_0, v) currently in T^* is maximal.

Thus, the only possible maximal directed spanning trees are formed by substitutions of zero slack edges for corresponding edges in T^*.

To prove assertion 1, suppose on the contrary that there is a maximal tree T^{**} with paths $\{P^{**}(w)\}$, to which (u, v) belongs. Since both T^* and T^{**} have maximal paths to u and to v, and since $S(u, v) > 0$, we can write:

$$
\begin{aligned}
C(P^*(v)) &> C(P^*(u)) + c(u, v) \\
&= C(P^{**}(u)) + c(u, v) \\
&= C(P^{**}(v))
\end{aligned}
$$

which contradicts optimality of T^{**}; thus assertion 1 is established.

The proof of assertion 2 is left to the reader as Exercise 5. ∎

In summary, part (a) of the theorem says that our search for all maximal paths may be confined to a search for all maximal trees. Part (b) implies that we may find all maximal trees simply by starting with one, say T^*, generated by the Critical Path Algorithm, and exhausting all substitutions of slack zero edges not in T^* for edges that point to the same vertex in T^*. Note, incidentally, that Theorem 2 yields a necessary and sufficient condition for uniqueness of a maximal path, namely, the condition that the slack $S(u, v) > 0$ for all edges (u, v) not in T^*.

It may have already occurred to you that the same *Mathematica* tools that we used in the last section for minimal directed spanning trees can be used here too. DirectedSpanningTreeFirstStep shows initial slack values for a given initial tree. In DirectedSpanningTreeOneStep it is our choice which edge to substitute in; so instead of choosing the one with the largest (positive) slack, for the maximum problem we choose the one with the smallest (negative) slack. Again there is a tool in KnoxOR`Graphs` that executes the whole algorithm, as described in the documentation below, but I suggest that you run through the step-by-step procedure to solve problems, for the most educational benefit.

```
? MaximalDirectedSpanningTree
```

MaximalDirectedSpanningTree[theGraph, initialTree,
    theRoot, opts] takes a given initial spanning
    tree of a given directed, quasi-connected graph,
    both in adjacency matrix form, and the vertex
    number of the root of the tree. It performs the
    full maximal directed spanning tree algorithm,
    displaying all intermediate graphs unless the
    option ShowTree is set to False, and returns   the
    maximal spanning tree in adjacency matrix form. The
    display options of DisplayGraph may be passed in.

EXAMPLE 1. The table below shows the tasks necessary to proceed from the raw ingredients of a turkey dinner, with stuffing, mashed potatoes, and gravy, to the finished product. (Invaluable technical advice for this example was provided by Mrs. Lois L. Hastings.) Our goal is to find the most time-consuming sequence of tasks.

Task	Predecessor	Time (min.)
A. Defrost turkey	none	480
B. Break bread for stuffing	none	15
C. Peel potatoes	none	15
D. Remove and boil innards	*A*	60
E. Make stuffing	*B*, *D*	20
F. Stuff turkey	*E*	10
G. Boil potatoes	*C*	40
H. Bake turkey	*F*	240
I. Mash potatoes	*G*	10
J. Make gravy	*H*	15
K. Remove stuffing from turkey	*H*	5
L. Slice turkey	*K*	15
M. Serve potatoes	*I*	1
N. Serve turkey	*L*	1
P. Serve stuffing	*K*	1
Q. Put gravy on potatoes	*J*, *M*	1

Some experimental artwork is usually necessary to obtain a convenient graphical representation of the dependencies among tasks. A good place to begin is to draw a root, out of which point edges corresponding to tasks that have no predecessors. For us, these are *A*, *B*, and *C*. Examine tasks one by one, drawing an edge that points away from the vertex to which its predecessor points. For instance, in the graph below, *D* follows *A*, and *G* follows *C*. For an edge like *E*, which has two predecessors *D* and *B*, the predecessors must both be drawn so as to point into the vertex away from which *E* points.

With some care, we obtain the graph in Figure 1.34. The *Mathematica* name of the overall graph is turkey, and for the display options we have used vertpos34 for VertexPositions, vlabpos34 for VertexLabelPositions, and turkeytasklabelpositions for the EdgeLabelPositions, as you may see by opening the closed cell below this paragraph.

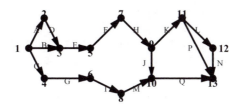

Figure 1.34 – Project graph for making a turkey dinner

We select an initial directed spanning tree, compute the path costs using the completion times in the table, and we compute the slack values for the unused edges. These are shown in Figure 1.35. The edge of minimum slack is $D = \{2, 3\}$, which we substitute for edge $B = \{1, 3\}$ in the spanning tree to produce the new tree in Figure 1.36.

Two similar edge substitutions, first $N = \{12, 13\}$ for $Q = \{10, 13\}$, then $J = \{9, 10\}$ for $M = \{8, 10\}$, produce the next tree in Figure 1.37 and the final tree in Figure 1.38. Since all slacks are non-negative and there are no unused edges of slack zero, this is the unique maximal tree. The critical sequence of tasks is therefore A, D, E, F, H, K, L, N. Most of these involve direct operations on the turkey, which is no surprise to anyone who has practical experience with this problem. Additional information we can gather from the final tree, for example, is that the sequence of operations C, G, I, M involving the potatoes can be delayed for as long as 759 minutes without delaying the meal. Also, it requires 831 minutes from the time when the defrosting starts until the time when the meal is ready. ∎

```
initturkeytree =
  {{0, 480, 15, 15, 0, 0, 0, 0, 0, 0, 0, 0, 0},
   {0, 0, 0, 0, 0, 0, 0, 0, 0, 0, 0, 0, 0},
   {0, 0, 0, 0, 20, 0, 0, 0, 0, 0, 0, 0, 0},
   {0, 0, 0, 0, 0, 40, 0, 0, 0, 0, 0, 0, 0},
   {0, 0, 0, 0, 0, 0, 10, 0, 0, 0, 0, 0, 0},
   {0, 0, 0, 0, 0, 0, 0, 10, 0, 0, 0, 0, 0},
   {0, 0, 0, 0, 0, 0, 0, 0, 240, 0, 0, 0, 0},
   {0, 0, 0, 0, 0, 0, 0, 0, 0, 1, 0, 0, 0},
   {0, 0, 0, 0, 0, 0, 0, 0, 0, 0, 5, 0, 0},
   {0, 0, 0, 0, 0, 0, 0, 0, 0, 0, 0, 0, 1},
   {0, 0, 0, 0, 0, 0, 0, 0, 0, 0, 0, 15, 0},
   {0, 0, 0, 0, 0, 0, 0, 0, 0, 0, 0, 0, 0},
   {0, 0, 0, 0, 0, 0, 0, 0, 0, 0, 0, 0, 0}};
DirectedSpanningTreeFirstStep[turkey,
  initturkeytree, 1, AspectRatio → .5,
  VertexPositions → vertpos34,
  VertexLabelPositions → vlabpos34,
  EdgeLabelPositions → turkeytasklabelpositions];
```

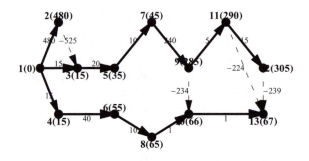

Figure 1.35 – Turkey dinner, step 1

```
currturkeytree = DirectedSpanningTreeOneStep[
   turkey, initturkeytree, 1, {2, 3},
   AspectRatio → .5, VertexPositions → vertpos34,
   VertexLabelPositions → vlabpos34,
   EdgeLabelPositions →
    turkeytasklabelpositions];
```

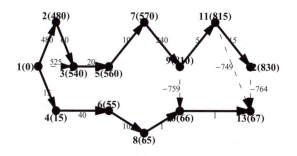

Figure 1.36 – Turkey dinner, step 2

```
currturkeytree = DirectedSpanningTreeOneStep[
   turkey, currturkeytree, 1, {12, 13},
   AspectRatio → .5, VertexPositions → vertpos34,
   VertexLabelPositions → vlabpos34,
   EdgeLabelPositions →
    turkeytasklabelpositions];
```

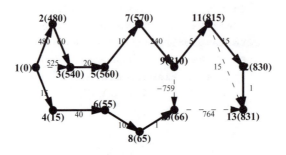

Figure 1.37 – Turkey dinner, step 3

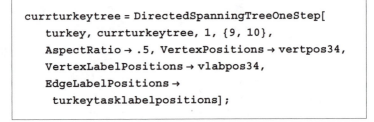

```
currturkeytree = DirectedSpanningTreeOneStep[
    turkey, currturkeytree, 1, {9, 10},
    AspectRatio → .5, VertexPositions → vertpos34,
    VertexLabelPositions → vlabpos34,
    EdgeLabelPositions →
     turkeytasklabelpositions];
```

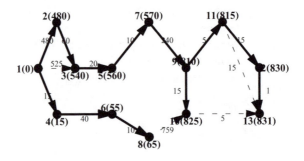

Figure 1.38 – Turkey dinner, step 4 and last

Activity 3 – Make up your own project example, and try to produce the project graph.

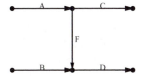

Figure 1.39 – Inserting a dummy edge into a project graph

REMARK 1. In forming the original graph, a special situation can arise that bears mentioning. If task A must precede each of tasks C and D, but task B precedes D only, then in order to represent these constraints, a dummy edge F may be inserted, as shown in Figure 1.39. We need to then take steps to be sure that the dummy edge, which is not really a part of the original project, is not included in any maximal directed spanning tree. One way to ensure this is by observing that the slack of an edge (u, v) is found by subtracting $c(u, v) + C(P(u))$ from $C(P(v))$, so that if the cost $c(u, v)$ of a dummy edge is taken to be a large magnitude negative number, larger in magnitude than any path cost $C(P(u))$ could be, then a negative number is subtracted from $C(P(v))$, so that a positive slack must be attached to the dummy edge. Thus, it will never be swapped in.

REMARK 2. Tasks may also be represented on vertices instead of edges, in which case the technical difficulties mentioned in Remark 1 do not arise. A solution algorithm can be developed, but we do not discuss it here. The project with the dependencies in Figure 1.39 above, with A preceding both C and D, and B preceding D, would be represented with tasks on vertices as in Figure 1.40.

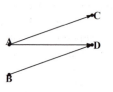

Figure 1.40 – Project graph with tasks on vertices

EXAMPLE 2. Several years ago there was a major renovation of the small library in my office building. The library was on two levels, and the major work was the replacement of the rickety spiral stairway leading to the second level by a large new stairway and front desk unit, and the construction of built-in study carrells on both levels. Some wood trim work was also done, the carpet was replaced, and the walls were painted. To do all this however, the library had to be emptied and old fixtures torn out. The 17 tasks into which the whole project breaks down are listed below, as well as time estimates, in days, of how long the tasks take. You should carefully go over the list of predecessor tasks to see that they make logical sense.   For instance, workers cannot tear up the old carpet until the old shelving fixtures are moved out, the new carpet should not be laid until the staining and painting is done, etc. Let us use the critical path method to find how many days the project will take, and make note of possible delay-causing tasks.

Task	Time	Predecessor
A. box books	8	none
B. move books	2	*A*
C. move old fixtures	1	*B*
D. tear up old carpet	3	*C*
E. remove spiral staircase	2	*C*
F. do upstairs carpentry	15	*C*
G. build new staircase	20	*D, E*
H. build new front desk	5	*G*
I. build new downstairs carrells	8	*D, E*
J. build new upstairs carrells	8	*F*
K. paint and stain downstairs	3	*H, I*
L. paint and stain upstairs	3	*G, J*
M. lay downstairs carpet	5	*K*
N. lay upstairs carpet	5	*L*
O. move old fixtures back	1	*M, N*
P. move books back	2	*O*
Q. unbox books	8	*P*

Figure 1.41 displays the graph of the project. Several points are worth noting. Since task *L* depends on both *G* and *J*, but task *H* depends only on *G*, we have had to insert a dummy edge as in Remark 1 connecting vertex 8 to

vertex 9. A second dummy edge, used this time to avoid multiple edges between a pair of vertices, connects vertices 6 and 5, since task *G* depends on both of *D* and *E*, which both emanate from the same vertex. This indicates that in the original task breakdown we could have simplified matters by combining tasks *D* and *E* into one task of duration 3 days. The dummy edges have been given weight of −200, a good choice given the magnitudes of the edge costs in the table. The name of the graph and the option values controlling the graph display are as shown in the command. These are defined in the closed cell above the DisplayGraph command.

```
DisplayGraph[librarygraph, GraphType → Directed,
   AspectRatio → .5, VertexPositions → libraryvpos,
   VertexLabelPositions → libraryvlabpos,
   EdgeLabels → libraryelab,
   EdgeLabelPositions → libraryelabpos];
```

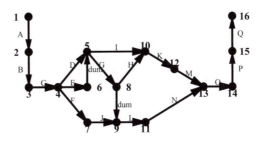

Figure 1.41 – Project graph for library renovation

This time let us use the MaximalDirectedSpanningTree command to complete the computation. An initial directed spanning tree generated by breadth-first search uses edges *A*, *B*, *C*, *D*, *E*, *F*, *I*, *G*, *J*, *K*, *L*, *M*, *O*, *P*, and *Q*. The adjacency matrix of that tree is defined in the closed cell immediately below as librarytree.

```
DirectedSpanningTreeFirstStep[librarygraph,
   librarytree, 1, AspectRatio → .5,
   VertexPositions → libraryvpos,
   VertexLabelPositions → libraryvlabpos,
   EdgeLabelPositions → libraryelabpos];
MaximalDirectedSpanningTree[librarygraph,
   librarytree, 1, AspectRatio → .5,
   VertexPositions → libraryvpos,
   VertexLabelPositions → libraryvlabpos,
   EdgeLabelPositions → libraryelabpos];
```

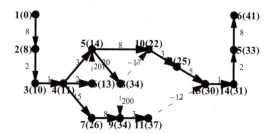

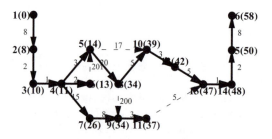

Figure 1.42 – Initial and final spanning trees for library renovation

As Figure 1.42 illustrates, only one edge substitution, namely $H = (8, 10)$ in place of $I = (5, 10)$, was necessary to finish the problem. All slacks are strictly positive, and so we have found the unique maximal spanning tree and critical path

$$A, B, C, D, G, H, K, M, O, P, Q.$$

Any task on this path will, if delayed, cause a delay in the completion time of the project. With no delays, the length of the critical path from 1 to 16 is 58 days, which is how long the project will take. The dummy edges are not in the final spannning tree. In view of the positive slack value of edge (5, 10), task *I*, the building of the downstairs carrells, may be delayed by as much as 17 days without delaying the whole project. ∎

Activity 4 – Are there any other tasks in the project of Example 2 that can be delayed, and if so, by how much?

Exercises 1.4

1. Call a directed graph double quasi-connected if each pair of vertices has not only a common ancestor, but also a common descendant. Show that a double quasi-connected graph has both a root and a terminus.

2. (*Mathematica*) Find all maximal trees and maximal paths for the graph of Figure 1.2.

3. Find all critical paths for the graph of Exercise 12 of Section 1.3 whose adjacency matrix is as below. Do this by hand, rather than with *Mathematica*.

$$\begin{pmatrix} - & 4 & 3 & 4 & 2 & - & - & - & - & - \\ - & - & - & - & - & 3 & - & - & - & - \\ - & - & - & - & - & 6 & 2 & 2 & - & - \\ - & - & - & - & - & - & - & 1 & - & - \\ - & - & - & - & - & - & - & 5 & 2 & - \\ - & - & - & - & - & - & - & - & - & 4 \\ - & - & - & - & - & - & - & - & - & 2 \\ - & - & - & - & - & - & - & - & - & 3 \\ - & - & - & - & - & - & - & - & - & 4 \\ - & - & - & - & - & - & - & - & - & - \end{pmatrix}$$

4. (*Mathematica*) A job requires ten stages of work. The completion times for each are in the table below. Also listed in the table is the information of which stages cannot begin until other stages are complete, e.g., stage *D*

requires both stages A and B to be finished before it can begin. Find all critical sequences of job stages.

Task	Immediate predecessor	Completion time (hr.)
A.	none	2
B.	none	6
C.	none	3
D.	A, B	4
E.	C	5
F.	D	1
G.	D	1
H.	E, F	4
I.	G	3
J.	H	5

5. Finish the proof of Theorem 2 by showing assertion 2: If (u, v) is an omitted edge and $S(u, v) = 0$, then the spanning tree created by substituting (u, v) for the edge (u_0, v) currently in T^* is maximal.

6. A large computer program is to be tested and debugged in modules, some of which require other modules to be completely tested before testing on them can proceed. The table below shows the dependencies, and the times required to finish the testing and debugging of each module. How long does it take until the entire group of modules is debugged? Do this by hand, rather than with *Mathematica*.

Module	Time (days)	Immediate predecessor
A	1	none
B	2	none
C	2	A
D	3	B
E	1	B
F	2	C, D

7. Intuitively, it is clear what we mean when we say that a graph is a "line of vertices" (see below). Give a set-theoretic definition of a line of vertices, and show that if a directed network is not a line of vertices, then its underlying graph must have an undirected cycle.

Exercise 7

8. (*Mathematica*) An office wants to install an information system. The main tasks are below, with time estimates in days and task dependencies indicated. Find the amount of time required to get the system up and running, and find the set of tasks which could delay the project if they were delayed.

Task	Time (days)	Predecessor
A. Run wiring	1	none
B. Research hardware	2	none
C. Research software	2	none
D. Purchase hardware	4	*B*
E. Purchase software	2	*C*
F. Install hardware	1	*D, A*
G. Install software	1	*E, F*
H. Install network facilities	1	*F*
I. Set up database	4	*G*
J. Train employees	5	*H*

9. (*Mathematica*) An advertising agency has contracted to prepare a commercial. The main tasks, time estimates, and task dependencies are shown in the table below. How many days will it take to produce this commercial?

Task	Time (days)	Predecessor
A. Write script	3	none
B. Consult with client	1	*A*
C. Revise script	2	*B*
D. Hire actors	3	*A*
E. Produce special effects	6	*C*
F. Film studio scenes	2	*C, D*
G. Film outdoor scenes	3	*F*
H. Second consult with client	1	*E, G*
I. Do voiceover work	1	*H*
J. Final editing	4	*I*

10. For the project in Exercise 4, form a project graph with tasks on vertices.

11. A variation on the critical path problem is the *task scheduling problem*. In this problem, unlike the critical path problem, explicit attention is paid to how many workers are available to do tasks, and the goal is to assign and schedule tasks among workers so that predecessor conditions are satisfied and the project is finished as quickly as possible. Even though a project graph such as those in this section might indicate that a task is ready to be done, there may not be a free worker to do it, so that it might have to wait longer than the critical path algorithm expects. Suppose that there are just two workers available to do the project with tasks listed below. Develop a work schedule for the workers that gets the project done in as little time as possible.

Task	A	B	C	D	E	F	G	H	I
Time	2	3	2	5	1	4	2	3	3
Predecessor	none	none	none	A	B	D	D	F	G

(Hint: A useful device for the scheduling problem is a bar graph that contains one horizontal row for each worker. To schedule tasks consists of placing bars, one per task, of width equal to the time required for the task, end-to-end in the worker rows, obeying predecessor constraints. For instance, in the picture below, there are three workers 1–3 and 6 tasks A–F taking 4, 2, 6, 3, 1, and 2 minutes respectively. The only requirements are that task D must be done after task A, and task E must be done after task C. The work schedule shows worker 1 doing task A, then beginning task F immediately on completion. Worker 2 does task B, then waits until task A is finished, and then does task D. Worker 3 does tasks C and E in succession. Is this the most efficient schedule?)

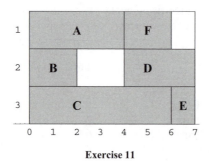

Exercise 11

1.5 Maximal Flow Problems

Problem Description

We now consider a different maximum weight problem for directed graphs in which edges have *flow capacities*, and we wish to use part or all of the available capacity on edges. Specifically, we wish to put a weighting on the graph such that:

1. Each edge has non-negative weight less than or equal to the capacity for that edge.
2. For each vertex except the source and the sink, the total weight of edges directed into the vertex equals the total weight of edges directed out of the vertex.
3. The total weight of edges directed out of the source is the largest among all weightings satisfying conditions 1 and 2.

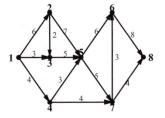

Figure 1.43 – A flow graph

One may think of a vertex as an intersection, an edge as a street, and a capacity as the maximum possible number of vehicles that can pass on the particular edge per unit time. The problem is to find traffic flows per unit time on each street so that the rate at which vehicles enter each intersection equals the rate at which they depart (an equilibrium, or conservation condition expressed by condition 2), the capacities for the streets are not exceeded, and the total outflow from the source of the traffic is as large as possible.

Like the critical path problem, the maximal flow problem will be solved by a policy improvement strategy. We begin with an initial feasible flow, and modify it step-by-step until a termination condition comes into effect (which we must prove characterizes a maximal flow). This approach is called the *Ford–Fulkerson maximal flow algorithm*, and it is contained in several sources, but here we follow the presentation of Swamy and Thulasiraman [57].

To define the problem more carefully, let $G = (V, E)$ be a directed network of n vertices. We suppose that the vertices are denoted $v_1, v_2, ..., v_{n-1}, v_n$. There is a *source* vertex v_1 and a *sink* vertex v_n, where the words "source" and "sink" have the same meaning as in Definition 1 of Section 1.4. Vertices 1 and 8, respectively, in Figure 1.43 are the source and sink for that network. Let c be a function from the set of edges to the non-negative real numbers; $c(v, w)$ is the *capacity* of edge (v, w). A *flow* is a function $f : E \to \mathbb{R}_+$ such that

> (a) $0 \le f(e) \le c(e)$ for all edges $e \in E$
>
> (b) $\sum_j f(v_i, v_j) = \sum_j f(v_j, v_i)$ for $i = 2, ..., n - 1$ (1)

where the sums in the second equation are taken over those vertices v_j such that an edge of the stated form does exist in E. These properties are direct translations of constraints 1 and 2 mentioned above: the first equation says that the flow on every edge cannot exceed the capacity of that edge, and the second equation states that the total flow out of vertex v_i equals the total flow into v_i, for all intermediate vertices v_i.

Activity 1 – Why can there be no edges pointing into the source vertex, nor can there be any edges pointing out of the sink?

The *value of flow f* is the total flow out of the source vertex, i.e.,

$$V(f) = \sum_j f(v_1, v_j) \qquad (2)$$

Shortly, we will show that the conservation condition (1b) implies that the value of a flow also equals the total inflow to the sink. A *maximal flow f** is such that for all other flows f

$$V(f^*) \ge V(f) \qquad (3)$$

The improvement of a given flow involves locating a path from the source to the sink such that the flow along each edge in the path can be increased by a fixed positive amount. The improvement procedure stops when there does not exist such an "augmenting path." Our main theoretical result is Theorem 1 below, which says that when the stopping condition becomes true, the maximal flow has been reached. To prove this result, we must introduce the idea of a *cut* in a graph.

DEFINITION 1. Let $G = (V, E)$ be a graph, let V_0 be a subset of vertices of G, and denote by V_0^c the complement of V_0 in V. The *cut* corresponding to V_0, denoted by $K = K(V_0)$, is the set of all edges that have one vertex in V_0 and one in V_0^c. A cut $K(V_0)$ in a network *separates* the source and sink if the source is in V_0 and the sink is in V_0^c.

EXAMPLE 1. In the graph of Figure 1.43, let $V_0 = \{1, 2, 5, 6\}$. Then $V_0^c = \{3, 4, 7, 8\}$, and the cut $K(V_0)$ corresponding to this set of vertices consists of the edges: (1, 3), (1,4), (2,3), (3,5), (4,5), (5,7), (6,8), and (7,6). Note that this cut separates the source and sink. The edges of the cut are displayed in Figure 1.44. ■

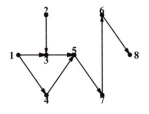

Figure 1.44 – A cut for the graph of Figure 1.43

DEFINITION 2. The *capacity* $c(K)$ of a cut $K = K(V_0)$ is

$$c(K) = \sum c(e)$$

where the sum is taken only over those edges whose initial vertex is in V_0 and whose terminal vertex is in V_0^c. A cut K^* separating the source and sink is a *minimum cut* if

$$c(K^*) \leq c(K)$$

for all cuts K that separate source and sink.

For example, if $K = K(V_0)$ is the cut in Example 1, then the capacity of K is $3 + 4 + 2 + 5 + 8 = 22$, using edges (1, 3), (1,4), (2,3), (5,7), and (6,8). Notice that we do not include edges pointing from V_0^c into V_0.

Activity 2 – If $V_0 = (1, 2)$ in the graph of Figure 1.43, find the capacity of the cut $K = K(V_0)$.

One further piece of notation: if f is a flow, and V_1 and V_2 are subsets of vertices, we denote by

$$f(V_1, V_2) = \sum f(v_i, v_j) \qquad v_i \in V_1, \ v_j \in V_2$$

the total flow along edges pointing from V_1 into V_2. For example, if we put the flow indicated in Figure 1.45 on the graph of Figure 1.43, with $V_0 = \{1, 2, 5, 6\}$ then $f(V_0, V_0^c) = 1 + 1 + 0 + 2 + 3 = 7$ and $f(V_0^c, V_0) = 1 + 1 + 1 = 3$ (edges (3, 5), (4, 5), and (7, 6)). Note that the value of the flow is 4, which is the same as the difference $f(V_0, V_0^c) - f(V_0^c, V_0)$. This is generally true, as shown by the first lemma in the following subsection.

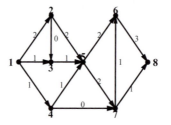

Figure 1.45 – A flow on the graph of Figure 1.43

Main Results and Algorithm

LEMMA 1. Let f be any flow on a graph $G = (V, E)$ and let V_0 be a set of vertices containing the source vertex v_1 but not the sink v_n. Then

$$V(f) = f(V_0, V_0^c) - f(V_0^c, V_0) \qquad (4)$$

Consequently,

$$V(f) \leq c(K) \tag{5}$$

where K is any cut seperating the source and the sink.

Proof. Note that since nothing flows into the source, the difference between the flow out of V_0 and the flow into V_0 is

$$\sum_{v \in V_0} \sum_{w \in V} f(v, w) - \sum_{v \in V_0} \sum_{w \in V} f(w, v)$$
$$= \sum_{w \in V} f(v_1, w) + \sum_{v \in V_0 - v_1} \sum_{w \in V} f(v, w)$$
$$- \sum_{v \in V_0 - v_1} \sum_{w \in V} f(w, v) \tag{6}$$
$$= V(f) + 0 = V(f)$$

by the conservation condition in (1). For the edges (v, w) such that w is in V_0, $f(v, w)$ appears in both terms in the difference on the left hand side of (6), hence it subtracts away. Thus, (6) may be rewritten as

$$V(f) = \sum_{v \in V_0} \sum_{w \in V_0^c} f(v, w) - \sum_{v \in V_0} \sum_{w \in V_0^c} f(w, v)$$
$$= f(V_0, V_0^c) - f(V_0^c, V_0)$$

The first assertion is proved.

Since flows are non-negative, we have

$$V(f) \leq f(V_0, V_0^c) = \sum_{v \in V_0} \sum_{w \in V_0^c} f(v, w)$$
$$\leq \sum_{v \in V_0} \sum_{w \in V_0^c} c(v, w) = c(K)$$

which establishes the second assertion. ∎

REMARK. Applying Lemma 1 to the cut $K(V_0)$ whose vertex set is $V_0 = V - \{v_n\}$, we see that since $V_0^c = \{v_n\}$,

$$V(f) = f(V_0, V_0^c) - f(V_0^c, V_0)$$
$$= \sum_{w \in V_0} f(w, v_n) - 0$$
$$= \text{total flow into sink}$$

because no edges are directed out of the sink. In other words, as anticipated earlier, the value of a flow is not only equal to the total flow out of the source, but is also equal to the total flow into the sink.

Next we have a lemma that is a stepping stone to the characterization of a maximal flow.

LEMMA 2. If a flow f and a cut $K = K(V_0)$ separating the source and the sink can be found such that

$$V(f) = c(K)$$

then f is a maximum flow and K is a minimum cut. ∎

You are asked for a proof in Exercise 3. (Hint: Use the second assertion of Lemma 1, i.e., the value of any flow is dominated by the capacity of any cut.)

Below is a precise definition of the idea of a flow augmenting path that was mentioned earlier. First, it should be explained that we are now thinking of *undirected* paths $P : v_1, v_2, v_3, ..., v_n$ from the source to the sink in the *underlying* graph. An edge in such a path is called *forward* if it is oriented in the same direction as in the original directed graph, and *reverse* otherwise.

DEFINITION 3. A simple path P is called an *augmenting path* for a flow f if for all edges e in the path

        (i) $f(e) < c(e)$ if e is a forward edge
        (ii) $f(e) > 0$ if e is a reverse edge

Figure 1.46 – Left: initial graph; right: augmented graph

For example, consider the (undirected) path v_1, a, b, c, d, v_n, in Figure 1.46, which is a small segment of some larger graph. The edge labels are the (capacity, flow) pairs. It is possible to increase the flows on the forward edges (v_1, a), (a, b), (c, d), and (d, v_n) by one unit, and to decrease the flow on the reverse edge (c, b) by one unit, while satisfying the capacity constraint and not changing net outflow from any intermediate vertices. This flow increase was arrived at by taking the smaller of the minimum "slack" (i.e., capacity minus flow) among all forward edges, and the minimum flow along all reverse edges.

Activity 3 – Find an augmenting path for the flow in Figure 1.45.

We can now prove the main theorem.

THEOREM 1. A flow is maximal if and only if it has no augmenting path.

Proof. To show the forward part of the equivalence, we will show its contrapositive: if there is an augmenting path for the flow f, then f is not maximal.

Let P be an augmenting path for the flow f. Define

$$\epsilon_1 = \min \{c(e) - f(e) \mid e \text{ is a forward edge in } P\} \qquad (7)$$

$$\epsilon_2 = \min \{f(e) \mid e \text{ is a reverse edge in } P\} \qquad (8)$$

$$\epsilon = \min \{\epsilon_1, \epsilon_2\} \qquad (9)$$

and define a new function f_ϵ on the edge set of the graph by

$$f_\epsilon(e) = \begin{cases} f(e) + \epsilon & \text{if } e \text{ is a forward edge of } P \\ f(e) - \epsilon & \text{if } e \text{ is a reverse edge of } P \\ f(e) & \text{otherwise} \end{cases} \qquad (10)$$

We claim that f_ϵ is a feasible flow, and that f_ϵ has strictly greater value than f, and consequently f is not maximal.

We leave it to the reader to check that for all edges, the capacity and nonnegativity constraints (1a) are satisfied by f_ϵ, (see Exercise 6). The first edge in path P must point out of the source v_1, hence it is a forward edge. The flow along that edge has been increased by $\epsilon > 0$, by the definition of an augmenting path and the construction of f_ϵ, so that f_ϵ is a strict improvement of f. So the only part of the claim left to show is that f_ϵ satisfies the conservation condition (1b) at interior vertices. Those interior vertices that are not on the augmenting path P satisfy the conservation condition because no flows on edges incident to them have changed. The path is also to be simple, thus there are four possible orientations of edges incident to a vertex u on the path P, as shown in Figure 1.47. In case (a), both total inflow to u and total outflow from u increase by ϵ, and therefore the balance is maintained. In case (b), there is no change to total outflow at all and the net change to total inflow is zero, since flow along one edge is increased by ϵ, while flow along the other is decreased by ϵ. The reader can easily check the conservation of flow condition for cases (c) and (d). Therefore we have found a feasible flow that is better than f, as desired. The forward half of the equivalence is shown.

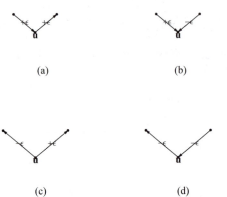

(a)   (b)

(c)   (d)

Figure 1.47 – Four orientations of directed edges at a vertex

To show the reverse implication, let f be a flow for which there is no augmenting path. Let V_0 be the set of vertices to which the source v_1 has an augmenting path. Then v_1 is in V_0, but by assumption the sink v_n is not in V_0. Consider the cut $K(V_0)$, which separates source and sink. If $e = (v, w)$ is an edge in this cut for which $v \in V_0$ and $w \in V_0^c$, then it must be true that $f(e) = c(e)$. To see this, observe that since $v \in V_0$, there is an augmenting path from v_1 to v, and if $f(e) < c(e)$, then edge e could be adjoined to this augmenting path to create an augmenting path to w. This cannot be, since w is not in V_0. Furthermore, if $e = (w, v)$ is an edge in the cut for which $v \in V_0$ and $w \in V_0^c$, then it must be true that $f(e) = 0$. Otherwise, there would be an augmenting path from v_1 to v, which can be extended along the reverse edge e to an augmenting path to w, contrary to the assumption that $w \in V_0^c$.

The argument of the last paragraph shows that

$$f(V_0, V_0^c) = \sum_{v \in V_0} \sum_{w \in V_0^c} f(v, w)$$
$$= \sum_{v \in V_0} \sum_{w \in V_0^c} c(v, w)$$
$$= c(K)$$

Also,

$$f(V_0^c, V_0) = \sum_{v \in V_0} \sum_{w \in V_0^c} f(w, v) = 0$$

Consequently, by Lemma 1,

$$V(f) = f(V_0, V_0^c) - f(V_0^c, V_0) = c(K)$$

which implies, in view of Lemma 2, that f is a maximum flow. ∎

The proof of the last theorem shows exactly how to improve a given flow, if an augmenting path can be found. Compute the increment ϵ by formulas (7)–(9). Note that ϵ is the smallest number among the slacks $c(e) - f(e)$ of forward edges and the flows $f(e)$ on reverse edges of the augmenting path. Then add this increment to all forward edges on the path, and subtract it from all reverse edges. One may repeat this augmentation process until an augmenting path can no longer be found, at which point we know that the current flow is optimal. The main problem, therefore, is to find an augmenting path from the source to the sink in a flow.

An augmenting path for a given flow can be found using a vertex labeling algorithm, which labels vertices with three pieces of information. First, the source is given a label. Other vertices receive labels successively from some vertex that is already labeled. In the vertex v about to be labeled, we record:

1. The preceding vertex u from which v receives its label.
2. A forward (+) designation if (u, v) is a forward edge, and otherwise a reverse designation (–).
3. The *slack* $S(v)$ of v, that is, if u is the vertex from which it receives its label,

$$S(v) = \begin{cases} \min\{S(u), c(u, v) - f(u, v)\} & \text{if } (u, v) \text{ is a forward edge} \\ \min\{S(u), f(u, v)\} & \text{if } (u, v) \text{ is a reverse edge} \end{cases} \quad (11)$$

Also, a vertex v can only receive a label from a vertex u if the slack defined by (11) is strictly positive. For example, in Figure 1.48, the label on u indicates that it received its label from a vertex named 6, along a forward edge pointing from vertex 6 to u, and the slack of u is 4. The flow capacities and current flow values of the two edges are displayed. Now vertex v, which is presently unlabeled, has slack equal to $\min\{4, 3 - 1\} = 2$, and can therefore receive the label $(u, +, 2)$. The edge (u, w) is a reverse edge. The slack that would be given to w if it could be labeled from u is $\min\{4, 0\} = 0$. According to the labeling rule cited above, vertex w cannot receive a label from u.

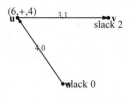

Figure 1.48 – Vertex labeling

The labeling algorithm to find an augmenting path for a flow f proceeds roughly as follows. Label the source $(0, -, \infty)$. Fan outward in an attempt to label vertices, using a breadth-first search plan that scans unlabeled neighbors of labeled vertices, in the order in which the labeled vertices received their labels. Stop when the sink v_n is labeled. If indeed the sink can be labeled, then the definition of slack, formula (11), clearly implies that there is a chain of vertices leading from the source to the sink, such that the sink has the smallest slack among the vertices in the chain. Moreover, if ϵ is the slack, then $c(u, v) - f(u, v) \geq \epsilon$ for all forward edges on this path, i.e., there are ϵ units of unused flow capacity on every forward edge of the path. Hence, ϵ units of flow may be added to each forward edge without violating the capacity constraint. Also, $f(u, v) \geq \epsilon$ for all reverse edges on the path, so that ϵ units of flow may be subtracted from each reverse edge without causing negative flow. The reason for the requirement that slack must be strictly greater than zero for a labeling to occur should now be clear.

Therefore, if the sink can be labeled, we can trace the augmenting path back from the sink to the source, and use it in the way already described to improve the current flow. We leave it to the reader to show (see Exercise 7) that if it is not possible to label the sink using this scheme, then there is no augmenting path, and hence, by Theorem 1, the current flow is optimal. This discussion is the heart of the proof that the following algorithm will locate a maximal flow. We omit the details.

MAXIMAL FLOW ALGORITHM
  1. Initialize flow $f(e) = 0$ for all edges e.
  2. Repeat steps 3–6 until the sink v_n cannot be labeled.
        {Steps 3–5 are the labeling algorithm.}
     3. Erase all previous labels.
     4. Label the source v_1 as $(0, -, \infty)$.
     5. Do a breadth-first search for vertices to label, until the
        sink is labeled, or no further labeling can be
        done.
     {Find the augmenting path, and augment the flow.}
     6. If the sink has been labeled, then:
     a. Let ϵ be the slack of the sink.
     b. Let P be the path of labeled vertices from v_1 to v_n.
     c. Set the new flow f equal to the flow f_ϵ defined by
        (10).

There are two commands in the KnoxOR`Graphs` package that can help you carry out the algorithm one step at a time. To set up the problem, form a matrix like an adjacency matrix that contains the flow capacities, and another matrix of flows on edges, initialized to zero. On each step, first search for an augmenting path. The function FindAugmentingPath returns the list {augmenting path, epsilon}, where the path is a list of vertex numbers, and epsilon is the amount of flow by which edges on the path can be augmented. There is one boolean option, ShowLabels, which if left at its initial value of True displays a table of vertex labels found by the labeling algorithm described above.

```
(***   FindAugmentingPath[
    capacities,flows,source,sink,opts]   ***)
```

```
Options[FindAugmentingPath]
```

{ShowLabels → True}

The second function takes the capacity matrix, the current flow matrix, and the augmenting path and epsilon values that are returned by FindAugmenting-Path, and returns the matrix of the augmented flow, which can then be used in the next augmentation step. The user can then use DisplayGraph to see the results of the augmentation.

```
(*** AddFlow[capacities,
        flows,augmentingpath,epsilon] ***)
```

Examples

EXAMPLE 2. To illustrate the application of the Maximal Flow Algorithm, consider the directed network whose capacities are depicted in Figure 1.49(a). This might represent the floor plan of a manufacturing plant in which pieces of heavy equipment are to be assembled. The nodes are work stations, and the capacities are numbers of pieces of equipment that can be moved from one station to another per half hour. Vertex 1 is the initial location of the parts, and vertex 6 is the shipping area. (In the closed cell below that generated the graph, the capacity matrix is defined as capacities49, and the graph option values are vpos49, vlabelpos49, and elabelpos49 as seen below in the other DisplayGraph commands.)

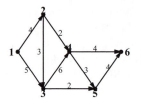

Figure 1.49(a) – A flow capacity graph

***

Activity 4 – We will use the algorithm and the *Mathematica* tools below to find the optimal flow on this graph, but it is rather easy on such a small graph to find it intuitively by hand. Try this.

***

Let the initial flow be 0 along all edges, as defined in the flows49 variable below. We first label vertex 1 as (0, –, ∞). We are able to label vertices 2 and 3 from the source, since the source has infinite slack. The slack along edge (1, 2) is the same as the unused capacity in that edge, namely 4. Similarly, the slack along edge (1, 3) is 5. The information is displayed in the table below the input cell. From vertex 2 we can label vertex 4, from vertex 3 we can label 5, and from vertex 4 the sink vertex can be labeled. The slack of the sink is 2, and hence a flow of two units can be added to each edge of the path 1, 2, 4, 6. The new flow is in Figure 1.49(b).

```
flows49 = {{0, 0, 0, 0, 0, 0}, {0, 0, 0, 0, 0, 0},
   {0, 0, 0, 0, 0, 0}, {0, 0, 0, 0, 0, 0},
   {0, 0, 0, 0, 0, 0}, {0, 0, 0, 0, 0, 0}};
{augpath, epsilon} = FindAugmentingPath[
   capacities49, flows49, 1, 6]
flows49 = AddFlow[capacities49,
      flows49, augpath, epsilon];
DisplayGraph[capacities49, GraphType → Directed,
   AspectRatio → .7, VertexPositions → vpos49,
   VertexLabelPositions → vlabelpos49,
   EdgeLabels → flows49,
   EdgeLabelPositions → elabelpos49];
```

```
vertex      label
1           0   -   ∞
2           1   +   4
3           1   +   5
4           2   +   2
5           3   +   2
6           4   +   2

{{1, 2, 4, 6}, 2}
```

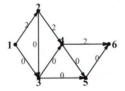

Figure 1.49(b) – First augmentation

The next breadth-first scan labels vertices 2 and 3 from the source. Note that since the flow on edge $(1, 2)$ is now 2, the slack induced on vertex 2 is now $4 - 2 = 2$. This time, vertex 4 cannot be labeled from vertex 2, since the slack is

$$\min\{S(2),\ c(2, 4) - f(2, 4)\} = 0,$$

but vertex 4 can be labeled from vertex 3. The sink is then labeled from vertex 4, and since there are only 2 unused units of flow capacity on edge (4, 6), the slack of the sink is 2. The augmenting path is {1, 3, 4, 6}. Throughout these steps, all edges used for labeling are forward edges.

```
{augpath, epsilon} =
  FindAugmentingPath[capacities49, flows49, 1, 6]
flows49 = AddFlow[capacities49,
    flows49, augpath, epsilon];
DisplayGraph[capacities49, GraphType → Directed,
  AspectRatio → .7, VertexPositions → vpos49,
  VertexLabelPositions → vlabelpos49,
  EdgeLabels → flows49,
  EdgeLabelPositions → elabelpos49];
```

```
vertex      label
1           0  -  ∞
2           1  +  2
3           1  +  5
4           3  +  5
5           3  +  2
6           4  +  2

{{1, 3, 4, 6}, 2}
```

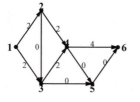

Figure 1.49(c) – Second augmentation

The rest of the computation is displayed in Figures 1.49(d)–(f). If you open the closed cell you will see that we use the ShowLabels->False option to suppress the tables of vertex labels, but you should check the labelings yourself. After the last augmentation, the sink vertex 6 cannot be labeled from either of its predecessors 4 or 5 because there is no unused capacity in either of the edges involved. We therefore know that the current flow among assembly stations is maximal. The maximum output per half hour is eight pieces. ∎

{{1, 3, 5, 6}, 2}

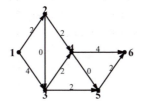

Figure 1.49(d) – Third augmentation

{{1, 3, 4, 5, 6}, 1}

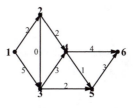

Figure 1.49(e) – Fourth augmentation

`{{1, 2, 3, 4, 5, 6}, 1}`

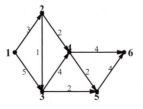

Figure 1.49(f) – Fifth augmentation

There is a command in KnoxOR`Graphs` that carries out the complete Maximal Flow Algorithm, as listed below. Its arguments are the capacity matrix and the vertex numbers of the source and sink, and the final matrix of flows is returned. It accepts the ShowLabels option to display the vertex labels found in each step by the breadth-first search process, and it has a new option of its own ShowSteps, initialized to True, which shows intermediate steps. If ShowSteps is set to False, then MaximalFlow goes immediately to the final flow matrix. It also accepts the display options of DisplayGraph.

```
(* MaximalFlow[capacities,source,sink,opts] *)
```

```
Options[MaximalFlow]
```

```
{ShowSteps → True, ShowLabels → True,
 GraphType → Undirected, VertexLabels → Automatic,
 VertexPositions → Automatic,
 VertexLabelPositions → Automatic,
 EdgeLabels → Automatic,
 EdgeLabelPositions → Automatic,
 EdgeStyle → Thickness[0.005], EdgeSeparation → 0.01,
 DisplayFunction → (Display[$Display, #1] &),
 AspectRatio → 1,
 LoopPositions → Automatic, LoopSize → 0.05}
```

The heart of the algorithm is below, which essentially automates the procedure we went through in the last example. After initializing the matrix of flows to the 0 matrix, while there is still an augmenting path to be found (indicated by the fact that the boolean variable *done* has the value False), if the ShowSteps value *ssteps* is True, then the current flow graph is displayed. An augmenting path and amount of flow called *newflow* are found by calling on FindAugmentingPath. If steps are to be shown, then the augmenting path and new flow are printed. Then if the augmenting path was empty, we mark the variable *done* as True, else we add the new flow to produce the updated flow matrix.

```
done = False;
While[Not[done],
  If[ssteps, DisplayGraph[capacities,
    EdgeLabels → flows, dispopts]];
  {augpath, newflow} = FindAugmentingPath[
    capacities, flows, source,
    sink, ShowLabels → slabels];
  If[ssteps, Print["Augmenting path: ",
    augpath, "   New flow: ", newflow]];
  If[augpath == {}, done = True,
    flows = AddFlow[capacities,
      flows, augpath, newflow]]];
```

EXAMPLE 3. Consider the simple traffic flow system in Figure 1.50(a), in which edge (4, 3) may be thought of as a side street leading to another side street (3, 6), leading to the entrance to an expressway at node 6. Streets (1, 4), (4, 5), and (5, 6) are wider streets capable of supporting more traffic per unit time. We execute the maximal flow algorithm again, this time calling on the MaximalFlow function.

```
capacity50 = {{0, 4, 0, 7, 0, 0}, {0, 0, 4, 0, 0, 0},
    {0, 0, 0, 0, 0, 2}, {0, 0, 2, 0, 8, 0},
    {0, 0, 0, 0, 0, 8}, {0, 0, 0, 0, 0, 0}};
vpos50 = {{0, 0}, {1, 0}, {1.5, 1},
    {1.5, -1}, {2, 0}, {3, 0}};
vlabelpos50 = {ToLeft, Above, Above,
    Below, Above, ToRight};
elabelpos50 = {{0, Above, 0, Below, 0, 0},
    {0, 0, ToLeft, 0, 0, 0}, {0, 0, 0, 0, 0, Above},
    {0, 0, ToLeft, 0, ToRight, 0},
    {0, 0, 0, 0, 0, Below}, {0, 0, 0, 0, 0, 0}};
DisplayGraph[capacity50, GraphType → Directed,
    AspectRatio → .7, VertexPositions → vpos50,
    VertexLabelPositions → vlabelpos50,
    EdgeLabels → capacity50,
    EdgeLabelPositions → elabelpos50];
```

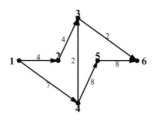

Figure 1.50(a) – A traffic flow system

In the first stage, the augmenting path 1, 2, 3, 6 is found, and the augmenting flow is 2 units. Then we can augment the flow by 7 units along path 1, 4, 5, 6. In the last step, vertex 2 can be labeled from 1, and vertex 3 can be labeled from 2, but the remaining vertices cannot be labeled, no augmenting path is found, and the algorithm ends with a maximal flow of 9 units. ■

```
MaximalFlow[capacity50, 1, 6,
  GraphType → Directed, AspectRatio → .7,
  VertexPositions → vpos50,
  VertexLabelPositions → vlabelpos50,
  EdgeLabelPositions → elabelpos50]
```

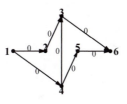

vertex	label		
1	0	–	∞
2	1	+	4
3	2	+	4
4	1	+	7
5	4	+	7
6	3	+	2

Augmenting path: {1, 2, 3, 6}   New flow: 2

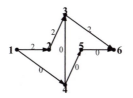

vertex	label		
1	0	–	∞
2	1	+	2
3	2	+	2
4	1	+	7
5	4	+	7
6	5	+	7

Augmenting path: {1, 4, 5, 6}   New flow: 7

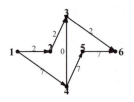

```
vertex        label
1             0   -   ∞
2             1   +   2
3             2   +   2
4             0   -   ∞
5             0   -   ∞
6             0   -   ∞

Augmenting path: {}    New flow: 0
```

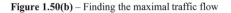

```
{{0, 2, 0, 7, 0, 0}, {0, 0, 2, 0, 0, 0},
 {0, 0, 0, 0, 0, 2}, {0, 0, 0, 0, 7, 0},
 {0, 0, 0, 0, 0, 7}, {0, 0, 0, 0, 0, 0}}
```

Figure 1.50(b) – Finding the maximal traffic flow

Exercises 1.5

1. (a) Consider the directed network below, whose edge capacities are indicated. For each of the vertex sets {1, 2}, {1, 2, 3}, and {1, 4}, list the edges in the cut corresponding to the set and compute the capacity of the cut.

   (b) Assume it is the case that the vertex set {1, 2} determines a minimum cut. Use your intuition to find a maximum flow without executing the algorithm (Hint: see Lemma 2).

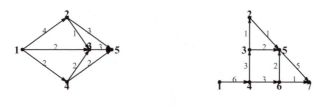

Exercise 1   **Exercise 2**

2. For the graph with flows as indicated, and $V_0 = \{1, 3, 5\}$, check the veracity of the first assertion of Lemma 1.

3. Prove Lemma 2.

4. For the capacity graph of Figure 1.43, find the capacity of the cut corresponding to the vertex set $\{1, 4, 7\}$. Is this cut a minimum cut?

5. Paths (a) and (b) are each paths in some larger network. In each case, decide whether the path is an augmenting path, and if so, use the method suggested by (7)–(10) to augment the path.

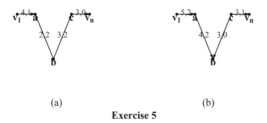

(a)   (b)

Exercise 5

6. Check the constraint (a) of (1) for the augmented flow f_ϵ, defined by (10).

7. Show that if, in the maximal flow algorithm, the breadth–first search cannot label the sink, then there is no augmenting path from source to sink. (Hint: Suppose that one did exist. Consider the first vertex without a label on this path.)

8. Find the optimal vehicular flow for the traffic network with capacities below. Do this problem by hand, rather than in *Mathematica*.

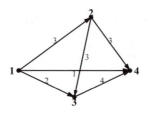

Exercise 8

9. Let the intermediate nodes on the graph of Exercise 1 represent switching locations at a busy train station located at node 5, to which trains are arriving from node 1. The edge capacities represent the number of parallel train tracks connecting switching locations. Use the maximal flow algorithm to decide the most efficient way of routing incoming trains. Do this problem by hand, rather than in *Mathematica*.

10. (*Mathematica*) The graph of Exercise 15 of Section 1.2 modeling a forced-air heat distribution system is displayed again below, with one additional edge. This time we suppose the fully connected system exists and due to pipe diameter differences, there are individual maximum airflow capacities on edges as shown in the graph. Find the maximal flow on the network from the furnace at vertex 1 to the vent at vertex 9.

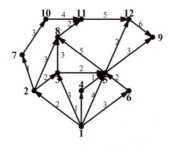

Exercise 10

11. (*Mathematica*) The diagram below represents the lubrication system of a machine; the lubricant flows from a source area at node 1, through components 2–6, which require lubrication, and collects at node 7. Edge capacities are maximum allowable flow rates from one position to another. Find the feasible flow that maximizes the total flow of lubricant through the machine.

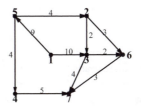

Exercise 11

12. Devise a graph in which at some step the maximal flow algorithm will reduce the flow along a reverse edge.

13. (*Mathematica*) Implement your own version of the AddFlow command described in this section.

14. In the problem of Example 3, suppose that in breadth-first search, vertex 4 is labeled first, before vertex 2, so that vertex 3 will be labeled by vertex 4 rather than vertex 2. Carry out the maximal flow algorithm by hand and note where a reverse edge arises in the algorithm.

1.6 Maximum Matching Problems

Definitions and Problem Description

We now look at the *optimal assignment problem*, which is a special case of the class of problems known as *maximum matching problems*. In the introduction, we saw an example that foreshadowed the discussion in this section. Referring again to Figure 1.3, we have a group of nodes on the left representing workers, and another group of nodes on the right representing tasks. The weight of an edge connecting a worker to a task represents the worker's effectiveness at that task. We are looking for a way of matching workers uniquely with tasks in order to maximize the total effectiveness of all workers.

Throughout this section we consider only undirected graphs. We again give a presentation close to that of Swamy and Thulasiraman ([57], Chapters 8 and 15). The basic definitions pertinent to the problem are as follows.

DEFINITION 1. (a) A graph $G = (V, E)$ is called *bipartite* if there exist disjoint subsets V_1 and V_2 of V such that each edge has one endpoint in V_1 and the other in V_2. We refer to V_1 and V_2 as the *sides* of the graph.

(b) A *matching* M in a bipartite graph $G = (V, E)$ is a collection of edges in E, no two of which share a common vertex.

(c) A vertex in a matching M is called *saturated* by M if it is an endpoint of an edge in the matching.

(d) A matching M in a bipartite graph with sides V_1 and V_2 is *complete* if all vertices in V_1 are saturated.

```
Needs["KnoxOR`Graphs`"];
```

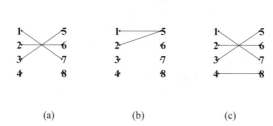

(a)   (b)   (c)

Figure 1.51 – (a) partial matching; (b) not a matching; (c) complete matching

In Figure 1.51(a), $M = \{\{1, 7\}, \{2, 6\}, \{3, 5\}\}$ is a matching, in which vertices 1, 2, and 3 of side $V_1 = \{1, 2, 3, 4\}$ are saturated. The set of edges $\{\{1, 5\}, \{2, 5\}\}$ in Figure 1.51(b) is not a matching since the two edges share vertex 5. The set $M^* = \{\{1, 7\}, \{2, 6\}, \{3, 5\}, \{4, 8\}\}$ in Figure 1.51(c) is a complete matching.

We specialize the general problem of finding complete, maximum weight matchings in graphs in the following ways. We assume:

(a) The graph G is a weighted, undirected bipartite graph with sides V_1 and V_2.

(b) Both V_1 and V_2 have n vertices.

(c) Edges $e = \{v, w\}$ exist between each vertex $v \in V_1$ and $w \in V_2$.

Together, these assumptions imply the existence of $n!$ complete matchings. (See Exercise 1). Our goal is to find one among these many matchings of maximum total weight. From the point of view of applications, (c) is not a very restrictive assumption, since a worker may be given an effectiveness measure of zero at a task for which he is unsuitable. We will denote the vertices of V_1 by v_1, v_2, ..., v_n and those of V_2 by w_1, w_2, ..., w_n. We also write W_{ij} for the weight of the edge connecting v_i and w_j. These weights form an $n \times n$ *weight matrix* $W = (W_{ij})$.

DEFINITION 2. A complete matching M^* is *maximal* if, for all other complete matchings M,

$$\sum_{\{v_i, w_j\} \in M^*} W_{ij} \geq \sum_{\{v_i, w_j\} \in M} W_{ij}$$

Activity 1 – Must there be a unique solution to the maximal matching problem? What if the weights of all edges are different?

There is a tool in the KnoxOR`Graphs` package that easily sketches bipartite graphs given the weight matrix of the graph, which is a *Mathematica* matrix in the form of a list of row lists, that has a row for each left-side vertex and a column for each right-side vertex. The command DisplayBipartiteGraph given below has the weight matrix as its only argument, and takes options ShowWeights → True to show the edge weights, Labeling, which can be set to a list of vertex labels in the matching algorithm to be described below, and Matching, which can be set to a list of edges in a matching. Those edges will be shown solid, and edges not in the matching will be shown dashed. Other options are those of DisplayGraph, except that the EdgeLabels option is overridden. For your convenience, the VertexPositions option and VertexLabelPositions options have been initialized so as to produce a good-looking bipartite graph, although you may change them if you wish.

```
(* DisplayBipartiteGraph[weightmatrix,opts] *)
```

```
Options[DisplayBipartiteGraph]
```

```
{ShowWeights → False,
 Labeling → Automatic, Matching → None,
 DisplayFunction → (Display[$Display, #1] &),
 GraphType → Undirected, VertexLabels → Automatic,
 VertexPositions → Automatic,
 VertexLabelPositions → Automatic,
 EdgeLabels → Automatic,
 EdgeLabelPositions → Automatic,
 EdgeStyle → Thickness[0.005], EdgeSeparation → 0.01,
 DisplayFunction → (Display[$Display, #1] &),
 AspectRatio → 1,
 LoopPositions → Automatic, LoopSize → 0.05}
```

In the graph with the weight matrix below, it is easy to check by enumerating all matchings that $M^* = \{\{1, 4\}, \{2, 6\}, \{3, 5\}\}$ is maximal. This matching is shown in Figure 1.52.

$$W = \begin{pmatrix} 10 & 4 & 6 \\ 1 & 3 & 8 \\ 2 & 7 & 0 \end{pmatrix}$$

```
wmatrix52 = {{10, 4, 6}, {1, 3, 8}, {2, 7, 0}};
vlabel52 = {"1", "2", "3", "4", "5", "6"};
DisplayBipartiteGraph[wmatrix52,
   VertexLabels → vlabel52, AspectRatio → .7,
   Matching → {{1, 4}, {2, 6}, {3, 5}}];
```

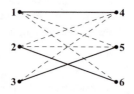

Figure 1.52 – A maximal matching

We will develop an algorithm that successively improves an initial matching until a maximal matching is found. This algorithm proceeds in phases. In each phase, we construct a (proper) subgraph of the original graph. In each step within a phase, there is a matching in the subgraph. If, at any step, the matching is complete, then because of the way in which the subgraph will be chosen, the matching is maximal. Otherwise, we locate an *augmenting path* and use it to create a new matching with one more edge. If an augmenting path cannot be found and we are not yet done, then a new phase is entered. A new subgraph is generated that contains the previous matching, and an augmenting path may exist within the new subgraph. The definition of an augmenting path is as follows.

DEFINITION 3. An *augmenting path* for a matching M is a path with no repeated edges that connects two vertices not saturated by M in such a way that edges in $E - M$ alternate with edges in M.

EXAMPLE 1. In the graph of Figure 1.53(a), the solid lines are edges of a matching M and the broken lines are edges not in M. The path v_3, w_3, v_2, w_2 is an augmenting path. Notice that if we delete from M the edge $\{v_2, w_3\}$ on the augmenting path, and adjoin to M the edges $\{v_3, w_3\}$ and $\{v_2, w_2\}$ on the augmenting path, we obtain a matching M' that has one more edge than M. ∎

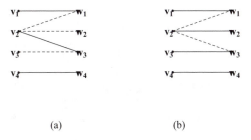

(a) (b)

Figure 1.53 – (a) Matching M; (b) augmented matching M'

Activity 2 – Try to justify that augmentation must always increase the number of edges in the current matching by exactly 1. Then compare your justification to the one in the proof of Theorem 1 below.

Matching Algorithm

We now show that the last observation of Example 1 is true in general, i.e., the exclusive union of a matching with an augmenting path improves the matching.

THEOREM 1. Let P be an augmenting path for a matching M, and define

$$M' = (M - P) \cup (P - M)$$

Then M' is a matching with one more edge than M.

Proof. The path P begins and ends in an unsaturated vertex, hence it must begin and end with an edge not in the matching M. Edges that are "out" of M alternate in the path with edges that are "in" M, hence P has an odd number $m = 2k + 1$ of edges as follows:

$$\{w_0, w_1\} \quad \{w_1, w_2\} \quad \{w_2, w_3\} \quad \cdots \quad \{w_{m-1}, w_m\}$$

$$e_1 \qquad\quad e_2 \qquad\quad e_3 \qquad \cdots \qquad e_m$$

$$\text{out} \qquad \text{in} \qquad \text{out} \qquad \cdots \qquad \text{out}$$

Since P has no repeated edges, we see easily that the vertices w_i in the path are all distinct, and among them, w_0 and w_m are the only vertices not saturated by the matching M.

Note that

$$P - M = \{e_1, e_3, ..., e_m\}$$

and

$$M - P = M - \{e_2, e_4, ..., e_{m-1}\}$$

and these two sets of edges are disjoint. Using $|\ |$ to represent the cardinality of a set, we have

$$
\begin{aligned}
|M'| = |(P - M) \cup (M - P)| &= |P - M| + |M - P| \\
&= (k + 1) + (|M| - k) \\
&= |M| + 1
\end{aligned}
$$

It remains to show that M' is a matching. We leave this to the reader as Exercise 5. ∎

EXAMPLE 2. The location of an augmenting path can be done by a quick scan of the graph for small problems. Consider the graph of Figure 1.54, in which the solid edges are those in a matching M and the broken edges are in the graph, but are not used by M. Begin at an unsaturated vertex on the left side, and follow an unused edge to a vertex on the right side. Then follow a used edge to a vertex on the first side, then an unused edge, etc. Stop as soon as an unsaturated vertex on the second side is found. For instance, starting with vertex 4 on the left side, we trace the path 4, 8, 3, 9. Addition of edges {4, 8} and {3, 9} and deletion of edge {3, 8} augments the matching. ∎

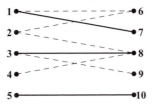

Figure 1.54 – Path 4, 8, 3, 9 is an augmenting path

Activity 3 – Try to find other augmenting paths in Figure 1.54.

The transition to a new phase of the main matching algorithm requires us to introduce the notion of *vertex labeling* for a bipartite graph.

DEFINITION 4. A real-valued function L on the vertex set V of a bipartite graph G with weight matrix W is a *feasible labeling* of V if, for all edges {v_i, w_j},

$$L(v_i) + L(w_j) \geq W_{ij} \quad (*)$$

The subgraph $G(L)$ of G whose edge set consists of all edges such that equality occurs in (*) is the *equality subgraph* of G generated by L.

EXAMPLE 3. The following defines a feasible vertex labeling; in fact, it will be the starting point of the matching algorithm:

$$L(v_i) = \max_{k=1,\ldots,n} W_{ik} \quad i = 1, 2, \ldots, n$$
$$L(w_j) = 0 \qquad\qquad j = 1, 2, \ldots, n \tag{1}$$

Since

$$L(v_i) + L(w_j) = \max_{k=1,\ldots,n} W_{ik} \geq W_{ij}$$

for all i and j, (1) satisfies the defining condition in the definition. Because $L(w_j) = 0$ for all j, the equality subgraph consists of those edges $e = \{v_i, w_j\}$ such that

$$W_{ij} = \max_{k=1,\ldots,n} W_{ik}$$

At least one such edge exists for each vertex v_i in side V_1. For example, let the weight matrix of a bipartite graph be as below. Recall that the rows refer to the left-side vertices v_1, v_2, and v_3, and the columns refer to the right-side vertices w_1, w_2, and w_3.

$$W = \begin{pmatrix} 3 & 2 & 3 \\ 1 & 6 & 4 \\ 4 & 5 & 3 \end{pmatrix}$$

The maximum weights in the three rows are 3, 6, and 5, respectively. In row 1, the maximum is taken on jointly at w_1, and w_3, whereas the maximum is unique in the other two rows. Therefore the equality subgraph has edges $\{v_1, w_1\}$, $\{v_1, w_3\}$, $\{v_2, w_2\}$, and $\{v_3, w_2\}$. ∎

Activity 4 – Find, and sketch, the equality subgraph of the bipartite graph with the weight matrix below, using labeling strategy (1).

$$W = \begin{pmatrix} 4 & 5 & 4 & 3 \\ 2 & 0 & 3 & 3 \\ 7 & 3 & 6 & 4 \\ 2 & 4 & 4 & 1 \end{pmatrix}$$

EXAMPLE 4. Given a labeling L_m, we can create a new labeling L_{m+1} in the following way. Let S be a given subset of vertices in V_1 and let T be the set of all vertices in V_2 that are adjacent to any vertex in S in the equality subgraph $G(L_m)$. Define

$$\Delta = \min_{v_i \in S, w_j \notin T} \{L_m(v_i) + L_m(w_j) - W_{ij}\} \tag{2}$$

No edge $\{v_i, w_j\}$ referred to above can be in $G(L_m)$, since $w_j \notin T$, hence the quantity in braces must be strictly positive for each such edge. It follows that Δ is strictly positive. Define L_{m+1} by

$$L_{m+1}(v_i) = \begin{cases} L_m(v_i) - \Delta & \text{if } v_i \in S \\ L_m(v_i) & \text{otherwise} \end{cases}$$

$$L_{m+1}(w_j) = \begin{cases} L_m(w_j) + \Delta & \text{if } w_j \in T \\ L_m(w_j) & \text{otherwise} \end{cases} \tag{3}$$

The reader is asked in Exercise 10 to check the following four claims about the new labeling L_{m+1}:

If $v_i \in S$ and $w_j \in T$, then $L_{m+1}(v_i) + L_{m+1}(w_j) = W_{ij}$;
consequently the feasibility condition is satisfied, and edge $\{v_i, w_j\}$ (4)
belongs to $G(L_{m+1})$.

If $v_i \in S$ and $w_j \notin T$, then the feasibility condition is satisfied.
Also, there exists an edge of this type that is in $G(L_{m+1})$ but not (5)
$G(L_m)$.

If $v_i \notin S$ and $w_j \in T$, then the feasibility condition is satisfied. (6)

If $v_i \notin S$ and $w_j \notin T$, then the feasibility condition is satisfied.
Also, if such an edge is in $G(L_m)$, then it is also in $G(L_{m+1})$. (7)

In all cases, L_{m+1} is seen to satisfy the feasibility condition (*), and consequently it is a feasible vertex labeling. ∎

Our algorithm will move from one phase to the next by changing labelings in this way. When a search for an augmenting path is unsuccessful, there will be a set S of left-side vertices as in Example 4 that we have encountered while searching for the path. Then the set T of all right-side vertices adjacent to vertices in S can be found. In the partial graph in Figure 1.55, our augmenting path search may have started at v_1, and gone to w_1, then v_2, w_2, and ended at v_3. In this case, and if there are no other edges coming out of v_1, v_2, or v_3 that are not shown, then $S = \{v_1, v_2, v_3\}$ and $T = \{w_1, w_2\}$. The next phase of the algorithm begins by resetting the graph to be the equality subgraph of the new labeling determined by S and T, in which the amount Δ is subtracted from the labels in S and added to the labels in T. We then try to find an augmenting path in the new graph. Observations (4) and (5) above give us hope because the edges used in the unsuccessful augmenting path are still in the new equality subgraph, and there must be at least one new edge as well from a vertex in S to a vertex not in T. There is not a guarantee of a new augmenting path in the next phase, however, and notice that we could actually lose edges pointing into T in the case of (6) when changing labeling. The reason that we cannot be sure of an augmenting path is that it is possible that the new edge points to a right-side vertex that was already matched with a left-side vertex, from which the augmenting path search could stall. But we also do not lose ground, because edges in the current matching that were not on the attempted augmenting path have left-side vertices that are not in S and right-side vertices that are not in T (else like w_2 and v_3 in the diagram the companion left-side vertex would have been a part of the augmenting path in S). By observation (7), the edge connecting them will still be in the new equality subgraph. If no augmenting path is found under the new labeling, we must relabel and search again.

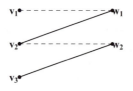

Figure 1.55 – Forming the sets S and T in the matching algorithm

Exercise 11 leads you through an argument that eventually the label changing algorithm must produce an equality subgraph that has a complete matching. The following theorem shows that once we have produced a complete matching in the equality subgraph of a feasible vertex labeling, then the computation may cease.

THEOREM 2. Suppose that M^* is a complete matching in the equality subgraph $G(L)$ of a feasible vertex labeling L. Then M^* is a maximal matching.

Proof. Recall that for arbitrary edges $e = \{v_i, w_j\}$ in the original graph,

$$L(v_i) + L(w_j) \geq W_{ij}$$

and for edges $\{v_i, w_j\}$ in the equality subgraph, particularly those in the matching M^*,

$$L(v_i) + L(w_j) = W_{ij}$$

In a complete matching, all vertices are saturated. Thus, if M is any other complete matching in G,

$$\sum_{e \in M} W_{ij} \le \sum_{i=1}^{n} L(v_i) + \sum_{j=1}^{n} L(w_j) = \sum_{e \in M^*} W_{ij} \qquad (8)$$

This proves that M^* has at least as much weight as M, and since M was an arbitrary complete matching, M^* is optimal. ∎

Below is a statement of the matching algorithm for the optimal assignment problem.

MAXIMAL MATCHING ALGORITHM
    1. Initialize $M = \phi$, labeling L as in formula (1), and $G =$ equality subgraph of L.
    2. Repeat steps 3–8 until the size $|M|$ of the matching is n.
    3. Repeat steps 4 and 5 until an augmenting path cannot be found.
        4. Search for an augmenting path in G.
        5. If the path P is found, then replace M by
          $(M - P) \cup (P - M)$.
    6. Let S be the set of vertices on the left found on the unsuccessful augmenting path, and let T be the set of all their right-side neighbors in G.
    7. Redefine labeling L by formula (3).
    8. Let G be the equality subgraph of the new L.

We will be using a few new commands and data structures in order to carry out the matching algorithm in *Mathematica*. The basic input to the algorithm is the weight matrix of the bipartite graph described earlier. A vertex labeling will be stored as a list of numbers, one for each vertex. For the weight matrix of Example 3, the initial labeling would be $\{3, 6, 5, 0, 0, 0\}$. The DisplayBipartiteGraph command can be used to display the graph, and the option Labeling can be set to the list of labels in order to display the labeling. We must represent matchings and augmenting paths, which we choose to do as lists of edges. The matching that pairs vertex 1 to 6, 2 to 5, and 3 to 4 would be stored as $\{\{1, 6\}, \{2, 5\}, \{3, 4\}\}$, for example. All edges will be written with vertices referred to by number, not name, and with the left-side vertices first and the right-side vertices second. The following tools are available in KnoxOR`Graphs`:

```
(* AugmentMatching[matching,augmentingpath] *)
```

AugmentMatching returns a new matching as in Theorem 1, which uses the given augmenting path and the given matching. (Both are lists of edges as described above.)

```
(* EqualitySubgraph[weightmatrix,labeling] *)
```

EqualitySubgraph produces the weight matrix of the equality subgraph as in Definition 4 of the given labeling, for the given weight matrix of the full bipartite graph.

```
(* ReviseLabeling[weightmatrix,labeling,S,T] *)
```

ReviseLabeling returns the pair $\{\Delta$, newlabeling$\}$, where newlabeling is the list of vertex labels defined by formula (3), given the weight matrix of the original bipartite graph, the current labeling, and the sets S and T of vertices described in the algorithm associated with the unsuccessful search for an augmenting path.

So after entering the weight matrix of the original bipartite graph, an empty initial matching, and the initial labeling, we can use the output of EqualitySubgraph as input to DisplayBipartiteGraph to look at the graph to find an augmenting path. Then AugmentMatching can be called on to get a new matching, which can then in turn be given as an option to DisplayBipartiteGraph to set up the search for a new augmenting path. When it happens that an augmenting path cannot be found, the ReviseLabeling command can be called upon to give the new labeling, which can then be passed to Equality-Subgraph, and the process of finding an augmenting path can continue, until a complete matching is found at some phase in an equality subgraph.

Activity 5 – Use the EqualitySubgraph and DisplayBipartiteGraph commands to show the first equality subgraph for the graph of Activity 4, and if there is a maximal matching, use the Matching option to show it as well.

Examples

The matching algorithm is illustrated and clarified by the next examples.

EXAMPLE 5. A department chairman is to assign four faculty members to four courses on the basis of data obtained from student evaluations conducted in the past. The students were asked to rate the professors on a scale of 1–10, with 10 representing the highest rating. The average scores received by each professor in each course are recorded in the matrix below, in which rows correspond to faculty members and columns to courses. The chairman is to find a matching between professors and courses that achieves maximal total evaluation rating.

$$\begin{pmatrix} 9 & 8 & 2 & 9 \\ 4 & 6 & 4 & 5 \\ 2 & 3 & 3 & 3 \\ 4 & 5 & 3 & 5 \end{pmatrix}$$

We see that the maximum in row 1 is 9, taken on at columns 1 and 4 (that is, vertices 5 and 8 on the right side, the maximum in row 2 is 6, taken on only in column 2 (vertex 6), etc. So we obtain the initial labeling $L_1 = \{9, 6, 3, 5, 0, 0, 0, 0\}$ with the equality subgraph shown in Figure 1.56. (Notice that here we choose to override the default VertexLabelPositions because with the vertex labeling shown the label positions are better off above the vertices to avoid being cut off.)

```
wmatrix56 = {{9, 8, 2, 9},
    {4, 6, 4, 5}, {2, 3, 3, 3}, {4, 5, 3, 5}};
L1 = {9, 6, 3, 5, 0, 0, 0, 0};
initwt = EqualitySubgraph[wmatrix56, L1];
initmatch = {};
vlabelpos56 = {Above, Above, Above,
    Above, Above, Above, Above, Above};
DisplayBipartiteGraph[initwt, AspectRatio → .7,
    VertexLabelPositions → vlabelpos56,
    Labeling → L1, Matching → initmatch];
```

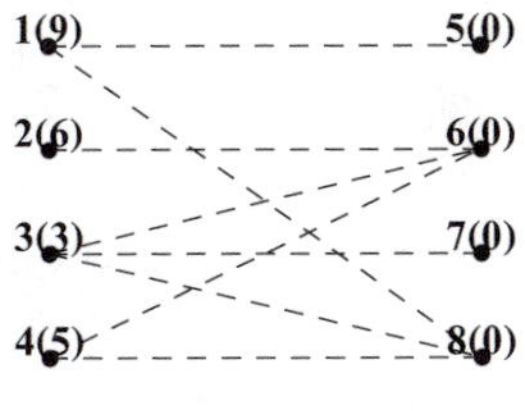

Figure 1.56 – Initial equality subgraph

Begin with the empty matching in the graph of Figure 1.56 and the unsaturated vertex 1. We find the augmenting path 1, 5 immediately, and add it to the matching. If we now choose the unsaturated vertex 3, we find the augmenting path 3, 6 immediately, and add it to the previous matching. The resulting graph is in Figure 1.57, where as usual the edges in the matching are solid, and the edges not in the matching are broken.

```
matching2 = AugmentMatching[initmatch, {{1, 5}}]
matching3 = AugmentMatching[matching2, {{3, 6}}]
DisplayBipartiteGraph[initwt,
    AspectRatio → .7, Matching → matching3];
```

{{1, 5}}

{{1, 5}, {3, 6}}

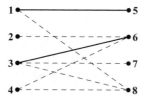

Figure 1.57 – Graph after two augmentations

To renew a search for an augmenting path, we pick the unsaturated vertex 2 on the left side. Its only neighbor is 6, which is saturated by the current matching. The partner of 6 is 3, and from 3 we find the unsaturated vertex 7. The augmenting path is therefore 2, 6, 3, 7, or in *Mathematica* format, {{2, 6}, {3, 6}, {3, 7}}. Augmentation leads us to delete edge {3, 6} from the matching, and add edges {2, 6} and {3, 7}. We now have the matching in Figure 1.58.

```
augpath = {{2, 6}, {3, 6}, {3, 7}};
matching4 = AugmentMatching[matching3, augpath]
DisplayBipartiteGraph[initwt,
   AspectRatio → .7, Matching → matching4];
```

{{1, 5}, {2, 6}, {3, 7}}

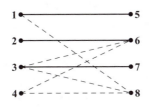

Figure 1.58 – Graph after three augmentations

In only one more step, we can match vertex 4 to 8, producing the maximal matching {{1, 5}, {2, 6}, {3, 7}, {4, 8}}. In this example, we were so fortunate as to have enough edges in the first equality subgraph that a change of subgraph was unnecessary, and we did not yet have to use the ReviseLabeling command. Looking at the rows of the original weight matrix, we see that we were able to match up left-side vertices with one of their most "favorite" right-side vertices, without having to sacrifice any weight. The purpose of using the initial labeling that we did is to check if that is possible, for in the first equality subgraph, only edges from vertices to their favorite mates are included. If a complete matching exists there, it must be optimal.
∎

EXAMPLE 6. In this example, a change of labeling will be required to find an augmenting path. A track coach must decide which of six runners to assign to which of six events. Knowing the skills of his runners, he estimates how many tenths of a second better than the competition is each possible runner on each possible event. These estimates are in the weight matrix below. Rows represent runners, and columns represent events. The problem is to assign runners to maximize the total time difference between the runners of his team and those of the competition.

$$W = \begin{pmatrix} 5 & 5 & 2 & 0 & 0 & 1 \\ 4 & 5 & 6 & 2 & 3 & 0 \\ 1 & 2 & 3 & 3 & 3 & 1 \\ 2 & 2 & 4 & 2 & 1 & 1 \\ 0 & 3 & 4 & 2 & 3 & 2 \\ 4 & 0 & 0 & 1 & 2 & 4 \end{pmatrix}$$

The initial labeling is easily checked to be $L_1 = \{5, 6, 3, 4, 4, 4, 0, 0, 0, 0, 0, 0\}$. The equality subgraph, i.e., those edges for which the sum of the labels of the endpoints equals the weight of the edge, is in Figure 1.59. We jump ahead a little in the algorithm and suppose that we have augmented an initial empty matching with edges $\{1, 7\}, \{2, 9\}, \{3, 10\}$, and $\{6, 12\}$.

```
wmatrix59 = {{5, 5, 2, 0, 0, 1}, {4, 5, 6, 2, 3, 0},
    {1, 2, 3, 3, 3, 1}, {2, 2, 4, 2, 1, 1},
    {0, 3, 4, 2, 3, 2}, {4, 0, 0, 1, 2, 4}};
L1 = {5, 6, 3, 4, 4, 4, 0, 0, 0, 0, 0, 0};
initwt59 = EqualitySubgraph[wmatrix59, L1];
initmatch59 = {{1, 7}, {2, 9}, {3, 10}, {6, 12}};
vpos59 =
    {{0, 5}, {0, 4}, {0, 3}, {0, 2}, {0, 1}, {0, 0},
    {2, 5}, {2, 4}, {2, 3}, {2, 2}, {2, 1}, {2, 0}};
vlabelpos59 = {Above, Above, Above,
    Above, Above, Above, Above, Above,
    Above, Above, Above, Above};
DisplayBipartiteGraph[initwt59,
    VertexPositions → vpos59,
    VertexLabelPositions → vlabelpos59,
    Labeling → L1, Matching → initmatch59,
    AspectRatio → .7];
```

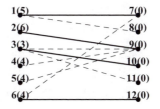

Figure 1.59 – No further augmenting path exists in this equality subgraph

At this point, following the algorithm we set $S = \{4\}$, which is one of only two unsaturated vertices on the left side. But 4 only has one neighbor, namely 9, and the vertex 2 to which 9 is connected in the matching produces no other candidates. So, with $S = \{2, 4\}$ and $T = \{9\}$ we pass out of loop 3–5 of the algorithm with an incomplete match, and we move to step 6, the change in labeling and subgraph.

```
S = {2, 4}; T = {9};
{delta, L2} = ReviseLabeling[wmatrix59, L1, S, T]
newwt59 = EqualitySubgraph[wmatrix59, L2];
DisplayBipartiteGraph[
   newwt59, VertexPositions → vpos59,
   VertexLabelPositions → vlabelpos59,
   Labeling → L2, Matching → initmatch59,
   AspectRatio → .7];
```

{1, {5, 5, 3, 3, 4, 4, 0, 0, 1, 0, 0, 0}}

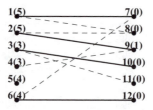

Figure 1.60 – Result of first relabeling

Notice that the number Δ of formula (2) is 1, which has been subtracted from the labels of vertices 2 and 4, and added to vertex 9. This Δ was found by considering rows 2 and 4 of the matrix, and columns other than column 3 (for vertex 9). The smallest difference between total vertex label and edge weight is 1, taken on in row 2 and column 2 (for vertex 8). Notice that the relabeling adds an edge {2, 8} that was not present before. Edges {5, 9} and {3, 9} were lost. But examination of the new graph shows an augmenting path 4, 9, 2, 8. We now augment the matching and look at the new graph.

```
newmatch59 = AugmentMatching[
   initmatch59, {{4, 9}, {2, 9}, {2, 8}}]
DisplayBipartiteGraph[newwt59,
   VertexPositions → vpos59,
   VertexLabelPositions → vlabelpos59,
   Labeling → L2, Matching → newmatch59,
   AspectRatio → .7];
```

{{1, 7}, {2, 8}, {3, 10}, {4, 9}, {6, 12}}

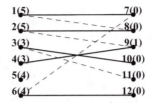

Figure 1.61 – New matching; relabeling must follow

Now the only unmatched left vertex is 5, but we reach an impasse because there is no edge at all incident to vertex 5. So we must relabel again, with $S = \{5\}$ and $T = \emptyset$.

```
S = {5}; T = {};
{delta, L3} = ReviseLabeling[wmatrix59, L2, S, T]
newwt59 = EqualitySubgraph[wmatrix59, L3];
DisplayBipartiteGraph[
   newwt59, VertexPositions → vpos59,
   VertexLabelPositions → vlabelpos59,
   Labeling → L3, Matching → newmatch59,
   AspectRatio → .7];
```

{1, {5, 5, 3, 3, 3, 4, 0, 0, 1, 0, 0, 0}}

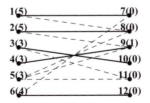

Figure 1.62 – A complete matching exists in this equality subgraph

Again Δ comes out to be 1; and since the minimum difference between vertex label total and edge weight in row 5 occurred in all of columns 2, 3, and 5, the equality subgraph gains new edges {5, 8}, {5, 9}, and {5, 11}. This permits us to augment with edge {5, 11}, for a final matching of {{1, 7}, {2, 8}, {3, 10}, {4, 9}, {5, 11}, {6, 12}}. The purpose of the relabeling and the meaning of Δ is now becoming clear: relabeling is a sacrifice of sorts. It is an admission that we will not be able to assign favorite right-side vertices to all left-side vertices. Some left-side vertex must be paired with a right vertex that is not a favorite. Relabeling as we have done finds a next best collection of potential assignments, and in fact Δ measures how much we have given up. Referring to the rows of the original weight matrix, two relabelings with $\Delta = 1$ have given us a situation where vertex 2 is paired with 8 (an edge weight of one less than 2's favorite weight), and vertex 5 is paired with 11 (an edge weight of one less than 5's favorite weight). All other left vertices are matched with right vertices that maximize the weight among all edges incident to them. The tale of the algorithm is now complete. ∎

Exercises 1.6

1. Argue using combinatorics and mathematical induction that, under assumptions (a)–(c) listed at the start of the section, there are $n!$ total possible complete matchings.

2. Verify that the matching in Figure 1.52 is maximal by computing the total weight of each possible matching.

3. Find an augmenting path for the matching below, and use it to produce a new matching with more edges.

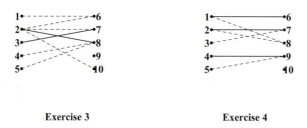

Exercise 3   Exercise 4

4. Repeat Exercise 3 for the matching above.

5. Finish the proof of Theorem 1, i.e., show that M' is a matching.

6. (*Mathematica*) Write your own versions of the *Mathematica* functions: (a) AugmentMatching; (b) ReviseLabeling; (c) EqualitySubgraph.

7. Let $G = (V, E)$ be a bipartite graph with sides V_1 and V_2, each of n vertices. Show that if there is a complete matching of V_1 to V_2, then for every subset S of V_1,

$$|S| \le |A(S)|$$

where $A(S)$ is the set of all vertices in V_2 adjacent to some vertex in S. (This is one half of a double implication called *Hall's Theorem*.)

8. (*Mathematica*) Consider the weight matrix, displayed below, of a bipartite graph.
   (a) Compute the feasible labeling L_1 of formula (1), and sketch the equality subgraph of L_1.
   (b) If $S = \{1, 3\}$ compute the feasible labeling L_2 defined by (3), and sketch its equality subgraph.

$$\begin{pmatrix} 3 & 5 & 1 & 0 & 0 & 2 \\ 6 & 4 & 3 & 2 & 5 & 4 \\ 1 & 4 & 2 & 2 & 1 & 2 \\ 1 & 2 & 3 & 3 & 3 & 1 \\ 2 & 1 & 3 & 2 & 4 & 2 \\ 3 & 2 & 5 & 4 & 6 & 6 \end{pmatrix} \qquad \begin{pmatrix} 4 & 1 & 0 & 2 & 3 \\ 1 & 3 & 3 & 2 & 1 \\ 4 & 5 & 5 & 2 & 1 \\ 0 & 0 & 3 & 2 & 0 \\ 2 & 1 & 2 & 6 & 6 \end{pmatrix}$$

Exercise 8   **Exercise 9**

9. Repeat Exercise 8, with the weight matrix above and $S = \{2, 3, 4\}$. This time, do the problem by hand, rather than in *Mathematica*.

10. Verify claims (4)–(7) about the change of labeling.

11. Show that the label changing algorithm must produce an equality subgraph that has a complete matching, by arguing as follows:
   (a) Upon changing labeling, since by claim (5) there is a new edge from S to T^c, show that there will either be an augmenting path in the new equality subgraph, or else the set S must become strictly larger.
   (b) In the worst case, a labeling will be reached where S is the whole left side. Then show that further relabelings can lose no edges and must gain edges. Therefore conclude that a complete matching must be reached.

12. A dishonest politician has four candidates for four patronage jobs. Each candidate has agreed to bribe the politician to obtain each job, by amounts shown in the matrix below (units of thousands of dollars). Find two different ways of assigning the candidates to jobs, each of which maximizes the politician's total profit. Do this problem by hand, and not in *Mathematica*.

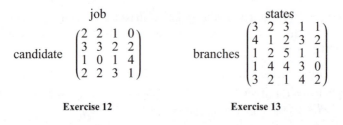

candidate $\begin{array}{c}\text{job}\\\begin{pmatrix}2 & 2 & 1 & 0\\3 & 3 & 2 & 2\\1 & 0 & 1 & 4\\2 & 2 & 3 & 1\end{pmatrix}\end{array}$

branches $\begin{array}{c}\text{states}\\\begin{pmatrix}3 & 2 & 3 & 1 & 1\\4 & 1 & 2 & 3 & 2\\1 & 2 & 5 & 1 & 1\\1 & 4 & 4 & 3 & 0\\3 & 2 & 1 & 4 & 2\end{pmatrix}\end{array}$

Exercise 12   **Exercise 13**

13. (*Mathematica*) A company is planning to locate five branches in five states. After studying various factors related to the local economies and tax laws of the states, the company has managed to quantify how beneficial each branch would be if located in each state. These ratings are in the matrix above. How should the company allocate the branches among the states, with no more than one per state, so as to maximize the overall benefit?

14. (*Mathematica*) Find a maximal matching for the graph of Exercise 8.

15. (*Mathematica*) Find a maximal matching for the graph of Exercise 9.

16. (*Mathematica*) A sales manager must assign each of eight salespeople to one of eight different regions. He has asked the salespeople to rate their choice of regions in order, with 8 representing their most preferred choice and 1 representing their least preferred. The results are shown in the matrix below, with rows corresponding to salespeople and columns to regions. How should the sales manager assign the salespeople to maximize the total rating?

salespeople $\begin{array}{c}\text{regions}\\\begin{pmatrix}7 & 8 & 4 & 5 & 3 & 6 & 2 & 1\\6 & 8 & 7 & 5 & 4 & 2 & 3 & 1\\5 & 4 & 8 & 7 & 6 & 3 & 1 & 2\\3 & 6 & 4 & 8 & 5 & 7 & 2 & 1\\1 & 5 & 6 & 7 & 4 & 8 & 3 & 2\\4 & 3 & 2 & 1 & 7 & 8 & 6 & 5\\2 & 3 & 4 & 1 & 6 & 5 & 7 & 8\\1 & 5 & 3 & 4 & 2 & 8 & 6 & 7\end{pmatrix}\end{array}$

1.7 Other Problems of Graph Theory

We are nearing the end of our study of algorithmic graph theory, and it might now be helpful to summarize the main ideas. The chapter began with an investigation of issues related to connectivity of directed and undirected graphs. We saw how to find connected components of a graph and how to use the adjacency matrix to count the number of paths of a given length. We then devised algorithms for finding spanning trees, i.e., connected subgraphs with the fewest possible connections. Then we gave costs to the edges, and proceeded in Section 1.3 to find spanning trees of minimal cost. Such trees indicate the most efficient way to maintain communications among several stations. The algorithm in the undirected case was a simple extension of the algorithm constructed in the previous section for trees without cost. A different algorithm, based on the idea of dynamic programming, was given for minimal directed spanning trees. This algorithm was adapted in Section 1.4 to find paths of maximum weight in a directed network. The idea was that such a path constitutes the most time-consuming sequence of operations in a large job whose prerequisite structure was described by the graph.

In the maximal flow problem, we were to find non-negative weights on the edges of a directed network, smaller than or equal to corresponding edge capacities, which maximize the total weight of edges pointing out of the source. As in the directed minimal cost tree problem, the strategy was to begin with an arbitrary feasible solution, and improve it step-by-step until a condition indicating optimality became true. This improvement required the location of an augmenting path. The augmenting path idea also arose in the matching problem of Section 1.6, in which we were to match vertices on two sides of a bipartite graph to maximize the total weight of edges used in the matching. An extra complication arose when it was no longer possible to find an augmenting path but the matching was not yet complete. A vertex labeling method was used to produce a new graph in which augmentation of the matching could occur.

Much of the material in Sections 1.1–1.4 was based on the presentation of Dierker and Voxman [19], with additional information on quasi-connectivity taken from Mott, Kandel, and Baker [46]. As mentioned before, Swamy and Thulasiraman [57] was the main source for the material in Sections 1.5 and 1.6. For further information, the reader may consult many books on discrete mathematics and graph theory. Of particular help may be Gibbons [26] and Minieka [44].

We have not attempted to analyze the efficiency of the algorithms given here. The texts on algorithmic graph theory mentioned above [26, 44, and 57] examine this issue. Efficiency can depend strongly on the proper choice of data structures used in the program implementation. The reader is encouraged to refer to a good data structures and algorithm analysis book such as

Aho, Hopcraft, and Ullman [1] for information on this subject.

Due to space limitations, we have reluctantly chosen to omit full discussion of several very well-known operations research problems related to graphs. It is hoped that if these are described briefly here, then the reader's appetite will be whetted for independent study.   The descriptions and references below can serve as the jumping-off point for an expository writing assignment as well.

Graph Coloring Problem

One of the most famous of graph theoretic problems involves the partitioning of the vertices into sets called "colors," or, less formally, the "graph coloring problem." Consider an intersection of streets such as that of Figure 1.63(a).   The street connecting points A and C is a one-way street, and the street connecting B and D is a two-way street. Some turns can be made simultaneously without causing accidents, and thus such turns can all be permitted to occur on a single phase (color) of a traffic light.   For example, the right turn denoted by AD from A to the right lane of D can occur at the same time as turn BC from B to the right lane of C. Other turns must be prevented when this color is lit, and allowed to occur only when some other color is lit. A car cannot, for instance, be permitted to make the left turn BC at the same time another car is passing through the intersection from A to C, which we denote by "turn" AC.   The problem is to find a way of assigning turns to colors, utilizing as few colors as possible, such that no collisions can occur. If we view the turns as vertices of a graph and connect those that cannot be allowed to proceed at the same time, as in Figure 1.63(b), then we are searching for a minimal coloring of vertices such that adjacent vertices have different colors.

(a)   (b)
Figure 1.63 – Graph coloring and street intersections

Activity 1 – Explain how the following situation can be modeled as a graph coloring problem. Five hazardous materials are to be transported by train to a distant underground repository. Legal safety regulations require that material 1 cannot be transported in the same car with materials 3 and 4, material 2 cannot be in the same car with materials 4 or 5, and material 3 cannot ride with material 1 or 5. What is the minimum number of cars needed to transport all the materials?

Several elementary results are easy to see. For example, a complete graph on n vertices requires n colors, since every vertex is adjacent to every other, hence no pair can share the same color. A less obvious result is that if the graph has no cycles with odd length, then two colors suffice. To show this, we can use a variation of the breadth-first search technique. (See Figure 1.64.) Start at a vertex and give it a color, say blue. Fan out to its children and color them red. Next, fan out to the children of these recent vertices and color them blue, if they are not already colored. If any of these new children was adjacent to the original blue starting vertex, then there would be a cycle from the start to the child, to the grandchild to the start, but this cycle would have length 3, which is forbidden. We can continue the process of fanning out to children, coloring them red if their parents are blue, and blue if their parents are red, and it is easy to give a convincing argument that a blue vertex can never be adjacent to a blue vertex, or a red to a red, else there would be an odd length cycle, in contradiction to assumption. It is not hard to write a *Mathematica* function to implement this algorithm. (Try it.)

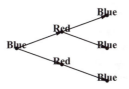

Figure 1.64 – Coloring with two colors

You should not be too disappointed if you cannot develop a complete solution to the graph coloring problem in one afternoon. Many researchers have spent many decades on the coloring problem. One of the great mathematical results of the twentieth century was the exhaustive, computer-intensive verification that no more than four colors will be necessary if the graph is a planar graph, which means, roughly speaking, if vertices are able to be arranged so that all edges can be drawn without crossing each other. For further information on graph coloring, see Aho, Hopcraft and Ullman [1], Gibbons [26], Busacker and Saaty [11], or any other good graph theory text.

Shortest Paths Problem

Figure 1.65 reproduces the computer network of Figure 1.24 in Section 1.3. At that point in our study, we were interested in minimal total cost networks, and we did not particularly care whether a minimal spanning tree gave a shortest possible path from one vertex to another. But in many problems we are interested in finding shortest paths; for instance, if messages are to be broadcast from vertex 1 throughout the network, we would need a way of finding shortest paths from 1 to each other vertex.

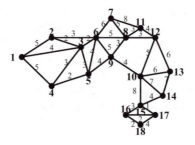

Figure 1.65 – Finding shortest paths

An efficient algorithm, due to Dijkstra, does exist for this shortest paths problem. The idea is that one keeps a set A of vertices at each step to which shortest paths have been found, and in the next step add to A the vertex not already in A that would result in the shortest path from 1. We keep track of the predecessors of each vertex in the path from 1 as we go along. In the diagram above, we would initialize $A = \{1\}$. The set of vertices adjacent to A is now $\{2, 3, 4\}$, and the closest vertex to 1 is vertex 3 at a distance of 4 units. Thus, we add it in to get $A = \{1, 3\}$, and note that vertex 3 has a permanent predecessor of vertex 1. Now vertex 1 is adjacent to vertices 2 and 4, and path costs from 1 for these vertices would each be 5. Also, vertex 3 is adjacent to 2, 4, 5, and 6. Since it costs 4 units already to get to vertex 3, if we choose to add in vertex 2 from 3, the path cost from 1 to 2 is 6, and the possible path costs from 1 through 3 to the other vertices 4, 5, and 6 are 7, 7, and 6, respectively. Among these path costs, the direct edge from either 1 to 2 or from 1 to 4 would be the least costly choice. So add vertex 2 to A with predecessor 1. Vertex 4 would be added next, also with predecessor 1. The vertex adjacent to one of the vertices in $A = \{1, 2, 3, 4\}$ that creates the next shortest path is vertex 6, with predecessor 3 and path cost 6. After adding 6 to the set A, the shortest paths that we have found so far are: $\{1, 2\}, \{1, 3\}, \{1, 4\}, \{1, 3, 6\}$. Continue in this way until all vertices are included in A.

Activity 2 – What are the next three vertices added to the set A in the example above?

Most texts on Discrete Mathematics and Graph Theory have discussions of the shortest path problem. In particular, you will find a nice discussion in the book by Dossey, Otto, Spence, and Vanden Eynden [20].

Traveling Salesman Problem

Another problem of prominence in operations research has the provocative name of the "traveling salesman" problem. Mathematically, the statement of the problem is very simple. Given a weighted graph, either directed or undirected, find a cycle of smallest total weight that passes through all of the vertices. To see the reason for the name, consider again the graph of Figure 1.1 reproduced here as Figure 1.66. Suppose now that the vertices are cities to be visited by a salesman, and the weights are distances between cities. The traveling salesman must visit all of the cities at least once, beginning and ending at some home base, while minimizing the total distance traveled. Several different approaches to the problem can be found in Minieka [44] and Gibbons [26], and a formulation as an integer program (see Chapter 3) is given in Winston [61]. We will be content to comment on the computational difficulty of the problem, and to talk about some *heuristic* approaches, that is, methods that common sense indicates should give good solutions but which are not guaranteed to yield optimal solutions.

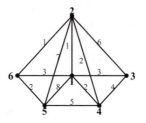

Figure 1.66 – The traveling salesman problem

First observe that each cycle that passes through all vertices of a graph exactly once is in 1–1 correspondence with a permutation of the vertices. The cycle 1, 2, 3, 4, 5, 6, 1 in Figure 1.66 corresponds to the permutation (1, 2, 3, 4, 5, 6) of the first six integers. It seems that there could be as many as 6!, or in general for a graph of n vertices $n!$, different cycles to check. An exhaustive algorithm would have to systematically examine each, add the total cost of the edges in the cycle, and update the best cycle and best cost if necessary. There are roughly $n - 1$ additions to be made, and a comparison, for each cycle to be examined, hence there are $n \cdot n!$ operations to be done. In the case of, say, $n = 30$ cities, even an extremely fast computer that could do a billion operations per second, at 3600 seconds per hour, 24 hours per day, and 365 days per year, would take the following number of years to do the job:

$$N[30 * (30!) / (10^9 * 3600 * 24 * 365)]$$

2.52333×10^{17}

By changing the 30 above to 40 or 50, you can see easily the problem with exhaustive checking.

Actually the situation is not quite that bad. First, some potential cycles can be terminated early because not all edges may be available in the graph. The path 1, 3, 5, 6, 2, 4, 1 is not legal in the graph of Figure 1.66 because edge {3, 5} does not exist. Second, we may choose to start and end our cycle at an arbitrary, but fixed, vertex such as vertex 1. The cycle 1, 2, 3, 4, 5, 6, 1 has the same cost and traverses the vertices in the same order as the cycle 2, 3, 4, 5, 6, 1, 2, so it is not necessary to examine both of these cycles. In that case, the cycles that we need to examine fix the first member of the permutation, and just permute the last $n - 1$ vertices. So there are a mere $(n - 1)!$ different vertices instead of $n!$. Moreover, a cycle like 1, 2, 3, 4, 5, 6, 1 is the same as the cycle in the reverse order 1, 6, 5, 4, 3, 2, 1. So we may cut the number of cycles in half. But you must agree that even using these devices and dividing the number above by $30\,(2) = 60$ does not change the essential explosive nature of the problem.

Activity 3 – For the graph of Figure 1.66, how many cycles at most can there be? List the cycles that begin 1, 2, ... and compute their path costs.

In doing Activity 3, you may have devised a systematic approach to itemizing cycles somewhat along the following lines. Form a tree with vertex 1 at the top, and child vertices 2, 3, etc. under it corresponding to the possible second vertices that can be visited from vertex 1 as the cycle begins. Under each of these children, place the possible third vertices on the cycle, etc., until the depth of the tree is equal to the number of vertices, and then adjoin vertex 1 at the bottom as the last vertex in the cycle. A part of the corresponding tree for Figure 1.66 is in Figure 1.67. We show only the cycles that begin with 1, 2. Some paths get stuck because an edge to an unvisited vertex no longer exists in the graph. For all paths that reach the bottom of the tree, the edge costs along the path can be added to find the cost of the cycle. In the part of the tree shown here, only two of the six paths shown form complete cycles.

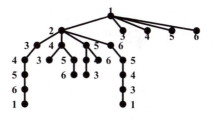

Figure 1.67 – Enumerating all cycles on a tree

One heuristic method that has been effective is the *nearest-neighbor* heuristic. Suppose for simplicity that we have a complete graph, with edges existing between every pair of vertices, such as the small four-vertex graph in Figure 1.68. We construct a cycle by starting from a given vertex, and following the strategy that whatever vertex we are currently at, we next visit the vertex that is closest (in terms of smallest edge cost) to the current vertex among those that have not been visited yet. So from vertex 1 in this graph, we visit 3 next because its edge cost of 4 is smaller than the edge costs 8 and 6 to vertices 2 and 4, respectively. From 3 we visit 2 next, and then 4. The total cost of the cycle 1, 3, 2, 4, 1 is $4 + 6 + 3 + 6 = 19$. Try to use the exhaustive tree method to verify that this is the smallest possible cost.

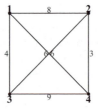

Figure 1.68 – Nearest -neighbor heuristic

Activity 4 – Another well-known heuristic method for the traveling salesman problem is the *sorted edges* heuristic. Much as in Kruskal's algorithm, you first sort the edges in increasing order of cost. Then you attempt to build a cycle by selecting edges in the list in order, as long as they neither finish a cycle that does not go through all of the vertices, nor give any vertex degree 3, which cannot be in a cycle. Try this approach on the graph of Figure 1.68 and show that you find the same cycle as the nearest-neighbor heuristic found. Does the approach work on the graph of Figure 1.67?

Our purpose has been to show the mutually beneficial interplay between theory and practical results in graph theory, as well as to solve a few special operations research problems related to networks. Any algorithm must be proved to do the job that it is meant to do, and some interesting mathematics arises in doing so, as in Kruskal's algorithm and the results characterizing optimality in the maximal flow and maximal matching problems. We will see often the interrelation of mathematics and algorithm development as we move on to study linear programming, stochastic processes, and dynamic programming in the remaining chapters.

Linear Programming

Introduction

Mathematical programming is the area of mathematics that is concerned with optimizing an objective function of several variables subject to constraints on those variables. We have already encountered such a problem in Chapter 1. Consider again the minimal cost spanning tree problem as illustrated by Figure 1.1. A subgraph of the original communications network can be represented by a matrix $A = (x_{ij})_{i,j=1,\ldots,n}$ in which $x_{ij} = 1$ if edge $\{v_i, v_j\}$ is in the subgraph and $x_{ij} = 0$ otherwise. In finding a tree of minimal cost, we are really solving the problem: minimize the total cost of all edges in the subgraph subject to the condition that there is exactly one path in the subgraph from each vertex to each other vertex. Since $A^k(i, j)$ is the number of paths of length k from v_i to v_j, we see that we can write the problem as:

$$\text{minimize: } g(x_{11}, x_{12}, \ldots, x_{n,n-1}, x_{nn}) = \sum_{i<j} \sum c(i, j) x_{ij}$$

$$\text{subject to: } \sum_{k=1}^{n-1} A^k(i, j) = 1 \text{ for } i \neq j, \quad x_{ij} = 0 \text{ or } 1$$

where n is the number of vertices and $c(i, j)$ is the cost of edge $\{v_i, v_j\}$.

The objective function g is linear in the variables x_{ij}, but the first constraint is highly non-linear, and the second constraint forbids the variables from taking on arbitrary real values in some interval. Both of these conditions make the problem difficult. Fortunately, we developed other techniques for the spanning tree problem.

The problems that we will study in this chapter are more tractable, though they are still non-trivial and have wide applications. The underlying idea of the problem is to optimally allocate limited resources. The objective function will be linear, the constraints will be linear, and the variables will usually take values in subsets of the non-negative half of the real line. Such problems belong to the area of mathematical programming called *linear programming*.

The following example is a good illustration. A winery makes three kinds of wine: red, white, and rosé. A gallon of red wine yields a profit of $1.25 and requires 2 bushels of type I grapes, 0 bushels of type II grapes, 2 lbs. of sugar, and 2 labor hours to produce. The corresponding numbers for white wine are \$1.50 , 0 bushels, 2 bushels, 1 lb., and 1 labor hour, and those for rosé wine are \$2, 1 bushel, 1 bushel, 1.5 lbs., and 2 labor hours. If in a week, the winery has available 200 bushels of type I grapes, 150 bushels of type II grapes, 90 lbs. of sugar, and 250 labor hours, then how much of each wine should be made to maximize total profit?

We define x_1, x_2, and x_3, respectively, to be the numbers of gallons of red, white, and rosé wine to be made in this week. The total weekly profit is then $1.25\,x_1 + 1.50\,x_2 + 2.00\,x_3$, in units of dollars. Each variable x_i is clearly greater than or equal to 0. We may use no more than 200 bushels of type I grapes. We will require $2\,x_1$ bushels of these for red wine, $0\,x_2$ for white wine, and $1\,x_3$ bushels for rosé. Thus, we obtain the first constraint below. The other three constraints reflect the limitations on type II grapes, sugar, and labor hours in that order. We formulate the linear programming (LP) problem as follows:

$$\text{maximize: } f(x_1,\ x_2,\ x_3) = 1.25\,x_1 + 1.50\,x_2 + 2.00\,x_3$$

$$
\begin{array}{rcrcrcr}
2\,x_1 & & & + & x_3 & \le & 200 \\
& & 2\,x_2 & + & x_3 & \le & 150 \\
\text{subject to: } 2\,x_1 & + & x_2 & + & 1.5\,x_3 & \le & 90 \\
2\,x_1 & + & x_2 & + & 2\,x_3 & \le & 250 \\
& & & & x_1,\ x_2,\ x_3 & \ge & 0
\end{array}
$$

This problem is said to be in *standard maximum* form; that is, it can be written in matrix notation as:

$$\text{maximize: } f = \mathbf{c}' \cdot \mathbf{x}$$

$$\text{subject to: } A\,\mathbf{x} \le \mathbf{b}$$
$$\mathbf{x} \ge \mathbf{0}$$

where $\mathbf{b}$ is at least 0 in every component. Here $\mathbf{c}'$ is the row vector of coefficients of the objective function, $\mathbf{x}$ is the column vector of variables, A is the matrix of constraint coefficients, and $\mathbf{b}$ is the column vector of constants on the right sides of the inequalities. Specifically, for the winery problem:

$$\mathbf{c'} = (1.25, \ 1.50, \ 2.00) \qquad \mathbf{x} = \begin{pmatrix} x_1 \\ x_2 \\ x_3 \end{pmatrix}$$

$$A = \begin{pmatrix} 2 & 0 & 1 \\ 0 & 2 & 1 \\ 2 & 1 & 1.5 \\ 2 & 1 & 2 \end{pmatrix} \qquad \mathbf{b} = \begin{pmatrix} 200 \\ 150 \\ 90 \\ 250 \end{pmatrix}$$

In Section 1, we consider the graphical solution of LP problems in two variables. We use the geometric intuition developed there to derive in Section 2 some of the general theory that leads to an approach to higher dimensional problems. Section 3 introduces the so-called *simplex algorithm* to solve the standard maximum problem described above. We use the important notion of *duality* between problems to extend the approach to standard minimum problems in Section 4. Our study of linear programming continues in Chapter 3; in particular, we will see how to solve problems that are in neither standard maximum nor standard minimum form.

Activity 1 – What aspect of the vectors and matrix above changes if a new formulation of the white wine requires only 1.5 bushels of type II grapes? If there are 110 lbs. of sugar available? If the profit on red wine is $1.10 per gallon?

2.1 Two-Variable Problems

We begin our study of linear programming by looking at problems involving two variables in an informal way, noting as we proceed several features of the problem that generalize to higher dimensions. Consider the problem:

$$\text{maximize:} \quad f(x_1, x_2) = x_1 + 2 x_2 \qquad (1)$$

$$\text{subject to:} \qquad \begin{aligned} x_2 &\le 2 \\ x_1 + x_2 &\le 3 \\ 2 x_1 + x_2 &\le 5 \\ x_1, x_2 &\ge 0 \end{aligned} \qquad (2)$$

The function f in (1) to be optimized is called the *objective function*, the inequalities in (2) are called *constraints*, and the simultaneous solution set of the constraints is called the *feasible region* of the problem. Figure 2.1 is a sketch of the three boundary lines associated with the three inequalities, which we produced in a special way using *Mathematica*. The *Mathematica* package called KnoxOR`LinearProgramming` that we will be using heavily in this chapter automatically loads the standard *Mathematica* package Graphics`ImplicitPlot`, which contains the useful command

ImplicitPlot[listofequations, plotdomain1,plotdomain2]

ImplicitPlot can plot the graphs of one or more equations in two variables. The first argument is the list of equations, and the other arguments are the plot domains for the two variables. ImplicitPlot also accepts the usual kind of plot options to control the appearance of the graph. Here is how it works on our three linear equations.

```
Needs["KnoxOR`LinearProgramming`"]
```

```
ImplicitPlot[
    {x₂ == 2, x₁ + x₂ == 3, 2 x₁ + x₂ == 5}, {x₁, 0, 3},
    {x₂, 0, 3}, AspectRatio -> 1, TextStyle ->
    {FontFamily → "Times", FontSize → 8}];
```

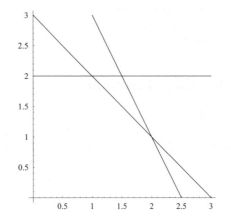

Figure 2.1 – Constraint boundary lines

Since the inequalities are all of ≤ form, the solution sets all lie to the southwest of the lines. The commands below solve for the intersection points that are corners of the polygonal solution region. These corners are: the *y*-intercept of the first constraint line, the intersection of the first two constraints, the intersection of the second and third constraints, and the *x*-intercept of the third constraint.

```
Solve[{x₁ == 0, x₂ == 2}, {x₁, x₂}]
Solve[{x₂ == 2, x₁ + x₂ == 3}, {x₁, x₂}]
Solve[{x₁ + x₂ == 3, 2 x₁ + x₂ == 5}, {x₁, x₂}]
Solve[{2 x₁ + x₂ == 5, x₂ == 0}, {x₁, x₂}]
```

$\{\{x_1 \to 0, \ x_2 \to 2\}\}$

$\{\{x_1 \to 1, \ x_2 \to 2\}\}$

$\{\{x_1 \to 2, \ x_2 \to 1\}\}$

$\left\{\left\{x_1 \to \dfrac{5}{2}, \ x_2 \to 0\right\}\right\}$

KnoxOR`LinearProgramming` has a command for plotting feasible regions of two-variable LP problems. The syntax is as follows:

```
(*  PlotFeasibleRegion[constraints,
    xdomain,ydomain,corners,objective]   *)
```

```
Options[PlotFeasibleRegion]
```

```
{DisplayFunction → (Display[$Display, #1] &),
 ObjectiveLines → Automatic, ShowTable → True,
 ShadingStyle → GrayLevel[0.7],
 ObjectiveLineStyle → {RGBColor[0, 0, 0]},
 AspectRatio → Automatic, Axes → Automatic,
 AxesLabel → None, AxesOrigin → Automatic,
 AxesStyle → Automatic, Background → Automatic,
 ColorOutput → Automatic, DefaultColor → Automatic,
 Epilog → {}, Frame → False, FrameLabel → None,
 FrameStyle → Automatic, FrameTicks → Automatic,
 GridLines → None, PlotLabel → None,
 PlotPoints → 39, PlotRange → Automatic,
 PlotRegion → Automatic, PlotStyle → Automatic,
 Prolog → {}, RotateLabel → True,
 Ticks → Automatic, DefaultFont :→ $DefaultFont,
 DisplayFunction :→ $DisplayFunction,
 FormatType :→ $FormatType,
 TextStyle :→ $TextStyle, ImageSize → Automatic}
```

PlotFeasibleRegion takes the list of equality constraints, plot domains on both the horizontal and vertical axis variables, the list of corners of the feasible region, and the name of the objective function, and it returns the graph of the feasible region. The corners should be listed in either clockwise or counterclockwise order, starting with any corner. Besides inheriting the options of ImplicitPlot, PlotFeasibleRegion has a few options of its own. ShowTable is a boolean option that, if set to true, displays a table of values of the objective function at the corners of the feasible region. Objective-Lines may be set to a list of constant values (see the next paragraph) for the plotting of c-level sets. And ShadingStyle and ObjectiveLineStyle can be used to apply a fill style to the feasible region and a style to the objective lines, respectively.

```
f[x_, y_] := x + 2 y;
PlotFeasibleRegion[
   {x₂ == 2, x₁ + x₂ == 3, 2 x₁ + x₂ == 5},
   {x₁, 0, 3}, {x₂, 0, 3},
   {{0, 0}, {0, 2}, {1, 2}, {2, 1}, {2.5, 0}},
   f, ObjectiveLines → {3, 4, 5, 6},
   ObjectiveLineStyle → RGBColor[1, 0, 0],
   ShowTable → True, TextStyle ->
     {FontFamily → "Times", FontSize → 8}];
```

x	y	objective
0	0	0
0	2	4
1	2	5
2	1	4
2.5	0	2.5

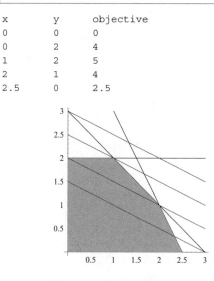

Figure 2.2 – The feasible region and c-level sets

The set of points (x_1, x_2) at which the objective function f takes on a constant value c is called the *c-level set* of f. The maximum LP problem essentially is to find the largest c such that the c-level set intersects the feasible region, and to find the point (or points) of intersection. Then c is the optimal value, and the point of intersection is the point at which f achieves its optimum. Figure 2.2 shows several level sets for c values of 3, 4, 5, and 6 proceeding from southwest to northeast. The sets in this case are lines of slope $-1/2$, because the equation for a general level set in this example is $x_1 + 2x_2 = c \Longleftrightarrow x_2 = (-1/2)x_1 + c/2$. As c increases, the lines move upward until finally the $c = 5$ line intersects the feasible region at the corner point $(1, 2)$. For all $c > 5$, the c-level set does not intersect the feasible region. This means that the maximum value of f is 5, taken on at $x_1 = 1$, and $x_2 = 2$. In the electronic version of the text, you can execute the cell below this paragraph, and then select and animate the graphics to watch the c-level sets move as c increases.

```
f[x_, y_] := x + 2 y;
Table[PlotFeasibleRegion[
    {x₂ == 2, x₁ + x₂ == 3, 2 x₁ + x₂ == 5}, {x₁, 0, 3},
    {x₂, 0, 3}, {{0, 0}, {0, 2}, {1, 2}, {2, 1},
    {2.5, 0}}, f, ObjectiveLines → {c},
    ObjectiveLineStyle → RGBColor[1, 0, 0],
    ShowTable → False,
    TextStyle -> {FontFamily → "Times",
    FontSize → 12}], {c, 3, 5, .1}];
```

Activity 2 – Use the *Mathematica* tools described above to plot the feasible region of the problem below, to find the corner points and their objective function values, and to plot c-level sets for values 4, 6, 8, and 10.

maximize: $f(x, y) = 4x - 2y$
subject to: $x + y \le 3$
$2x - y \le 4$
$x, y \ge 0$

Suppose the objective function had been $f(x_1, x_2) = x_1 + x_2$. Then the c-level sets are lines parallel to one of the constraint boundaries, namely $x_1 + x_2 = 3$. The largest c for which the c-level set intersects the feasible region is $c = 3$, and the intersection is the line segment connecting $(1, 2)$ and $(2, 1)$, as shown in Figure 2.3. The optimal value of this new objective is 3, taken on at all points on the segment between the corner points (1, 2) and (2, 1).

```
g[x_, y_] := x + y;
PlotFeasibleRegion[
    {x₂ == 2, x₁ + x₂ == 3, 2 x₁ + x₂ == 5},
    {x₁, 0, 3}, {x₂, 0, 3},
    {{0, 0}, {0, 2}, {1, 2}, {2, 1}, {2.5, 0}}, g,
    ObjectiveLines → {2, 3, 4}, ObjectiveLineStyle →
    RGBColor[1, 0, 0], TextStyle ->
    {FontFamily → "Times", FontSize → 8}];
```

x	y	objective
0	0	0
0	2	2
1	2	3
2	1	3
2.5	0	2.5

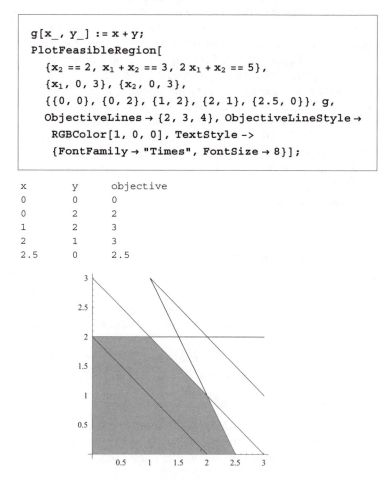

Figure 2.3 – Level sets parallel to a constraint boundary

In both of the examples above, the feasible region was a bounded polygon. Care must be exercised when the constraints yield an unbounded region. Examine the constraints

$$2 x_1 + x_2 \geq 3$$
$$x_1 + x_2 \geq 2$$
$$x_1, x_2 \geq 0$$

which generate the unbounded feasible region in Figure 2.4. The feasible corner points are at (0, 3), (1, 1), and (2, 0). (Check these.) Consider three problems relative to this feasible region:

1. minimize: $f_1(x_1, x_2) = 3\,x_1 + 2\,x_2$
2. minimize: $f_2(x_1, x_2) = x_1 - x_2$
3. maximize: $f_3(x_1, x_2) = x_1 + 3\,x_2$

For objective f_1, we search for the smallest value of c such that the c-level set intersects the feasible region. By graphing level sets as above, it is easy to see that the smallest such c is 5, taken on at the corner point (1, 1), since for all smaller c, the c-level set does not intersect the feasible region. (Notice in the command that we have had to pretend that the points (0, 5), (5, 5), and (5, 0) were corners in order to get *Mathematica* to gray out the whole polygon.)

```
f1[x_, y_] := 3 x + 2 y;
PlotFeasibleRegion[
  {2 x₁ + x₂ == 3, x₁ + x₂ == 2}, {x₁, 0, 5}, {x₂, 0, 5},
  {{1, 1}, {0, 3}, {0, 5}, {5, 5}, {5, 0}, {2, 0}},
  f1, ObjectiveLines → {5, 6, 7},
  ObjectiveLineStyle → RGBColor[1, 0, 0],
  ShowTable → False, AspectRatio → 1, TextStyle ->
  {FontFamily → "Times", FontSize → 8}];
```

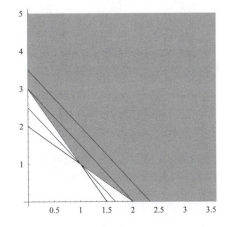

Figure 2.4 – An unbounded feasible region, with a minimum value

We conclude that there may be a solution to an LP problem even if the feasible region is unbounded. However, consider objective function f_2. The

c-level sets shown in Figure 2.5(a) correspond to smaller c values as they move to the northwest. If we hold x_1 at 0 and send x_2 to infinity, we observe that we can achieve arbitrarily small values of the objective within the feasible region. This problem has no optimal solution. Similarly, in Figure 2.5(b) for the maximum problem, the c-level sets correspond to larger c values as they move to the northeast, and so the objective function f_3 has no maximum value. The latter two problems are called *unbounded problems*. Note that it is not the feasible region alone that makes a problem unbounded. The key feature is whether or not the *objective function* is unbounded on the feasible region.

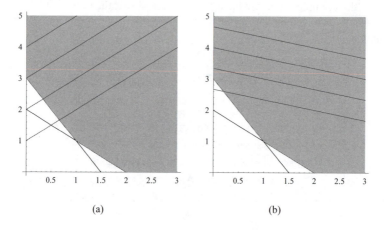

(a)   (b)

Figure 2.5 – (a) An unbounded minimum problem; (b) an unbounded maximum problem

Another difficulty that can arise involves constraints that are inconsistent with one another. For instance, if the constraints are

$$x_1 + x_2 \le 1$$
$$4x_1 + 2x_2 \ge 8$$
$$x_1, x_2 \ge 0$$

then the solution sets of the individual inequalities do not overlap. The feasible region is empty, so that no matter what the objective function is, the problem can have no solution. Such a problem is called *infeasible*.

Activity 3 – Sketch a graph to check the infeasibility of the problem above. Why, algebraically, are the first and second constraints together incompatible with the third constraint?

Our experience with the examples above leads to a methodology for approaching two-variable LP problems.

1. Sketch the feasible region. If it is empty, then the problem is infeasible and has no solution.

2. If the feasible region is unbounded, look carefully at the objective function to see if it too is unbounded in the feasible region. If the objective is unbounded below for a minimization problem, or unbounded above for a maximization problem, then there is no solution.

3. Otherwise, there is at least one optimal solution, taken on at some corner point of the feasible region. One may inspect the *c*-level sets, or simply compute the coordinates and functional values of all corner points, and pick out the largest or smallest.

4. If two corner points both achieve the optimal value of the objective, then so do all points on the line segment connecting them.

In order for the reader to gain some intuition into the general theory of solutions to be presented in the next section, we would like to make a few more geometric and algebraic observations. These will be the basis for an algorithm that is successful in solving higher-dimensional problems.

5. The feasible region of a two–variable LP problem is an intersection of half-planes, and is therefore either empty or *convex*. By the latter, we mean that given any two points in the region, the line segment connecting the points lies entirely in the feasible region.

6. The corner points do not lie on any line segment connecting two other feasible points. Corners are called *extreme points*.

7. Any feasible point may be expressed as a convex combination of extreme points. A vector x is a *convex combination* of vectors $x_1, x_2, \dots, x_n$ if there exist coefficients $t_1, t_2, \dots, t_n$ in $[0, 1]$ such that

$$\Sigma_{i=1}^{n} t_i = 1 \quad \text{and} \quad x = \Sigma_{i=1}^{n} t_i x_i$$

This property of feasible points is more subtle and difficult to see, so let us illustrate how to find such coefficients for the point $(1, 1)$ in the feasible region of Figure 2.2. The extreme points are $(0, 2)$, $(1, 2)$, $(2, 1)$, $(5/2, 0)$, and $(0, 0)$. We look for numbers $t_1, t_2, t_3, t_4,$ and t_5 in $[0, 1]$ whose sum is 1, such that:

$$\begin{pmatrix} 1 \\ 1 \end{pmatrix} = t_1 \begin{pmatrix} 0 \\ 2 \end{pmatrix} + t_2 \begin{pmatrix} 1 \\ 2 \end{pmatrix} + t_3 \begin{pmatrix} 2 \\ 1 \end{pmatrix} + t_4 \begin{pmatrix} 5/2 \\ 0 \end{pmatrix} + t_5 \begin{pmatrix} 0 \\ 0 \end{pmatrix}$$

Thus, we must solve the system:

$$1 = t_2 + 2t_3 + \frac{5}{2}t_4$$
$$1 = 2t_1 + 2t_2 + t_3$$
$$1 = t_1 + t_2 + t_3 + t_4 + t_5$$

There is a *Mathematica* command in the KnoxOR`LinearProgramming` package called Dictionary, whose syntax is below.

```
(***  Dictionary[constraints,
       basiclist,nonbasiclist]  ***)
```

Dictionary takes as its first argument the list of constraint equations, as its second argument a list of variables to be solved for simultaneously in those equations, and as its third argument the list of remaining variables. Dictionary returns an aligned display in which the designated variables are solved for in terms of the others. Returning to our system, there are five unknowns and only three equations, so we look for an equivalent system in which three variables, say the first three, are represented in terms of the other two. The general solution, found by *Mathematica*, is

```
Dictionary[{1 == t₂ + 2 t₃ + (5 / 2) t₄,
            1 == 2 t₁ + 2 t₂ + t₃,
            1 == t₁ + t₂ + t₃ + t₄ + t₅},
  {t₁, t₂, t₃}, {t₄, t₅}]
```

$$t_1 = 1 - \frac{1}{2}t_4 - 3t_5$$
$$t_2 = -1 + \frac{3}{2}t_4 + 4t_5$$
$$t_3 = 1 - 2t_4 - 2t_5$$

We see that there are infinitely many choices of coefficients using which $(1, 1)$ can be expressed as a convex combination, subject to the restriction that all t_i are between 0 and 1. Choosing $t_4 = 1/4$ and $t_5 = 1/4$, for instance, gives values $t_1 = 1/8$, $t_2 = 3/8$, $t_3 = 0$ for the remaining coefficients.

8. Referring to constraints (2), there exist variables $s_1, s_2, s_3 \geq 0$ such that

$$x_2 + s_1 \qquad\qquad = 2$$
$$x_1 + x_2 \qquad + s_2 \qquad = 3$$
$$2x_1 + x_2 \qquad\qquad + s_3 = 5$$

These new variables are called *slack variables* because they "take up the slack" in the original constraint inequalities. Had there been an inequality of the form $d x_1 + e x_2 \geq b$, there would exist a *surplus variable $s \geq 0$* such that $d x_1 + e x_2 - s = b$. The introduction of slack (or surplus) variables to transform the constraints to equality form gives this problem five variables. After stating a more precise definition of convexity in the next section, we will see that the convexity of the feasible region is preserved under this enlargement of dimension.

9. At each extreme point (at least) two of the variables x_1, x_2, s_1, s_2, s_3 are 0 and (at most) three are non-zero. Note that two is the number of original variables in the problem, and three is the number of constraints. For instance, for the constraints above, when slack variables s_1 and s_2 are zero, we obtain the values 1, 2, and 1, respectively, for the other variables x_1, x_2, and s_3, as we see by forcing s_1 and s_2 to equal zero in the following output of the Dictionary command:

```
Dictionary[
    {x2 + s1 == 2, x1 + x2 + s2 == 3, 2 x1 + x2 + s3 == 5,
       s1 == 0, s2 == 0}, {x1, x2, s3}, {s1, s2}]
```

```
x1 = 1 +   0 s1 +   0 s2
x2 = 2 +   0 s1 +   0 s2
s3 = 1 +   0 s1 +   0 s2
```

We see that the designation of zero values for s_1 and s_2 puts us at the extreme point $(x_1, x_2) = (1, 2)$. Similarly, we can generate the following complete table of extreme points and variable values for the constraints in Remark 8:

Extreme point	x_1	x_2	s_1	s_2	s_3
(0, 0)	0	0	2	3	5
(0, 2)	0	2	0	1	3
(1, 2)	1	2	0	0	1
(2, 1)	2	1	1	0	0
$(\frac{5}{2}, 0)$	$\frac{5}{2}$	0	2	$\frac{1}{2}$	0

More than two variables can be zero; e.g., if all three constraint lines had intersected at the same point, then the slack variables for each constraint would have been zero. (Try making up a problem that exhibits this behavior.)

It is the task of the next section to prove these observations for LP problems of arbitrary dimension.

Exercises 2.1

1. A town post office has $800,000 available for the purchase of delivery vehicles. There are two models, a Jeep style and a van style, under consideration. Each Jeep costs $8000 and each van costs $10,000. Estimated annual maintenance costs per vehicle are $800 and $600, respectively, for jeeps and vans. The town will allocate $80,000 annually for maintenance. If the jeep achieves 25 miles per gallon of gasoline and the van achieves 20 miles per gallon, how many of each type of vehicle should the town buy in order to maximize the total among all vehicles of miles per gallon of gasoline? Can the problem of maximizing average gas mileage per vehicle be treated by the methods of this chapter?

2. Solve the LP problem:

$$\text{minimize: } g = x_1 + x_2$$

$$\text{subject to: } \begin{array}{rrrrl} 2x_1 & + & 3x_2 & \geq & 6 \\ 4x_1 & + & 3x_2 & \geq & 12 \\ 6x_1 & + & x_2 & \geq & 6 \\ x_1, & & x_2 & \geq & 0 \end{array}$$

3. Find the complete solution set of the problem:

$$\text{maximize: } f = 4000\,x_1 + 4000\,x_2$$

$$\text{subject to: } \begin{array}{rrrrl} x_1 & + & 2x_2 & \leq & 5 \\ x_1 & + & x_2 & \leq & 3 \\ 2x_1 & + & x_2 & \leq & 5 \\ x_1, & & x_2 & \geq & 0 \end{array}$$

4. In a psychology experiment on conditioning, an experimenter places mice and rats into two types of conditioning boxes, I and II. Each mouse spends 20 minutes per day and each rat spends 40 minutes per day in box I. Each mouse spends 40 minutes per day and each rat spends 20 minutes per day in box II. Suppose box I is available to the experimenter for 640 minutes per day and box II is available for 800 minutes per day. How many rats and

mice should be used to maximize the total number of animals in the experiment in a day?

5. Find the solution set of the problem:

minimize: $2x_1 + x_2$

$$x_2 \leq -\frac{1}{2}x_1 + 1$$

subject to:
$$
\begin{aligned}
x_1 + 2x_2 &\geq 4 \\
2x_1 + x_2 &\geq 6 \\
x_1, \quad x_2 &\geq 0
\end{aligned}
$$

6. A hospital patient is required to have at least 90 units of drug I and 120 units of drug II. The drugs are both contained in two substances S_1 and S_2. Suppose that a gram of S_1 contains 6 units of drug I and 4 units of drug II, and a gram of S_2 contains 3 units of drug I and 3 units of drug II. But in addition, each gram of S_1 contains 2 units of a mildly toxic drug and each gram of S_2 contains 1 unit of this other undesirable drug. How much of each substance should be given to the patient to achieve the medication requirements with minimal dosage of the toxin? How much of the toxin does the patient receive with this optimal mixture?

7. Find the optimal solution, if it exists, of the problem:

maximize: $f = x_1 - x_2$

subject to:
$$
\begin{aligned}
-x_1 + 2x_2 &\leq 3 \\
x_1 + x_2 &\geq 1 \\
x_2 - x_1 &\leq 1 \\
x_1 - 3x_2 &\leq 1 \\
x_1, \quad x_2 &\geq 0
\end{aligned}
$$

8. (a) Given the feasible region below, find the associated set of constraints.
(b) For what set of non-negative coefficients c_1 and c_2 will (1, 2) be the maximum point of the objective function $f = c_1 x_1 + c_2 x_2$?
(c) For what set of non-negative coefficients will the points (1, 2) and (2, 0) both be maximum points?

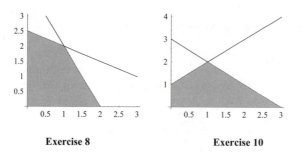

Exercise 8   **Exercise 10**

9. For the feasible region of Exercise 8, express (a) the point $(3/2, 1)$ and (b) the point $(1, 1)$ as a convex combination of extreme points.

10. Repeat Exercise 8 (a) and (b) for the feasible region sketched above.

11. Repeat Exercise 9 for the feasible region of Exercise 10.

12. Introduce slack variables into the following constraints, and give the values of all variables at each vertex of the feasible region.

$$
\begin{aligned}
x_1 &- x_2 \le 2 \\
2x_1 &+ x_2 \le 10 \\
x_1 &+ x_2 \le 8 \\
x_1, \ x_2 &\ge 0
\end{aligned}
$$

13. Suppose that the objective function to be maximized is the piecewise linear function:

$$
f = \begin{cases} 2x_1 + x_2 & \text{if } x_1 \le 1/2 \\ x_1 + x_2 + 1/2 & \text{otherwise} \end{cases}
$$

Sketch the c-level sets carefully in order to maximize f over the feasible region of Exercise 8.

14. Consider the feasible region of Exercise 8 and the objective function $f = 3x_1 + 2x_2$. Beginning at a point (x_1, x_2) in the feasible region, in what direction does f increase most rapidly? If we begin at $(0, 0)$ and move through the interior of the feasible region in the direction of most rapid increase of f, at what point on the boundary do we land?

2.2 Geometry of Linear Programming

We now go about the task of deriving the key theoretical results about linear programming problems, which will lead in the next section to a solution algorithm.

Since slack variables may be appended to the objective function as long as they are given zero coefficients, it is clear that the standard maximum linear programming problem may be written with its constraints in equality form as follows:

$$\text{maximize:} \quad f = \mathbf{c}' \cdot \mathbf{x}$$

$$\text{subject to:} \quad A\mathbf{x} = \mathbf{b}, \ \mathbf{x} \geq \mathbf{0}$$

Here $\mathbf{c} = (c_j)$ is an $n \times 1$ column vector of objective function coefficients, $\mathbf{x} = (x_j)$ is an $n \times 1$ column vector of variables, $\mathbf{A} = (a_{ij})$ is an $m \times n$ matrix of constraint coefficients, and $\mathbf{b} = (b_j)$ is an $m \times 1$ column vector of constants. We use the notation M' for the transpose of a matrix M. The problem listed at the beginning of Section 2.1, for instance, may be written in equality form as:

$$\text{maximize:} \quad f = (1 \ \ 2 \ \ 0 \ \ 0 \ \ 0) \begin{pmatrix} x_1 \\ x_2 \\ x_3 \\ x_4 \\ x_5 \end{pmatrix}$$

$$\text{subject to:} \quad \begin{pmatrix} 0 & 1 & 1 & 0 & 0 \\ 1 & 1 & 0 & 1 & 0 \\ 2 & 1 & 0 & 0 & 1 \end{pmatrix} \begin{pmatrix} x_1 \\ x_2 \\ x_3 \\ x_4 \\ x_5 \end{pmatrix} = \begin{pmatrix} 2 \\ 3 \\ 5 \end{pmatrix}, \quad \mathbf{x} \geq \mathbf{0}$$

For the sake of brevity, we prefer the matrix form of (1) and (2) as the description of the maximum problem, but for the reader's reference, the long-hand version is below.

$$\text{maximize:} \quad f = c_1 x_1 + c_2 x_2 + \cdots + c_n x_n$$

$$\text{subject to:} \quad \begin{aligned} a_{11} x_1 + a_{12} x_2 + \cdots + a_{1n} x_n &= b_1 \\ a_{21} x_1 + a_{22} x_2 + \cdots + a_{2n} x_n &= b_2 \end{aligned}$$

$$\vdots$$

$$a_{m1} x_1 + a_{m2} x_2 + \cdots + a_{mn} x_n = b_m$$
$$x_i \geq 0 \quad \text{for all } i = 1, \ldots, n$$

In this section, we will refer only to the maximum problem. We will study the minimum problem later.

DEFINITION 1. (a)   A *hyperplane* in $\mathbf{R}^n$ is the set of points $\mathbf{x} = (x_1, \ x_2, \ ..., \ x_n)$ satisfying a linear equation of the form:

$$a_1 x_1 + a_2 x_2 + \cdots + a_n x_n = b$$

(b) The *line segment* connecting points $\mathbf{x}_1$ and $\mathbf{x}_2$ in $\mathbf{R}^n$ is the set of all points $\mathbf{x}$ of the form:

$$\mathbf{x} = \lambda\, \mathbf{x}_1 + (1 - \lambda)\, \mathbf{x}_2 \,, \ \lambda \in [0, \ 1]\,.$$

(c) A subset S of $\mathbf{R}^n$ is *convex* if, given any two points $\mathbf{x}_1$ and $\mathbf{x}_2$ in S, the line segment connecting $\mathbf{x}_1$ and $\mathbf{x}_2$ is contained in S.

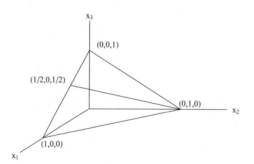

Figure 2.6 – The plane $x_1 + x_2 + x_3 = 1$

EXAMPLE 1. The reader is probably familiar with hyperplanes (or simply, planes) in $\mathbf{R}^3$. The plane associated with the equation $x_1 + x_2 + x_3 = 1$ forms a triangular region when restricted to the first octant, as shown in Figure 2.6. It is clear from the picture that the line segment connecting points $(0, 1, 0)$ and $(1/2, 0, 1/2)$ is entirely in the plane, which leads us to a guess that planes are convex. This guess is substantiated in Theorem 1 below. The points on this segment are of the form:

$$x = \lambda \cdot (0, 1, 0) + (1 - \lambda) \cdot (1/2, \ 0, \ 1/2)\,, \ \lambda \in [0, 1]$$

When $\lambda = 0$, we have $\mathbf{x} = (1/2, 0, 1/2)$; and when $\lambda = 1$, $\mathbf{x} = (0, 1, 0)$. As λ grows from 0 to 1, we may think of a path being traced out beginning at

$(1/2, 0, 1/2)$ and ending at $(0, 1, 0)$. When $\lambda = 1/4$, for example, the coordinates of the point on the segment are:

$$(1/4)\cdot(0, 1, 0) + (3/4)\cdot(1/2, 0, 1/2) = (3/8, 1/4, 3/8)$$

Conversely, given a point on the segment it is easy to solve algebraically for the corresponding λ by equating components; for the point $(1/4, 1/2, 1/4)$, we have

$$
\begin{aligned}
1/4 &= \lambda\cdot 0 + (1-\lambda)\cdot(1/2) \\
1/2 &= \lambda\cdot 1 + (1-\lambda)\cdot 0 \\
1/4 &= \lambda\cdot 0 + (1-\lambda)\cdot(1/2)
\end{aligned}
$$

and consequently $\lambda = 1/2$. If a given point is not on the segment, then it will not be possible to solve for λ such that all corresponding components are equal. See Activity 1 below.

Note finally that this restricted hyperplane is exactly the set of points in $\mathbb{R}^3$ obtained by insertion of a non-negative slack variable x_3 into an inequality constraint $x_1 + x_2 \leq 1$ in which both x_1 and x_2 are non-negative. The introduction of a slack variable has imbedded a two-dimensional feasible region into three dimensions. ∎

Activity 1 – Check whether the point $(1/4, 1/4, 1/4)$ is on the segment connecting $(1/2, 0, 1/2)$ and $(0, 1, 0)$.

Next we state the result on convexity of the feasible region that was mentioned earlier.

THEOREM 1. If the feasible region of a standard LP problem written in equality form (2) is not empty, then it is convex.

Proof. We claim first that a hyperplane is convex. Let $\mathbf{y}$ and $\mathbf{z}$ be two points on a hyperplane with equation $a_1 x_1 + a_2 x_2 + \cdots + a_n x_n = b$, which, in vector notation, is $\mathbf{a}' \cdot \mathbf{x} = b$. Consider the point $\lambda \mathbf{y} + (1 - \lambda)\mathbf{z}$, $\lambda \in [0, 1]$ on the line segment connecting $\mathbf{y}$ and $\mathbf{z}$. By the linearity of dot product,

$$
\begin{aligned}
\mathbf{a}' \cdot (\lambda \mathbf{y} + (1-\lambda)\mathbf{z}) &= \lambda\, \mathbf{a}' \cdot \mathbf{y} + (1-\lambda)\,\mathbf{a}' \cdot \mathbf{z} \\
&= \lambda b + (1-\lambda) b = b
\end{aligned}
$$

Hence $\lambda \mathbf{y} + (1 - \lambda)\mathbf{z}$ is also on the hyperplane, which proves the claim.

Let $S_1, S_2, \ldots, S_k$ be convex sets whose intersection S is non-empty. We ask the reader to show that S is convex in Exercise 3. Also, the feasible region is the intersection of m hyperplanes of the form:

$$a_{i1} x_1 + a_{i2} x_2 + \cdots + a_{in} x_n = b_i \qquad i = 1, \ldots, m$$

with the set $\{\mathbf{x} \in \mathbb{R}^n : x_i \geq 0 \text{ for } i = 1, \ldots, n\}$, which is clearly convex. By the exercise cited above, the feasible region is convex (if non-empty). ■

DEFINITION 2. A vector $\mathbf{x}$ in a convex set S is called an *extreme point* of S if it cannot be expressed as a convex combination $t\mathbf{y} + (1 - t)\mathbf{z}, t \in (0, 1)$ of any other pair of vectors $\mathbf{y}, \mathbf{z} \in S$.

Following are two important facts about extreme points.

THEOREM 2. (a) Suppose that S is a closed, bounded convex set in $\mathbb{R}^n$ with a finite number of extreme points $\mathbf{x}_1, \mathbf{x}_2, \ldots, \mathbf{x}_k$. Then any $\mathbf{x} \in S$ can be written as a convex combination of extreme points:

$$\mathbf{x} = \sum_{i=1}^{k} \lambda_i \mathbf{x}_i, \qquad \sum_{i=1}^{k} \lambda_i = 1 \qquad (5)$$

(b) The feasible region of an LP problem written in standard equality form, if non-empty, is a convex set with a finite number of extreme points. ■

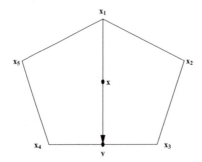

Figure 2.7 – Writing a point as the convex combination of extreme points

The full proof of Theorem 2 would take us somewhat farther into the subject of convex analysis than we wish to go, but the following discussion forms the intuitive basis for a proof. Regarding (a), consider a convex set in $\mathbb{R}^2$ with five extreme points $\mathbf{x}_1, \mathbf{x}_2, \mathbf{x}_3, \mathbf{x}_4,$ and $\mathbf{x}_5$ as shown in Figure 2.7. Let $\mathbf{x}$ be in the set, not itself an extreme point. Because the set is closed, bounded, and convex, the ray connecting $\mathbf{x}_1$ to $\mathbf{x}$ will intersect a side of the set at some point, say $\mathbf{y}$, between $\mathbf{x}_3$ and $\mathbf{x}_4$. Then we have, for some $t, s \in [0, 1]$,

$$\mathbf{x} = t \cdot \mathbf{x}_1 + (1 - t) \cdot \mathbf{y}$$
$$\mathbf{y} = s \cdot \mathbf{x}_3 + (1 - s) \cdot \mathbf{x}_4$$
$$\Longrightarrow \mathbf{x} = t \cdot \mathbf{x}_1 + (1 - t) \cdot s \cdot \mathbf{x}_3 + (1 - t) \cdot (1 - s) \cdot \mathbf{x}_4$$

It is easy to check that the sum of the coefficents of $\mathbf{x}_1$, $\mathbf{x}_3$, and $\mathbf{x}_4$ above is one, thus we have expressed $\mathbf{x}$ as a convex combination of these extreme points. In the case where the dimension is higher than two, the idea is roughly the same. We successively project $\mathbf{x}$ down onto lower dimensional boundaries of the feasible region, until eventually we find a point lying on a segment between two extreme points. The reader may consult Rockafellar [50] for details. Referring to part (b), the convexity of the feasible region has already been noted. The finite cardinality of the set of extreme points can be proved by identifying extreme points with particular solutions $\mathbf{x}$ of the system $A\mathbf{x} = b$ in which $n - m$ components of $\mathbf{x}$ are set to 0 (see Theorem 3 below). Since the system has m equations in n unknowns, such a choice of $n - m$ zero components determines the remaining components of $\mathbf{x}$ uniquely. There are only a finite number of ways of choosing $n - m$ components from among n components, hence there can only be a finite number of extreme points.

Activity 2 – Write the point $(1/2, 1/4)$, which is in the interior of the square whose corners are $(0, 0)$, $(0, 1)$, $(1, 1)$, and $(1, 0)$, as a convex combination of the corners. Draw a picture similar to Figure 2.7, showing the projection idea.

The discussion in the last paragraph leads to the following definition. First, recall from linear algebra that vectors $\mathbf{z}_1, \mathbf{z}_2, \ldots, \mathbf{z}_k$ are called *linearly independent* if it cannot be the case that

$$\sum_{i=1}^{k} t_i \mathbf{z}_i = 0$$

$(3)-(4)$

unless all coefficients t_i are zero.

DEFINITION 3. A *basic feasible solution* of the LP problem (1)–(2) is a vector $\mathbf{y}$, satisfying the constraints, such that (at least) $n - m$ of its components are zero, and in addition the system $A\mathbf{x} = b$ can be rewritten equivalently so that the columns of A corresponding to the remaining non-zero components of $\mathbf{y}$ are linearly independent. A basic feasible solution $\mathbf{y}$ is called *non-degenerate* if *exactly* $n - m$ of its components are zero, otherwise it is *degenerate*.

EXAMPLE 2. To illustrate the definition, consider again the problem of the beginning of the section, written in equality form:

$$\text{maximize: } f = (1 \quad 2 \quad 0 \quad 0 \quad 0) \begin{pmatrix} x_1 \\ x_2 \\ x_3 \\ x_4 \\ x_5 \end{pmatrix}$$

$$\text{subject to: } \begin{pmatrix} 0 & 1 & 1 & 0 & 0 \\ 1 & 1 & 0 & 1 & 0 \\ 2 & 1 & 0 & 0 & 1 \end{pmatrix} \begin{pmatrix} x_1 \\ x_2 \\ x_3 \\ x_4 \\ x_5 \end{pmatrix} = \begin{pmatrix} 2 \\ 3 \\ 5 \end{pmatrix}, \quad \mathbf{x} \geq 0$$

The vector $\mathbf{y} = (0 \ 0 \ 2 \ 3 \ 5)$ is a basic feasible solution. To see this, note that $n = 5$ is the number of variables, $m = 3$ is the number of constraints, and thus $n - m = 2$. By setting two components, namely x_1 and x_2, equal to zero, the constraint equations easily yield $x_3 = 2$, $x_4 = 3$, and $x_5 = 5$, hence $\mathbf{y}$ is feasible. Also, the columns corresponding to the non-zero variables x_3, x_4, and x_5 are just unit coordinate vectors in $\mathbf{R}^3$, and are therefore linearly independent. Suppose that we now subtract row 1 from row 2, and subtract row 1 from row 3 in the constraint equations. The following equivalent system of constraints results:

$$\begin{pmatrix} 0 & 1 & 1 & 0 & 0 \\ 1 & 0 & -1 & 1 & 0 \\ 2 & 0 & -1 & 0 & 1 \end{pmatrix} \begin{pmatrix} x_1 \\ x_2 \\ x_3 \\ x_4 \\ x_5 \end{pmatrix} = \begin{pmatrix} 2 \\ 1 \\ 3 \end{pmatrix}$$

Set x_1 and x_3 equal to zero, and it is easy to see that the remaining components x_2, x_4, and x_5 are 2, 1, and 3, respectively. Moreover, the columns corresponding to these non-zero components are again unit coordinate vectors, so that $\mathbf{y} = (0 \ 2 \ 0 \ 1 \ 3)$ is a basic feasible solution of the problem. ∎

   The following theorem provides the crucial connection between the geometry of linear programming problems and the algebraic solution of systems of linear equations, which is the heart of the algorithm to solve the problem. The proof we give here follows along the lines of ([38], Theorem 2.8).

THEOREM 3. Assume that the feasible region S of the LP problem (1)–(2) is non-empty, and that there exist m columns of A that are linearly independent. Then every extreme point of S is a basic feasible solution. Conversely, every basic feasible solution of (1)–(2) is an extreme point of S.

Proof. First, let $\mathbf{x} = (x_1, x_2, \ldots, x_n)'$ be an extreme point of S. If all n components of $\mathbf{x}$ are zero, then $\mathbf{x}$ is a basic feasible solution by the assumption that A has at least m independent columns. Otherwise, we may relabel the coordinates so that $\mathbf{x} = (x_1, x_2, \ldots, x_r, 0, 0, \ldots, 0)'$, where $1 \leq r \leq n$ and $x_i > 0$ for $i = 1, \ldots, r$. Denote the columns of A under the new labeling system by $\mathbf{A}^j$, $j = 1, \ldots, n$. Because of the form of $\mathbf{x}$, one can see for example that

$$a_{11} x_1 + a_{12} x_2 + a_{13} x_3 + \cdots + a_{1r} x_r = b_1$$

and similarly for the other rows of $\mathbf{b}$. In vector form, the constraint equation (2) can therefore be written:

$$\sum_{j=1}^{r} x_j \cdot \mathbf{A}^j = \mathbf{b} \tag{6}$$

We would like to show that the m-component column vectors $\mathbf{A}^1, \ldots, \mathbf{A}^r$ are linearly independent. Suppose on the contrary that there exist constants $t_1, t_2, \ldots, t_r$ that are not all equal to zero such that

$$\sum_{j=1}^{r} t_j \cdot \mathbf{A}^j = \mathbf{0} \tag{7}$$

Since $x_1, \ldots, x_r$ are strictly positive, there exists a small positive number ϵ such that both $x_j + \epsilon t_j$ and $x_j - \epsilon t_j$ are positive for all $j = 1, \ldots, r$. Multiplying equation (7) by ϵ, then respectively adding and subtracting it from equation (6) yields the two equations:

$$\sum_{j=1}^{r} (x_j + \epsilon t_j) \mathbf{A}^j = \mathbf{b} \qquad \sum_{j=1}^{r} (x_j - \epsilon t_j) \mathbf{A}^j = \mathbf{b} \tag{8}$$

Thus, the following two vectors

$$\mathbf{y} = (x_1 + \epsilon t_1, \ldots, x_r + \epsilon t_r, 0, 0, \ldots, 0)'$$
$$\mathbf{z} = (x_1 - \epsilon t_1, \ldots, x_r - \epsilon t_r, 0, 0, \ldots, 0)'$$

are feasible, and $\mathbf{x} = (1/2)\mathbf{y} + (1/2)\mathbf{z}$. This is a contradiction of the assumption that $\mathbf{x}$ is an extreme point. Therefore, columns $\mathbf{A}^1, \ldots, \mathbf{A}^r$ are linearly independent vectors (of length m).

A standard theorem from linear algebra states that there can be no more than m independent vectors in $\mathbb{R}^m$. Therefore $r \leq m$, so that $\mathbf{x}$ has at least $n - m$ zero components, and the columns of A corresponding to the non-zero components of $\mathbf{x}$ are independent. This means that $\mathbf{x}$ is a basic feasible solution, and the first half of the theorem is established.

To show the second statement, let $\mathbf{x}$ be a basic feasible solution. If $\mathbf{x} = (0, 0, \dots, 0)$, then it is impossible for $\mathbf{x}$ to be written as $\mathbf{x} = t\mathbf{y} + (1 - t)\mathbf{z}$ for two other feasible vectors $\mathbf{y}, \mathbf{z} \geq \mathbf{0}$. That is, $\mathbf{0}$ is an extreme point. If $\mathbf{x}$ is not the zero vector, then $\mathbf{x}$ can be written:

$$\mathbf{x} = (x_1, \dots, x_r, 0, 0, \dots, 0) \tag{9}$$

with $x_i > 0$ for $i = 1, \dots, r$ by suitably relabeling the variables. We know that $1 \leq r \leq m$, and that the columns $\mathbf{A}_1, \dots, \mathbf{A}_r$ of A must be linearly independent, by the definition of basic feasible solution.

Assume that $\mathbf{x}$ is not an extreme point. Then there exists vectors $\mathbf{y}$ and $\mathbf{z}$ in S such that $\mathbf{y} \neq \mathbf{z}$ and

$$\mathbf{x} = t\mathbf{y} + (1 - t)\mathbf{z} \text{ for some } t \in (0, 1)$$

Since $\mathbf{y}, \mathbf{z} \geq \mathbf{0}$ and $\mathbf{x}$ has the form (9), $\mathbf{y}$ and $\mathbf{z}$ must also be zero after their r^{th} components. Because $\mathbf{y}$ and $\mathbf{z}$ are feasible,

$$\sum_{j=1}^{r} y_j \mathbf{A}^j = \mathbf{b} , \quad \sum_{j=1}^{r} z_j \mathbf{A}^j = \mathbf{b} \implies \sum_{j=1}^{r} (y_j - z_j) \mathbf{A}^j = \mathbf{0}$$

This contradicts the linear independence of the columns $\mathbf{A}^1, \dots, \mathbf{A}^r$, and completes the proof. ∎

The importance of extreme points, and therefore basic feasible solutions, is that the optimal value of the objective function of a linear program can be found at one of them, if there is a solution at all. The *simplex algorithm*, to be discussed in the next section, will step from one basic feasible solution (corner point) to another in such a way that the objective function is improved. Thanks to Theorem 3, the procedure is a routine algebraic matter of manipulating systems of linear equations; and because of the next theorem, we will eventually find the optimal value at one of the basic feasible solutions.

THEOREM 4. Assume that the feasible region S is non-empty and bounded. Then the maximum value of f is taken on at an extreme point. If the maximum value occurs at several points $\mathbf{y}_1, \mathbf{y}_2, \dots, \mathbf{y}_k$, then it also occurs at all convex combinations of these points.

Proof. The feasible region

$$S = \{\mathbf{x} \in \mathbf{R}^n : A\mathbf{x} = \mathbf{b}, \mathbf{x} \geq \mathbf{0}\}$$

is clearly closed, so that we may appeal to a well-known result of analysis, which says that the maximum of a continuous function on a closed and bounded non-empty set is attained at some point, say $\mathbf{x}_0 \in S$. If $\mathbf{x}_0$ is an extreme point, then the first assertion of the theorem is proved. If $\mathbf{x}_0$ is not an extreme point, then by Theorem 2 it can be written as a convex combination

$$\mathbf{x}_0 = \sum_{i=1}^{k} \lambda_i \, \mathbf{x}_i$$

where the $\mathbf{x}_i$ are the extreme points, necessarily finite in number, of the feasible region. Suppose that the maximal objective value among the extreme points is taken on at $\mathbf{x}^*$. Then, since $\mathbf{x}_0$ was the optimal point, $f(\mathbf{x}_0) \geq f(\mathbf{x}^*)$. But also,

$$
\begin{aligned}
f(\mathbf{x}_0) = \mathbf{c}' \cdot \mathbf{x}_0 &= \mathbf{c}' \cdot (\sum_{i=1}^{k} \lambda_i \, \mathbf{x}_i) \\
&= \sum_{i=1}^{k} \lambda_i \, \mathbf{c}' \cdot \mathbf{x}_i \\
&= \sum_{i=1}^{k} \lambda_i \, f(\mathbf{x}_i) \\
&\leq \sum_{i=1}^{k} \lambda_i \, f(\mathbf{x}^*) \\
&= (\sum_{i=1}^{k} \lambda_i) \, f(\mathbf{x}^*) = f(\mathbf{x}^*)
\end{aligned}
$$

Thus, $f(\mathbf{x}_0) = f(\mathbf{x}^*)$, and the extreme point $\mathbf{x}^*$ is also an optimal point.

Now suppose that $\mathbf{y}_1, \ldots, \mathbf{y}_k$ are points whose objective value is optimal. Consider the functional value of a convex combination of the $\mathbf{y}_i$. We have, since f is linear,

$$f(\sum_{i=1}^{k} t_i \, \mathbf{y}_i) = \sum_{i=1}^{k} t_i \, f(\mathbf{y}_i) = (\sum_{i=1}^{k} t_i) \, f^* = f^*$$

where f^* is the common functional value of the $\mathbf{y}_i$. This proves the second part of the theorem. ∎

We will not attempt to do a comprehensive theoretical study of the case where S is unbounded, but the algorithm itself will indicate when there is no optimal value. If the optimal value is attainable, the algorithm will yield an extreme point optimal solution. Except for the existence problem, the assumption of boundedness is not really needed in Theorem 4. Several results can be seen almost immediately in the unbounded, as well as the bounded case:

1.   The second statement of Theorem 4 remains true when S is unbounded.

2. If there is a unique optimal point, it must occur at an extreme point. For, let $\mathbf{x}$ be the point in question. If $\mathbf{x}$ is not an extreme point, then it can be written $\mathbf{x} = t\mathbf{y} + (1 - t)\mathbf{z}$ for some non-identical feasible points $\mathbf{y}$, $\mathbf{z}$ and a number $t \in (0, 1)$. By linearity of the objective function,

$$f(\mathbf{x}) = t\,f(\mathbf{y}) + (1 - t)\,f(\mathbf{z})$$

Either $f(\mathbf{x})$, $f(\mathbf{y})$, and $f(\mathbf{z})$ are all the same, which cannot be since $\mathbf{x}$ was the unique optimum point, or exactly one of $f(\mathbf{y})$ or $f(\mathbf{z})$ exceeds $f(\mathbf{x})$. In both cases, we have a contradiction; thus $\mathbf{x}$ is an extreme point.

3. The maximum of the objective function can never occur in the interior of the feasible region. As usual, let $f = \mathbf{c}' \cdot \mathbf{x}$, and suppose that $\mathbf{x}^*$ is in the interior of the feasible region. Then there is $\epsilon > 0$ small enough that the point $\mathbf{y}^* = \mathbf{x}^* + \epsilon\mathbf{c}$ is still in the feasible region. Then $\mathbf{y}^*$ has functional value

$$f(\mathbf{y}^*) = \mathbf{c}' \cdot \mathbf{y}^* = \mathbf{c}' \cdot \mathbf{x}^* + \epsilon\mathbf{c}'\mathbf{c} = f(\mathbf{x}^*) + \epsilon \parallel \mathbf{c} \parallel^2 \, > f(\mathbf{x}^*)$$

This says that we can increase the value of the objective by moving away from $\mathbf{x}^*$ in the direction $\mathbf{c}$; hence $\mathbf{x}^*$ cannot be optimal.

4. If $\mathbf{x}^*$ is a locally optimal solution, then it is also a globally optimal solution. This result will tell us when to terminate the simplex algorithm. You are asked for a proof of this fact in Exercise 9.

Activity 3 – Trace through the proof of Theorem 4, and check the veracity of claim 1 above.

Exercises 2.2

1. Consider a bounded standard maximum problem in two variables whose feasible region is of the form:

$$a_{11}\,x_1 \;+\; a_{12}\,x_2 \;\leq\; b_1$$
$$\vdots$$
$$a_{m1}\,x_1 \;+\; a_{m2}\,x_2 \;\leq\; b_m$$
$$x_1, \, x_2 \geq 0$$

where the constants b_i are non-negative. Give a geometric argument that a feasible point can be written as the convex combination of at most three corner points.

2. (a) In Example 1, show that the line segment connecting $(1/2, 0, 1/2)$ and $(1/2, 1/2, 0)$ is entirely in the triangular region.

  (b) Is the point $(1/2, 1/8, 3/8)$ on this segment?

  (c) Under what conditions on x_2 and x_3 is the point $(1/2, x_2, x_3)$ on this segment?

3. (a) By using the definition of convexity only, and not Theorem 1, show that the set of points (x_1, x_2, x_3) such that:

$$
\begin{aligned}
x_1 + x_2 + x_3 &\le 1 \\
2x_1 + x_2 + 2x_3 &\le 5 \\
x_1, x_2, x_3 &\ge 0
\end{aligned}
$$

is convex.

  (b) Show that the intersection of a finite number of convex sets, if non-empty, is convex.

4. If f is the objective of any feasible, bounded linear program with a given system of constraints, show that the optimal value of f is taken on at the same point as the optimal value of $c \cdot f$, where c is a positive constant.

5. A function $f : \mathbb{R}^n \to \mathbb{R}$ is called *convex* if for any $\mathbf{x}, \mathbf{y} \in \mathbb{R}^n$ and $t \in [0, 1]$,

$$
f(t \cdot \mathbf{x} + (1 - t) \cdot \mathbf{y}) \le t \cdot f(\mathbf{x}) + (1 - t) \cdot f(\mathbf{y})
$$

Show that a non-constant convex function defined on a bounded convex set cannot take on its maximum value in the interior of the convex set.

6. Construct a counterexample to Theorem 2 part (a) if the set S is not bounded.

7. Show that $(4/3, 10/3, 0, 0)$ is a basic feasible solution of the system:

$$
\begin{aligned}
2x_1 + x_2 + x_3 \quad\quad &= 6 \\
x_1 + 2x_2 \quad\quad + x_4 &= 8 \\
x_i \geq 0 \quad \text{for all } i &
\end{aligned}
$$

8. Express the constraints below in equality form by inserting slack variables, then find two basic feasible solutions and argue that they do satisfy the definition of basic feasible solution.

$$
\begin{aligned}
x_1 - x_2 + x_3 &\leq 2 \\
x_1 + x_2 \quad\quad &\leq 4 \\
x_1 \quad\quad + x_3 &\leq 6 \\
x_1, \ x_2, \ x_3 &\geq 0
\end{aligned}
$$

9. Prove that if a feasible solution to a maximum problem in equality form is locally optimal, then it is optimal. (A solution is *locally optimal* if its objective value exceeds those of all feasible points in some neighborhood of this solution.)

2.3 Simplex Algorithm for the Standard Maximum Problem

The Simplex Algorithm

In this section, we use the theoretical results of Section 2 to develop an algorithm to find the optimal solution to s standard maximum problem, if the solution exists. Recall that a linear programming problem is in *standard maximum* (inequality) *form* if it can be written:

$$\text{maximize:} \quad f = \mathbf{c}' \cdot \mathbf{x}$$

$$\text{subject to:} \quad A\mathbf{x} \le \mathbf{b}, \ \mathbf{x} \ge 0$$

where $\mathbf{b}$ is a column vector, all of whose entries are non-negative. If there are m constraints, then $\mathbf{b}$ has m entries, and A has m rows. We have already seen two examples of such problems, namely the winery problem of the introduction and Example 2 of Section 2.2.

Because $\mathbf{0}$ itself clearly satisfies the constraint system (2), the problem is feasible. Depending on the entries of A, the feasible region may be unbounded. You are asked to show in Exercise 6 that if at least one constraint, say the ith, is such that all of its coefficients $a_{i1}, a_{i2}, \ldots, a_{in}$ are strictly positive, then the region is bounded. The algorithm will have the ability to detect unboundedness of the objective function, and to find optimal solutions when they exist, even in problems with unbounded feasible regions. We illustrate later the difficulty of degenerate basic feasible solutions.

To illustrate the algorithm, consider again the winery problem:

$$\text{maximize:} \quad f = \tfrac{5}{4} x_1 + \tfrac{3}{2} x_2 + 2 x_3$$

$$
\begin{array}{rcrcrcrcl}
2 x_1 & + & & & x_3 & & & \le & 200 \\
& & 2 x_2 & + & x_3 & & & \le & 150 \\
\text{subject to:} \quad 2 x_1 & + & x_2 & + & \tfrac{3}{2} x_3 & & & \le & 90 \\
2 x_1 & + & x_2 & + & 2 x_3 & & & \le & 250 \\
& & x_1, & & x_2, & & x_3 & \ge & 0
\end{array}
$$

Slack variables x_4, x_5, x_6, and x_7 may be introduced into the constraints (4) to write the problem in equality form. Let us think of f as a variable as well, and write an equivalent version of the problem as a system of linear equations:

$$
\begin{array}{rcrcrcrcrcrcl}
2 x_1 & + & & & x_3 & + & x_4 & & & & & = & 200 \\
& & 2 x_2 & + & x_3 & & & + & x_5 & & & = & 150 \\
2 x_1 & + & x_2 & + & \tfrac{3}{2} x_3 & & & & & + & x_6 & = & 90 \\
2 x_1 & + & x_2 & + & 2 x_3 & & & & & & + x_7 & = & 250 \\
\tfrac{5}{4} x_1 & + & \tfrac{3}{2} x_2 & + & 2 x_3 & & & & & & & = & f
\end{array}
$$

Remember the *Mathematica* command Dictionary that was introduced in Section 2.1. We can use it to express the enlarged equality constraint system

(5) in a way that is convenient for the algorithm, first by solving for the slack variables and f in terms of the other variables.

```
Needs["KnoxOR`LinearProgramming`"];
```

```
constraints = {2 x₁ + x₃ + x₄ == 200,

    2 x₂ + x₃ + x₅ == 150, 2 x₁ + x₂ + 3/2 x₃ + x₆ == 90,

    2 x₁ + x₂ + 2 x₃ + x₇ == 250, 5/4 x₁ + 3/2 x₂ + 2 x₃ == f};

Dictionary[constraints, {x₄, x₅, x₆, x₇, f},
    {x₁, x₂, x₃}]
```

$$x_4 = 200 - 2 \; x_1 + 0 \; x_2 - 1 \; x_3$$
$$x_5 = 150 + 0 \; x_1 - 2 \; x_2 - 1 \; x_3$$
$$x_6 = 90 \; - 2 \; x_1 - 1 \; x_2 - \tfrac{3}{2} \, x_3$$
$$x_7 = 250 - 2 \; x_1 - 1 \; x_2 - 2 \; x_3$$
$$f \; = 0 \quad + \tfrac{5}{4} \, x_1 + \tfrac{3}{2} \, x_2 + 2 \; x_3$$

The first four equations express x_4, x_5, x_6, and x_7 as linear functions of x_1, x_2, and x_3. We say that the former are the *basic* variables for this system and the latter are the *non-basic* variables. At this point, the system still has infinitely many solutions. But suppose we set the three non-basic variables equal to zero. Then the collection of values

$$x_1 = 0, \; x_2 = 0, \, x_3 = 0, \; x_4 = 200, \, x_5 = 150, \, x_6 = 90, \, x_7 = 250 \qquad (6)$$

is a basic feasible solution, therefore an extreme point of the feasible region, by Theorem 3 of Section 2.2. To see this, notice that the constraint equations in system (5) can be written in matrix form as $A\,\mathbf{x} = \mathbf{b}$, where:

$$A = \begin{pmatrix} 2 & 0 & 1 & 1 & 0 & 0 & 0 \\ 0 & 2 & 1 & 0 & 1 & 0 & 0 \\ 2 & 1 & \frac{3}{2} & 0 & 0 & 1 & 0 \\ 2 & 1 & \frac{1}{2} & 0 & 0 & 0 & 1 \end{pmatrix}, \; \mathbf{x} = \begin{pmatrix} x_1 \\ x_2 \\ x_3 \\ x_4 \\ x_5 \\ x_6 \\ x_7 \end{pmatrix}, \; \mathbf{b} = \begin{pmatrix} 200 \\ 150 \\ 90 \\ 250 \end{pmatrix} \qquad (7)$$

In the notation of Section 2, $m = 4$ and $n = 7$. We have set $n - m = 3$ variables x_1, x_2, x_3 equal to zero, and the columns corresponding to the remaining variables x_4, x_5, x_6, and x_7 are unit coordinate vectors in $\mathbb{R}^4$, hence they are linearly independent. Thus, the list (6) represents a basic feasible solution, and a corner point of the feasible region.

So far, we have merely identified one extreme point. Looking back to (5), we see that the current value of f is zero, since x_1, x_2, and x_3 are equal to zero. Because each of these three variables has a positive coefficient, we can increase the value of f by making one of them positive. We would like to substitute a non-basic variable, say x_2, as a newly entering basic variable to replace one of the old basic variables in such a way that the resulting point is still feasible. To do this, we will pick one of the first four equations in (5), solve for x_2, then substitute for x_2 in all of the other equations. To decide which of the four to pick, we ask: by how much can x_2 be increased so that the entire solution is still feasible, i.e., all variables remain non-negative? In the first equation, x_2 does not appear. In the second, if $x_2 \leq 150/2 = 75$, then x_5 remains non-negative. Similarly, in the third equation, if $x_2 \leq 90$, then x_6 remains non-negative; and in the fourth equation, if $x_2 \leq 250$, then x_7 remains non-negative. The second equation is therefore the most restrictive, or *binding* equation. In it, we solve for x_2 using the Dictionary command, which amounts to replacing the *departing basic variable* x_5 by the new *entering basic variable* x_2, to get

```
Dictionary[constraints,
    {x₄, x₂, x₆, x₇, f}, {x₁, x₃, x₅}]
```

$$x_4 = 200 - 2\ x_1 - 1\ x_3 + 0\ x_5$$
$$x_2 = 75 + 0\ x_1 - \tfrac{1}{2}\ x_3 - \tfrac{1}{2}\ x_5$$
$$x_6 = 15 - 2\ x_1 - 1\ x_3 + \tfrac{1}{2}\ x_5$$
$$x_7 = 175 - 2\ x_1 - \tfrac{3}{2}\ x_3 + \tfrac{1}{2}\ x_5$$
$$f = \tfrac{225}{2} + \tfrac{5}{4}\ x_1 + \tfrac{5}{4}\ x_3 - \tfrac{3}{4}\ x_5$$

Notice that the way in which we managed this operation is by manipulating the order of appearance of x_2 in the basic list in the second argument, so that x_2 becomes the basic variable in the second equation. The order of appearance of variables in the non-basic list controls the columns that the non-basic variables appear in. Once again, setting the non-basic variables x_1, x_3, and x_5 to zero gives a basic feasible solution as listed below, and the value of f has increased to 225/2.

$$x_1 = 0,\ x_2 = 75,\ x_3 = 0,\ x_4 = 200,\ x_5 = 0,\ x_6 = 15,\ x_7 = 175 \qquad (8)$$

We ask the reader to show in Exercise 2 that (8) is indeed a basic feasible solution. Throughout the procedure, the solution $(x_1, x_2, \ldots, x_7)$ that we generate will continue to have the proper number of zeros, and the columns of the coefficient matrix corresponding to the basic variables will continue to be independent, hence the definition of basic feasible solution will be satisfied by the solution. Henceforth, we give this fact no special mention.

The value of f may still be increased by increasing either x_1 or x_3, but not x_5, since its coefficient in the bottom (i.e., the objective) row of the current system is negative. Introduce x_3 as the entering basic variable. In order to maintain non-negativity of the old basic variables, four conditions must be satisfied: $x_3 \leq 200$ from the first equation, $x_3 \leq 75/(1/2) = 150$ from the second, $x_3 \leq 15$ from the third, and $x_3 \leq 175/(3/2) = 350/3$ from the fourth equation. The third equation is the binding one, so that the departing basic variable is x_6. We solve for x_3 in the third equation to obtain the following new system:

```
Dictionary[constraints,
    {x₄, x₂, x₃, x₇, f}, {x₁, x₅, x₆}]
```

$$x_4 = 185 + 0 \ x_1 - \tfrac{1}{2} \ x_5 + 1 \ x_6$$

$$x_2 = \tfrac{135}{2} + 1 \ x_1 - \tfrac{3}{4} \ x_5 + \tfrac{1}{2} \ x_6$$

$$x_3 = 15 - 2 \ x_1 + \tfrac{1}{2} \ x_5 - 1 \ x_6$$

$$x_7 = \tfrac{305}{2} + 1 \ x_1 - \tfrac{1}{4} \ x_5 + \tfrac{3}{2} \ x_6$$

$$f = \tfrac{525}{4} - \tfrac{5}{4} \ x_1 - \tfrac{1}{8} \ x_5 - \tfrac{5}{4} \ x_6$$

$$x_1 = 0, \ x_2 = 135/2, \ x_3 = 15, \ x_4 = 185, \ x_5 = 0, \ x_6 = 0, \ x_7 = 305/2 \quad (9)$$

The objective function now has the value $f = 525/4$.

In the objective equation of the current system, the objective function f is expressed as a linear function of the non-basic variables, all of whose coefficients are negative. Therefore, f cannot be improved by increasing any of these variables, and the current solution (9) is locally optimal, hence it is also globally optimal. In terms of the original winery problem, we have shown that a maximum profit of $525/4 = 131.25$ dollars is obtained by making red, white, and rosé wine in amounts of 0, 135/2, and 15 gallons, respectively. The slack variable $x_4 = 185$ indicates the number of unused bushels of type I grapes, and the slack variable $x_7 = 305/2$ gives the number of unused labor hours. Since $x_5 = x_6 = 0$, all type II grapes and sugar are used.

Activity 1 – Use the Dictionary command to select a different sequence of entering basic variables, and see if you arrive at the same optimal solution.

To summarize, the algorithm to solve standard maximum LP problems with inequality constraints is as follows.

SIMPLEX ALGORITHM FOR STANDARD MAXIMUM PROBLEMS

1. a. {Set up initial system} By introducing slack variables, write the constraints in equality form.
   b. Solve for the slack variables, and adjoin the objective function.
2. While there are still variables with positive coefficients in the objective function row of the system of equations, do step 3.
3. a. Choose an entering basic variable with a positive coefficient on the objective function row.
   b. Choose a departing basic variable by finding the row, among those in which the entering basic variable selected in (a) has a negative coefficient, which has the smallest ratio of the constant term divided by the absolute value of that coefficient. If two rows share the smallest such ratio, pick either one; {The basic variable currently solved for in this row will depart}
   c. Solve for the new list of basic variables in the constraints and objective equation.
4. {The optimal solution has been reached.} Return values of 0 for the non-basic variables in the final system of equations, and return the appropriate constant terms in the system for the values of the basic variables, and of f.

The rule for selection of the departing basic variable in step 3b is new and needs some discussion. Suppose, without loss of generality, that at a certain step of the algorithm, we have basic variables $x_1,\ldots, x_m$ written as functions of non-basic variables $x_{m+1}, \ldots, x_n$ as below, and we wish to introduce x_{m+1} as a basic variable.

$$x_1 = c_1 + a_{1,m+1} x_{m+1} + a_{1,m+2} x_{m+2} + \cdots + a_{1,n} x_n$$
$$x_2 = c_2 + a_{2,m+1} x_{m+1} + a_{2,m+2} x_{m+2} + \cdots + a_{2,n} x_n$$

$$\vdots$$

$$x_m = c_m + a_{m,m+1} x_{m+1} + a_{m,m+2} x_{m+2} + \cdots + a_{m,n} x_n$$

We locate the binding constraint by solving the system of inequalities:

$$c_1 + a_{1,m+1} x_{m+1} \geq 0$$
$$c_2 + a_{2,m+1} x_{m+1} \geq 0$$
$$\cdot \qquad (10)$$
$$\cdot$$
$$\cdot$$
$$c_m + a_{m,m+1} x_{m+1} \geq 0$$

If $a_{i,m+1} \geq 0$, then there is no restriction on how large x_{m+1} can be, and the ith row can be passed by. Otherwise, the restriction imposed by the ith row is $x_{m+1} \leq c_i / |a_{i,m+1}|$, and the binding constraint is the row with the minimal such ratio.

Activity 2 – Is there any restriction on the choice of the entering basic variable? Can you conceive of a good strategy for choosing it?

In the exercises, you are led through a proof that, in the absence of degeneracies, the simplex algorithm returns either an optimal solution, or the information that the problem is unbounded (see Exercises 14–16).

Special Behavior

The simplex algorithm as stated above is not a finished product, because of the problems of unboundedness and degeneracy, which we examine in the examples below. We would also like to show how the existence of more than one optimal solution can be read from the final simplex system. For expositional ease, we use two-variable problems as examples, but the principles are the same for larger problems.

EXAMPLE 1. Consider the problem

$$\text{maximize}: \quad f = x_1 + x_2$$

$$\begin{array}{rcrcl} -x_1 & + & x_2 & \leq & 1 \\ & & x_2 & \leq & 2 \\ \end{array}$$
$$\text{subject to}: \qquad x_1, \ x_2 \ \geq \ 0$$

The sketch of the feasible region, which is unbounded, is shown in Figure 2.8.

```
constraints28 = {-x₁ + x₂ == 1, x₂ == 2};
f1[x_ , y_] := x + y;
PlotFeasibleRegion[
    constraints28, {x₁, 0, 3}, {x₂, 0, 3},
    {{0, 0}, {0, 1}, {1, 2}, {3, 2}, {3, 0}},
    f1, ShowTable → False, TextStyle →
    {FontFamily -> "Times", FontSize → 8}];
```

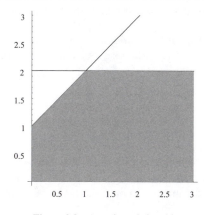

Figure 2.8 – An unbounded problem

Begin the execution of the algorithm by introducing slack variables x_3 and x_4 as basic variables:

```
equalityconstraints =
    {-x₁ + x₂ + x₃ == 1, x₂ + x₄ == 2, f == x₁ + x₂};
Dictionary[equalityconstraints,
    {x₃, x₄, f}, {x₁, x₂}]
```

$$x_3 = 1 + 1\ x_1 - 1\ x_2$$
$$x_4 = 2 + 0\ x_1 - 1\ x_2$$
$$f\ \ = 0 + 1\ x_1 + 1\ x_2$$

The current solution is $x_1 = 0$, $x_2 = 0$, $x_3 = 1$, $x_4 = 2$, $f = 0$. Choose x_1 as the entering basic variable. Since x_1 has a positive coefficient in the first constraint equation, and does not appear in the second equation, it may be increased without bound, while all problem variables remain non-negative and satisfy the other constraints. Since the objective function increases without bound as x_1 does, we see that the problem is unbounded. Physically, it is clear from Figure 2.8 that to increase x_1 while holding x_2 at zero means to move to the right along the x_1-axis, all the while staying in the feasible region. As you do so, the objective $f = x_1 + x_2$ becomes arbitrarily large. ∎

The implication of the last example is that *to check for unboundedness, the simplex algorithm should add a check at the beginning of step 2 for non-basic variable columns in which all constraint coefficients are non-negative and the objective function coefficient is positive. If such a column exists, the problem is unbounded.* (Exercise 14 asks for a general proof of this fact.) Another important observation is that the algorithm encounters no problem when the *feasible region is unbounded* but *the objective function is not.* (See the example in the discussion of the tableau method in the next subsection.)

EXAMPLE 2. *When multiple solutions are present, the final system of equations indicates this fact.* Suppose we wish to maximize $f = 4x_1 + 2x_2$ subject to the constraints

$$
\begin{aligned}
x_1 \ + \ x_2 &\leq 2 \\
2x_1 \ + \ x_2 &\leq 3 \\
x_1, \ x_2 &\geq 0
\end{aligned}
$$

The feasible region is sketched in Figure 2.9, and we note that the c-level sets of the objective funciton are parallel to the segment connecting $(1, 1)$ and $(3/2, 0)$. Thus, the maximum of f should be taken on at all points along this segment.

```
constraints29 = {x₁ + x₂ == 2, 2 x₁ + x₂ == 3};
f2[x_, y_] := 4 x + 2 y;
PlotFeasibleRegion[
  constraints29, {x₁, 0, 3}, {x₂, 0, 3},
  {{0, 0}, {0, 2}, {1, 1}, {3/2, 0}}, f2,
  ShowTable → False, ObjectiveLines → {3, 4, 5, 6},
  ObjectiveLineStyle → RGBColor[1, 0, 0],
  AspectRatio → .8, TextStyle →
    {FontFamily -> "Times", FontSize → 8}];
```

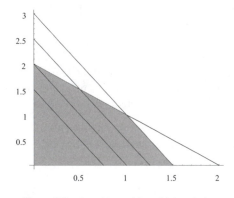

Figure 2.9 – A problem with multiple solutions

With slack variables x_3 and x_4 inserted, the initial simplex system is

```
equalityconstraints2 =
  {x₁ + x₂ + x₃ == 2, 2 x₁ + x₂ + x₄ == 3, f == 4 x₁ + 2 x₂};
Dictionary[equalityconstraints2,
  {x₃, x₄, f}, {x₁, x₂}]
```

$$x_3 = 2 - 1\,x_1 - 1\,x_2$$
$$x_4 = 3 - 2\,x_1 - 1\,x_2$$
$$f\ = 0 + 4\,x_1 + 2\,x_2$$

Hence, $x_1 = 0$, $x_2 = 0$, $x_3 = 2$, $x_4 = 3$, and $f = 0$. It is always useful to remember that a simplex system like the one above represents a basic feasible solution (when non-basic variables are set to zero), and in turn a basic feasible solution is an extreme point of the feasible region. The current system represents the origin, which is a corner of the feasible region in Figure 2.9. When x_1 is introduced to replace x_4 as a basic variable, we obtain

```
Dictionary[equalityconstraints2,
  {x₃, x₁, f}, {x₂, x₄}]
```

$$x_3 = \tfrac{1}{2} - \tfrac{1}{2}\,x_2 + \tfrac{1}{2}\,x_4$$
$$x_1 = \tfrac{3}{2} - \tfrac{1}{2}\,x_2 - \tfrac{1}{2}\,x_4$$
$$f\ = 6 + 0\,x_2 - 2\,x_4$$

For this new system, $x_1 = 3/2$, $x_2 = 0$, $x_3 = 1/2$, $x_4 = 0$, and $f = 6$. The extreme point represented by this system is the corner $(3/2, 0)$ of the feasible region. Since all coefficients on the bottom row are now non-positive, the solution $(3/2, 0)$ is optimal. But also, x_2 has coefficient 0 in the objective row. This means that x_2 can be increased without altering the value of f. A check of the ratios of constant terms to coefficients of x_2 in the constraint equations reveals that x_2 can be increased as far as 1 (with x_3 as departing basic variable) without violating feasibility. At this point, x_1 becomes 1, x_3 becomes 0, and x_4 remains 0; hence we obtain $(x_1, x_2) = (1, 1)$ as another optimal solution, as expected. ■

```
Dictionary[equalityconstraints2,
    {x₂, x₁, f}, {x₃, x₄}]
```

$$x_2 = 1 - 2\,x_3 + 1\,x_4$$
$$x_1 = 1 + 1\,x_3 - 1\,x_4$$
$$f\ = 6 + 0\,x_3 - 2\,x_4$$

The last example points out that *when a non-basic variable has a coefficient of 0 in the objective row of the final simplex system, that variable may be made basic, and another variable non-basic, thereby yielding an alternative optimal solution.*

EXAMPLE 3. The final special behavior to illustrate is degeneracy of basic feasible solutions. We will see that this occurs when more than two constraints intersect at a single extreme point. From a computational point of view, this is the stickiest sort of problem to handle because the number of steps required to reach the solution may increase greatly, and in the most extreme case, an infinite loop can be entered in the algorithm. Consider the standard LP problem

$$\text{maximize: } f = x_1 + 3\,x_2$$

$$\text{subject to: } \begin{array}{rcrcl} x_1 & + & 2\,x_2 & \le & 3 \\ 2\,x_1 & + & x_2 & \le & 3 \\ x_1 & + & x_2 & \le & 2 \\ x_1, & & x_2 & \ge & 0 \end{array}$$

whose feasible region is depicted in Figure 2.10. Here, all three constraint boundaries intersect at the point $(1, 1)$.

```
constraints210 =
   {x₁ + 2 x₂ == 3, 2 x₁ + x₂ == 3, x₁ + x₂ == 2};
f3[x_, y_] := x + 3 y;
PlotFeasibleRegion[constraints210, {x₁, 0, 2},
   {x₂, 0, 2}, {{0, 0}, {0, 3/2}, {1, 1}, {3/2, 0}},
   f3, ShowTable → False,
   ObjectiveLines → {3.5, 4, 4.5, 5},
   ObjectiveLineStyle → RGBColor[1, 0, 0],
   AspectRatio → .8, TextStyle →
      {FontFamily -> "Times", FontSize → 8}];
```

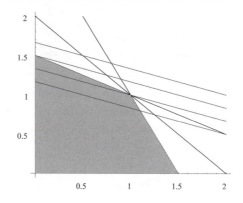

Figure 2.10 – A degenerate problem

```
equalityconstraints3 = {x₁ + 2 x₂ + x₃ == 3,
   2 x₁ + x₂ + x₄ == 3, x₁ + x₂ + x₅ == 2, f == x₁ + 3 x₂};
Dictionary[equalityconstraints3,
   {x₃, x₄, x₅, f}, {x₁, x₂}]
```

$$x_3 = 3 - 1\, x_1 - 2\, x_2$$
$$x_4 = 3 - 2\, x_1 - 1\, x_2$$
$$x_5 = 2 - 1\, x_1 - 1\, x_2$$
$$f\ = 0 + 1\, x_1 + 3\, x_2$$

Graphically, it is easy to see that we could reach the optimal point in one step by introducing x_2 into the basic variable list, but suppose we choose to let x_1 be the entering basic variable instead. The minimum ratio criterion indicates that x_4 should depart, and we obtain:

```
Dictionary[equalityconstraints3,
    {x₃, x₁, x₅, f}, {x₂, x₄}]
```

$$x_3 = \frac{3}{2} - \frac{3}{2} x_2 + \frac{1}{2} x_4$$

$$x_1 = \frac{3}{2} - \frac{1}{2} x_2 - \frac{1}{2} x_4$$

$$x_5 = \frac{1}{2} - \frac{1}{2} x_2 + \frac{1}{2} x_4$$

$$f = \frac{3}{2} + \frac{5}{2} x_2 - \frac{1}{2} x_4$$

This dictionary system represents the basic feasible solution $x_1 = 3/2$, $x_2 = 0$, $x_3 = 3/2$, $x_4 = 0$, $x_5 = 1/2$, and $f = 3/2$. Next, x_2 should become basic, but we have a tie between ratios in the first and third rows of the constraint system. Activity 3 below this example asks you to check that if x_3 is chosen as the departing basic variable in row 1, only one further step is necessary. Let us see what happens if we choose x_5 as the departing basic variable instead.

```
Dictionary[equalityconstraints3,
    {x₃, x₁, x₂, f}, {x₄, x₅}]
```

$$x_3 = 0 - 1 x_4 + 3 x_5$$

$$x_1 = 1 - 1 x_4 + 1 x_5$$

$$x_2 = 1 + 1 x_4 - 2 x_5$$

$$f = 4 + 2 x_4 - 5 x_5$$

We are now at the point of degeneracy where all the boundary lines intersect: $x_1 = 1$, $x_2 = 1$, $x_3 = 0$, $x_4 = 0$, $x_5 = 0$, $f = 4$. In this system, the slack variable x_3 is basic, yet it has the value 0 because the first constraint in the original problem is an equality at $(x_1, x_2) = (1, 1)$. This collection forms a degenerate solution, since more than $n - m = 5 - 3 = 2$ variables equal zero. In the next step, x_4 must enter, and x_3 departs the list of basic variables because 0/1 is the smallest coefficient ratio.

```
Dictionary[equalityconstraints3,
   {x₄, x₁, x₂, f}, {x₃, x₅}]
```

$$x_4 = 0 - 1\, x_3 + 3\, x_5$$
$$x_1 = 1 + 1\, x_3 - 2\, x_5$$
$$x_2 = 1 - 1\, x_3 + 1\, x_5$$
$$f\ = 4 - 2\, x_3 + 1\, x_5$$

The basic feasible solution is now $x_1 = 1$, $x_2 = 1$, $x_3 = 0$, $x_4 = 0$, $x_5 = 0$, and $f = 4$. There are several things worth noting about this system. The feasible solution is still degenerate, since the basic variable $x_4 = 0$. Geometrically, this system represents the same corner $(1, 1)$ of the feasible region, and the value of the objective has not increased. In theory (and in fact such examples have been constructed), it is conceivable that a sequence of entering and departing basic variables might be chosen so that the physical point remains the same indefinitely; only the roles of basic and non-basic variables are interchanged. Fortunately, this is rather rare, and in this problem x_5 now enters, x_1 departs, and the final system is given by

```
Dictionary[equalityconstraints3,
   {x₄, x₅, x₂, f}, {x₁, x₃}]
```

$$x_4 = \frac{3}{2} - \frac{3}{2}\, x_1 + \frac{1}{2}\, x_3$$
$$x_5 = \frac{1}{2} - \frac{1}{2}\, x_1 + \frac{1}{2}\, x_3$$
$$x_2 = \frac{3}{2} - \frac{1}{2}\, x_1 - \frac{1}{2}\, x_3$$
$$f\ = \frac{9}{2} - \frac{1}{2}\, x_1 - \frac{3}{2}\, x_3$$

The optimal solution is $x_1 = 0$, $x_2 = 3/2$, $x_3 = 0$, $x_4 = 3/2$, $x_5 = 1/2$, and the maximum value of the objective function is $f = 9/2$. ∎

Activity 3 – Use the Dictionary command to check that you avoid the degeneracy completely by letting x_2 be the first entering basic variable; and also to check that if you take the approach of the example up to the second step, but then let x_3 instead of x_5 be the departing basic variable when x_2 enters, you need only one more step.

It turns out that the difficulty illustrated in the last example can be avoided by a more judicious choice of entering and departing basic variables, though we would rather not discuss the problem in any greater detail in this introduction to the subject. For more information, the reader can study Papadimitriou and Steiglitz [47], or Gass [24].

Tableau Method

The method used in the examples above is called the *dictionary method*, because all variables and equations are explicitly listed at every step. There is another way of executing the simplex algorithm that is equivalent to the dictionary method. Only the coefficients of the equations are really necessary as long as we identify at each step which variables are basic. The alternative *tableau* method takes advantage of this observation by performing a procedure very similar to matrix Gaussian elimination. Gaussian elimination chooses a sequence of matrix elements about which to "pivot" (meaning to zero all entries beneath the pivot row in the pivot column by row operations), namely the diagonal elements, so as to transform the matrix to upper triangular form. The simplex algorithm does almost the same thing, except that it chooses its sequence of pivot elements according to the rules of step 3 of the simplex algorithm for selecting entering and departing basic variables.

To illustrate, consider a problem with the same constraints as the problem of Example 1, but with an objective function that turns out to be bounded on the feasible region of Figure 2.8. The problem statement and its equality form are shown below:

$$\text{maximize:} \qquad f = -2\,x_1 + x_2$$

$$\text{subject to :} \qquad \begin{aligned} -x_1 \;+\; x_2 &\le 1 \\ x_2 &\le 2 \\ x_1, \; x_2 &\ge 0 \end{aligned} \qquad\qquad (11)$$

$$\begin{aligned} -x_1 \;+\; x_2 \;+\; x_3 &= 1 \\ x_2 \;+\; x_4 &= 2 \\ -2\,x_1 \;+\; x_2 &= f \end{aligned}$$

(Note: Some authors, and some computer programs, display the objective function on the top row of the system.) Write the augmented matrix for this system, labeling each constraint row with the basic variable that it represents. This matrix is called the *initial tableau*, and is shown in Figure 2.11(a), to the right of the initial dictionary system to which it is equivalent.

$$\begin{array}{ll}
\begin{array}{rl}
x_3 = & 1 + x_1 - x_2 \\
x_4 = & 2 \qquad\;\; - x_2 \\
f = & \;\;\;\;\;\; -2x_1 + x_2
\end{array}
&
\begin{array}{c}
\begin{array}{cccc}
x_1 & x_2 & x_3 & x_4
\end{array} \\
\begin{array}{c}
x_3 \\ x_4
\end{array}
\left[\begin{array}{cccc|c}
-1 & 1 & 1 & 0 & 1 \\
0 & 1 & 0 & 1 & 2 \\
\hline
-2 & 1 & 0 & 0 & f-0
\end{array}\right]
\end{array}
\end{array}$$

Figure 2.11(a) – Initial dictionary system and simplex tableau

$$\begin{array}{ll}
\begin{array}{rl}
x_2 = & 1 + x_1 - x_3 \\
x_4 = & 1 - x_1 + x_3 \\
f = & 1 - x_1 - x_3
\end{array}
&
\begin{array}{c}
\begin{array}{cccc}
x_1 & x_2 & x_3 & x_4
\end{array} \\
\begin{array}{c}
x_2 \\ x_4
\end{array}
\left[\begin{array}{cccc|c}
-1 & 1 & 1 & 0 & 1 \\
1 & 0 & -1 & 1 & 1 \\
\hline
-1 & 0 & -1 & 0 & f-1
\end{array}\right]
\end{array}
\end{array}$$

Figure 2.11(b) – Second dictionary system and simplex tableau

Choose a new basic variable by choosing a column such that the coefficient in the objective row is positive. Thus, we must choose the x_2 column. Determine the departing basic variable by finding the minimal ratio of the constant right-hand side to the x_2 coefficient. In Figure 2.11(a), we see that the top row, i.e., the x_3 row, is the one to select. Designate the new basic variable for the first row to be x_2. Note that in the tableau we look for the row with a *positive* entry in the x_2 column such that the aforementioned ratio is minimal. Use the pivot element just selected, here it is in row 1 and column 2, to "pivot away" to zero all other entries in its column, i.e., to make them zero. The suitable row operations are:

$$\begin{array}{l}
\text{row } 2 := \text{row } 2 - \text{row } 1 \\
\text{row } 3 := \text{row } 3 - \text{row } 1
\end{array} \tag{12}$$

We obtain the new tableau in Figure 2.11(b), which is equivalent to the system of equations resulting from the dictionary method on the left of this tableau. In general, the tableau algorithm stops when there are no positive entries in the bottom row, which is the case in Figure 2.11(b). The maximum value of 1 is the number subtracted from f in the lower right-hand corner, the values of the non-basic variables are zero, and the values of the basic variables are the constants on the right-hand side of the tableau, as you can see from the final dictionary system. (Note that the latter is only true if the pivot elements are made equal to one by multiplication of the pivot row by a suitable constant.)

Activity 4 – Justify carefully why, in the tableau version of the simplex algorithm, you select as the entering basic variable a column that has a positive entry in the objective row; why the departing basic variable is found in the row, among those with a positive entry in the pivot column, such that the right side to entry ratio is smallest; and why the values of the basic variables are found in the constant column of the tableau, as long as the pivot entry is 1. Given a tableau, how can you tell what variable is considered basic in each row, and what variables are non-basic?

There is a *Mathematica* command in the KnoxOR`LinearProgramming` package that saves you the trouble of doing row operations yourself, but only requires you to select the entering and departing basic variables, based on the current tableau. It is called SimplexOneStep:

```
(** SimplexOneStep[tableau,varlist,
      entering,departing,basiclist]  **)
```

Its first argument is the current simplex tableau in the usual *Mathematica* form for a matrix, and its other arguments involve the variable names. "Varlist" is a list of all the variables in column order, "entering" is the name of the entering basic variable, "departing" is the name of the departing basic variable, and "basiclist" is the list of basic variable names in the row order they have in the tableau. SimplexOneStep pivots in the column of the entering basic variable and the row of the departing basic variable, shows the new tableau, then returns the pair {newtableau, newbasiclist} for use in the next step.

Here, for example, is how the command works on the initial tableau of Figure 2.11(a). We give the tableau and the variable list names and values first, then we ask SimplexOneStep to let x_2 replace x_3 in the basic list. It gives the printout of the tableau in Figure 2.11(b), with basic variables labeled on the left. Notice here that the bottom right corner will always be of the form $-c$, and you should interpret the entry as $f - c$, where c is the current value of the objective function.

```
tableau11 = {{-1, 1, 1, 0, 1},
   {0, 1, 0, 1, 2}, {-2, 1, 0, 0, 0}};
varlist = {x₁, x₂, x₃, x₄};
SimplexOneStep[tableau11,
  varlist, x₂, x₃, {x₃, x₄}]
```

```
      x₁  x₂  x₃  x₄
x₂   -1   1   1   0   1
x₄    1   0  -1   1   1
obj  -1   0  -1   0  -1

{{{-1, 1, 1, 0, 1}, {1, 0, -1, 1, 1},
   {-1, 0, -1, 0, -1}}, {x₂, x₄}}
```

We further illustrate the tableau form of the simplex algorithm in the next example.

EXAMPLE 4. A coal mining company owns two neighboring mines, whose coal outputs differ somewhat in quality and accessibility. Men and equipment can be shifted back and forth easily between the two mines. Suppose that each day there are 12 available mining hours. Each mine produces high, middle, and low quality coal. Because of storage restrictions, the company may mine no more than 60 tons, 90 tons, and 80 tons, respectively, of high, middle, and low quality coal in a day. An hour spent digging in mine I produces 2 tons, 4 tons, and 6 tons, respectively, of the three kinds of coal; similarly, an hour in mine II yields 3, 1, and 5 tons of the three coal types. Profits per ton are $500, $400, and $300 for high, middle, and low quality coal, respectively. How many hours should be spent in each of the two mines in order to maximize total profit per day?

First we need to identify the problem variables, then construct the objective function, and then translate the constraints. The optimization problem implicit in the question suggests that the variables are

$$x_1 = \text{number of hours spent in mine I}$$
$$x_2 = \text{number of hours spent in mine II}$$

What is the daily profit if these amounts of time are spent in the two mines? By the problem statement, mine I produces tonnages of

$$2\,x_1 \quad \text{(high)} \qquad 4\,x_1 \quad \text{(middle)} \qquad 6\,x_1 \quad \text{(low)}$$

and mine II will yield

$$3\,x_2 \quad \text{(high)} \qquad x_2 \quad \text{(middle)} \qquad 5\,x_2 \quad \text{(low)}$$

tons of the three varieties of coal. For the high quality coal we obtain a profit of $500 $(2 x_1 + 3 x_2)$, and we can calculate the profit for the other two types of coal similarly. The total profit for all three types is

$$f = 4400 x_1 + 3400 x_2 \tag{13}$$

There is a constraint on the total number of hours, and there are three other constraints on total tonnage of the three types of coal that can be mined. It is easy to check that these translate as

$$\begin{aligned}
x_1 + x_2 &\leq 12 \\
2 x_1 + 3 x_2 &\leq 60 \\
4 x_1 + x_2 &\leq 90 \\
6 x_1 + 5 x_2 &\leq 80
\end{aligned} \tag{14}$$

and as usual, both x_1 and x_2 are non-negative. The objective function in (13) and the constraints in (14) specify the LP problem of interest to the mining company. To construct the initial simplex tableau, we insert slack variables $x_3, x_4, x_5,$ and x_6 into the constraints:

$$\begin{aligned}
x_1 + x_2 + x_3 &= 12 \\
2 x_1 + 3 x_2 + x_4 &= 60 \\
4 x_1 + x_2 + x_5 &= 90 \\
6 x_1 + 5 x_2 + x_6 &= 80 \\
4400 x_1 + 3400 x_2 &= f - 0
\end{aligned} \tag{15}$$

The initial tableau below follows immediately:

```
tableau4 =
    {{1, 1, 1, 0, 0, 0, 12}, {2, 3, 0, 1, 0, 0, 60},
     {4, 1, 0, 0, 1, 0, 90}, {6, 5, 0, 0, 0, 1, 80},
     {4400, 3400, 0, 0, 0, 0, 0}};
TableForm[tableau4, TableHeadings →
    {None, {"x₁", "x₂", "x₃", "x₄", "x₅", "x₆", " "}}]
```

x_1	x_2	x_3	x_4	x_5	x_6	
1	1	1	0	0	0	12
2	3	0	1	0	0	60
4	1	0	0	1	0	90
6	5	0	0	0	1	80
4400	3400	0	0	0	0	0

In the first step, x_2 can be chosen as the entering basic variable, and the minimum ratio criterion implies that it should replace x_3 in the first row.

```
varlist = {x₁, x₂, x₃, x₄, x₅, x₆};
{tableau4, basiclist} = SimplexOneStep[
    tableau4, varlist, x₂, x₃, {x₃, x₄, x₅, x₆}];
```

	x_1	x_2	x_3	x_4	x_5	x_6	
x_2	1	1	1	0	0	0	12
x_4	−1	0	−3	1	0	0	24
x_5	3	0	−1	0	1	0	78
x_6	1	0	−5	0	0	1	20
obj	1000	0	−3400	0	0	0	−40800

And in the second and last step, x_1 replaces x_2 in the basic list.

```
{tableau4, basiclist} = SimplexOneStep[
    tableau4, varlist, x₁, x₂, basiclist];
```

	x_1	x_2	x_3	x_4	x_5	x_6	
x_1	1	1	1	0	0	0	12
x_4	0	1	−2	1	0	0	36
x_5	0	−3	−4	0	1	0	42
x_6	0	−1	−6	0	0	1	8
obj	0	−1000	−4400	0	0	0	−52800

We observe from the final tableau that the optimal values of x_1 and x_2 are 12 and 0, respectively, i.e., all time should be spent in mine I. The maximum profit for this activity is \$52,800 per day. Since the slack variables x_4, x_5, and x_6 are strictly positive in the optimal solution, it follows that the storage constraints are not binding. ∎

In what remains of this chapter, and in the next chapter, we will use the dictionary approach when clarity and ease of understanding are paramount, and the more concise tableau approach when efficiency of presentation is more important. You should become accustomed to them both.

Exercises 2.3

1. A small software production company wants to maximize the benefit of the time and money spent by its staff in working on development projects. It produces software that is roughly classified as one of three types, increasing in value and also difficulty from one type to the next. Type 1 projects are

half as valuable as type 2 projects, which are in turn half as valuable as type 3 projects. Each type 1 project requires two programmers, and 40 hours of planning and documentation over the span of a month; each type 2 project requires five programmers and 60 hours of time, and each type 3 project requires eight programmers and 90 hours of time. How should the company allocate its resources for a month's work if it has 30 programmers and 300 hours of time? Can non-integer solutions make sense in this problem; and if so, how?

2. Verify that the list in formula (8) constitutes a basic feasible solution for the winery problem (3)–(4).

3. Solve the following standard maximum problem by the simplex algorithm:

maximize: $f = 2x_1 - x_2 - x_3 + 4x_4$

subject to:

$$
\begin{array}{rcrcrcrcl}
x_1 & + & x_2 & & & & & \le & 3 \\
& & & & x_3 & + & x_4 & \le & 6 \\
x_1 & + & 2x_2 & + & x_3 & + & 2x_4 & \le & 10
\end{array}
$$

$x_i \ge 0$ for all i

4. An enterprising farmer wants to devote some of his land to the raising of hogs, chickens, and ostriches. He will use no more than 1000 square yards for this purpose. After deducting the cost of feed, he decides that he can profit by \$500 per hog, \$50 per chicken, and \$1000 per ostrich. Each ostrich, however, requires at least 100 square yards, each hog 20 square yards, and each chicken 5 square yards of his property. Also, he wishes to spend no more than 500 hours during the season tending to the animals, and each hog requires 10 hours, each ostrich 20 hours, and each chicken 5 hours of his time. What combination of hogs, chickens, and ostriches should he choose to raise, and what is the maximum profit he can achieve?

5. Solve the problem below. Show that the feasible region is unbounded.

maximize: $f = -2x_1 + x_2 + x_3$

subject to:

$$
\begin{array}{rcrcrcl}
-x_1 & + & x_2 & + & x_3 & \le & 2 \\
x_1 & - & x_2 & + & x_3 & \le & 2
\end{array}
$$

$x_1, \ x_2, \ x_3 \ge 0$

6. Show that if there exists at least one constraint in the standard maximum problem, say the ith, such that a_{ij} is strictly greater than zero for all j, then the problem is bounded.

7. A construction contractor builds single-family dwellings and apartment buildings. The contractor can make $5000 profit on each house and $50,000 on each apartment. It takes 400 hours to build a house and 1000 hours to build an apartment. Each house requires 1600 cubic feet of concrete, and each apartment requires 3000 cubic feet of concrete to lay the foundation. Bricks are the only other limiting resource; houses require 20,000 square feet and apartments require 100,000 square feet of brickface. A total of 10,000 hours, 60,000 cubic feet of concrete, and 1,000,000 square feet of brick are available. Find the numbers of single-family dwellings and apartments that should be made to maximize the contractor's profit.

8. Show that the following linear program is unbounded, using the dictionary implementation of the simplex algorithm.

$$\text{maximize: } f = x_1 + x_2 + x_3$$

$$\text{subject to: } \begin{array}{rcrcrcl} x_1 & + & x_2 & - & x_3 & \leq & 4 \\ 2x_1 & - & x_2 & + & x_3 & \leq & 6 \\ & & x_1, & x_2, & x_3 & \geq & 0 \end{array}$$

9. Redo the winery problem (3)–(4) using the tableau implementation of the simplex algorithm. Note the connection between the tableau you obtain at each step, and the corresponding system of equations in Section 3.

10. Solve Exercise 1 using the tableau method.

11. (a) What would you look for in a simplex tableau in order to conclude that a problem is unbounded?
   (b) What would you look for in a simplex tableau in order to detect a degeneracy?
   (c) What characteristic of a final simplex tableau indicates multiple solutions?

12. You may have thought of trying the method of Lagrange multipliers to find optimal solutions, since, after the introduction of slack variables into the standard maximum problem, the problem has the form:

$$\begin{aligned} &\text{maximize: } f(\mathbf{x}) \\ &\text{subject to: } g_1(\mathbf{x}) = 0 \\ &\qquad\qquad \vdots \\ &\qquad\quad g_m(\mathbf{x}) = 0 \\ &\qquad \mathbf{x} \geq 0 \end{aligned}$$

Try this on the problem below.

maximize: $f = x_1 + x_2 + x_3$

subject to:
$$
\begin{array}{rcrcrcl}
x_1 & & & + & x_3 & \leq & 10 \\
x_1 & + & x_2 & & & \leq & 5 \\
& & 2x_2 & + & x_3 & \leq & 8 \\
x_1, & & x_2, & & x_3 & \geq & 0
\end{array}
$$

(The trouble here lies in the fact that the slack variables do not appear in the objective function. Even when the problem is in the standard equality form, this method is inefficient at best. For more information, see Dantzig [16].)

13. Consider the problem:

maximize: $f = x_1 + x_2$

subject to:
$$
\begin{array}{rcrcl}
2x_1 & + & x_2 & \leq & 6 \\
x_1 & + & 2x_2 & \leq & 4 \\
x_1, & & x_2 & \geq & 0
\end{array}
$$

(a) Sketch the feasible region, find the coordinates of the corner points, and find the optimal value.

(b) Repeat (a) if the right-hand side constants are changed to $6 + h_1$ and $4 + h_2$, respectively. Under what conditions on h_1 and h_2 are x_1 and x_2 still basic in the optimal solution?

(c) Perform the simplex algorithm on the problem in part (b), noting the connection between the conditions on h_1 and h_2 derived there, and the choice of entering basic variables.

Exercises (14)–(16) step the reader through a proof of the simplex algorithm. In these problems, we suppose that we begin with a standard maximum problem (1)–(2) whose constraints have been turned into equality constraints by the introduction of slack variables. Refer to the statement of the simplex algorithm. We amend step 2 in the following way: include into the entrance condition for loop 3 a check for a non-basic variable column with all non-negative coefficients. If this condition causes loop termination, then in place of step 4, return a message that the problem is unbounded.

14. Show that if there is a non-basic variable all of whose coefficients are non-negative at some stage of execution, then the problem is unbounded, as the message described above claims.

15. Show inductively that at each pass through the loop, the next system represents a basic feasible solution. Show in addition that if there are no degeneracies, the value of the objective increases strictly.

16. Prove that if no degeneracies are encountered at any stage, then the algorithm terminates in finitely many steps with with an unboundedness message or an optimal solution.

2.4 Duality and the Standard Minimum Problem

In the last section, we used the simplex algorithm to solve the standard maximum problem: max: $f = c' \cdot x$ subject to: $A x \leq b, x \geq 0$. We could adapt the procedure directly to minimum problems, but in this section, we take a different approach in order to introduce the important notion of *duality* in linear programming. The class of minimum problems to be solved is illustrated by the following model problem.

EXAMPLE 1. A publishing company owns two printing facilities, F_1 and F_2, each of which prints a different publication. Facility F_1 uses 20 units of paper and 5 units of ink per copy, and F_2 uses 15 units of paper and 10 units of ink per copy. In a certain period, F_1 must produce at least 10,000 copies of its publication in order to stay in business, and F_2 must produce at least 20,000 copies. Also, on the average, F_1 requires 0.2 units of electricity per unit of paper in order to execute its printing and F_2 requires 0.4 units of electricity per unit of paper. In order to receive a reduced rate from the power company, the publisher wants to use at least 400,000 total units of electricity in its two facilities during the time period in question. It costs \$1 per unit of paper and \$.50 per unit of ink to pay for and deliver the raw materials to F_1, and \$2 per unit of paper and \$1 per unit of ink to supply F_2. How many units of each raw material should the firm purchase for each facility in order to minimize the total supply cost while satisfying the stated constraints?

The variables are $y_1 = \#$ units of paper to F_1 (in thousands), $y_2 = \#$ units of ink to F_1, $y_3 = \#$ units of paper to F_2, and $y_4 = \#$ units of ink to F_2. The constraint that F_1 must produce at least 10,000 copies says that F_1 needs at least $20(10,000) = 200,000$ units of paper and $1/4$ units of ink for each unit of paper. Similarly, F_2 needs at least $15(20,000) = 300,000$ units of paper and $2/3$ units of ink per unit of paper. That is,

$$
\begin{aligned}
y_1 &\geq 200 \\
y_2 &\geq (1/4)\, y_1 \\
y_3 &\geq 300 \\
y_4 &\geq (2/3)\, y_3
\end{aligned}
$$

In addition, the total electricity required must exceed 400,000, which yields

$$0.2\, y_1 + 0.4\, y_3 \geq 400$$

The costs listed in the last paragraph indicate that the following objective function is to be minimized:

$$\text{minimize: } g = y_1 + (1/2)\, y_2 + 2\, y_3 + y_4$$

By moving terms involving the variables to the left side of the inequalities in (1), we may write the problem expressed by (1)–(3) in matrix form as

$$\text{minimize:} \qquad g = \mathbf{b}' \cdot \mathbf{y} \qquad (4)$$

$$\text{subject to:} \qquad A'\, \mathbf{y} \geq \mathbf{c}, \;\; \mathbf{y} \geq \mathbf{0} \qquad (5)$$

where

$$
\mathbf{b} = \begin{pmatrix} 1 \\ 1/2 \\ 2 \\ 1 \end{pmatrix}
\qquad
\mathbf{y} = \begin{pmatrix} y_1 \\ y_2 \\ y_3 \\ y_4 \end{pmatrix}
\qquad
A' = \begin{pmatrix}
1 & 0 & 0 & 0 \\
-1/4 & 1 & 0 & 0 \\
0 & 0 & 1 & 0 \\
0 & 0 & -2/3 & 1 \\
0.2 & 0 & 0.4 & 0
\end{pmatrix}
$$

$$
\mathbf{c} = \begin{pmatrix} 200 \\ 0 \\ 300 \\ 0 \\ 400 \end{pmatrix}
$$

We will show how to solve a problem of the form (4)–(5) by converting it to a corresponding standard maximum problem. ∎

DEFINITION 1. (a) A problem of the form (4)–(5) for which $\mathbf{b} \geq \mathbf{0}$ is said to be a *standard minimum problem*.

(b) The *dual problem* of (4)–(5) is the standard maximum problem:

maximize: $f = \mathbf{c}' \cdot \mathbf{x}$
subject to: $A\mathbf{x} \leq \mathbf{b},\ \mathbf{x} \geq \mathbf{0}$

So to go from the minimum problem to its dual standard maximum problem, the matrix of constraint coefficients is transposed, the inequalities are reversed, and the roles of the vectors $\mathbf{b}$ and $\mathbf{c}$ are interchanged. Duality is a pairwise idea; the minimum problem of (4)–(5) and the maximum problem of part (b) of Definition 1 are duals *of each other*. We will show how the solution of the standard maximum problem by the simplex method gives immediately the solution to its dual standard minimum problem. At the end of the section, we give an economic explanation of why the two problems are connected in this way.

For example, the following two problems are duals of each other.

$$\text{maximize } f = 3x_1 + x_2 + x_3$$

$$\text{subject to: } \begin{array}{rrrrr} & x_2 & + & 2x_3 & \leq 5 \\ x_1 & - \ 3x_2 & + & x_3 & \leq 3 \\ x_1, & x_2, & x_3 & \geq 0 \end{array}$$

$$\text{minimize } g = 5y_1 + 3y_2 \tag{6}$$

$$\text{subject to: } \begin{array}{rrrr} & y_2 & \geq 3 \\ y_1 & - \ 3y_2 & \geq 1 \\ 2y_1 & + \ y_2 & \geq 1 \\ y_1, & y_2 & \geq 0 \end{array}$$

Activity 1 – Form the dual maximum problem of the problem expressed by (1)–(3).

THEOREM 1 (Weak Duality). If $\mathbf{x}$ is feasible for the standard maximum problem, and $\mathbf{y}$ is feasible for its dual standard minimum problem, then

$$\mathbf{b}' \cdot \mathbf{y} \geq \mathbf{c}' \cdot \mathbf{x} \tag{7}$$

Suppose that feasible solutions $\mathbf{x}^*$ and $\mathbf{y}^*$ as above can be found such that equality holds in (7). Then $\mathbf{x}^*$ is an optimal solution of the maximum problem, $\mathbf{y}^*$ is optimal for its dual minimum problem, and the maximum value of f equals the minimum value of g.

Proof. By feasibility of $\mathbf{y}$,

$$\mathbf{c}' \le (A' \, \mathbf{y})' = \mathbf{y}' \, A$$

Thus,

$$\begin{aligned}
\mathbf{b}' \cdot \mathbf{y} - \mathbf{c}' \cdot \mathbf{x} &\ge \mathbf{b}' \cdot \mathbf{y} - \mathbf{y}' A \mathbf{x} \\
&= \mathbf{y}' \cdot \mathbf{b} - \mathbf{y}' A \mathbf{x} \\
&= \mathbf{y}' \cdot (\mathbf{b} - A \mathbf{x}) \ge 0
\end{aligned}$$

The first line uses the fact that $\mathbf{x} \ge \mathbf{0}$, and the last inequality follows from the feasibility of $\mathbf{x}$. We therefore have (7).

Now suppose that $\mathbf{x}^*$ and $\mathbf{y}^*$ are as described in the second statement of the theorem. Given any feasible $\mathbf{x}$ for the maximum problem, (7) implies:

$$\mathbf{c}' \cdot \mathbf{x} \le \mathbf{b}' \cdot \mathbf{y}^* = \mathbf{c}' \cdot \mathbf{x}^*$$

hence $\mathbf{x}^*$ is optimal for the maximum problem. Similarly, given any feasible $\mathbf{y}$ for the minimum problem,

$$\mathbf{b}' \cdot \mathbf{y} \ge \mathbf{c}' \cdot \mathbf{x}^* = \mathbf{b}' \cdot \mathbf{y}^*$$

hence $\mathbf{y}^*$ is optimal for the minimum problem. Also,

$$\max f = \mathbf{c}' \cdot \mathbf{x}^* = \mathbf{b}' \cdot \mathbf{y}^* = \min g$$

which completes the proof. ∎

The next lemma helps to prove the strong duality theorem later, and also enables us to see something of the origin of the form of the minimum problem.

LEMMA 1. At any stage of execution of the simplex algorithm for the standard maximum problem, there exist numbers $y_1, y_2, \ldots, y_m$ such that the objective function row has the form:

$$f = \sum_{k=1}^{m} y_k b_k + \left(c_1 - \sum_{k=1}^{m} y_k a_{k1} \right) x_1 + \cdots + \left(c_n - \sum_{k=1}^{m} y_k a_{kn} \right) x_n - \tag{8}$$
$$y_1 x_{n+1} - \cdots - y_m x_{n+m}$$

where $x_1, \ldots, x_n$ are the original problem variables, and $x_{n+1}, \ldots, x_{n+m}$ are slack variables. In addition, if the slack variable x_{n+k} is basic for the simplex system, then its coefficient $y_k = 0$; and if the problem variable x_i is basic, then its coefficient $(c_i - \sum y_k a_{ki}) = 0$.

Proof. We give only an informal argument. It is convenient to think of the simplex algorithm in its tableau implementation, in order to see exactly what happens to the objective row. Choosing an equation in which to solve for a variable, then substituting into the remaining equations in the system (i.e., the dictionary method) is equivalent to subtracting a certain multiple of the pivot row from each row other than itself. (Exercise 16 asks the reader to show this for the first step of the simplex algorithm.) The starting point is the initial tableau, depicted below. It is easy to see inductively that, at any step of the algorithm, the net effect of the row operations up to that step is that each row of the current tableau is obtained by subtracting from the corresponding row in the initial tableau some net multiple y_1 times row 1 of the initial tableau, subtracting y_2 times row 2, etc., and lastly subtracting a multiple y_m of row m of the initial tableau.

x_1	x_2	$\cdots$	x_n	x_{n+1}	x_{n+2}	$\cdots$	x_{n+m}		
a_{11}	a_{12}	$\cdots$	a_{1n}	1	0	$\cdots$	0	$\mid$	b_1
a_{21}	a_{22}	$\cdots$	a_{2n}	0	1	$\cdots$	0	$\mid$	b_2
$\vdots$	$\vdots$	$\ddots$	$\vdots$	$\vdots$	$\vdots$	$\ddots$	$\vdots$	$\mid$	$\vdots$
a_{m1}	a_{m2}	$\cdots$	a_{mn}	0	0	$\cdots$	1	$\mid$	b_m
								$\mid$	
c_1	c_2	$\cdots$	c_n	0	0	$\cdots$	0	$\mid$	$f - 0$

Consider the initial tableau. Since x_{n+1} appears only in the first row with coefficient 1, the coefficient of x_{n+1} in the new objective row at a later step of the algorithm is just $-y_1$, where y_1 is as described in the last paragraph. More generally, for $k = 1, \ldots, m$, the coefficient of x_{n+k} is the negative of the net multiple y_k of row k that has been subtracted from the objective row. For $i = 1, \ldots, n$, the coefficient of x_i in the objective row after these subtractions must be

$$c_i - \sum_{k=1}^{m} y_k a_{ki}$$

as desired. The term $\sum_{k=1}^{m} y_k b_k$ is the total of all subtractions from f. The last statement of the lemma follows immediately from the fact that basic variables are guaranteed by the algorithm to have coefficient zero in the objective row at every step, because in order to have been made basic, their columns must have been converted to unit coordinate vectors with a zero in the last entry. ∎

This lemma supplies a great deal of insight about how the form of the dual minimum problem is obtained. The intuition is given by equation (8) and the following observation. We look for a minimum problem whose feasibility condition means optimality for the maximum problem, and whose minimum value equals the maximum value of f. Recall that termination of the algorithm occurs when all x_i have non-positive coefficients in the objective row. In view of (8), this requires that

$$c_i - \sum_{k=1}^{m} y_k \, a_{ki} \leq 0 \quad \text{for } i = 1, \ldots, n,$$
$$\text{and} \quad y_k \geq 0 \quad \text{for } k = 1, \ldots, m$$

This is precisely the feasibility condition (5) for the minimum problem. Also, as we have seen a number of times, the fact that the basic variables do not appear in the objective row implies that the constant $\sum y_k \, b_k$ is the value of the objective function. This helps to motivate the choice of coefficients for the minimum problem.

Activity 2 – For the maximum problem in (6), use the tableau method to perform one pivoting step by hand. Keep careful track of the multiples y_k of the pivot row that are subtracted from the other rows, and see how the objective row depends on those multiples, to gain a more concrete understanding of Lemma 1.

The following theorem shows how the optimal solution of the minimum problem may be extracted from the final simplex system of the maximum problem. The theorem also gives information about the existence of optimal solutions to the two dual problems.

THEOREM 2 (Strong Duality). (a) If the standard minimum problem has a feasible solution, then both the standard minimum and its dual standard maximum problem have optimal solutions, and min g = max f.
(b) Assume that the minimum problem (4)–(5) has a feasible solution. For $k = 1, \ldots, m$, let y_k^* be the negative of the coefficient of slack variable x_{n+k} in the objective row of the final simplex system for the dual maximum problem. Then the vector $\mathbf{y}^* = (y_k^*)$ is optimal for the minimum problem.

Proof. (a) Since the minimum problem is feasible, the weak duality theorem implies that, for any vector $\mathbf{y}$ that is feasible for the minimum problem, $\mathbf{b}' \cdot \mathbf{y}$ is an upper bound for the optimal value of f. By the assumption that $\mathbf{b} \geq \mathbf{0}$, the origin is feasible for the standard maximum problem. Therefore, we see that the maximum problem is both feasible and bounded. Thus, the simplex algorithm will produce an optimal extreme point solution to the maximum problem. The fact that the minimum problem has an optimal solution will follow from part (b), and as a by-product we will show that for the two optimal solutions $\mathbf{x}^*$ and $\mathbf{y}^*$, $\mathbf{b}' \cdot \mathbf{y}^* = \mathbf{c}' \cdot \mathbf{x}^*$. From this, it follows that $\min g = \max f$.

(b) Let y_k^*, $k = 1, \ldots, m$ be as in the statement of the theorem, and consider the objective row of the final simplex system for the dual maximum problem. By Lemma 1, for $j = 1, \ldots, n$, the coefficient c_j^* of the jth coordinate x_j^* of the optimal extreme point $\mathbf{x}^*$ is

$$c_j^* = c_j - \sum_{i=1}^{m} y_i^* a_{ij} \qquad\qquad j = 1, \ldots, n \qquad\qquad (9)$$

Since this is the final system, all c_j^* must be non-positive. In matrix notation, (9) can be written as:

$$\mathbf{c}^* = \mathbf{c} - A' \cdot \mathbf{y}^* \le \mathbf{0} \Longrightarrow \mathbf{c} \le A' \cdot \mathbf{y}^*$$

Also, the y_i^* themselves must be non-negative, by the algorithm termination condition, hence $\mathbf{y}^*$ is feasible for the minimum problem. Moreover, since all basic variables have zero coefficients in the objective row, and non-basic variables are 0, (8) implies that

$$0 = f - \sum_{k=1}^{m} y_k^* b_k \Longrightarrow f(\mathbf{x}^*) = \mathbf{b}' \cdot \mathbf{y}^* = g(\mathbf{y}^*)$$

By the weak duality theorem, $\mathbf{x}^*$ and $\mathbf{y}^*$ are optimal for their respective problems. ∎

In Exercises 5–8, some general results about unboundedness, infeasibility, and multiple solutions are given.

EXAMPLE 2. Let us solve the problem stated in Example 1 of supplying the two printing facilities at minimum cost. The technique is to dualize the minimum problem, execute the simplex algorithm on the resulting maximum problem, and then read the solution from the final system.

The minimum problem is:

$$\text{minimize:} \qquad g = y_1 + \tfrac{1}{2} y_2 + 2 y_3 + y_4 \qquad\qquad (10)$$

subject to:

$$
\begin{aligned}
y_1 & & & & & & &\ge 200 \\
-\tfrac{1}{4} y_1 &+ y_2 & & & & & &\ge 0 \\
& & y_3 & & & & &\ge 300 \\
& & -\tfrac{2}{3} y_3 &+ y_4 & & & &\ge 0 \\
\tfrac{1}{5} y_1 & &+ \tfrac{2}{5} y_3 & & & & &\ge 400 \\
y_1, & y_2, & y_3, & y_4 & & & &\ge 0
\end{aligned}
$$

We obtain the dual problem by transposing the coefficient matrix of the constraint system, and interchanging roles of the constraint constants and the objective function coefficients:

$$\text{maximize:} \qquad f = 200\, x_1 + 300\, x_3 + 400\, x_5 \qquad\qquad (11)$$

subject to:

$$
\begin{aligned}
x_1 - \tfrac{1}{4} x_2 \qquad\qquad\qquad + \tfrac{1}{5} x_5 &\le 1 \\
x_2 \qquad\qquad\qquad\qquad\qquad &\le 1/2 \\
x_3 - \tfrac{2}{3} x_4 + \tfrac{2}{5} x_5 &\le 2 \\
x_4 \qquad\qquad &\le 1 \\
x_1,\ x_2,\ x_3,\ x_4,\ x_5 &\ge 0
\end{aligned}
$$

Notice that the number of variables in the dual maximum problem equals the number of constraints in the original (or *primal*) minimum problem. Similarly, the number of constraints in the maximum problem equals the number of variables in the minimum problem.

Introduce slack variables x_6, x_7, x_8, x_9 into the constraints of problem (11) to obtain the initial simplex tableau:

	x_1	x_2	x_3	x_4	x_5	x_6	x_7	x_8	x_9	constant
x_6	1	$-\tfrac{1}{4}$	0	0	$\tfrac{1}{5}$	1	0	0	0	1
x_7	0	1	0	0	0	0	1	0	0	$\tfrac{1}{2}$
x_8	0	0	1	$-\tfrac{2}{3}$	$\tfrac{2}{5}$	0	0	1	0	2
x_9	0	0	0	1	0	0	0	0	1	1
	200	0	300	0	400	0	0	0	0	$f - 0$

```
Needs["KnoxOR`LinearProgramming`"];
```

We enter it into *Mathematica*, and then call on the SimplexOneStep command to carry out the computation. The entering basic variables x_1, x_2, x_3, x_4, x_5 were introduced in that order to obtain the final tableau for the maximum problem below. You may run the commands in the electronic version of the text to see the intermediate tableaux.

```
inittableau = {{1, -1 / 4, 0, 0, 1 / 5, 1, 0, 0, 0, 1},
    {0, 1, 0, 0, 0, 0, 1, 0, 0, 1 / 2},
    {0, 0, 1, -2 / 3, 2 / 5, 0, 0, 1, 0, 2},
    {0, 0, 0, 1, 0, 0, 0, 0, 1, 1},
    {200, 0, 300, 0, 400, 0, 0, 0, 0, 0}};
vlist = {x₁, x₂, x₃, x₄, x₅, x₆, x₇, x₈, x₉};
basicvarlist = {x₆, x₇, x₈, x₉};
{newtableau, newbasiclist} = SimplexOneStep[
    inittableau, vlist, x₁, x₆, basicvarlist];
```

	x_1	x_2	x_3	x_4	x_5	x_6	x_7	x_8	x_9	
x_1	1	$-\frac{1}{4}$	0	0	$\frac{1}{5}$	1	0	0	0	1
x_7	0	1	0	0	0	0	1	0	0	$\frac{1}{2}$
x_8	0	0	1	$-\frac{2}{3}$	$\frac{2}{5}$	0	0	1	0	2
x_9	0	0	0	1	0	0	0	0	1	1
obj	0	50	300	0	360	−200	0	0	0	−200

```
{newtableau, newbasiclist} = SimplexOneStep[
    newtableau, vlist, x₂, x₇, newbasiclist];
```

	x_1	x_2	x_3	x_4	x_5	x_6	x_7	x_8	x_9	
x_1	1	0	0	0	$\frac{1}{5}$	1	$\frac{1}{4}$	0	0	$\frac{9}{8}$
x_2	0	1	0	0	0	0	1	0	0	$\frac{1}{2}$
x_8	0	0	1	$-\frac{2}{3}$	$\frac{2}{5}$	0	0	1	0	2
x_9	0	0	0	1	0	0	0	0	1	1
obj	0	0	300	0	360	−200	−50	0	0	−225

```
{newtableau, newbasiclist} = SimplexOneStep[
    newtableau, vlist, x₃, x₈, newbasiclist];
```

	x_1	x_2	x_3	x_4	x_5	x_6	x_7	x_8	x_9	
x_1	1	0	0	0	$\frac{1}{5}$	1	$\frac{1}{4}$	0	0	$\frac{9}{8}$
x_2	0	1	0	0	0	0	1	0	0	$\frac{1}{2}$
x_3	0	0	1	$-\frac{2}{3}$	$\frac{2}{5}$	0	0	1	0	2
x_9	0	0	0	1	0	0	0	0	1	1
obj	0	0	0	200	240	−200	−50	−300	0	−825

```
{newtableau, newbasiclist} = SimplexOneStep[
    newtableau, vlist, x4, x9, newbasiclist];
```

	x_1	x_2	x_3	x_4	x_5	x_6	x_7	x_8	x_9	
x_1	1	0	0	0	$\frac{1}{5}$	1	$\frac{1}{4}$	0	0	$\frac{9}{8}$
x_2	0	1	0	0	0	0	1	0	0	$\frac{1}{2}$
x_3	0	0	1	0	$\frac{2}{5}$	0	0	1	$\frac{2}{3}$	$\frac{8}{3}$
x_4	0	0	0	1	0	0	0	0	1	1
obj	0	0	0	0	240	-200	-50	-300	-200	-1025

```
{newtableau, newbasiclist} = SimplexOneStep[
    newtableau, vlist, x5, x1, newbasiclist];
```

	x_1	x_2	x_3	x_4	x_5	x_6	x_7	x_8	x_9	
x_5	5	0	0	0	1	5	$\frac{5}{4}$	0	0	$\frac{45}{8}$
x_2	0	1	0	0	0	0	1	0	0	$\frac{1}{2}$
x_3	-2	0	1	0	0	-2	$-\frac{1}{2}$	1	$\frac{2}{3}$	$\frac{5}{12}$
x_4	0	0	0	1	0	0	0	0	1	1
obj	-1200	0	0	0	0	-1400	-350	-300	-200	-2375

According to the strong duality theorem, the minimum cost agrees with the maximum of f, which is 2,375,000. The slack coefficients in the bottom row give the optimal solution:

$$y_1^* = 1{,}400{,}000 \quad y_2^* = 350{,}000 \quad y_3^* = 300{,}000 \quad y_4^* = 200{,}000$$

The economic interpretation is that facility F_1 operates well over its minimum capacity, printing $1{,}400{,}000/20 = 70{,}000$ copies and using (as one would expect) only enough ink to cover this. Facility F_2 works at its minimum capacity of $300{,}000/15 = 20{,}000$ copies, using the minimal amount of ink and paper for the printing. ■

Activity 3 – Try a different sequence of entering basic variables in the last example. For instance, if you use the heuristic of always selecting the highest magnitude entry in the objective row to determine the pivot column, can you shorten the computation?

A general class of standard minimum problems called *mixing problems* can be described as follows. Let there be *m substances*, each composed of some combination of *n elementary ingredients*. We would like to find a mixture of the substances that achieves minimal requirements on the ingredients at a minimal cost. One example of this situation is in problems of diet, for which there are m possible foods to include in the diet, and each food is made up of some combination of n possible nutrients.

Denote by a_{ji} the number of units of ingredient i per unit of substance j. Suppose that it is required to have at least c_i units of ingredient i in the mixture. The cost per unit of substance j is b_j. Then, denoting by y_j the number of units of substance j in the mixture, $b_j y_j$ is the cost of the total amount of substance j and $a_{ji} y_j$ is the total amount of ingredient i that is contributed by substance j. The problem of minimizing the cost of the mixture subject to the minimal requirements of each ingredient takes the form:

$$\text{minimize: } b_1 y_1 + \cdots + b_m y_m \tag{12}$$

subject to:
$$
\begin{array}{ccccccc}
a_{11} y_1 & + & a_{21} y_2 & + & \cdots & + & a_{m1} y_m & \geq & c_1 \\
& & & & \vdots & & & & \\
a_{1n} y_1 & + & a_{2n} y_2 & + & \cdots & + & a_{mn} y_m & \geq & c_n \\
\end{array}
$$
$$\text{all } y_i \geq 0$$

EXAMPLE 3. A cat food manufacturer wishes to design a meat product composed of tuna, liver, and kidney such that minimum total amounts of protein and carbohydrates, 3 ounces and 6 ounces respectively, are present. In addition, the cost of the mixture should be minimized. Suppose that an ounce of tuna has 0.5 ounces of carbohydrates, 0.2 ounces of protein, and costs 2 cents. An ounce of liver has 0.4 ounces of carbohydrates, 0.3 ounces of protein, and costs 1.5 cents, and an ounce of kindey has 0.3 ounces of carbohydrates, 0.2 ounces of protein, and costs 1 cent. To formulate the problem of finding an optimal mixture, introduce variables y_1, y_2, and y_3 to represent, respectively, the number of ounces of tuna, liver, and kidney in the mixture. By the information on costs given above, the objective is clearly:

$$\text{minimize: } \quad g = 2 y_1 + 1.5 y_2 + y_3 \tag{13}$$

The total amount of protein must be at least 3 ounces, and the total amount of carbohydrate must be at least 6 ounces. The given compositions of the three meats yield the constraints:

$$
\begin{aligned}
0.2 y_1 + 0.3 y_2 + 0.2 y_3 &\geq 3 \quad \text{(protein)} \\
0.5 y_1 + 0.4 y_2 + 0.3 y_3 &\geq 6 \quad \text{(carbohydrate)} \\
y_1, y_2, y_3 &\geq 0
\end{aligned}
\tag{14}
$$

The dual maximum problem to the minimum problem expressed by (13)–(14) is:

$$\text{maximize:} \qquad f = 3\,x_1 + 6\,x_2$$

$$
\text{subject to:} \qquad
\begin{aligned}
0.2\,x_1 &+ 0.5\,x_2 &\le\; 2 \\
0.3\,x_1 &+ 0.4\,x_2 &\le\; 1.5 \\
0.2\,x_1 &+ 0.3\,x_2 &\le\; 1 \\
x_1, \quad x_2 &\;\ge\; 0
\end{aligned}
\qquad\qquad (15)
$$

After inserting slack variables x_3, x_4, and x_5 as usual, the initial simplex tableau below results. One pivot operation, the introduction of x_2 into the basic list replacing x_5, gives the final simplex tableau for the maximum problem.

```
tableau3 =
   {{.2, .5, 1, 0, 0, 2}, {.3, .4, 0, 1, 0, 1.5},
    {.2, .3, 0, 0, 1, 1}, {3, 6, 0, 0, 0, 0}};
MatrixForm[tableau3]
```

$$
\begin{pmatrix}
0.2 & 0.5 & 1 & 0 & 0 & 2 \\
0.3 & 0.4 & 0 & 1 & 0 & 1.5 \\
0.2 & 0.3 & 0 & 0 & 1 & 1 \\
3 & 6 & 0 & 0 & 0 & 0
\end{pmatrix}
$$

```
SimplexOneStep[tableau3,
   {x₁, x₂, x₃, x₄, x₅}, x₂, x₅, {x₃, x₄, x₅}];
```

	x_1	x_2	x_3	x_4	x_5	
x_3	-0.133333	0.	1	0	-1.66667	0.333333
x_4	0.0333333	0.	0	1	-1.33333	0.166667
x_2	0.666667	1.	0	0	3.33333	3.33333
obj	-1.	0.	0	0	-20.	-20.

The negatives of the coefficients of the slack variables in the objective row are the optimal values for the original minimum problem, i.e., $y_1 = 0$, $y_2 = 0$, $y_3 = 20$, and the minimum cost is the same as the maximum of f, namely 20. The conclusion is that the mixture should be entirely of kidney in order to minimize cost. ∎

Let us close the section by exploring the intuitive connection between a maximum problem and its dual minimum problem. Consider the winery problem discussed in the introduction. There are four available resources: type I grapes, type II grapes, sugar, and labor hours, which are used to make three products: red, white, and rosé wine. The general setup is depicted in the table below, in which we have chosen to leave the coefficients general rather than insert the numerical values given for them in the introduction.

	Red	White	Rosé	Bounds on Resources	Resources
Units of	a_{11}	a_{12}	a_{13}	b_1	Type I grapes
resource per	a_{21}	a_{22}	a_{23}	b_2	Type II grapes
unit of wine	a_{31}	a_{32}	a_{33}	b_3	Sugar
	a_{41}	a_{42}	a_{43}	b_4	Hours
Profit per					
unit of wine	c_1	c_2	c_3		

Consider the problem faced by an individual who wishes to purchase all of the winery's resources. A decision must be reached by this buyer about the following four offering prices:

$$y_1 = \text{price per unit of type I grapes}$$
$$y_2 = \text{price per unit of type II grapes}$$
$$y_3 = \text{price per unit of sugar}$$
$$y_4 = \text{price per unit of hours}$$

The buyer wishes to minimize the total purchase price. Since the winery has respectively b_1, b_2, b_3, and b_4 units of these resources on hand, the total price to the buyer will be

$$g = b_1 \cdot y_1 + b_2 \cdot y_2 + b_3 \cdot y_3 + b_4 \cdot y_4 \tag{16}$$

which is the objective function of the dual minimum problem. Also, the buyer must offer prices at least as large as the profit that the winery could derive from using the resources to make and sell wine. For instance, in the case of red wine, if the winery used a_{11}type I grapes, a_{21}type II grapes, a_{31}units of sugar, and a_{41}labor hours, the winery could make one unit of red wine for a profit of c_1. The buyer's total offer for this much of the four resources must exceed c_1, i.e.,

$$a_{11}\, y_1 + a_{21}\, y_2 + a_{31}\, y_3 + a_{41}\, y_4 \geq c_1 \tag{17}$$

This is the first constraint of the dual standard minimum problem, and similar analyses of profits for white wine and rosé wine yield the other two constraints of the minimum problem. The dual variables y_i, which are values per unit of the resources, are often called the *shadow prices* of the resources.

In a similar manner we can intuitively relate a given standard minimum problem to its dual maximum problem. Let us work in the context of Example 3 on cat food. The problem information is summarized in the table below, and again we leave the coefficients general in order that the reader may more easily recognize the dual problem.

	Tuna	Liver	Kidney	Amounts Required	Nutrients
Amount nutrient per	a_{11}	a_{21}	a_{31}	c_1	Protein
unit of ingredient	a_{12}	a_{22}	a_{32}	c_2	Carbohydrate
Cost per unit of ingredient	b_1	b_2	b_3		

Consider now an individual who is willing to sell all of the nutrients directly to the cat food manufacturer. This supplier would choose prices:

$$x_1 = \text{price per unit of protein}$$
$$x_2 = \text{price per unit of carbohydrate}$$

in order to maximize his total revenue from the sale. Since the manufacturer must purchase c_1 units of protein and c_2 units of carbohydrate, this revenue is

$$f = c_1 x_1 + c_2 x_2$$

which is the objective function of the dual standard maximum problem. Moreover, the prices offered by the supplier must be competitive with the price paid by the cat food manufacturer for enough units of tuna, liver, and kidney to satisfy the requirements. For example, a unit of tuna, at a cost of b_1, supplies a_{11} units of protein and a_{12} units of carbohydrate. Therefore the sale price offered by our supplier for this combination of nutrients must be less than b_1, i.e.,

$$a_{11} x_1 + a_{12} x_2 \le b_1$$

We recognize this as the first constraint of the dual standard maximum problem. Similar analyses performed on the liver and kidney ingredients produce the other two maximum constraints.

Exercises 2.4

1. (*Mathematica*) A sheep farmer will blend three types of feed for his sheep, costing $1 per pound, $2 per pound, and $3 per pound, respectively. Feed 1 consists of 50% fat and 50% protein, feed 2 is 25% fat and 75% protein, and feed 3 is 40% fat and 60% protein. Sheep are to receive at least 2 lbs. of fat per week and 3 lbs. of protein. Minimize the weekly cost of feed per sheep.

2. (*Mathematica*) Solve the standard minimum problem:

$$\text{minimize:} \quad g = y_1 + y_2$$

$$\text{subject to:} \quad \begin{aligned} y_1 &+ y_2 &\geq 2 \\ 4\,y_1 &+ y_2 &\geq 8 \\ y_1 &+ 4\,y_2 &\geq 8 \\ y_1, \quad y_2 &\geq 0 \end{aligned}$$

3. Verify that constraints (1)–(2) are equivalent to (5).

4. (*Mathematica*) A woman operating her own business is trying to plan her weekly sales activity schedule to produce the most valuable sales results in the least possible time. She can make personal visits, do phone calls, or work on mass mailings. She estimates that each hour spent on personal visits nets her $50, each hour on phone calls earns $40, and each hour on mass mailings earns $20. To limit her transportation and phone costs, and to reach a wider audience, she will spend at least twice as much time on the phone as she spends making personal visits, and at least twice as much time doing mass mailings as she spends making phone calls. If her goal is to make at least $1000 per week, how should she allocate her time so as to spend as few hours as possible?

5. Suppose that in the final simplex system for a dual maximum problem of a given minimum problem, there is a degenerate basic slack variable x_j. In the equation to which x_j belongs is some non-basic variable x_k with a positive coefficient; and furthermore in the objective row, x_k has a strictly negative coefficient c_k. Show that if x_k is made basic, the resulting vector $\mathbf{y}$ of negatives of slack coefficients also achieves the minimum of the objective function of the minimum problem. Need $\mathbf{y}$ be feasible for the minimum problem?

6. Suppose that in the final simplex system for a dual maximum problem of a given minimum problem, there is a non-basic variable in the objective row with coefficient zero. Recall that this indicates the presence of an alternative

optimal solution for the maximum problem. Show that if this variable is made basic, the same solution to the minimum problem results. (Together, Exercises 5 and 6 point out the duality between degeneracy in one problem, and non-uniqueness in the other.)

7. Prove that if a standard maximum problem is unbounded, then its dual standard minimum problem is infeasible.

8. Prove that if a standard minimum problem is unbounded, then its dual standard maximum problem is infeasible.

9. Solve the standard minimum problem below.

$$\text{minimize:} \qquad g = 2\,y_1 + y_2 + 3\,y_3$$

$$\text{subject to:} \quad \begin{array}{rcrcrcrcl} y_1 & + & y_2 & + & y_3 & \geq & 10 \\ y_1 & - & y_2 & + & 6\,y_3 & \geq & 15 \\ & & y_1, & y_2, & y_3 & \geq & 0 \end{array}$$

10. Give an economic interpretation of the dual maximum problem for the problem of Exercise 1.

11. Give an economic interpretation of the dual minimum problem for the problem of Exercise 7 of Section 2.3.

12. Let $A\mathbf{x} = \mathbf{b}$ be the system of equality constraints for a standard maximum problem, after the introduction of slack variables. There are n variables and m constraints. A *basic solution* is a vector $\mathbf{x}$ that satisfies the equality constraints, and has $n - m$ of its components equal to 0, but is not required to satisfy the non-negativity constraint. For each basic solution of the maximum problem, there is a *complementary basic solution* (again, not necessarily feasible) for its dual minimum problem, obtained by reading the negatives of the slack coefficients in the objective row of the corresponding maximum simplex system. To see what basic solutions represent in two dimensions, consider the problem:

$$\text{maximize:} \qquad f = x_1 + 5\,x_2$$

$$\text{subject to:} \quad \begin{array}{rcrcl} 2\,x_1 & + & x_2 & \leq & 4 \\ x_1 & + & 2\,x_2 & \leq & 6 \\ x_1, & x_2 & & \geq & 0 \end{array}$$

(a) Write the dual problem, and sketch the feasible regions for both problems. In each case find the coordinates of all corner points, whether feasible or infeasible.

(b) Write a simplex system for each basic solution of the maximum problem, and find the point on each graph to which the system corresponds.

13. Use duality and two-dimensional geometry to solve the following problem without recourse to the simplex algorithm:

$$\text{minimize:} \quad y_1 + 3\,y_2 + 4\,y_3$$

$$\text{subject to:} \quad \begin{array}{rcccrcr} -y_1 & + & & & y_3 & \geq & 2 \\ y_1 & - & y_2 & & & \geq & -1 \\ & & y_1, & y_2, & y_3 & \geq & 0 \end{array}$$

14. We studied duality only in the context of standard maximum and minimum problems. The concept can be extended to arbitrary problems using the following correspondences:

Max problem		Min problem
Objective coefficient c_i	$\Longleftrightarrow$	Constraint constant c_i
Constraint constant b_i	$\Longleftrightarrow$	Objective coefficient b_i
Variable $x_i \geq 0$	$\Longleftrightarrow$	Constraint i of $\geq$ form
Constraint j of $\leq$ form	$\Longleftrightarrow$	Variable $y_j \geq 0$
Constraint j of $=$ form	$\Longleftrightarrow$	Variable y_j unrestricted
Variable x_j unrestricted	$\Longleftrightarrow$	Constraint i of $=$ form
Constraint coefficient A	$\Longleftrightarrow$	Constraint coefficient A'

Notice that the first four correspondences are known to us already from our study of standard problems, and the last three show how to dualize in non-standard problems.

(a) Use the correspondences given in the table above to find the dual problem of the problem:

$$\text{minimize:} \quad g = 2\,y_1 - y_2 + y_3$$

$$\text{subject to:} \quad \begin{array}{rcrcrcr} y_1 & & & + & 5\,y_3 & \geq & 10 \\ 2\,y_1 & + & 4\,y_2 & & & = & 7 \\ & & y_1, & & y_3 & \geq & 0 \end{array}$$

(b) Assuming that it is still true that dual problems share the same optimal objective value, find the minimum value of g for the problem in part (a).

15. Referring to Exercise 14,
(a) Write the general form of the dual of the problem:

$$\begin{aligned} \text{minimize:} \quad & g = \mathbf{b}' \cdot \mathbf{y} \\ \text{subject to:} \quad & A' \, \mathbf{y} = \mathbf{c} \\ & \mathbf{y} \geq \mathbf{0} \end{aligned}$$

(b) Write the general form of the dual of the problem:

$$\begin{aligned} \text{maximize:} \quad & f = \mathbf{c}' \cdot \mathbf{x} \\ \text{subject to:} \quad & A \, \mathbf{x} = \mathbf{b} \\ & \mathbf{x} \geq \mathbf{0} \end{aligned}$$

(c) For each of the problems in (a) and (b), show that the dual of the dual problem is the original (i.e., *primal*) problem. This illustrates again the paired nature of dual problems.

16. Our proof of the Strong Duality Theorem, as well as the presentation of the tableau method in Section 3, depended on the fact that the dictionary and tableau methods were equivalent. More specifically, solving for an entering basic variable in a constraint equation, and then substituting the resulting expression into all other equations, is equivalent to pivoting about the element in the row of the departing basic variable and the column of the entering basic variable in the related simplex tableau. Consider a general standard maximum problem in equality form, and show for the first step of the algorithm that this is indeed the case.

3

Further Topics in Linear Programming

Introduction

There is a great deal more to linear programming than we have discussed in Chapter 2. In particular, we have only developed a method for solving a limited range of problems. The standard maximum problem takes the form:

$$\text{max: } f = \mathbf{c}' \mathbf{x}$$
$$\text{subject to: } A \mathbf{x} \leq \mathbf{b}, \mathbf{x} \geq \mathbf{0}$$

where the vector $\mathbf{b}$ of constant right-hand sides was assumed to be non-negative. We would like to investigate problems in which constraint constants are allowed to be negative, and constraints may be of the form "=" or "≥" in the initial problem. (We could also generalize to problems in which the variables are allowed to be negative, but for the sake of simplicity we do not thoroughly discuss this problem in this book apart from a brief remark below.) Similarly, we would like to be able to solve minimization problems that are not in standard form, hence not able to be dualized to standard maximum problems.

As an example of an interesting application leading to a non-standard problem, consider the following. An individual has a fixed amount of money to divide among several possible investment objects. For each dollar invested in each object, the investor can estimate the number of dollars at the end of the investment period. Due to the fact that some investment objects are riskier than others and the investor may be averse to risk, certain other constraints may be put on the allocation of wealth. To fix some numbers, suppose that the investor has \$20,000 to invest in either stocks, municipal bonds, or a savings account. The expected rate of return on stocks is 8% per year, the rate on bonds is 7%, and the savings account interest is 5%. Because the investor is averse to risk, he wishes to invest no more in stocks than in bonds, and no more than four times as much in bonds as in the savings account. In addition, he wants to keep at least \$5000 in savings for emergencies. Let x_1, x_2, and x_3, respectively, denote the numbers of dollars invested in stocks, bonds, and savings. Then it is easy to see that the LP problem to be solved is:

max: $f = 1.08x_1 + 1.07x_2 + 1.05x_3$

$$
\begin{array}{rcl}
x_1 + x_2 + x_3 &=& 20000 \\
x_1 - x_2 &\leq& 0 \\
\text{subject to:} \quad x_2 - 4x_3 &\leq& 0 \\
x_3 &\geq& 5000 \\
x_1, \ x_2, \ x_3 &\geq& 0
\end{array}
$$

which is a non-standard problem. Non-standard problems are covered in Section 1.

Another direction of generalization involves the domain of values that can be taken on by the problem variables. We have heretofore assumed that the variables x_i could take on all real values permitted by the constraints, but in applications where a variable represents the number of some indivisible item that are to be processed, only integer values are legal. *Integer programming* is a large subarea of operations research in its own right, and we cannot hope to do it justice here, but we will study an algorithm called the *transportation algorithm* for the special integer programming problem described roughly as follows. There are several locations, called *sources*, which are able to supply needed items to several other locations, called *destinations*, which require the items in some known quantities. The number of items on hand at each source is known, as are the costs per item to supply each destination from each source. The problem is to find a supply schedule that determines how many items are to be dispatched from each source to each destination, in order to minimize total shipment cost while satisfying demand. We show how to solve this transportation problem in Section 2.

The third extension to be discussed was foreshadowed in Exercise 13 of Section 2.3. Suppose that the operations research analyst has been presented with a linear programming problem, and has computed the solution, but the originator of the problem returns to say that some of the coefficients of the problem were mistaken. Or, perhaps the problem originator would like to see how the optimal solution changes if a higher profit per item is made for one or more of the quantities that have given rise to the problem variables. Yet another change to the problem might result if new levels of resources, not originally included in the right-hand side constraints, unexpectedly appear. An entirely new problem constraint might even enter the picture. In all of these cases, some change has been made to the original problem coefficients, and one is interested in the new optimal solution. Will it be necessary to solve the problem from the beginning with the new data? The answer is: not necessarily. We will see in Section 3 how to analyze the sensitivity of the optimal solution to changes in problem parameters, without repeating the entire computation. Some new computing will be necessary, but it is less time consuming than to solve the problem again from scratch.

3.1 Non-Standard Problems

Thus far, we have seen how to solve a problem of standard maximum form by the simplex algorithm, and we have seen how to solve a standard minimum problem by solving the dual maximum problem and applying the Strong Duality Theorem. In this section, we discuss a technique for non-standard LP problems. The approach requires two phases of computation. The first phase involves reformulating the original objective function and performing the simplex algorithm on the new problem. Its goal is to produce a basic feasible solution. The second phase uses the final system of the first phase, together with the original objective function, to produce an optimal solution by the ordinary simplex method.

The following theorem, whose proof is requested in Exercise 5, allows us to treat maximum problems only.

THEOREM 1. Suppose that $g(\mathbf{y}) = \mathbf{b}' \cdot \mathbf{y}$ is the objective function for a minimum problem with some non-empty feasible region F. Define $f(\mathbf{x}) = -g(\mathbf{x}) = -(\mathbf{b}' \cdot \mathbf{x})$. The problem of maximizing f over F has an optimal solution $\mathbf{x}^*$ if and only if $\mathbf{x}^*$ is also optimal for the problem of minimizing g, and in addition,

$$\min g = -(\max f) \quad \blacksquare$$

Therefore, given a minimum problem, we may solve instead the problem of maximizing the negative of the objective over the same feasible region. It should be pointed out that there are other ways of solving minimum problems directly, and it may not be computationally most efficient to translate all minimum problems into maximum problems in this way. But this proposition does allow us to focus our attention only on solving non-standard maximum problems, and therefore it has the advantage of simplifying the exposition.

EXAMPLE 1. Consider the two-variable problem:

$$\max: f = 4x_1 + x_2$$

$$\text{subject to:} \quad \begin{aligned} -x_1 + x_2 &\ge -1 \\ x_1 + x_2 &\ge 2 \\ 2x_1 + x_2 &\le 8 \\ x_1, \ x_2 &\ge 0 \end{aligned} \tag{1}$$

This is not of standard form, because of the presence of a negative number on the right side of the first constraint, and inequalities in the $\ge$ direction in the first two constraints. The feasible region is depicted in Figure 3.1.

Notice that the origin is not feasible. This is in contrast to the situation for standard maximum problems, in which the origin is always feasible. Recall also that in the simplex algorithm, the initial system represented the origin, which was feasible, and at each succeeding step another basic feasible solution was produced. The main difficulty with non-standard linear programs is the need to produce an initial simplex system that represents a basic feasible solution. Once that is done, the simplex algorithm can finish the problem. ∎

```
Needs["KnoxOR`LinearProgramming`"];
```

```
constraints =
   {-x₁ + x₂ == -1, x₁ + x₂ == 2, 2 x₁ + x₂ == 8};
corners = {{0, 2}, {0, 8}, {3, 2}, {3 / 2, 1 / 2}};
f[x_, y_] := 4 x + y;
PlotFeasibleRegion[constraints,
   {x₁, 0, 4}, {x₂, 0, 8}, corners, f,
   ShowTable → False, AspectRatio → .8, TextStyle →
   {FontFamily -> "Times", FontSize → 8}];
```

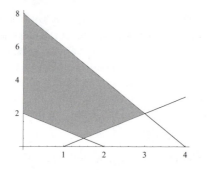

Figure 3.1 – Feasible region of a non-standard problem

Activity 1 – Verify that the feasible region for Example 1 is as in Figure 3.1. Find algebraically the coordinates of the feasible corner points, and make sure they are consistent with the picture.

The first step of our Phase 1 algorithm is to remove negative signs on the constant constraint bounds by multiplying through by −1, if necessary. Referring to the problem of Example 1, the first constraint in (1) becomes:

$$x_1 - x_2 \le 1$$

Now introduce slack variables into all "$\le$" constraints. At the same time, introduce *surplus variables* into all "$\ge$" constraints; e.g., in the second constraint of (1), there exists a non-negative surplus variable x_4 such that $x_1 + x_2 - x_4 = 2$. We should note here that if a constraint is already written in "$=$" form, then no slack or surplus variables are necessary, and the constraint may simply be left untouched. Thus, the first step of Phase 1 produces the following system of equality constraints for Example 1:

$$
\begin{array}{rcrcrcrcrcr}
x_1 & - & x_2 & + & x_3 & & & & & = & 1 \\
x_1 & + & x_2 & & & - & x_4 & & & = & 2 \\
2x_1 & + & x_2 & & & & & + & x_5 & = & 8
\end{array}
\tag{2}
$$
$$x_i \ge 0 \text{ for all } i$$

REMARK. We have generalized the standard maximum problem by allowing negative entries in the vector **b**, allowing "$\ge$" constraints, and allowing "$=$" constraints. There is one other direction that we could take, namely to permit negative values of the variables x_i. We show briefly how to convert such a problem into the current form. If a certain variable x_i is bounded from below by some number $L < 0$, then

$$x_i \ge L \Longrightarrow x_i - L \ge 0$$

We see that if we change variables in the constraints and the objective function by $x_i' = x_i - L$, then the new variable x_i' is constrained to be non-negative. If no such lower bound L is present, then x_i may be split into two non-negative variables x_i^+ and x_i^- by

$$x_i = x_i^+ - x_i^- \tag{3}$$

where

$$x_i^+ = \begin{cases} x_i & \text{if } x_i \ge 0 \\ 0 & \text{otherwise} \end{cases} \quad \text{and} \quad x_i^- = \begin{cases} 0 & \text{if } x_i \ge 0 \\ -x_i & \text{otherwise} \end{cases} \tag{4}$$

This introduces one extra variable into the system for every such x_i that is unbounded below, and increases the complexity of the computation. We will not discuss this issue further here; for more information, the reader may see Hillier and Lieberman [31] or Gribik and Kortanek [27]. ∎

Activity 2 – Verify that x_i does decompose as in formula (3). What happens to a constraint such as the third one in (1) if $x_2 \geq 0$ but x_1 is unconstrained?

The second step in Phase 1 is to solve a different maximization problem. To be specific, insert a new artificial variable a_i into each equality constraint of the current system, and define a new objective function to be the sum of the negatives of the a_i's. In the problem of Example 1, we produce the new problem:

$$\max: \ g = -a_1 - a_2 - a_3$$

subject to:

$$
\begin{array}{rcccccccccl}
x_1 & - & x_2 & + & x_3 & & & + & a_1 & & & = & 1 \\
x_1 & + & x_2 & & & - & x_4 & & & + & a_2 & = & 2 \\
2x_1 & + & x_2 & & & + & x_5 & & & + & a_3 & = & 8
\end{array}
\qquad (5)
$$

$$x_i, a_i \geq 0 \text{ for all } i$$

The reason for doing this is given in the following theorem.

THEOREM 2. Let A be an $m \times n_0$ matrix with $n_0 \geq m$, let $\mathbf{c}$ be a vector of n_0 entries, let $\mathbf{b}$ be a vector of m non-negative components, let I be the $m \times m$ identity matrix, and denote by $\mathbf{1}$ the vector all of whose m entries are equal to 1. Consider the following two LP problems:

(LP1) $\max_x: \qquad \mathbf{c}' \cdot \mathbf{x}$ (LP2) $\max_{x,a}: \quad \mathbf{0} \cdot \mathbf{x} + (-1) \cdot \mathbf{a}$

$\qquad$ subject to: $A\mathbf{x} = \mathbf{b}$ $\qquad\qquad$ subject to: $A\mathbf{x} + I\mathbf{a} = \mathbf{b}$

$\qquad\qquad\qquad \mathbf{x} \geq \mathbf{0}$ $\qquad\qquad\qquad\qquad\qquad \mathbf{x}, \mathbf{a} \geq \mathbf{0}$

We have the following.

(a) Problem (LP2) is both feasible and bounded, hence the simplex algorithm produces an optimal solution $(\mathbf{x}^*, \mathbf{a}^*)$.

(b) The optimum value of (LP2) is strictly less than zero if and only if (LP1) is infeasible.

(c) If the optimum value of (LP2) is zero, taken on at $\mathbf{a}^* = \mathbf{0}$, then $\mathbf{x}^*$ is a basic feasible solution of (LP1). Also, the final simplex system of constraints for (LP2), with $\mathbf{a} = \mathbf{0}$, is equivalent to the initial constraints $A\mathbf{x} = \mathbf{b}$ for (LP1).

Proof. (a) Since we have assumed that $b \geq 0$, the vector $(x, a) = (0, b)$ satisfies the feasibility conditions for (LP2). Also, since $a \geq 0$, the objective function of (LP2) is bounded above by 0. Together these observations prove part (a).

(b) Suppose the optimum value of (LP2) is strictly less than zero. We wish to show that (LP1) is infeasible. If, on the contrary, there exists a feasible vector x for (LP1), then $x \geq 0$ and

$$A x + I \cdot 0 = A x = b \qquad (6)$$

so that $(x, 0)$ is feasible for (LP2). But this is a contradiction, because $(x, 0)$ has objective value 0 and the maximum for (LP2) was assumed to be less than zero. You are asked to show the converse in Exercise 8.

(c) By (6), applied at the optimal solution $(x^*, 0)$ of (LP2), x^* is feasible for (LP1). We also are required to show that x^* is basic, i.e., x^* has at least $n_0 - m$ zero components and the columns of the coefficient matrix corresponding to the remaining components are linearly independent. Now (LP2) is a problem with $n_0 + m$ variables. At each stage of execution of the simplex algorithm on (LP2), m of these variables are basic and at least $n_0 + m - m = n_0$ variables equal 0. In particular, in the final simplex system, all of the m a_i^* terms have the value zero, there are m basic variables in the system, and at least $n_0 - m$ of the remaining x_i^* terms are zero. Those x_i^* terms that are non-zero are basic for the final system, hence the corresponding columns of the final coefficient matrix A^* are linearly independent. The final system for (LP2) may be written in the form:

$$A_1^* x + A_2^* a = b^* \qquad (7)$$

where A_1^* is an $m \times n_0$ matrix and A_2^* is an $m \times m$ matrix. Furthermore, since this system was obtained from the initial system by elementary row operations, the solution spaces of the two are exactly the same. In particular, the subset of the solution space of (7) consisting of all (x, a) such that $a = 0$ is the same as the subset of the feasible region of (LP2) for which $a = 0$. But that is just the feasible region of (LP1), therefore the system $A_1^* x = b$ is equivalent to the system $A x = b$, which proves the second statement of part (c). ∎

EXAMPLE 1 (cont.). Returning to the linear program of Example 1, notice that objective function and constraints are in the form of (LP1) as a result of our initial manipulations; here $m = 3$, $n_0 = n + m = 5$, and

$$A = \begin{pmatrix} 1 & -1 & 1 & 0 & 0 \\ 1 & 1 & 0 & -1 & 0 \\ 2 & 1 & 0 & 0 & 1 \end{pmatrix}, \quad b = \begin{pmatrix} 1 \\ 2 \\ 8 \end{pmatrix}, \quad c = \begin{pmatrix} 4 \\ 1 \\ 0 \\ 0 \\ 0 \end{pmatrix}$$

The introduction of the three artificial variables in (5) defines a new problem of the type (LP2), with $n_0 + m = 8$ variables: x_1, ..., x_5 and a_1, a_2, a_3. The next step of Phase 1 is to perform the simplex algorithm on the new problem, starting with the artificial variables as basic. Notice that the initial vector $(\mathbf{x}, \mathbf{a}) = (\mathbf{0}, \mathbf{b})$ is feasible for (LP2). Part (a) of the theorem says that we will get an optimal solution. Part (b) of the theorem says that if it is not the case that all a_i's are zero in that optimal solution, then the original problem was infeasible, and so we should stop. If indeed all of the a_i's are zero in the final system of the new problem, then delete all references to the a_i's in the constraint equations of that final system and the resulting system is equivalent to the constraints for the original problem. Moreover, it represents a basic feasible solution for (LP1). From there, we may proceed with the simplex algorithm and the original objective from (1).

Activity 3 – Look at the constraint system (2) again. The main difficulty with non-standard problems seems to be to obtain an initial basic feasible solution. Why are the aritificial variables necessary, that is, why not just let the slack and surplus variables be the initial basic variables?

Let us perform the operations described above on the problem of Example 1. We start with system (5), and we can use the Dictionary command again to do the tedious algebra. Observe that there is one slight variation in (5) on the usual simplex format. Currently, the objective is written as a function of the basic variables a_1, a_2, and a_3. The objective should be expressed in terms of the non-basic variables. Dictionary has no trouble alleviating this difficulty for us by simply substituting for these variables using the constraint equations:

```
phase1system =
    {x₁ - x₂ + x₃ + a₁ == 1, x₁ + x₂ - x₄ + a₂ == 2,
     2 x₁ + x₂ + x₅ + a₃ == 8, g == -a₁ - a₂ - a₃};
Dictionary[phase1system, {a₁, a₂, a₃, g},
    {x₁, x₂, x₃, x₄, x₅}]
```

```
a₁ = 1    -  1 x₁ +  1 x₂ -  1 x₃ +  0 x₄ +  0 x₅
a₂ = 2    -  1 x₁ -  1 x₂ +  0 x₃ +  1 x₄ +  0 x₅
a₃ = 8    -  2 x₁ -  1 x₂ +  0 x₃ +  0 x₄ -  1 x₅
g  = -11 +  4 x₁ +  1 x₂ +  1 x₃ -  1 x₄ +  1 x₅
```

Continuing in the usual way, with entering basic variables x_2 replacing a_2, x_5 replacing a_3, and x_3 replacing a_1, gives the final simplex system for the altered problem, listed below. (You can delete the semicolons to display the output of the intermediate steps in the electronic version of the text.)

```
Dictionary[phase1system,
  {a₁, x₂, a₃, g}, {x₁, x₃, x₄, x₅, a₂}];
Dictionary[phase1system, {a₁, x₂, x₅, g},
  {x₁, x₃, x₄, a₂, a₃}];
Dictionary[phase1system, {x₃, x₂, x₅, g},
  {x₁, x₄, a₁, a₂, a₃}]
```

```
x₃ = 3 -  2 x₁ +  1 x₄ -  1 a₁ -  1 a₂ +  0 a₃
x₂ = 2 -  1 x₁ +  1 x₄ +  0 a₁ -  1 a₂ +  0 a₃
x₅ = 6 -  1 x₁ -  1 x₄ +  0 a₁ +  1 a₂ -  1 a₃
g  = 0 +  0 x₁ +  0 x₄ -  1 a₁ -  1 a₂ -  1 a₃
```

Since the current values of the x variables are $x_1 = 0$, $x_2 = 2$, $x_3 = 3$, $x_4 = 0$, and $x_5 = 6$, we are now at the feasible corner point (0, 2) in Figure 3.1, and all a_i's are zero. Drop the a_i's from the constraint equations and return the original objective to the bottom row. Again, we see that the expression for f contains a basic variable x_2, but Dictionary eliminates it for us.

```
phase2system = {2 x₁ + x₃ - x₄ == 3,
  x₁ + x₂ - x₄ == 2, x₁ + x₄ + x₅ == 6, f == 4 x₁ + x₂};
Dictionary[phase2system, {x₃, x₂, x₅, f}, {x₁, x₄}]
```

```
x₃ = 3 -  2 x₁ +  1 x₄
x₂ = 2 -  1 x₁ +  1 x₄
x₅ = 6 -  1 x₁ -  1 x₄
f  = 2 +  3 x₁ +  1 x₄
```

Phase 1 is complete.

Phase 2 simply performs the simplex algorithm on this new system. If we introduce x_1 to replace x_3, then x_4 to replace x_5, we obtain in two steps the following final system.

```
Dictionary[phase2system,
   {x₁, x₂, x₅, f}, {x₃, x₄}];
Dictionary[phase2system, {x₁, x₂, x₄, f}, {x₃, x₅}]
```

$$x_1 = 3 - \frac{1}{3}x_3 - \frac{1}{3}x_5$$

$$x_2 = 2 + \frac{2}{3}x_3 - \frac{1}{3}x_5$$

$$x_4 = 3 + \frac{1}{3}x_3 - \frac{2}{3}x_5$$

$$f = 14 - \frac{2}{3}x_3 - \frac{5}{3}x_5$$

The maximum value of f is 14, taken on at the corner point (3, 2). ∎

Let us now summarize the steps of the Phase 1 algorithm for maximum problems with general linear constraints and non-negative variables. We assume that the input problem has already been identified as non-standard.

PHASE 1 SIMPLEX ALGORITHM FOR NON-STANDARD MAXIMUM PROBLEMS
1. {Prepare the initial system for the new problem.}
   (a) Multiply constraints in which the constant is negative by -1.
   (b) Introduce slack variables into "≤" constraints and surplus variables into "≥" constraints to convert to standard equality form.
2. Construct the problem of maximizing the sum of the negatives of artificial variables, where one such variable is inserted into each constraint equation.
3. Substitute for the artificial variables in the objective row of the new problem.
   {The initial system for the new problem is now ready.}
4. Use the simplex algorithm to solve the new problem.
5. If the optimal value of the new problem is less than zero, the original problem is infeasible, so stop.
Otherwise, do steps 6–7
{Set up the tableau for Phase 2}
6. Delete the artificial variables from the constraint equations and restore the original objective to the bottom row.
7. By substitution, if necessary, eliminate basic variables from the objective row.
{The system is now ready for Phase 2, the ordinary simplex algorithm.}

After execution of the Phase 1 algorithm, the optimal solution to the problem can be computed as in the standard maximum case. By the way, you may have already noticed that in practice you can save a few variables by using slack (but not surplus) variables as basic variables in their constraint equations, and not including artificial variables for those equations. This is justifiable; the only reason we advised at the outset to insert artificial variables into each constraint row was to simplify the proof of Theorem 2. We try this approach in the next example.

EXAMPLE 2. Let us now illustrate how the Phase 1–Phase 2 approach proceeds in the tableau implementation of the method. The idea is the same as in the standard maximum problem, i.e., only coefficients are preserved, the constraint constants are isolated on the right side of the equations, and the rows are labeled with the variable that is currently basic in the equation.

We work in the context of the investment problem discussed in the introduction. For ease of reference it is reproduced below. Recall that x_1, x_2, and x_3 represent the investment amounts in stocks, bonds, and the savings account, respectively.

$$\text{max: } f = 1.08x_1 + 1.07x_2 + 1.05x_3$$

$$\text{subject to: } \begin{array}{rcrcrcl} x_1 & + & x_2 & + & x_3 & = & 20000 \\ x_1 & - & x_2 & & & \leq & 0 \\ & & x_2 & - & 4\,x_3 & \leq & 0 \\ & & & & x_3 & \geq & 5000 \end{array}$$

$$x_1, x_2, x_3 \geq 0$$

There are no negative right-hand sides to remove. Slack variables x_4 and x_5, respectively, are to be inserted into the second and third constraints, and a surplus variable x_6 is needed in the fourth constraint. Insert artificial variables a_1 and a_2 into the first and fourth constraints, and the following system of equations and initial tableau results.

$$\begin{array}{rrrrrrr} x_1 & +x_2 & +x_3 & & +a_1 & = & 20000 \\ x_1 & -x_2 & & +x_4 & & = & 0 \\ & x_2 & -4\,x_3 & & +x_5 & = & 0 \\ & & x_3 & & -x_6 & +a_2 & = & 5000 \\ & & & & -a_1 & -a_2 & = & g \end{array}$$

```
phase1tableau = {{1, 1, 1, 0, 0, 0, 1, 0, 20000},
    {1, -1, 0, 1, 0, 0, 0, 0, 0},
    {0, 1, -4, 0, 1, 0, 0, 0, 0},
    {0, 0, 1, 0, 0, -1, 0, 1, 5000},
    {0, 0, 0, 0, 0, 0, -1, -1, 0}};
MatrixForm[phase1tableau]
```

$$\begin{pmatrix} 1 & 1 & 1 & 0 & 0 & 0 & 1 & 0 & 20000 \\ 1 & -1 & 0 & 1 & 0 & 0 & 0 & 0 & 0 \\ 0 & 1 & -4 & 0 & 1 & 0 & 0 & 0 & 0 \\ 0 & 0 & 1 & 0 & 0 & -1 & 0 & 1 & 5000 \\ 0 & 0 & 0 & 0 & 0 & 0 & -1 & -1 & 0 \end{pmatrix}$$

Note that the basic variables a_1, a_2 have non-zero coefficients in the bottom row. These may be removed by pivoting about the 1's in the artificial variable columns (by letting the a_i replace themselves in the SimplexOne-Step command), the result of which is the tableau that follows.

```
varlist = {x₁, x₂, x₃, x₄, x₅, x₆, a₁, a₂};
{newtableau, newbasiclist} =
   SimplexOneStep[phase1tableau,
      varlist, a₁, a₁, {a₁, x₄, x₅, a₂}];
{newtableau, newbasiclist} = SimplexOneStep[
   newtableau, varlist, a₂, a₂, newbasiclist];
```

	x_1	x_2	x_3	x_4	x_5	x_6	a_1	a_2	
a_1	1	1	1	0	0	0	1	0	20000
x_4	1	-1	0	1	0	0	0	0	0
x_5	0	1	-4	0	1	0	0	0	0
a_2	0	0	1	0	0	-1	0	1	5000
obj	1	1	1	0	0	0	0	-1	20000

	x_1	x_2	x_3	x_4	x_5	x_6	a_1	a_2	
a_1	1	1	1	0	0	0	1	0	20000
x_4	1	-1	0	1	0	0	0	0	0
x_5	0	1	-4	0	1	0	0	0	0
a_2	0	0	1	0	0	-1	0	1	5000
obj	1	1	2	0	0	-1	0	0	25000

Introduce x_3 to replace a_2 in the basic list, then x_2 to replace a_1, and the tableaux below result.

```
{newtableau, newbasiclist} = SimplexOneStep[
    newtableau, varlist, x₃, a₂, newbasiclist];
```

	x₁	x₂	x₃	x₄	x₅	x₆	a₁	a₂	
a₁	1	1	0	0	0	1	1	-1	15000
x₄	1	-1	0	1	0	0	0	0	0
x₅	0	1	0	0	1	-4	0	4	20000
x₃	0	0	1	0	0	-1	0	1	5000
obj	1	1	0	0	0	1	0	-2	15000

```
{newtableau, newbasiclist} = SimplexOneStep[
    newtableau, varlist, x₂, a₁, newbasiclist];
```

	x₁	x₂	x₃	x₄	x₅	x₆	a₁	a₂	
x₂	1	1	0	0	0	1	1	-1	15000
x₄	2	0	0	1	0	1	1	-1	15000
x₅	-1	0	0	0	1	-5	-1	5	5000
x₃	0	0	1	0	0	-1	0	1	5000
obj	0	0	0	0	0	0	-1	-1	0

This is the final tableau of Phase 1. Note that the artificial variables are non-basic, and the basic variables have positive, feasible values. (Check that the current solution satisfies the original problem constraints.)

Delete the artificial variable columns from this final Phase 1 tableau, and restore the original objective function f to the bottom row. This gives the initial tableau for Phase 2. Since x_2 and x_3 are basic, we must first pivot away the coefficients in the bottom row. After this is done, the tableau below results, and Phase 2, the simplex algorithm, is ready to be carried out.

```
phase2tableau = {{1, 1, 0, 0, 0, 1, 15000},
    {2, 0, 0, 1, 0, 1, 15000}, {-1, 0, 0, 0,
      1, -5, 5000}, {0, 0, 1, 0, 0, -1, 5000},
    {1.08, 1.07, 1.05, 0, 0, 0, 0}};
MatrixForm[phase2tableau]
varlist2 = {x₁, x₂, x₃, x₄, x₅, x₆};
{newtableau, newbasiclist} =
  SimplexOneStep[phase2tableau,
    varlist2, x₂, x₂, {x₂, x₄, x₅, x₃}];
{newtableau, newbasiclist} = SimplexOneStep[
    newtableau, varlist2, x₃, x₃, newbasiclist];
```

$$\begin{pmatrix} 1 & 1 & 0 & 0 & 0 & 1 & 15000 \\ 2 & 0 & 0 & 1 & 0 & 1 & 15000 \\ -1 & 0 & 0 & 0 & 1 & -5 & 5000 \\ 0 & 0 & 1 & 0 & 0 & -1 & 5000 \\ 1.08 & 1.07 & 1.05 & 0 & 0 & 0 & 0 \end{pmatrix}$$

	x_1	x_2	x_3	x_4	x_5	x_6	
x_2	1	1	0	0	0	1	15000
x_4	2	0	0	1	0	1	15000
x_5	-1	0	0	0	1	-5	5000
x_3	0	0	1	0	0	-1	5000
obj	0.01	0.	1.05	0	0	-1.07	-16050.

	x_1	x_2	x_3	x_4	x_5	x_6	
x_2	1	1	0	0	0	1	15000
x_4	2	0	0	1	0	1	15000
x_5	-1	0	0	0	1	-5	5000
x_3	0	0	1	0	0	-1	5000
obj	0.01	0.	0.	0	0	-0.02	-21300.

A basic feasible solution with $x_1 = 0$, $x_2 = 15000$, and $x_3 = 5000$ has been reached. We next let x_1 enter the basic list, replacing x_4 by the minimum ratio rule:

```
{newtableau, newbasiclist} = SimplexOneStep[
    newtableau, varlist2, x₁, x₄, newbasiclist];
```

	x_1	x_2	x_3	x_4	x_5	x_6	
x_2	0	1	0	$-\frac{1}{2}$	0	$\frac{1}{2}$	7500
x_1	1	0	0	$\frac{1}{2}$	0	$\frac{1}{2}$	7500
x_5	0	0	0	$\frac{1}{2}$	1	$-\frac{9}{2}$	12500
x_3	0	0	1	0	0	-1	5000
obj	0.	0.	0.	-0.005	0	-0.025	-21375.

This is the last tableau, because all objective row entries are non-positive. Therefore, the amounts $x_1 = 7500$, $x_2 = 7500$, and $x_3 = 5000$ are the optimal investments in stocks, bonds, and savings. ∎

Exercises 3.1

1. (*Mathematica*) Solve the non-standard problem:

max: $f = x_1 + x_2$

$$\text{subject to:} \quad \begin{array}{rcrcl} & & x_2 & \geq & 1 \\ x_1 & - & x_2 & \geq & -2 \\ x_1 & + & x_2 & \leq & 6 \\ & & x_1, \ x_2 & \geq & 0 \end{array}$$

2. (*Mathematica*) What happens in Phase 1 of the investment problem of Example 2 if the first entering basic variable is chosen to be x_1 instead of x_3? Do you get the same basic feasible solution at the start of Phase 2?

3. (*Mathematica*) Find the optimal solution of:

max $f = x_1 + x_2 + x_3$

$$\text{subject to:} \quad \begin{array}{rcrcrcl} x_1 & & & & & \geq & 3 \\ & & x_2 & + & x_3 & \leq & 6 \\ x_1 & - & x_2 & + & x_3 & = & 5 \\ & & x_1, \ x_2, \ x_3 & & & \geq & 0 \end{array}$$

4. (*Mathematica*) A bakery employs a skilled pastry chef, who should work at least 6 hours per day. An oven suitable for the use of the chef is available 8 hours per day. Three types of pastry are to be made; each batch requires labor time (in hours) by the chef and time in the oven as below:

	Pastry type		
	1	2	3
Chef time :	1	1	2
Oven time :	2	1	2

Suppose that the profit per batch is $10 for type 1, $5 for type 2, and $10 for type 3 pastry. How many batches of each pastry type should be made to maximize profit?

5. Prove Theorem 1.

6. (*Mathematica*) Find the minimum value of $2x_1 - x_2$ subject to

$$
\begin{aligned}
2x_1 &+ x_2 &\geq 4 \\
x_1 &+ x_2 &\leq 5 \\
x_1,\ x_2 &\ &\geq 0
\end{aligned}
$$

7.   Solve the following non-standard problem without recourse to the simplex algorithm.

maximize: $f = 2x_1 + 3x_2$

subject to:
$$
\begin{aligned}
x_1 &- x_2 &\geq 1 \\
2x_1 &+ x_2 &\leq 6 \\
x_1 &+ 2x_2 &= -1 \\
x_1,\ x_2 &\ &\geq 0
\end{aligned}
$$

8.   Show the converse of Theorem 2(b); i.e., show that if problem (LP1) is infeasible, then the optimum value of problem (LP2) is strictly less than zero.

9. (*Mathematica*) Express the following problem in non-standard form with all variables constrained to be non-negative. Then solve the problem by the Phase 1–Phase 2 approach. Sketch the feasible region.

maximize: $f = -x_1 + 2x_2$

subject to:
$$
\begin{aligned}
x_1 &- x_2 &\geq 1 \\
&\ x_2 &\leq 5
\end{aligned}
$$

x_1 unconstrained, $x_2 \geq -2$

10.   (*Mathematica*) A woman beginning a small business will borrow $10,000.   There are three possible lenders; one is an in-town bank who charges an effective annual interest rate of 10%, the second is a savings and loan whose interest rate is 8%, and the third is a major out-of-town bank, whose interest rate is also 8%. Because she wishes to establish a significant credit history at the in-town bank where she will do most of her banking, she will borrow at least $5000 from this institution. Of the remaining money, she will borrow at least as much from the out-of-town bank as from the savings and loan. How much should she borrow from each institution to minimize the yearly interest she pays?

11. There is an alternative method for solving problems with mixed inequality constraints, which can result in computational savings, called the "Big M" method. Instead of introducing an artificial variable into every constraint as

Phase 1 does, introduce it only into the "≥" constraints, i.e., those with surplus variables. Then maximize the original objective minus a large but unspecified number M times the sum of the artificial variables.

(a) Solve Exercise 1 by the Big M method.

(b) Argue that if the original problem has an optimal solution, then at the end of the Big M procedure all artificial variables will be zero and the final system will represent the solution. (Which variables should be basic in the initial system?)

(c) Examine what happens to this procedure when it is applied to the following infeasible problem:

maximize: $f = x_1 + x_2$

$$
\begin{array}{rccccc}
 & 2x_1 & + & x_2 & \leq & 4 \\
\text{subject to:} & x_1 & + & x_2 & \geq & 5 \\
 & & x_1, & x_2 & \geq & 0
\end{array}
$$

12. (*Mathematica*) A maker of bird seed will use three ingredients, labeled A, B, and C, to form boxes of exactly 100 grams of seed. It has been determined that the profit per gram of A is 5, and the profits per gram of B and C are 4 each. It is desired to achieve a threshhold value of at least 260 units of protein in the mixture, while limiting the fat content to no more than 80 units. Suppose that each gram of ingredient A has 2 units of protein and 1 unit of fat, each gram of B has 3 units of protein but no fat, and each gram of C has 4 units of protein and 1 unit of fat. Formulate the problem as a non-standard linear programming problem, and solve it by: (a) the Phase 1–Phase 2 approach, and (b) the Big M method (see Exercise 11).

13. Formulate as a non-standard linear program, but do not solve, the maximal flow problem of Example 2 of Section 1.5.

3.2 Transportation Problem

We now introduce a class of problems known as *transportation problems* and a streamlined algorithm to solve them. Suppose that there are m sources of supply and that supplies must be transported to n destinations. For simplicity, we will assume that the total available supply exactly equals the total demand by these destinations. The ith source has a quantity s_i of supplies available, and the jth destination needs a quantity d_j. There is a transportation cost per item of c_{ij} for a move from source i to destination j. The problem may be depicted by a bipartite, weighted, directed graph as in Figure 3.2, in which vertices 1, 2, 3 are sources and vertices 4, 5, 6 are destinations. We are to determine how many items must be sent from each source to each destination to meet the demands at minimum total cost.

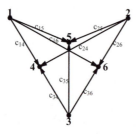

Figure 3.2 – A transportation problem

Denote by x_{ij} the number of items sent from source i to destination j. Then $c_{ij} x_{ij}$ is the cost of this particular shipment. Also, notice that

$$\sum_{j=1}^{n} x_{ij}$$

is the total amount of supplies shipped out by source i, and

$$\sum_{i=1}^{m} x_{ij}$$

is the total amount received by destination j. It follows that the transportation problem can be formulated as a linear program by:

$$\text{minimize:} \quad \sum_{i=1}^{m} \sum_{j=1}^{n} c_{ij} x_{ij} \tag{1}$$

$$\text{subject to:} \quad
\begin{aligned}
\sum_{j=1}^{n} x_{ij} &= s_i \quad \text{for each } i = 1, 2, \dots, m \\[2mm]
\sum_{i=1}^{m} x_{ij} &= d_j \quad \text{for each } j = 1, 2, \dots, n \\[2mm]
\text{all } x_{ij} &\geq 0
\end{aligned} \tag{2}$$

We observe that there are mn variables x_{ij}, m supply constraints and n demand constraints, comprising a total of $m + n$ constraints. However, since the total demand Σd_j equals the total supply Σs_i, the sum of the supply constraints in (2) equals the sum of the demand constraints, which gives the system a dependency. Because of this, one constraint is superfluous, which implies that the number of basic variables in the modified version of the simplex algorithm to be discussed will be $m + n - 1$, rather than $m + n$.

The constraints of (2) are already in standard equality form, and the origin is not feasible. Recall that we may solve the minimum problem by solving instead:

$$\text{maximize: } -\sum_{i=1}^{m} \sum_{j=1}^{n} c_{ij} x_{ij} \tag{3}$$

The optimal value of the minimum cost transportation problem will be the negative of the optimal value of (3). So, we have a non-standard problem of the type that could be treated by the Phase 1 algorithm of Section 1. However, that algorithm, whose purpose is to find a basic feasible solution, requires the introduction of many new artificial variables. The result is that an already large problem becomes even larger. Because of the special form of the constraints, it should be possible to replace Phase 1 by another routine that does not increase the size of the problem. Several strategies have been used; here we list one that is not necessarily the most efficient, but is easy to understand and is consistent in style with our earlier discussion of the simplex algorithm. For alternative treatments, you may refer to Rao [49], or Walker [59].

To gain an appreciation of how special the constraints are, let us write in full the system (2).

$$
\begin{array}{llll}
x_{11}+x_{12}+\cdots+x_{1n} & & & =s_1 \\
& x_{21}+x_{22}+\cdots+x_{2n} & & =s_2 \\
& & \cdot & \cdot \\
& & \cdot & \cdot \\
& & x_{m1}+x_{m2}+\cdots+x_{mn}=s_m
\end{array}
$$

$$
\begin{array}{llll}
x_{11} & + x_{21}+ & \cdots & +x_{m1} & =d_1 \\
x_{12} & +x_{22}+ & \cdots & + x_{m2} & =d_2 \\
\cdot & \cdot & & \cdot & \cdot \\
x_{1n} & +x_{2n}+ & \cdots & +x_{mn}=d_n
\end{array}
$$

Figure 3.3 – The system of constraints for the transportation problem

All coefficients are 1 initially, and each variable appears exactly once in a supply constraint and exactly once in a demand constraint. If, as in the simplex algorithm, we imagine choosing a variable to be basic, then there will only be two equations to be tested for the binding constraint; and once the proper equation is located, there will only be one constraint equation, and the objective row, into which to substitute the expression for the new basic variable. We will see that this property remains true, not only in the first step, but throughout the transportation algorithm.

Activity 1 – Is there an easy way of knowing which of the two constraint equations in which an entering basic variable appears is the binding one?

The strategy will be to identify one basic variable in a step. Any other variables in the equation in which it becomes basic are declared to be non-basic, not only for the next step but throughout the remainder of this phase of the algorithm. The choice of which variable is to become basic is made from among all variables not already declared to be either basic or non-basic. To attempt to steer the algorithm toward a good basic feasible solution, we select at each step a new basic variable from among undeclared variables that has the smallest current cost coefficient (equivalently, the largest coefficient in the objective row of the maximum problem (3)). We stop when there are $m + n - 1$ basic variables. The system of constraints will then represent a basic feasible solution, and Phase 2, the ordinary simplex algorithm, can be brought in to finish the problem.

The example below illustrates the procedure. We first do a problem longhand, and then later we will show how to use a command similar to the SimplexOneStep command to get *Mathematica* to carry out the computations.

EXAMPLE 1. An army commander must send tanks from two bases to three battle positions. The supply and demand requirements are given in the table below; also listed are the transportation times of moving a tank from each base to each battle position. How should the tanks be assigned so as to minimize the total transportation time?

		Battle Positions			
		1	2	3	Tanks Available
	1	5	10	12	$s_1 = 20$
Bases	2	6	8	8	$s_2 = 20$
	Required # Tanks	$d_1 = 8$	$d_2 = 12$	$d_3 = 20$	40

Let x_{ij} be the number of tanks to be sent from base i to position j. Then, for instance, the fact that base 1 has 20 tanks says that $x_{11} + x_{12} + x_{13} = 20$. The base 2 supply requirement, and the three battle position demand requirements may be obtained from the table similarly. Adjoining the objective of maximizing negative cost, we can write the initial transportation simplex system as follows.

$$
\begin{aligned}
0 &= 20 - x_{11} - x_{12} - x_{13} \\
0 &= 20 \qquad\qquad\qquad\qquad - x_{21} - x_{22} - x_{23}
\end{aligned}
$$

$$
\begin{aligned}
0 &= 8 - x_{11} \qquad\qquad - x_{21} \\
0 &= 12 \qquad - x_{12} \qquad\qquad - x_{22} \\
0 &= 20 \qquad\qquad - x_{13} \qquad\qquad\qquad - x_{23}
\end{aligned} \tag{4}
$$

$$
f = \quad -5 x_{11} - 10 x_{12} - 12 x_{13} - 6 x_{21} - 8 x_{22} - 8 x_{23}
$$

The form given above, in which 0's appear on the left, is used as a reminder that as yet no basic variables have been declared. The maximum coefficient in the objective row belongs to x_{11}. Since $8 < 20$, it is the first demand constraint that is binding. There, $x_{11} = 8 - x_{21}$. Declare x_{11} to be basic in row 3, declare x_{21} to be non-basic for the rest of transportation phase 1, and substitute the expression for x_{11} into the first supply constraint and the bottom row to get:

$$
\begin{aligned}
0 &= 12 - x_{12} - x_{13} + x_{21} \\
0 &= 20 \qquad\qquad\qquad - x_{21} - x_{22} - x_{23}
\end{aligned}
$$

$$
\begin{aligned}
x_{11} &= 8 \qquad\qquad\qquad - x_{21} \\
0 &= 12 - x_{12} \qquad\qquad - x_{22} \\
0 &= 20 \qquad\qquad - x_{13} \qquad\qquad\qquad - x_{23}
\end{aligned} \tag{5}
$$

$$
f = -40 - 10 x_{12} - 12 x_{13} - x_{21} - 8 x_{22} - 8 x_{23}
$$

$$
\text{basic: } x_{11} \qquad\qquad \text{non-basic: } x_{21}
$$

Among the undeclared variables, the largest objective coefficient belongs to both x_{22} and x_{23}. If we declare x_{22} basic, then the second demand constraint is binding, and x_{12} is declared non-basic. Substituting the expression for x_{22} from row 4 into the second row and the bottom row, we obtain:

$$
\begin{aligned}
0 &= 12 - x_{12} - x_{13} + x_{21} \\
0 &= 8 + x_{12} \qquad\qquad - x_{21} - x_{23}
\end{aligned}
$$

$$
\begin{aligned}
x_{11} &= 8 \qquad\qquad\qquad - x_{21} \\
x_{22} &= 12 - x_{12} \\
0 &= 20 \qquad\qquad - x_{13} \qquad\qquad - x_{23}
\end{aligned} \tag{6}
$$

$$
f = -136 - 2 x_{12} - 12 x_{13} - x_{21} - 8 x_{23}
$$

$$
\text{basic: } x_{11}, x_{22} \qquad\qquad \text{non-basic: } x_{21}, x_{12}
$$

Variable x_{23} is the next to be declared basic. The second supply constraint (row 2) is binding, and no new non-basic variables are declared. The next system is:

$$
\begin{array}{rrlll}
0 = & 12 & - \ x_{12} & - \ x_{13} & + \ x_{21} \\
x_{23} = & 8 & + \ x_{12} & & - \ x_{21} \\
\hline
x_{11} = & 8 & & & - \ x_{21} \\
x_{22} = & 12 & - \ x_{12} & & \\
0 = & 12 & - \ x_{12} & - \ x_{13} & + \ x_{21} \\
\hline
\multicolumn{5}{c}{f = -200 - 10\,x_{12} - 12\,x_{13} + 7\,x_{21}}
\end{array}
\tag{7}
$$

$$\text{basic: } x_{11}, x_{22}, x_{23} \qquad \text{non-basic: } x_{21}, x_{12}$$

Now we clearly see the degeneracy mentioned earlier. The first supply and third demand equations are the same. Delete the top row and declare x_{13}, the only remaining choice, to be basic in the third demand equation.

$$
\begin{array}{rrll}
x_{23} = & 8 & + \ x_{12} & - \ x_{21} \\
\hline
x_{11} = & 8 & & - \ x_{21} \\
x_{22} = & 12 & - \ x_{12} & \\
x_{13} = & 12 & - \ x_{12} & + \ x_{21} \\
\hline
\multicolumn{4}{c}{f = -344 + 2\,x_{12} - 5\,x_{21}}
\end{array}
\tag{8}
$$

$$\text{basic: } x_{11}, x_{22}, x_{23}, x_{13} \qquad \text{non-basic: } x_{21}, x_{12}$$

$$x_{11} = 8, \ x_{12} = 0, \ x_{13} = 12, \ x_{21} = 0, \ x_{22} = 12, \ x_{23} = 8, \ f = -344 \tag{9}$$

The revised Phase 1 for the transportation problem is now complete. The solution represented by (9) is feasible, and the ordinary simplex method leads us to the optimal solution in one more step by introducing x_{12} to replace x_{13} in the basic list (note the degeneracy):

$$
\begin{array}{rrll}
x_{23} = & 20 & - \ x_{13} & - \ x_{21} \\
\hline
x_{11} = & 8 & & - \ x_{21} \\
x_{22} = & 0 & + \ x_{13} & - \ x_{21} \\
x_{12} = & 12 & - \ x_{13} & + \ x_{21} \\
\hline
\multicolumn{4}{c}{f = -320 - 2\,x_{13} - 3\,x_{21}}
\end{array}
\tag{10}
$$

$$x_{11} = 8, \ x_{12} = 12, \ x_{13} = 0, \ x_{21} = 0, \ x_{22} = 0, \ x_{23} = 20, \ f = -320 \tag{11}$$

The solution means that we split the 20 tanks of base 1 among positions 1 and 2, and position 3 is supplied entirely by base 2 tanks. This solution is degenerate, since the basic variable x_{22} has the value zero. ∎

Activity 2 – Find all alternative optimal solutions, if any, in Example 1.

The algorithm below outlines the general procedure illustrated by the example. It requires the cost coefficients $-c_{ij}$ and the supply and demand constants s_i and d_j.

TRANSPORTATION PHASE 1 ALGORITHM
1. Initialize all x_{ij} as undeclared, and all supply and demand equations as unused.
2. Do a–b until $n + m - 1$ basic variables have been declared.
   a. Find an x_{ij} from the undeclared list whose corresponding coefficient in the objective row is maximal. Declare x_{ij} to be basic.
   b. Of the two equations, one supply and one demand, in which x_{ij} appears, select the binding equation, and
   i. Solve in this equation for x_{ij}.
   ii. Substitute into the other constraint equation in which x_{ij} appears, and into the objective equation.
   iii. For each variable that is currently undeclared in the equation in which x_{ij} became basic, declare it to be non-basic.
3. Delete the unused equation and return the others for use by the ordinary simplex algorithm.

Some other questions regarding the correctness of the algorithm arise, which we now address informally. Note that we have chosen to discard an equation at the end of the algorithm. Since there are only $n + m - 1$ independent constraints, discarding an equation produces no change in the feasible region. In Exercise 3 you are asked to show that, as step 2b requires, if a variable is currently undeclared, then it appears with its original coefficient in exactly one unused supply and exactly one unused demand equation.

How do we know that the algorithm will succeed at producing $n + m - 1$ basic variables? Refer to the constraints in Figure 3.3. If a variable is declared basic in a supply constraint, then since each unused demand equation has at most one variable in common with this supply equation, each such demand equation can lose at most one undeclared variable. Similarly, if a variable becomes basic in a demand equation, then this demand equation can have at most one variable in common with each unused supply constraint, so that each of those loses at most 1 undeclared variable. Separate supply constraints and demand constraints have no undeclared variables in common, so the declaration of a variable in a supply constraint does not change the count of undeclared variables in other supply constraints, and similarly for demand constraints. Consequently, a sequence of k_1 choices of basic variables in supply constraints and k_2 choices of basic variables in demand constraints results in at least $n - k_2$ undeclared variables per unused supply constraint in the $m - k_1$ unused supply constraints, and at least $m - k_1$ undeclared variables per unused demand constraint in the $n - k_2$ unused demand constraints. Each undeclared variable appears once in each group, hence (correcting for double counting) we have at least:

$$((n - k_2)(m - k_1) + (m - k_1)(n - k_2))/2 = (m - k_1)(n - k_2)$$

undeclared variables after this sequence of moves. As long as we do not exhaust all equations in either the supply or the demand group, then we know that this number of undeclared variables is at least 1, so that there will be a variable to declare as basic. This points up that perhaps we should add a check to the basic algorithm that once a group, either supply or demand, of equations comes within one equation of exhaustion, we should discard the last equation (which is dependent on the others anyway), and just finish by declaring exactly one basic variable in each remaining unused equation in the other group, for a total of $n + m - 1$ basic variables. (See, however, Exercise 11.)

Activity 3 – Explain how we know that the set of variable values at the end of Phase 1 forms a basic feasible solution.

```
Needs["KnoxOR`LinearProgramming`"];
```

The LinearProgramming package has a command that performs a step of the transportation algorithm in tableau form.

```
(** TransportationOneStep[tableau,
    varlist,entering,row,basiclist] **)
```

Like SimplexOneStep, it takes the current simplex tableau for the constraint system with the objective adjoined, the list of variables, the variable that is now entering as a declared basic variable, the number of the row in which it is to become basic, and the list of current basic variables. Note that the fourth argument differs from that of SimplexOneStep in that there is no departing variable for that row. Also, the list of current basic variables will have blanks in any row for which a basic variable has not yet been declared.

Also like SimplexOneStep, it returns the pair{newtableau, newbasiclist}for use as arguments in the next step.

EXAMPLE 2. Let us illustrate the tableau implementation of the transportation algorithm. Suppose that there are two beer distributors, owned by a common parent company, in a small city. Each week they supply four taverns with kegs of a limited-edition dark beer. The costs per keg to ship from each distributor to each tavern are in the table below. Also listed are the required number of kegs in a week for each tavern, and the supplies on hand at each distributor. Find a distribution scheme that minimizes the total shipping cost to the company, while fulfilling the needs of the taverns.

		\multicolumn{4}{c}{Taverns}	
		1	2	3	4	Kegs Available
Distributors	1	1	2	3	2	25
	2	2	3	1	2	22
Required # Kegs:		10	12	10	15	47

We let x_{ij} be the number of kegs shipped from distributor i to tavern j, for $i = 1, 2$; $j = 1, 2, 3, 4$. Then the following system of equations represents the constraints. The objective function f to be maximized is the negative of the sum of coefficients c_{ij} obtained from the table, times x_{ij}.

$$
\begin{array}{l}
x_{11} + x_{12} + x_{13} + x_{14} \hspace{4.5cm} = 25 \\
\hspace{2.3cm} x_{21} + x_{22} + x_{23} + x_{24} = 22 \\
\hline
x_{11} \hspace{2.7cm} + x_{21} \hspace{3.0cm} = 10 \\
\hspace{0.6cm} x_{12} \hspace{3.0cm} + x_{22} \hspace{1.9cm} = 12 \hspace{1cm} (12) \\
\hspace{1.4cm} x_{13} \hspace{3.3cm} + x_{23} \hspace{1.0cm} = 10 \\
\hspace{2.2cm} x_{14} \hspace{3.9cm} + x_{24} = 15 \\
\hline
-x_{11} - 2\,x_{12} - 3\,x_{13} - 2\,x_{14} - 2\,x_{21} - 3\,x_{22} - x_{23} - 2\,x_{24} = f - 0
\end{array}
$$

No variables are yet declared. This system gives rise to the initial tableau shown below.

```
tableau = {{1, 1, 1, 1, 0, 0, 0, 0, 25},
   {0, 0, 0, 0, 1, 1, 1, 1, 22},
   {1, 0, 0, 0, 1, 0, 0, 0, 10},
   {0, 1, 0, 0, 0, 1, 0, 0, 12}, {0, 0, 1, 0, 0,
      0, 1, 0, 10}, {0, 0, 0, 1, 0, 0, 0, 1, 15},
   {-1, -2, -3, -2, -2, -3, -1, -2, 0}};
MatrixForm[tableau]
```

$$
\begin{pmatrix}
1 & 1 & 1 & 1 & 0 & 0 & 0 & 0 & 25 \\
0 & 0 & 0 & 0 & 1 & 1 & 1 & 1 & 22 \\
1 & 0 & 0 & 0 & 1 & 0 & 0 & 0 & 10 \\
0 & 1 & 0 & 0 & 0 & 1 & 0 & 0 & 12 \\
0 & 0 & 1 & 0 & 0 & 0 & 1 & 0 & 10 \\
0 & 0 & 0 & 1 & 0 & 0 & 0 & 1 & 15 \\
-1 & -2 & -3 & -2 & -2 & -3 & -1 & -2 & 0
\end{pmatrix}
$$

The rule of thumb that says to select the largest entry in the objective row, which would correspond to the minimum cost coefficient, would say to let either x_{11} or x_{23} enter the basis. Choosing x_{11}, the first demand constraint, i.e., row 3, is binding. Now we set up and use TransportationOneStep.

```
varlist = {x₁₁, x₁₂, x₁₃, x₁₄, x₂₁, x₂₂, x₂₃, x₂₄};
basiclist = {"", "", "", "", "", ""};
{tableau, basiclist} = TransportationOneStep[
   tableau, varlist, x₁₁, 3, basiclist];
```

	x_{11}	x_{12}	x_{13}	x_{14}	x_{21}	x_{22}	x_{23}	x_{24}	
	0	1	1	1	-1	0	0	0	15
	0	0	0	0	1	1	1	1	22
x_{11}	1	0	0	0	1	0	0	0	10
	0	1	0	0	0	1	0	0	12
	0	0	1	0	0	0	1	0	10
	0	0	0	1	0	0	0	1	15
obj	0	-2	-3	-2	-1	-3	-1	-2	10

basic variables: x_{11} non-basic variables: x_{21}

Because the undeclared variable x_{21} was in the equation in which x_{11} was declared basic, it is placed into the non-basic list. Next to enter is x_{23}, and the binding constraint is the third demand constraint, in row 5. Since the undeclared variable x_{13} is in row 5, it is included into the non-basic list.

```
{tableau, basiclist} = TransportationOneStep[
    tableau, varlist, x₂₃, 5, basiclist];
```

	x_{11}	x_{12}	x_{13}	x_{14}	x_{21}	x_{22}	x_{23}	x_{24}	
	0	1	1	1	-1	0	0	0	15
	0	0	-1	0	1	1	0	1	12
x_{11}	1	0	0	0	1	0	0	0	10
	0	1	0	0	0	1	0	0	12
x_{23}	0	0	1	0	0	0	1	0	10
	0	0	0	1	0	0	0	1	15
obj	0	-2	-2	-2	-1	-3	0	-2	20

basic variables: x_{11}, x_{23} \qquad non-basic variables: x_{21}, x_{13}

Now we can choose any of $x_{12}, x_{14},$ or x_{24}. Using x_{24} for example, the second supply constraint (row 2) and the fourth demand constraint (row 6) are the competitors, and we see that row 2 is binding. Since x_{22} is an undeclared variable in that list, we must mark it as non-basic.

```
{tableau, basiclist} = TransportationOneStep[
    tableau, varlist, x₂₄, 2, basiclist];
```

	x_{11}	x_{12}	x_{13}	x_{14}	x_{21}	x_{22}	x_{23}	x_{24}	
	0	1	1	1	-1	0	0	0	15
x_{24}	0	0	-1	0	1	1	0	1	12
x_{11}	1	0	0	0	1	0	0	0	10
	0	1	0	0	0	1	0	0	12
x_{23}	0	0	1	0	0	0	1	0	10
	0	0	1	1	-1	-1	0	0	3
obj	0	-2	-4	-2	1	-1	0	0	44

basic variables: x_{11}, x_{23}, x_{24} \qquad non-basic variables: x_{21}, x_{13}, x_{22}

If we next let x_{12} enter, the second demand constraint (row 4) is binding and no new non-basic variables come in.

```
{tableau, basiclist} = TransportationOneStep[
    tableau, varlist, x₁₂, 4, basiclist];
```

	x_{11}	x_{12}	x_{13}	x_{14}	x_{21}	x_{22}	x_{23}	x_{24}	
	0	0	1	1	-1	-1	0	0	3
x_{24}	0	0	-1	0	1	1	0	1	12
x_{11}	1	0	0	0	1	0	0	0	10
x_{12}	0	1	0	0	0	1	0	0	12
x_{23}	0	0	1	0	0	0	1	0	10
	0	0	1	1	-1	-1	0	0	3
obj	0	0	-4	-2	1	1	0	0	68

basic variables: $x_{11}, x_{23}, x_{24}, x_{12}$
non-basic variables: x_{21}, x_{13}, x_{22}

To finish Phase 1, x_{14} enters the basis in either row 1 or row 6. We choose row 1.

```
{tableau, basiclist} = TransportationOneStep[
    tableau, varlist, x₁₄, 1, basiclist];
```

	x_{11}	x_{12}	x_{13}	x_{14}	x_{21}	x_{22}	x_{23}	x_{24}	
x_{14}	0	0	1	1	-1	-1	0	0	3
x_{24}	0	0	-1	0	1	1	0	1	12
x_{11}	1	0	0	0	1	0	0	0	10
x_{12}	0	1	0	0	0	1	0	0	12
x_{23}	0	0	1	0	0	0	1	0	10
	0	0	0	0	0	0	0	0	0
obj	0	0	-2	0	-1	-1	0	0	74

basic variables: $x_{11}, x_{23}, x_{24}, x_{12}, x_{14}$
non-basic variables: x_{21}, x_{13}, x_{22}

To pass to Phase 2, the ordinary simplex algorithm, we would delete the unneeded row of zeros and continue to pivot away positive entries in the objective row, but we see that in fact there are none, and so Phase 1 has already led us to the optimal solution. Because the objective row corresponds to the equation $f + 74 = -2x_{13} - x_{21} - x_{22}$ and these variables are non-basic, we observe that the maximal value of f is -74, hence the minimal cost is 74. The optimal values of the problem variables are:

$$x_{11} = 10, x_{12} = 12, x_{13} = 0, x_{14} = 3,$$
$$x_{21} = 0, x_{22} = 0, x_{23} = 10, x_{24} = 12$$

For your reference, to delete the unneeded row in *Mathematica* so as to begin using SimplexOneStep for phase 2 when you do need to do that, you can use the standard *Mathematica* command

```
Take[list, {m, n}]
```

which returns elements *m* through *n* of the given list. For the example above, we would keep rows 1–5 and the objective row to make a new tableau for phase 2 by the following:

```
phase2tableau = Join[
    Take[tableau, {1, 5}], Take[tableau, {7, 7}]];
MatrixForm[phase2tableau]
```

$$
\begin{pmatrix}
0 & 0 & 1 & 1 & -1 & -1 & 0 & 0 & 3 \\
0 & 0 & -1 & 0 & 1 & 1 & 0 & 1 & 12 \\
1 & 0 & 0 & 0 & 1 & 0 & 0 & 0 & 10 \\
0 & 1 & 0 & 0 & 0 & 1 & 0 & 0 & 12 \\
0 & 0 & 1 & 0 & 0 & 0 & 1 & 0 & 10 \\
0 & 0 & -2 & 0 & -1 & -1 & 0 & 0 & 74
\end{pmatrix}
$$

Activity 4 – Check in Example 2 that if you do not use the minimum cost rule to select entering basic variables, then after Phase 1 is executed there are still several steps of the ordinary simplex algorithm to execute. Try, for example, the sequence of entering basic variables $x_{13}, x_{22}, x_{14}, x_{21}, x_{24}$.

Exercises 3.2

1. (*Mathematica*) Solve the transportation problem with 3 sources and 3 destinations whose cost structure and supply and demand requirements are in the table below. (The table entries are costs per unit shipped.)

| | destination | | | |
source	1	2	3	available
1	10	8	10	100
2	12	15	20	200
3	20	10	20	100
required:	150	150	100	400

2. (*Mathematica*) A manufacturer of auto batteries has two plants, which are to supply four retailers. The plants have 1000 and 1500 batteries available,

respectively. The four retailers have ordered 800, 500, 400, and 800 batteries, respectively. The shipping costs, in cents per battery, from plant 1 to the four retailers are 25, 25, 10, and 15, and costs from plant 2 to the retailers are 30, 20, 10, and 20. Determine a supply strategy to minimize the total transportation costs.

3. Prove that, in reference to the Transportation Phase 1 algorithm, if a variable is currently undeclared in step 2a, then it appears with its original coefficient in exactly one unused supply and exactly one unused demand equation.

4. Prove that if the entries in any single row or column of the cost matrix (c_{ij}) of a transportation problem are all reduced by the same number, then the optimal solution does not change.

5. (*Mathematica*) Solve the transportation problem whose supply and demand requirements, and transportation costs are given in the table below.

source	destination 1	2	3	4	available
1	4	2	2	3	80
2	1	4	5	2	50
3	6	3	3	2	100
4	3	1	1	3	50
required:	60	100	80	40	280

6. Prove that under the assumptions of this section, Phase 1 of the Transportation Algorithm must result in an integer-valued feasible solution.

7. One alternative to the minimum cost selection rule for the transportation algorithm is the *Northwest Corner Rule*. In this approach, the chosen sequence of basic variables is simpler. Display the variables x_{ij} in an array as shown below:

$$
\begin{array}{cccc}
x_{11} & x_{12} & x_{13} & \cdots & x_{1n} \\
x_{21} & x_{22} & x_{23} & \cdots & x_{2n} \\
& & \vdots & & \\
x_{m1} & x_{m2} & x_{m3} & \cdots & x_{mn}
\end{array}
$$

First let x_{11} be basic (i.e., begin in the northwest corner of the array) and let all other variables in the binding constraint be non-basic. If the supply constraint was binding, then x_{1j} are declared non-basic for all $j = 1, \ldots, n$, and we may effectively delete the first row of the array. If the binding constraint was the demand constraint, then the first column may be deleted. Choose as the next entering basic variable the entry in the northwest corner of the reduced array. In the first case above, the next basic variable would be x_{21}, and in the second case it would be x_{12}. Continue in this manner until there are $m + n - 1$ basic variables.

(a) Why should you expect that in general this approach will not lead as quickly to an optimal solution as the minimum cost algorithm?

(b) (*Mathematica*) Redo Example 1 using the Northwest Corner Algorithm.

8. (*Mathematica*) Redo Example 2 using the Northwest Corner Algorithm (see Exercise 7).

9. Suppose that in Exercise 1, source 1 can only supply 90 items. Execute the Transportation Algorithm on the resulting problem, and explain what happens in the final system or tableau. (This time, do not discard an equation when it is the only one still unused in its group.)

10. Suppose that in Exercise 1, source 1 can now supply 110 items. As in Exercise 9, execute Phase 1 of the Transportation Algorithm and explain the result.

11. Consider a step in the Transportation Algorithm in which there remains exactly one supply constraint that has no basic variable corresponding to it, and there are two or more unused demand constraints. Show that it cannot be the case that when a new basic variable is selected, the final supply constraint is the binding constraint. (Hint: express the current constant in the unused supply equation in terms of the demand constants that have been subtracted from it up to this step, and argue by contradiction that this constant must exceed the demand constant in the demand constraint in which the entering basic variable appears.)

12. (*Mathematica*) Use the tableau version of the Transportation Phase 1 algorithm, and if necessary the Phase 2 simplex algorithm, to solve the following problem. A bakery has five trucks servicing the four supermarkets in a town. The trucks contain 10, 8, 7, 10, and 6 units of bread, respectively, and the supermarkets are demanding 12, 8, 14, and 7 units, respectively. The delivery costs per unit from each truck to each supermarket are shown in the table below. Devise an optimal delivery plan.

Supermarkets

	1	2	3	4
1	4	3	2	6
2	9	5	4	7
Trucks 3	5	3	5	2
4	3	5	4	8
5	3	5	6	4

13. We may view the optimal assignment problem of Chapter 1 as a transportation problem in the following way. Let a variable x_{ij} equal 1 if worker i is assigned to task j, and 0 otherwise. There is a cost c_{ij} of assigning worker i to task j.

(a) Formulate an objective function for the problem of minimization of total cost of assignment.

(b) Formulate constraints corresponding to the requirement that all workers have a distinct task.

(c) Solve the problem if the costs are as follows:
$$c_{11} = 6, \quad c_{12} = 4, \quad c_{13} = 3,$$
$$c_{21} = 6, \quad c_{22} = 5, \quad c_{23} = 8,$$
$$c_{31} = 8, \quad c_{32} = 2, \quad c_{33} = 4 .$$

14. Referring to the discussion of Exercise 13, solve Exercise 12 of Section 6 of Chapter 1 using the Transportation Algorithm.

3.3 Sensitivity Analysis

Discussion of the Problem

Up to this point in our study of linear programming it has been assumed that all parameters, including objective coefficients, constraint coefficients, and constraint constants, are perfectly known and unchanging. This might not be the case in practice, as indicated by another look at the winery problem of Chapter 2. For convenience, we display the problem and its final simplex system below.

maximize: $f(x_1, x_2, x_3) = 1.25\,x_1 + 1.50\,x_2 + 2.00\,x_3$

$$
\begin{array}{llllll}
& 2\,x_1 & & +x_3 & \leq & 200 & \text{(type 1 grapes)} \\
& & 2\,x_2 & +x_3 & \leq & 150 & \text{(type II grapes)} \\
\text{subject to:} & 2\,x_1 & +x_2 & +1.5\,x_3 & \leq & 90 & \text{(sugar)} \\
& 2\,x_1 & +x_2 & +2\,x_3 & \leq & 250 & \text{(labor time)} \\
& & & x_1,\ x_2,\ x_3 & \geq & 0 &
\end{array}
$$

$$
\begin{aligned}
x_4 &= 185 & & - \tfrac{1}{2}\,x_5 & + & x_6 \\
x_2 &= \tfrac{135}{2} + & x_1 & - \tfrac{3}{4}\,x_5 & + & \tfrac{1}{2}\,x_6 \\
x_3 &= 15 - & 2\,x_1 & + \tfrac{1}{2}\,x_5 & - & x_6 \\
x_7 &= \tfrac{305}{2} + & x_1 & - \tfrac{1}{4}\,x_5 & + & \tfrac{5}{4}\,x_6 \\
f &= \tfrac{524}{4} - & \tfrac{5}{4}\,x_1 & - \tfrac{1}{8}\,x_5 & - & \tfrac{5}{4}\,x_6
\end{aligned}
$$

$x_1 = 0,\ x_2 = 135/2,\ x_3 = 15,$
$x_4 = 185,\ x_5 = 0,\ x_6 = 0,\ x_7 = 305/2$

The standard maximum problem written in matrix notation is:

$$
\max:\ \mathbf{c}' \cdot \mathbf{x}, \quad \text{subject to } A\,\mathbf{x} \leq \mathbf{b},\ \mathbf{x} \geq \mathbf{0}
$$

Recall that the variables x_1, x_2, and x_3 represent production quantities in gallons of three types of wine: red, white, and rosé. The objective coefficient c_j is the profit per gallon of wine j, the constraint coefficient a_{ij} is the amount of resource i (type I grapes, type II grapes, sugar, or labor time) needed to make a gallon of wine j, and the constraint constant b_i is the available amount of resource i. In the optimal solution for these parameters, x_2 and x_3 are basic, and x_1 is non-basic, meaning that no red wine should be made.

It is reasonable to ask the question: by how much can the profit per unit on red wine be increased before it is no longer optimal to produce no red wine? Must we execute the simplex algorithm over and over again with different choices of c_1 until by very good luck we hit upon the critical profit at which red wine enters the basis? Another question involves changes in available resources. In the current optimal solution, the slack variable x_5 is non-basic, which means that there are no unused type II grapes. Suppose that the winery has an unexpectedly good harvest of these grapes, and therefore, there are more than the original 150 bushels available. A change in the constraint constant b_2 results. How does the optimal solution change? Yet another question pertains to changes in resource requirements. Suppose that a reduction in the amount of sugar in red wine is being contemplated.

This would alter the constraint coefficient a_{31}. How do the optimal production variables change? Might it become profitable for red wine to be manufactured?

All of these questions focus on what happens to the optimal solution when a parameter changes. This subject is referred to as *sensitivity* (or *post-optimality*) *analysis*. We will analyze the effect on the optimal solution of: (1) changing the objective function coefficients, (2) changing the constraint constants, and (3) changing the column of constraint coefficients for a non-basic variable. Our goal is to characterize the range of values of the parameter under study for which the current basic list remains optimal, and to recompute the optimal value of the objective function and the values of the basic variables, if indeed they change. This new computation can be done without repeating the simplex algorithm. We remark briefly at the end of the section on other forms of sensitivity analysis, and how to obtain a new optimal solution if the basic list becomes sub-optimal, or even infeasible. Our presentation follows that of Winston [61], and also draws somewhat from Hillier and Lieberman [31], and Rao [49].

Matrix-Geometric View of the Simplex Method

In order to have a concise way of expressing the final simplex tableau for a perturbed problem in terms of the original problem, it will be helpful to look once again at how a final tableau is obtained. We now adopt the point of view of matrix geometry. The initial tableau for the winery problem is displayed in Figure 3.4(a), and elementary row operations produce the final tableau in Figure 3.4(b). The initial and final tableaux for this problem, as well as any other standard maximum problem, can be written in block form as in Figure 3.5(a) and (b), respectively.

x_1	x_2	x_3		x_4	x_5	x_6	x_7			
2	0	1	|	1	0	0	0	|	200	x_4
0	2	1	|	0	1	0	0	|	150	x_5
2	1	3/2	|	0	0	1	0	|	90	x_6
2	1	2	|	0	0	0	1	|	250	x_7
5/4	3/2	2	|	0	0	0	0	|	$f - 0$	

(a)

x_1	x_2	x_3		x_4	x_5	x_6	x_7			
0	0	0	$\mid$	1	$1/2$	-1	0	$\mid$	185	x_4
-1	1	0	$\mid$	0	$3/4$	$-1/2$	0	$\mid$	$135/2$	x_2
2	0	1	$\mid$	0	$-1/2$	1	0	$\mid$	15	x_3
-1	0	0	$\mid$	0	$1/4$	$-3/2$	1	$\mid$	$305/2$	x_7
$-5/4$	0	0	$\mid$	0	$-1/8$	$-5/4$	0	$\mid$	$f - 525/4$	

(b)

Figure 3.4 – (a) Initial, and (b) final tableaux for winery problem

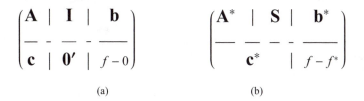

(a) (b)

Figure 3.5 – (a) Initial, and (b) final tableaux in block form

Below is a glossary of notation. Some of it is already familiar, and some is illustrated in Figure 3.5. The rest is necessary because we will need to express all relevant quantities in terms of the separate contributions of the basic variables and the non-basic variables for the final simplex system.

(a) $A = m \times n$ matrix of constraint coefficients for original problem in inequality form;

(b) $\mathbf{b} = m \times 1$ column vector of constraint constants for original problem;

(c) $I = m \times m$ identity matrix;

(d) $\mathbf{0}$ = vector of length m all of whose components are 0;

(e) $\mathbf{c} = 1 \times n$ row vector of objective function coefficients for original problem;

(f) f = objective function;

(g) $\mathbf{c}^* = 1 \times (n + m)$ row vector of objective function coefficients in final tableau;

(h) f^* = value of objective function in final tableau;

(i) $\mathbf{b}^* = m \times 1$ column vector of values of basic variables in final tableau;

(j) $A^* = m \times n$ matrix of constraint coefficients for objective variables in final tableau;

(k) $\mathbf{x} = (n + m) \times 1$ column vector of variables, including slacks, in (arbitrary) simplex system;

(l) $S = m \times m$ matrix of slack variable constraint coefficients in final tableau;

(m) β = list of basic variables in optimal simplex tableau, ordered according to the rows in which they are basic;

(n) v = list of non-basic variables in optimal simplex tableau, written in an arbitrary order;

(o) $B = m \times m$ submatrix of initial constraint coefficent matrix whose columns are the columns of the variables in β, in the order specified by β;

(p) $N = m \times n$ submatrix of initial constraint coefficent matrix whose columns are the columns of the variables in v, in the order specified by v;

(q) $\mathbf{c}_b = 1 \times m$ row vector of entries of $\mathbf{c}$ corresponding to β, in order;

(r) $\mathbf{c}_{nb} = 1 \times n$ row vector of entries of $\mathbf{c}$ corresponding to v, in order;

(s) $\mathbf{x}_b = m \times 1$ column vector of entries of $\mathbf{x}$ corresponding to β, in order;

(t) $\mathbf{x}_{nb} = n \times 1$ column vector of entries of $\mathbf{x}$ corresponding to v, in order.

The main idea here is to divide up the parts of the problem that have to do with the final list of basic variables from those that have to do with the non-basic variables. For example, in the winery problem whose initial and final tableaux are in Figure 3.4(a) and (b),

$$\beta = \{x_4, \, x_2, \, x_3, \, x_7\}, \quad v = \{x_1, \, x_5, \, x_6\},$$
$$\mathbf{c}_b = [\, 0, 3/2, 2, 0 \,], \quad \mathbf{c}_{nb} = [\, 5/4, \; 0, \; 0 \,] \tag{3}$$

> **Activity 1** – Identify all of the other quantities named in the glossary for the winery problem. Note particularly the submatrices B and N.

Recall that we wish to find ranges of values of the problem parameters in which the list β of basic variables does not change after perturbation of the parameters. The actual values of the basic variables, as well as the objective function and the objective coefficient vector $\mathbf{c}^*$ may change under the perturbation. It turns out that we can obtain closed-form expressions for these quantities in terms of the parameters of the original problem, and, perhaps surprisingly, in terms of the matrix S of slack variable constraint coefficients in the final tableau of the unperturbed problem.

Note that the standard maximum problem in equality form can be written, by separating the contributions of the basic and non-basic variables, as:

$$\max : \; f = \mathbf{c}_b \, \mathbf{x}_b + \mathbf{c}_{nb} \, \mathbf{x}_{nb}$$

$$\text{subject to} : \; B \, \mathbf{x}_b + N \, \mathbf{x}_{nb} = \mathbf{b}, \quad \mathbf{x} \geq 0 \tag{4}$$

(See Exercise 1, which asks you to check this for the winery problem.) When we use the subscripts b and nb, you should remember that we mean the basic and non-basic lists in the *optimal* tableau for the *unperturbed* problem. The main theorem is as follows.

THEOREM 1. (a) The matrix B, composed of columns of the initial system of constraints corresponding to the variables in β, and the matrix S, composed of slack variable coefficients in the final system of constraints, are inverses of one another.

(b) The values of the optimal basic variables are the components of the vector

$$\mathbf{b}^* = B^{-1} \cdot \mathbf{b} = S \cdot \mathbf{b}$$

in the order specified by β.

(c) The objective equation corresponding to the final row of the optimal tableau has the form

$$f = \mathbf{c}_b \cdot S \cdot \mathbf{b} + (\mathbf{c}_{nb} - \mathbf{c}_b \cdot S \cdot N) \cdot \mathbf{x}_{nb},$$

where N is the matrix composed of columns of the initial system of constraints corresponding to the variables in v. In particular, $\mathbf{c}_b \cdot S \cdot \mathbf{b}$ is the optimal objective value, and the vector coefficient $(\mathbf{c}_{nb} - \mathbf{c}_b \cdot S \cdot N)$ of $\mathbf{x}_{nb}$ in the second term on the right gives the non-basic variable coefficients in the order specified by v.

Proof. (a) Consider the basic variable submatrix B. Because of the way in which we have ordered the set β, the first member of β, corresponding to the first column of B, will be basic in the first row of the final tableau, the second member of β will be basic in the second row, etc. As we have seen many times, a column of a basic variable in the final tableau has an entry of 1 in the row in which it is basic, and 0 elsewhere. Thus, the elementary row operations performed in computing the final tableau have the effect of transforming B into I, the $m \times m$ identity matrix. Those same row operations transform the identity matrix, formed by the slack variable columns of the initial tableau, into S (see Figure 3.5). By the well-known procedure from linear algebra, S must be the inverse matrix of B.

(b) The system of constraints for problem (4) is equivalent to

$$\mathbf{x}_b + B^{-1} \cdot N \cdot \mathbf{x}_{nb} = B^{-1}\,\mathbf{b} \qquad\qquad (5)$$

But the values of the non-basic variables are zero in the final tableau, hence the second term in the sum on the left of the equation vanishes. Since $B^{-1} = S$ for the final tableau, the optimal values of the basic variables are contained in the column vector $S \cdot \mathbf{b}$.

   (c)   Substitution of expression (5) for the basic variables $\mathbf{x}_b$ into the objective function f in equation (4) yields

$$f = \mathbf{c}_b \cdot (B^{-1}\,\mathbf{b} - B^{-1} \cdot N \cdot \mathbf{x}_{nb}) + \mathbf{c}_{nb} \cdot \mathbf{x}_{nb}$$

The formula in (c) is a simple algebraic rearrangement of the above, using the fact that $B^{-1} = S$. ∎

EXAMPLE 1.   We can check the results of Theorem 1 on the winery example.   Refer to Figure 3.4.   Recall that $\beta = \{x_4, x_2, x_3, x_7\}$, and $v = \{x_1, x_5, x_6\}$, and that S is the slack variable part of the final tableau, and B is the basic variable part of the initial tableau in Figure 3.4.   We use *Mathematica* to check easily that $B\,S = I = S\,B$:

```
B = {{1, 0, 1, 0}, {0, 2, 1, 0},
     {0, 1, 3 / 2, 0}, {0, 1, 2, 1}};
S = {{1, 1 / 2, -1, 0}, {0, 3 / 4, -1 / 2, 0},
     {0, -1 / 2, 1, 0}, {0, 1 / 4, -3 / 2, 1}};
{MatrixForm[B.S], MatrixForm[S.B]}
```

$$\left\{\begin{pmatrix} 1 & 0 & 0 & 0 \\ 0 & 1 & 0 & 0 \\ 0 & 0 & 1 & 0 \\ 0 & 0 & 0 & 1 \end{pmatrix}, \begin{pmatrix} 1 & 0 & 0 & 0 \\ 0 & 1 & 0 & 0 \\ 0 & 0 & 1 & 0 \\ 0 & 0 & 0 & 1 \end{pmatrix}\right\}$$

Also, $\mathbf{b}^* = S\,\mathbf{b}$, as asserted by part (b) of the theorem.

```
b = {200, 150, 90, 250};
S.b
```

$$\left\{185,\ \frac{135}{2},\ 15,\ \frac{305}{2}\right\}$$

As in part (c), the optimal value of the objective function coincides with $\mathbf{c}_b \cdot S \cdot \mathbf{b}$.

```
cb = {0, 3 / 2, 2, 0};
cb.S.b
```

$$\frac{525}{4}$$

■

Activity 2 – Check, for the winery example, that the non-basic variable coefficients in the objective row of the final tableau are the components of the row vector $(\mathbf{c}_{nb} - \mathbf{c}_b \cdot S \cdot N)$.

Determining Sensitivity of Parameters

We are now in a position to find the range of values of a parameter for which the current list of basic variables remains feasible and optimal for a problem in which the parameter is perturbed. We will also be able to compute the new values of the basic variables and the objective function for the perturbed problem. There are a number of changes to a problem that could be antici-pated; here we discuss only three: (1) changing an objective function coeffi-cient; (2) changing a constant constraint bound; and (3) changing a con-straint coefficient for a variable that is non-basic in the optimal solution for the original problem.

Case 1: Perturbing c

Suppose that we change one (or more) of the entries of the coefficient vector **c** of the objective function of the original problem. This can be accom-plished by adding some increment vector $\Delta \mathbf{c}$ to **c**. Let $\Delta \mathbf{c}_b$ be the subvector of this increment vector that increments the basic variable coefficients, and let $\Delta \mathbf{c}_{nb}$ correspond similarly to the non-basic variable coefficients. The constraint constants **b** do not change. There is no change to the matrix B, which has to do only with constraint coefficients, nor to its inverse matrix S. Thus, part (b) of Theorem 1 allows us to conclude that:

The values of the basic variables do not change under perturbation
of the objective coefficient. (6)
In particular, the basic solution remains feasible.

Examination of the constant term in the linear equation for f listed in part (c) of Theorem 1 yields that:

The new value of the objective function under perturbation of the
objective coefficient is:

$$(\mathbf{c}_b + \Delta \, \mathbf{c}_b) \cdot S \cdot \mathbf{b} \tag{7}$$

This objective value is still the optimal value of the new f, as long as the optimality condition of the simplex algorithm is satisfied. By part (c) of the theorem again, this is the case if all non-basic variable coefficients in the new objective row are non-positive, i.e.:

The current basic solution is optimal iff:

$$\mathbf{c}_{nb} + \Delta \, \mathbf{c}_{nb} - (\mathbf{c}_b + \Delta \, \mathbf{c}_b) \cdot S \cdot N \le \mathbf{0} \tag{8}$$

Notice what the matrix-geometric approach has allowed us to accomplish. We have a check in (8) for optimality of the current solution, which requires only knowledge of the original objective coefficients, the increment vector, the slack variable portion S of the final tableau, and the non-basic variable portion N of the initial tableau. No further simplex computations are necessary. Also, by (7), we can compute the new optimal objective value knowing only the basic variable coefficients in the original objective function, the perturbation, the matrix S, and the initial constraint constant vector $\mathbf{b}$. The values of the basic variables themselves do not change under this kind of perturbation, and of course the values of the non–basic variables remain at zero.

EXAMPLE 2. Return to the winery problem. We analyze the effect of perturbing the profit coefficient of red wine by an amount Δ_1, white by Δ_2, and rosé by Δ_3. The vectors $\mathbf{c}_b$ and $\mathbf{c}_{nb}$ will become:

$$\mathbf{c}_b + \Delta \, \mathbf{c}_b = [\, 0, \, \tfrac{3}{2} + \Delta_2, \, 2 + \Delta_3, \, 0 \,] \,,$$
$$\mathbf{c}_{nb} + \Delta \, \mathbf{c}_{nb} = [\, \tfrac{5}{4} + \Delta_1, \, 0, \, 0 \,]$$

By part (c) of the theorem, the vector of new objective function coefficients in the final tableau will be

$$\mathbf{c}_{nb} + \Delta \, \mathbf{c}_{nb} - (\mathbf{c}_b + \Delta \, \mathbf{c}_b) \cdot S \cdot N \tag{9}$$

Using the definition of the slack variable submatrix S from before, and introducing the non-basic submatrix N, we get

```
cnb = {5 / 4, 0, 0}; Δcnb = {Δ₁, 0, 0};
Δcb = {0, Δ₂, Δ₃, 0};
Nmatrix =
   {{2, 0, 0}, {0, 1, 0}, {2, 0, 1}, {2, 0, 0}};
Expand[Simplify[cnb + Δcnb - (cb + Δcb).S.Nmatrix]]
```

$c_b = \{0, \frac{3}{2}, 2, 0\}$ ⟵

$$\left\{ -\frac{5}{4} + \Delta_1 + \Delta_2 - 2\,\Delta_3\,,\quad -\frac{1}{8} - \frac{3\,\Delta_2}{4} + \frac{\Delta_3}{2}\,,\quad -\frac{5}{4} + \frac{\Delta_2}{2} - \Delta_3 \right\}$$

Recall that the ordering is determined by v, hence these are the coefficients of the non-basic variables x_1, x_5, and x_6, respectively. The current solution remains optimal if and only if all of the three components of the vector above are less than or equal to zero. This system of inequalites forms a polyhedral region in three-dimensional space consisting of those points $(\Delta_1, \Delta_2, \Delta_3)$ whose corresponding perturbation does not change the optimal values of the variables. We will not attempt to sketch this set; rather, we will be content to determine the range of values of each Δ_i individually such that the current solution remains optimal, assuming the other profit coefficients are not changed. See Exercise 6 for the two at a time perturbations.

If Δ_1 is the only non-zero perturbation, then the *Mathematica* output shows that only the first component of the vector in (9) provides any restriction. To maintain optimality of the current solution, we must have

$$-\tfrac{5}{4} + \Delta_1 \le 0 \implies \Delta_1 \le \tfrac{5}{4}$$

Since we are not perturbing any basic variable coefficients, the optimal value of the objective function will still be 525/4. If more than 5/4 is added to the objective coefficient of x_1, then the first non-basic variable, namely x_1 itself, has a negative coefficient in the "final" tableau, and would be introduced into the basic solution. In other words, it would become profitable to make red wine.

If Δ_2 is the only non-zero perturbation, then all three components of the vector in (9) restrict Δ_2. The same basic feasible solution as in the unperturbed problem will occur if

$$-\tfrac{5}{4} + \Delta_2 \le 0$$
$$-\tfrac{1}{8} - \tfrac{3}{4}\Delta_2 \le 0$$
$$-\tfrac{5}{4} + \tfrac{1}{2}\Delta_2 \le 0$$

This system of inequalities is easily solved, to yield

$$-\tfrac{1}{6} \le \Delta_2 \le \tfrac{5}{4}$$

For this range of values of Δ_2, the solution $x_1 = 0$, $x_2 = 135/2$, $x_3 = 15$ is still optimal. By Theorem 1(c), the new optimal objective value is $(\mathbf{c}_b + \Delta\,\mathbf{c}_b)\cdot S\cdot \mathbf{b}$, which we compute as:

```
Δcb = {0, Δ2, 0, 0};
Expand[Simplify[(cb + Δcb).S.b]]
```

$$\frac{525}{4} + \frac{135\,\Delta_2}{2}$$

If Δ_3 is the only non-zero perturbation, then we obtain three restrictions on this perturbation in a similar way:

$$
\begin{aligned}
-\tfrac{5}{4} &- 2\Delta_3 &\le\ 0 \\
-\tfrac{1}{8} &+ \tfrac{1}{2}\Delta_3 &\le\ 0 \qquad &\Longrightarrow\quad -\tfrac{5}{8} \le \Delta_3 \le \tfrac{1}{4}\\
-\tfrac{5}{4} &- \Delta_3 &\le\ 0
\end{aligned}
$$

For this range of values of Δ_3, the old optimal solution remains optimal, and the new optimal objective value is

```
Δcb = {0, 0, Δ3, 0};
Expand[Simplify[(cb + Δcb).S.b]]
```

$$\frac{525}{4} + 15\,\Delta_3$$

∎

Activity 3 – Suppose that the profit coefficient of red wine was increased by 1/4. By how much can each of the individual coefficients on white wine and rosé be changed without changing the optimal solution?

Case 2: Perturbing b

Suppose next that we change one or more of the entries of the constraint constant vector $\mathbf{b}$, by adding some increment vector $\Delta\mathbf{b}$ to $\mathbf{b}$. By Theorem 1(c), we observe that:

Perturbation of **b** does not change the non-basic variable
coefficients in the objective row of the final system; thus the   (10)
current solution, if feasible, is still optimal.

We check for feasibility by inspecting the list of values of the basic variables. If all entries are non-negative, then the current solution is both feasible and optimal. Theorem 1(b) shows us how to verify this.

Feasibility of the solution under perturbation of **b** is equivalent to   (11)
$$S \cdot (\mathbf{b} + \Delta \mathbf{b}) \geq \mathbf{0}$$

If the old basis is indeed optimal, Theorem 1(c) enables us to compute the new optimal objective value.

The value of the objective function in the final simplex system   (12)
after perturbation of **b** is $f = \mathbf{c}_b \cdot S \cdot (\mathbf{b} + \Delta \mathbf{b})$.

As before, some simple matrix multiplications using data from the initial and final simplex tableaux for the unperturbed problem are sufficient to find the range of values of the perturbation vector $\Delta \mathbf{b}$ such that the old optimal basic variables are still basic in the new solution. It is also easy to recalculate new optimal values of the variables and the objective without resorting to another application of the simplex algorithm.

EXAMPLE 3. Now let us determine the effect of changes in the constraint constants on the solution of the winery problem. The problem is simply to find the range of values of the perturbation vector $\Delta \mathbf{b}$ such that the inequality in (11) holds. But by Theorem 1(b), this vector inequality is equivalent to

$$\mathbf{b}^* + S \cdot \Delta \mathbf{b} \geq \mathbf{0} \qquad (13)$$

The vector on the left side of inequality (13) gives the new values of the basic variables. Computing these in *Mathematica* gives

```
bstar = {185, 135/2, 15, 305/2}; Δb = {Δ₁, Δ₂, Δ₃, Δ₄};
Expand[Simplify[bstar + S.Δb]]
```

$$\left\{ 185 + \Delta_1 + \frac{\Delta_2}{2} - \Delta_3, \ \frac{135}{2} + \frac{3\,\Delta_2}{4} - \frac{\Delta_3}{2}, \right.$$
$$\left. 15 - \frac{\Delta_2}{2} + \Delta_3, \ \frac{305}{2} + \frac{\Delta_2}{4} - \frac{3\,\Delta_3}{2} + \Delta_4 \right\}$$

Treat the perturbation one parameter at a time, setting the others to zero, beginning with Δ_1. We see from the first element of the vector above that the current solution is still feasible and optimal if $185 + \Delta_1 \geq 0$, i.e., $\Delta_1 \geq -185$. In terms of the applied problem, the current solution remains feasible if no more than 185 bushels of type I grapes are subtracted from the original supply of 200. Note that since the optimal x_3 is 15 in the original problem, we need precisely 15 bushels of these grapes to make the rosé wine. Also, 185 is the value of the slack variable associated with the type I grape resource. The perturbation Δ_4 is also easy to analyze, since it appears only in the last element of $\mathbf{b}^* + S \cdot \Delta \mathbf{b}$. Feasibility of the current solution is preserved if $\Delta_4 \geq -305/2$. As long as no more than $305/2$ is subtracted from the available labor time (note that $305/2$ is the value of the slack variable associated to labor time in the optimal solution), the current list of basic variables is feasible. In both of these cases, the perturbation changes only the value of the basic variable in the row in which the perturbation appears. More precisely, the new value of basic variable x_4 under a perturbation of the constraint constant for type I grapes is $185 + \Delta_1$. Other basic variables, and all non-basic variables, do not change. Similarly, $305/2 + \Delta_4$ is the new value of basic variable x_7 under a perturbation of the labor time constraint constant, and no other variables change their value.

To see the effect of a change Δ_2 in the availability of type II grapes requires simultaneously setting all four elements of $\mathbf{b}^* + S \cdot \Delta \mathbf{b}$ greater than or equal to zero. You should verify that the range of values under which the current solution is still feasible is

$$-90 \leq \Delta_2 \leq 30$$

Similarly, the range of perturbations of the third constraint, involving sugar, is

$$-15 \leq \Delta_3 \leq 305/3$$

In both cases, since the perturbation terms Δ_i are contained in every element of $\mathbf{b}^* + S \cdot \Delta \mathbf{b}$, the values of all of the basic variables change. The new values are determined by substituting the values of the perturbations Δ_i into the formulas from *Mathematica* above.

We also have an expression for the new value of the objective function under the perturbation. As long as the perturbation terms indicate that the current list of basic variables is still feasible, the new optimal value is $f^* = \mathbf{c}_b \cdot S \cdot (\mathbf{b} + \Delta \mathbf{b})$, which we compute below.

```
Expand[Simplify[c_b.S.(b + Δb)]]
```

$$\frac{525}{4} + \frac{\Delta_2}{8} + \frac{5\,\Delta_3}{4}$$

The coefficients of the Δ_i's above give useful information; they are the rates of increase in profit per unit increase in the resources of type II grapes and sugar. Notice that they are the same as the negatives of the slack variable coefficients (i.e., the optimal dual variables) in the optimal simplex system (2) corresponding to these resources. In Section 4 of Chapter 2, we referred to these values as the *shadow prices* of the resources. We therefore have an interesting connection between duality theory and sensitivity theory, namely that the shadow price of a resource is the same as the rate of increase in the objective function per unit increase of the resource. (See Exercise 10 for a general result.) ∎

Activity 4 – What are the values of the basic variables if simultaneously there is an increase of 30 bushels of type II grapes and 10 lbs. of sugar? Is the old optimal solution still feasible and optimal?

Case 3: Perturbing constraint coefficient column of a non-basic variable

Finally, we will investigate changes in a single column of constraint coefficients in the original problem. We only consider the case in which the column variable is non-basic in the final simplex system. By Theorem 1(b) and (c), the only change induced in the final system is that the non-basic variable portion N of the constraint matrix will change in one column. Recall that S is the inverse of B, which is the basic variable portion of the constraint matrix; consequently, S will not change, nor are we changing the original objective coefficients or the constraint constants in this case. We observe immediately that

> A change in a non-basic variable constraint coefficient column leaves $\mathbf{b}^*$ untouched; consequently, the current solution is still feasible for the perturbed problem. $\qquad$ (14)

> The value of the objective function does not change under perturbation of a non-basic variable constraint coefficient column. $\qquad$ (15)

Consequently, it remains to check only whether the vector of coefficients of non-basic variables in the objective row of the final system is still entirely non-positive after the change to N. A perturbation of a column, say the jth, in N can be accomplished by adding a matrix ΔN to N, which is zero in every column except column j, and in that column has entries $\Delta_1, \Delta_2, \ldots, \Delta_m$.

The current solution remains optimal under a perturbation of a
non-basic variable constraint coefficient column iff   (16)
$$(\mathbf{c}_{nb} - \mathbf{c}_b \cdot S \cdot N) - \mathbf{c}_b \cdot S \cdot \Delta N \le \mathbf{0}$$

Note that the parenthesized term in the inequality of (16) is the old
vector of non-basic variable coefficients in the final objective row. It is
known to be non-positive, by the optimality of the solution to the unperturbed problem. From this is subtracted a vector which, since ΔN is non-
zero only in its jth column, could only be non-zero in its jth component.
Because of this, it is only necessary to check whether the new coefficient of
the jth non-basic variable has become positive, and if not, the current
solution is still optimal for the perturbed problem. This fact will come out
more clearly in the example below.

EXAMPLE 4. In the winery example, suppose we contemplate increasing
the amount of sugar required to make a gallon of red wine by an amount Δ.
A negative value of Δ is allowed, meaning that there is a reduction in the
required amount of sugar. We find the range of values of Δ such that it is
still not optimal for x_1 to enter the basis. Following (16), we compute the
vector $\mathbf{c}_b \cdot S \cdot \Delta N$, where ΔN has a column for each non-basic variable, but
whose only non-zero element is in row 3 (for sugar) and column 1 (for red
wine).

```
ΔN = {{0, 0, 0}, {0, 0, 0}, {Δ, 0, 0}, {0, 0, 0}};
MatrixForm[ΔN]
c_b.S.ΔN
```

$$\begin{pmatrix} 0 & 0 & 0 \\ 0 & 0 & 0 \\ \Delta & 0 & 0 \\ 0 & 0 & 0 \end{pmatrix}$$

$$\left\{ \frac{5\,\Delta}{4}, 0, 0 \right\}$$

Since the coefficient of x_1 in the objective row of the final tableau in Figure
3.4 is $-5/4$, we see that x_1 will remain non-basic iff

$$-\frac{5}{4} - \frac{5}{4}\Delta \le 0 \iff \Delta \ge -1$$

This means that if the sugar content of red wine is reduced by no more than
one unit per gallon, then it will still be optimal not to make any red wine. ∎

We have not covered every possible perturbation of a given linear programming problem. One noticeable absence is the change of a constraint coefficient column belonging to a variable that is basic in the optimal solution to the original problem. Theorem 1 still applies to this kind of perturbation, but the situation is somewhat more complicated than the cases that we have addressed. Referring to the formulas listed in parts (b) and (c) of the theorem, we see that since the perturbation changes B, it changes the inverse matrix S in a way that is not obvious. The values of the basic variables, the value of the objective function, and the coefficients of the non-basic variables in the objective equation all change as well; hence the new solution may not be feasible or optimal. Hillier and Lieberman [31] shed some light on this problem.

Perturbations of a different type involve the introduction into the problem of features that were not even present before. For example, suppose a new constraint was introduced. In the winery example, this could arise in the form of a new ingredient for the wines, which is available in limited amounts. One easy way to check for optimality of the current solution is to see whether it satisfies the new constraint. If the solution is feasible for the new constraint, then it is still optimal for the new problem, because the new problem maximizes the same objective over a smaller set of feasible points. A feasible solution with a strictly better objective value than the current solution could not exist. If the current optimal solution does not satisfy the constraint, Rao ([49], Sec. 4.5) suggests a Phase 1–Phase 2 approach to find the new optimal solution.

Another way of introducing an entirely new feature into the problem is to include a new variable. In the case of the winery problem, it might be that a new kind of wine is being considered. This wine has its own profit coefficient and requirements on resources. We might wish to find the range of values of these coefficients such that the new variable does not enter into the basic solution. A clever way of attacking this problem is to pretend that the variable was present as a non-basic variable in the original problem, but with all coefficients equal to zero. Then employ the other perturbation techniques that we have discussed to analyze the effect of moving the coefficients away from zero. The reader can find more on these issues in Winston [61], as well as Hillier and Lieberman [31].

We have also not discussed the problem of what to do when the old solution is no longer optimal. In the case where the constraint constant vector $\mathbf{b}^*$ does not become negative, the feasibility condition for the basic variables will not be violated. Consequently, even when the perturbations are extreme enough to destroy optimality, we can just begin the simplex algorithm again, with the perturbed final tableau representing an initial feasible solution. Until now, there has been little need to say anything about changes to the A^* block (see Figure 3.5) of the final tableau for the original problem. But if we are to restart the simplex algorithm after a perturbation,

we must know the entries in this block. The only case among those discussed earlier in which A^* will change is when there is a change to a constraint coefficient column for a non-basic objective variable. This will not
change $S = B^{-1}$, and Figure 3.5 makes it clear that the matrix A^* will be
obtained by multiplying the perturbed matrix A by S.

   If feasibility is lost under the perturbation, it is possible to continue the
simplex algorithm from the current tableau, rather than returning to the
beginning of the problem. As noted above, there is a two-phase approach in
which the first phase locates a feasible solution for the new problem and the
second phase is just the ordinary simplex method. But there also exists an
algorithm called the *dual simplex algorithm*, which steps through super-opti-
mal, non-feasible solutions using similar row operations to the ordinary
simplex algorithm, until a feasible solution, which is also optimal, is
reached. Hillier and Lieberman ([31], Sec. 9.2) give a nice discussion of this
algorithm.

Exercises 3.3

1. Check that the winery problem can be decomposed as in formula (4).

2. Express the LP problem of Example 2 of Chapter 2, Section 3 in the form
(4). (For your convenience, we were to maximize $f = 4x_1 + 2x_2$ subject to
the constraints below.)

$$
\begin{aligned}
x_1 &+ x_2 \le 2 \\
2x_1 &+ x_2 \le 3 \\
x_1, x_2 &\ge 0
\end{aligned}
$$

3. We return to the coal mining example, which is Example 4 of Chapter 2,
Section 3.

   (a) Identify the vectors $\mathbf{b}^*$ and $\mathbf{c}^*$ and the matrix S of Figure 3.5 for this
problem.

   (b) Express the problem in the form of formula (4).

   (c) Verify the equations in parts (b) and (c) of Theorem 1 for this
problem.

4. (*Mathematica*) For Example 4 of Chapter 2, Section 3:

   (a) Find the range of values of each component of a perturbation vector
$\Delta \mathbf{c} = (\Delta_1, \Delta_2)$ such that the basic solution depicted in the final tableau is still
optimal.

   (b) Sketch the set of all pairs (Δ_1, Δ_2) in the plane such that the old
optimal solution is still optimal.

   (c) If the profit coefficient of x_2 in the original problem is changed to
4500, obtain the entire new "final" tableau under this perturbation. Note that

it no longer represents an optimal solution. Use the simplex algorithm to obtain an optimal solution for the perturbed problem.

5. Prove that the non-basic variable columns of the matrix A^* of Figure 3.5(b) are the corresponding columns of $S \cdot N$. (Hint: Use formula (5).)

6. (*Mathematica*) In Example 2, sketch the regions in the: (a) $\Delta_1 - \Delta_2$ plane; (b) $\Delta_2 - \Delta_3$ plane; and (c) $\Delta_1 - \Delta_3$ plane, such that the optimal solution of the original problem remains optimal when the given pair of perturbations is imposed, holding the remaining perturbation at zero.

7. (*Mathematica*) Consider Exercise 4 of Chapter 2, Section 3 involving the farmer and his hogs, chickens, and ostriches.
    (a) Find a system of inequalities characterizing the set of all perturbation vectors $\Delta \mathbf{b} = (\Delta_1, \Delta_2)$ for $\mathbf{b}$ such that the optimal solution does not change under the perturbation.
    (b) Express the new value of the objective function in terms of Δ_1 and Δ_2.
    (c) Find the range of values of the two individual perturbations such that the current solution remains feasible and optimal.

8. (*Mathematica*) Repeat Exercise 7 for the problem of Exercise 1 of Chapter 2, Section 3.

9. (*Mathematica*) (a) Find the new optimal tableau for the winery problem if the constraint constant vector is perturbed by a vector $\Delta \mathbf{b}$ whose components are: $\Delta_1 = -75$, $\Delta_2 = 4$, $\Delta_3 = 16$, and $\Delta_4 = 91/2$.
    (b) For fixed $\Delta_1 = -75$, $\Delta_4 = -21/2$, graph the set of all (Δ_2, Δ_3) in the plane such that the current optimal solution remains unchanged.

10. Show in general the observation that was made in Example 3. That is, prove that the negatives of the slack variable coefficients in the objective row of a final simplex system (the optimal values of the dual variables) are the same as the coefficients of $\Delta_1, \Delta_2, \dots, \Delta_m$ in the expression for the objective function of the problem in which the constraint constant vector $\mathbf{b}$ is perturbed by these Δ_i.

11. In Example 4 on the winery, find a system of inequalities for the perturbations $\Delta_1, \Delta_2, \Delta_3$, and Δ_4 of the red wine column of constraint coefficients to characterize the set of such perturbations under which the current solution is still optimal.

12. Consider Exercise 1 of Chapter 2, Section 1 on allocation of city funds for the purchase of two types of vehicles. Suppose that the purchase price of vans is incremented by an amount Δ_1, and the maintenance cost per year for a van is changed by an amount Δ_2. Characterize the set of all such Δ_1 and

Δ_2, such that the old optimal solution computed in that problem is still optimal.

13. (*Mathematica*) This problem refers to Example 3 of Chapter 2, Section 3, which is repeated below.

(a) Find the set of perturbations of the form (Δ, Δ, Δ) to the column of x_1 constraint coefficients that do not change the optimal solution.

(b) If $\Delta = -2$, find the resulting perturbed final tableau, and use it as the initial tableau in the simplex method to find an optimal solution for the perturbed problem. (Hint: Use Exercise 5.)

$$\text{maximize: } f = x_1 + 3x_2$$

$$\text{subject to: }\begin{array}{rcrcl} x_1 & + & 2x_2 & \leq & 3 \\ 2x_1 & + & x_2 & \leq & 3 \\ x_1 & + & x_2 & \leq & 2 \\ x_1, & & x_2 & \geq & 0 \end{array}$$

4
Markov Chains

Introduction

Appendix A gives a review of key ideas and results from probability theory, one of the main focuses of which is the random variable. A single random variable is appropriate in experiments that are static. The experiment is performed, an outcome ω happens, and we observe a numerical value $X(\omega)$. But to model an experiment that progresses over time, we need a family of random variables (X_t), one for each time t in some time set. The random variable X_t is the numerical value observed at time t. (Though we usually suppress it in the notation, bear in mind that random variables are functions of the outcome ω.) We see that this family of random variables, called a *stochastic process*, can be thought of as a function of two arguments: $X = X(t, \omega)$. For fixed time t, X is a random variable. For fixed outcome ω, X is an ordinary function of the real variable t. The latter will probably be most helpful to your intuition, because you can begin to think of a stochastic process as a random graph in the plane. Figure 4.1 below shows the sketch of an integer-valued process for a typical outcome ω, in which the time set is the set of non-negative reals. The sequence of values taken on by this process is 0, 1, 3, 2, ..., and the process spends an interval of time at each value before moving to the next. For a different experimental outcome this graph, called the *path* of the process, might look different. We are led to the definition below.

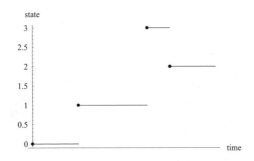

Figure 4.1 – A graph of $t \longrightarrow X_t(\omega)$ for fixed ω

DEFINITION. A *stochastic process* with *time set T* and *state space E* is a family of random variables $(X_t)_{t \in T}$ such that each random variable X_t takes values in the set E.

It is not hard to think of examples of stochastic processes. The following are only a few of many interesting models:

1. $X_t =$ price of a common stock at time t;
2. $X_t =$ level of water in a reservoir at time t;
3. $X_t =$ number of demands for service in a time-sharing computer system through time t;
4. $X_t =$ energy level of an electron within an atom at time t;
5. $X_t =$ number of customers waiting in a line at time t;
6. $X_t = 0$ or 1, respectively, according to the value of the bit in position t in a binary string.

Notice that in the last example the index set for the family of random variables represents position in a sequence rather than time. We mention this to indicate that the results we obtain are not confined only to problems of random motion through time, though this is the easiest way to think of a stochastic process.

We have not specified what kind of sets T and E are. Typically the time set T is either the non-negative integers $\mathbb{Z}_+$ or the non-negative real numbers $\mathbb{R}_+$, depending on the context of the problem. The space of states E is usually some subset of the real line or higher dimensional real space. Much useful and interesting work has been done on non-discrete state spaces; however, most of it requires measure theory and other mathematics that is somewhat beyond the level of this text. So, for our purposes, the state space will usually be some discrete subset of Euclidean space.

There are many probabilistic questions that can be asked about stochastic processes. As with single random variables, one can try to compute the distribution of the value of the process at a fixed time t; that is, $P[X_t \in B]$ for subsets B of the state space. We might also desire to know where the process tends to be in the long-run, i.e., $\lim_{t \to \infty} P[X_t \in B]$. Also, since time is a factor in the experiment, we may be able to observe the evolution of the process for awhile, and use the data to make predictions about the future. Thus, conditional probabilities like the following would be interesting:

$$P[X_t \in B \mid X_s \in A], \quad \text{for } t > s$$

which is the probability that the process is in set B at time t, given that it was in set A at the earlier time s.

In this chapter we study a class of memoryless discrete time processes with discrete state space, called *Markov chains*. After the basic definitions are given in Section 1, we will deduce the conditional and unconditional distributions of the state of the process at a fixed time n in Section 2. In the third section we will show an inductive method for the problem of finding the distribution of the time when a given state is first visited by the process. Sections 4, 5, and 6 give a rather detailed discussion of the limiting properties of Markov chains as time becomes infinite. Some chains can be shown to spend a stable fraction of time in each state, while others can be completely absorbed by a state.

This chapter will rely heavily on results and examples from two excellent books on the subject of stochastic processes: Cinlar [15] and Ross [52]. Other references include Volume 3 of the series by Hoel, Port, and Stone [34], and Volume 1 of the series by Karlin and Taylor [40].

4.1 Definitions and Examples

A Markov chain is a discrete process such that future motions are independent of the past, given the present state. More precisely, we have the following definition:

DEFINITION 1. A discrete time stochastic process $(X_n)_{n=0,1,2,\ldots}$ with finite or countable state space E is called a *Markov chain* if for each $n = 0, 1, 2, \ldots$, and subsets $B_0, B_1, \ldots, B_n$ of E,

$$P[X_{n+1} = j \mid X_0 \in B_0, X_1 \in B_1, \ldots, X_{n-1} \in B_{n-1}, X_n = i] = $$
$$P[X_{n+1} = j \mid X_n = i]$$

Furthermore, the Markov chain is called *time-homogeneous* if the conditional probabilities in the formula above do not depend on n. The *transition matrix* T of a time-homogeneous Markov chain is the matrix defined by:

$$T(i, j) = P[X_{n+1} = j \mid X_n = i] \quad i, j \in E$$

We may depict the state space E and the transition probabilities $T(i, j)$ together on a weighted directed graph called a *transition diagram* as in Figure 4.2. The chain hops in discrete time units from node to node on the transition diagram. For the chain to be Markov, at any time the distribution of the next state to be visited is conditionally independent of the past sequence of states, given the present state.

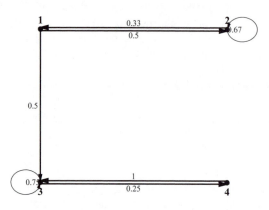

Figure 4.2 – Transition diagram of a Markov chain

For a time-homogeneous chain, the only kind that we will consider here, the conditional probability that the process goes to state j next, given that at the current time it is at state i, does not change as time progresses. The entry $T(i, j)$ in row i and column j of the transition matrix is this conditional probability. For example, in Figure 4.2, given that the current state is 4, the next state to be visited is 3 with probability 1. Thus, $T(4, 3) = 1$. Given that the present state is 1, the chain moves next to either state 2 or 3 with equal probability, stochastically independent of the past, and functionally independent of the clock time. So, $T(1, 2) = T(1, 3) = 1/2$. The complete transition matrix T is:

$$
T = \begin{array}{c} \\ \text{(current state)} \\ \\ \\ \\ \\ \end{array}
\begin{array}{c} \\ 1 \\ 2 \\ 3 \\ 4 \end{array}
\begin{pmatrix}
0 & 1/2 & 1/2 & 0 \\
1/3 & 2/3 & 0 & 0 \\
0 & 0 & 3/4 & 1/4 \\
0 & 0 & 1 & 0
\end{pmatrix}
$$

(next state)
(current state) 1   2   3   4

Activity 1 – What is the probability that the Markov chain whose transition diagram is as in Figure 4.2 goes in two steps from state 1 to state 4? From state 1 back to itself?

The world teems with examples that can be modeled by Markov chains, if one is willing to accept the assumption of independence of past and future. In the introduction to the chapter are a few examples; following are a few more that illustrate the diversity of the applications.

EXAMPLE 1. The state X_n of the chain at time n may be:
    (a) the salary of a worker in year n;
    (b) the number of items in inventory at time n;
    (c) the condition of a patient's health at time n;
    (d) the position of a case in the legal system on day n;
    (e) the level of the national debt in year n;
    (f) the condition of the weather at hour n;
    (g) the population of a town in year n.   ■

We consider three more examples in some detail, in order to illustrate how transition matrices are found.

EXAMPLE 2. Three companies are competing for the market in the gourmet frozen food industry. Company 1 is mounting an advertising campaign, and as a result the shares of the market possessed by each company change from day to day. Let us extrapolate individual behavior to mass behavior, e.g., to say that a randomly selected individual has probability 1/2 of favoring company 1 says that company 1 has 50% of the market. Let $X_n(\omega)$ be the company preferred by a randomly selected individual ω (an outcome in the sample space) at time n. If future preference is independent of the past given the present, then (X_n) forms a Markov chain with state space $E = \{1, 2, 3\}$. Suppose that in any time period, company 1 retains half of its customers, and the others are split equally among the other two companies. Half of the customers of company 2 are converted to company 1 in a time period, and the rest remain with company 2. Company 3 keeps 3/4 of its customers, and the rest change to company 2 in one time period. The transition matrix is below, and the transition diagram is shown in Figure 4.3. Recall that in the transition matrix, the rows represent current states and the columns are the states to be visited in the next time period.

$$T = \begin{pmatrix} 1/2 & 1/4 & 1/4 \\ 1/2 & 1/2 & 0 \\ 0 & 1/4 & 3/4 \end{pmatrix}$$

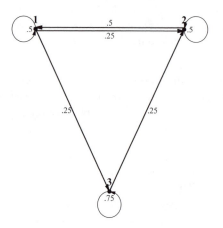

Figure 4.3 – Frozen food example

Company 1 would be interested in its share of the market on day n, which according to our extrapolation is $P[X_n = 1]$. Also of interest would be the limit of this probability as n approaches ∞, if that limit exists. This represents the long-run share of the market belonging to company 1. This company might also want to know the probability that it takes exactly k days to win over a customer initially favoring company 3. We will have the means to solve all of these problems later. ∎

Activity 2 – On the macroscopic level, what is the meaning of the probability that it takes k days for company 1 to win over a customer initially favoring company 3?

EXAMPLE 3. Successive customers to a discount store make their purchase decisions independently of one another, spending \$0, \$1, \$2, or \$3, respectively, with probabilities 1/2, 1/3, 1/12, and 1/12. Let us develop a Markov chain model for the process in which X_n is the cumulative dollar amount purchased by all customers through the nth.

Clearly the Markov property is satisfied, because X_{n+1} is the cumulative amount X_n spent up through the nth customer, plus the amount Z_{n+1} spent by the $n + 1^{st}$ customer, which is independent of the past before customer n. More formally,

$$
\begin{aligned}
P[X_{n+1} = j \mid X_n = i] &= P[X_n + Z_{n+1} = j \mid X_n = i] \\
&= P[Z_{n+1} = j - i \mid X_n = i] \\
&= P[Z_{n+1} = j - i]
\end{aligned}
$$

If the cumulative amount spent through customer n is i, then the amount spent through customer $n + 1$ is either i, $i + 1$, $i + 2$, or $i + 3$ with the probabilities above. The transition matrix of the chain is easy to derive from this observation.

$$
T = \begin{array}{c} \\ 0 \\ 1 \\ 2 \\ 3 \\ \vdots \end{array}
\begin{array}{cccccccc}
0 & 1 & 2 & 3 & 4 & 5 & \cdots \\
\left(1/2 \right. & 1/3 & 1/12 & 1/12 & 0 & 0 & \cdots \\
0 & 1/2 & 1/3 & 1/12 & 1/12 & 0 & \cdots \\
0 & 0 & 1/2 & 1/3 & 1/12 & 1/12 & \cdots \\
0 & 0 & 0 & 1/2 & 1/3 & 1/12 & \cdots \\
\vdots & \vdots & \vdots & \vdots & \vdots & \vdots & \left. \ddots \right)
\end{array}
$$ ■

EXAMPLE 4. Suppose that there are N residents of a college dormitory, all of whom start the week with the flu. Each day, a random positive number of those who had the flu the day before will recover. Let X_n be the number of sick individuals remaining on day n. Then $X_0 = N$, and for $n \geq 0$, the state X_{n+1} at day $n + 1$ is the state X_n at day n minus a random variable Z with the discrete uniform distribution on $1, \ldots, X_n$ which we suppose is independent of the past recovery history, in order that the Markov property is satisfied. Notice that under our conditions, once the number of infected individuals reaches either 1 or 0, the next state is 0 with certainty. This means that the transition matrix of the chain (X_n) is:

$$
T = \begin{array}{c} \\ 0 \\ 1 \\ 2 \\ 3 \\ \vdots \\ N \end{array}
\begin{array}{ccccccc}
0 & 1 & 2 & \cdots & N-1 & N \\
\left(1 \right. & 0 & 0 & \cdots & 0 & 0 \\
1 & 0 & 0 & \cdots & 0 & 0 \\
1/2 & 1/2 & 0 & \cdots & 0 & 0 \\
1/3 & 1/3 & 1/3 & \cdots & 0 & 0 \\
\vdots & \vdots & \vdots & \ddots & \vdots & \vdots \\
1/N & 1/N & 1/N & \cdots & 1/N & \left. 0 \right)
\end{array}
$$

Later we will calculate the expected value of the number of steps it takes, starting from the complete epidemic state N, to reach the healthy state 0. ■

Simulation

One way to obtain information about stochastic processes is to simulate them over many time periods and observe the facet of their behavior that is of interest. In the next two chapters we will be using some commands defined in the KnoxOR`StochasticProcesses` package that comes with this book. The command to load the package is in the closed cell just above Figure 4.1, which has already been executed if you initialized the electronic notebook. This package has been set up to load two of the *Mathematica* standard packages called Statistics`ContinuousDistributions` and Statistics`Discrete-Distributions`, which contain useful commands for simulating observations from given probability distributions. The following *Mathematica* functions are in both of these standard packages:

```
(**   Random[dist]      RandomArray[dist,n]  **)
```

The argument called dist can be set to one of the predefined distributions in *Mathematica*, such as

UniformDistribution[a,b]

which is the continuous uniform distribution on the interval [*a*, *b*]. Then the Random command returns a randomly sampled observation from the given distribution, and the next time that it is called, it samples another, typically different from the previous observation. For RandomArray, the second argument n can be set so that the command returns a list of such observations of length n. To repeat previously simulated random numbers, you can reinitialize the seed value from which the stream of random numbers comes using the following command:

```
(******   SeedRandom[seedvalue]  ******)
```

The seedvalue argument can be any integer; and if the argument is left blank, then the seed is reinitialized randomly.

Observe how *Mathematica*'s random number generation works in the sequence of commands below.

```
SeedRandom[4563];
Random[UniformDistribution[0, 2]]
Random[UniformDistribution[0, 2]]
Random[UniformDistribution[0, 2]]
```

1.91038

0.591598

0.983016

```
SeedRandom[4563];
Random[UniformDistribution[0, 2]]
Random[UniformDistribution[0, 2]]
SeedRandom[];
Random[UniformDistribution[0, 2]]
```

1.91038

0.591598

0.176409

When the seed is initialized to 4563, the sequence of numbers generated by the Random command begins with 1.91038, 0.591598,..., regardless of when you execute the commands. Unless the seed is reinitialized, the next random number in the sequence will be 0.983016. But in the second input cell above, we reinitialized the seed randomly, thus obtaining a different random number from the uniform (0,2) distribution. Like many random number generators, what *Mathematica* actually does is to create a stream of pseudo-random numbers deterministically as a function of the seed that nevertheless appear to have all the properties of truly random numbers. For more information, and for a concise review of the probability theory that is necessary for operations research, see Appendix A.

Let us try to write a Markov chain simulator that will take the transition matrix of the chain, a starting state, and a desired number of time steps, and that will output a simulated sequence of states that behaves as the Markov chain would. The heart of the algorithm will be the generation of a next state given the current state and the row of the transition matrix that forms the discrete probability distribution for the next state. There is a utility function in the KnoxOR`StochasticProcesses` package that has been set up to do this for you. Exercise 7 will ask you to program this.

```
(****** SimDiscreteDist[problist] ******)
```

SimDiscreteDist takes a list of numbers that forms a valid probability distribution on the integers $\{1, 2, ..., n\}$, and simulates a value taken from that distribution. It works by forming a list of cumulative probabilities and then searching for the first element of that list that exceeds a uniform(0,1) random number.

To simulate the Markov chain, we initialize a list of states of the chain with the given start state. Then for the desired number of steps, we append to the state list a newly simulated observation from the distribution determined by the row of the most recently added state in the transition matrix. Read the code below carefully. This command is also contained in the KnoxOR`StochasticProcesses` package.

```
SimMarkovChain[
    transmatrix_, start_, numsteps_] :=
  Module[{statelist},
    statelist = {start};
    Do[AppendTo[statelist,
      SimDiscreteDist[transmatrix[[
    Last[statelist]]]]], {numsteps}];
    statelist];
```

Here is a sample run of the simulator for the chain of Figure 4.3. Notice that, as the transition probabilities determine, the chain cannot make transitions from state 3 to state 1, nor from state 2 to state 3. Also, starting from state 3, the chain frequently stays at state 3.

```
graph43 =
  {{.5, .25, .25}, {.5, .5, 0}, {0, .25, .75}};
SeedRandom[91687];
states = SimMarkovChain[graph43, 3, 40]
ListPlot[states, PlotStyle → PointSize[.02],
  DefaultFont → {"TimesNewRoman", 8}];
```

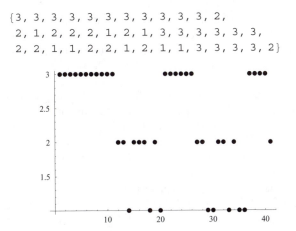

{3, 3, 3, 3, 3, 3, 3, 3, 3, 3, 3, 2,
 2, 1, 2, 2, 2, 1, 2, 1, 3, 3, 3, 3, 3, 3,
 2, 2, 1, 1, 2, 2, 1, 2, 1, 1, 3, 3, 3, 3, 2}

Later we will be able to analytically solve the problem of finding the long-run proportion of time that a Markov chain occupies each of its states. Simulation is able to give approximate answers to this problem too. Using *Mathematica*'s Frequencies command, we can simulate the chain over many time periods and tabulate the number of visits to each state as below. The output shows that among 10,000 time steps, each state seems to appear about equally often. You should reinitialize the seed randomly in the electronic text and execute the cell a few more times to see if this behavior is consistent.

```
SeedRandom[85091];
Frequencies[SimMarkovChain[graph43, 3, 10000]]
```

{{3312, 1}, {3435, 2}, {3254, 3}}

Exercises 4.1

1. A sales representative for a cosmetics firm makes calls in an area with four regions. If she is in region 1 this week, then she will be in region 2 with probability 60%, or region 4 with probability 40%, next week. If she is in region 2, then she goes to one of the other three regions next week with equal probability. The same property holds if she is in region 3 this week. If she is in region 4 this week, then next week she will be in region 2 with certainty. Define a Markov chain that models her travels, and find the transition matrix and transition diagram for this chain.

2. Compute, and interpret the meaning of, the row 1, column 4, entry of T^2 for the transition matrix of the chain of Exercise 1.

3. An $n \times n$ matrix is called a *Markov matrix* if its entries are non-negative and the sum of the entries in every single row is 1. Thus, the transition matrix of a Markov chain is a Markov matrix. Show that the product of two Markov matrices is a Markov matrix.

4. A television manufacturer inspects the TV sets that it makes before releasing them for sale. The inspection of a set results in classification into one of four categories: poor condition (P), fair (F), good (G), or excellent condition (E). Sets in excellent condition are sent off for sale, while those in poor condition are disposed of. Televisions classified as either fair or good are taken to the shop for adjustment, then reinspected. A set that had been in fair condition prior to the adjustment changes to poor condition afterward with probability 1/10, remains fair with probability 3/10, and changes to either good or excellent with equal probability. A set that was in good condition before the adjustment becomes good or excellent afterward with equal probability. The adjustments and reinspections continue until the set is either ready for sale or disposed of.

Define a Markov chain (X_n) that models the progression of a randomly selected set. Find the transition matrix and diagram of the chain. Compute $P[X_1 = F, \ X_2 = F, \ X_3 = E \mid X_0 = F]$.

5. Make up your own example of a Markov chain, and provide intuitive justification for the Markov property in Definition 1.

6. A model that is studied in theoretical computer science is the *finite state automaton*. This is a machine that reads input from a tape, one character at a time, and based on what it reads it moves from where it currently is to one of several other internal states. For instance, a machine that is built to recognize the pattern 000 in an input string of 0's and 1's can be designed as follows. State A is a start state, where we go if we have not seen a 0 yet, or have just seen a 1 so that the machine must try to look anew for the pattern. State B is a state we go to if we have seen a single 0, state C is a state we go to if we have seen two 0's in a row, and similarly state D is for three 0's in a row. The machine stops and returns a success message if it reaches state D; otherwise, if it does not before the input string runs out, then it returns an unsuccessful message.

(a) If the automaton described above is given the input string 1, 0, 0, 1, 0, 1, 0, 0, 0, what sequence of states does it occupy?

(b) Assuming that input strings are fed in such that characters are equally likely to be 0 or 1, independently of previous characters, argue that the sequence of automaton states forms a Markov chain, and find its transition matrix.

7. (*Mathematica*) Write your own version of the SimDiscreteDist function described in the section.

8. (*Mathematica*) Another problem that we will solve analytically later is the problem of finding the expected time that it takes a Markov chain to reach a state starting from another state. Consider the Markov chain with transition matrix and diagram below.

```
matrix8 = {{.375, .625, 0, 0, 0, 0, 0, 0, 0},
    {.375, 0, .625, 0, 0, 0, 0, 0, 0},
    {0, .375, 0, .625, 0, 0, 0, 0, 0},
    {0, 0, .375, 0, .625, 0, 0, 0, 0},
    {0, 0, 0, .375, 0, .625, 0, 0, 0},
    {0, 0, 0, 0, .375, 0, .625, 0, 0},
    {0, 0, 0, 0, 0, .375, 0, .625, 0},
    {0, 0, 0, 0, 0, 0, .375, 0, .625},
    {0, 0, 0, 0, 0, 0, 0, 0, 1}};
```

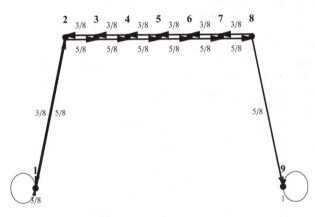

Exercise 8

Build a command to let you approximate the average number of transitions necessary to reach state 9, starting from a given one of the other states. Run your command for each of the initial states 6, 7, and 8.

4.2 Short-Run Distributions

As the examples in the last section suggest, the following quantities are important to know for a Markov chain (X_n):

$$P[X_n = j \mid X_0 = i] \tag{1}$$

$$\mathbf{p}^{(n)}(j) \equiv P[X_n = j] \tag{2}$$

The next theorem gives these, plus a little more. Its proof is a simple matter of manipulating conditional probabilities. It turns out that the powers of the transition matrix can be used to find both the conditional distribution of X_n given X_0 in (1), and the unconditional distribution of X_n in (2).

THEOREM 1. Let (X_n) be a time-homogeneous Markov chain with transition matrix T, and let $\mathbf{p}^{(0)}$ be the probability distribution of X_0, i.e., the row vector whose j^{th} component is $\mathbf{p}^{(0)}(j) \equiv P[X_0 = j]$. Then,
(a)   For all $n \geq 0$, $P[X_n = j \mid X_0 = i] = T^n(i, j)$ (i.e., the $i - j$ component of the nth power of T);
(b)   For all $n \geq 0$, $P[X_n = j] = \mathbf{p}^{(0)} \cdot T(j)$ (i.e., the jth component of the product $\mathbf{p}^{(0)} \cdot T$);
(c)   The conditional joint distribution of $X_{n+1}, \ldots, X_{n+m}$ given X_n is:

$$P[X_{n+1} = j_1, X_{n+2} = j_2, \ldots, X_{n+m} = j_m \mid X_n = i]$$
$$= T(i, j_1) T(j_1, j_2) \cdots T(j_{m-1}, j_m)$$

Proof.   It is convenient to prove part (c) first. By the multiplication rule for conditional probabilities and the Markov property:

$$P[X_{n+1} = j_1, X_{n+2} = j_2, \ldots, X_{n+m} = j_m \mid X_n = i]$$
$$= P[X_{n+1} = j_1 \mid X_n = i] \times P[X_{n+2} = j_2 \mid X_n = i, X_{n+1} = j_1] \cdots$$
$$\cdot P[X_{n+m} = j_m \mid X_n = i, X_{n+1} = j_1, \cdots, X_{n+m-1} = j_{m-1}]$$
$$= P[X_{n+1} = j_1 \mid X_n = i] \times P[X_{n+2} = j_2 \mid X_{n+1} = j_1] \cdots$$
$$\cdot P[X_{n+m} = j_m \mid X_{n+m-1} = j_{m-1}]$$
$$= T(i, j_1) T(j_1, j_2) \cdots T(j_{m-1}, j_m)$$

The last step follows from the time-homogeneity of the chain.
    To prove (a), we condition and uncondition on all states from times 1 to $n - 1$:

$$P[X_n = j \mid X_0 = i]$$

$$= \sum_{j_1 \in E} \sum_{j_2 \in E} \cdots \sum_{j_{m-1} \in E} P[X_1 = j_1, \cdots, X_{n-1} = j_{n-1}, X_n = j \mid X_0 = i]$$

$$= \sum_{j_1 \in E} \sum_{j_2 \in E} \cdots \sum_{j_{m-1} \in E} T(i, j_1) \, T(j_1, j_2) \cdots T(j_{n-1}, j)$$

$$= T^n(i, j)$$

The second line follows by part (c), and the third line is by the definition of matrix multiplication.

Part (b) now follows easily, since

$$P[X_n = j] = \sum_{i \in E} P[X_0 = i] \, P[X_n = j \mid X_0 = i]$$

$$= \sum_{i \in E} \mathbf{p}^{(0)}(i) \, T^n(i, j)$$

$$= \mathbf{p}^{(0)} \cdot T^n(j) \quad \blacksquare$$

Activity 1 – Considering the vectors $\mathbf{p}^{(n)}$ as a sequence of row vectors, show that they can be built up recursively by $\mathbf{p}^{(n)} = \mathbf{p}^{(n-1)} \cdot T$

EXAMPLE 1. Let us return to Example 2 of Section 4.1 on the frozen food companies to illustrate the application of Theorem 1. The transition matrix in that example was

$$T = \begin{pmatrix} 1/2 & 1/4 & 1/4 \\ 1/2 & 1/2 & 0 \\ 0 & 1/4 & 3/4 \end{pmatrix}$$

First, by Theorem 1(c), the conditional probability that a consumer will use company 1 at times 1 and 2 given that he was initially a company 2 customer is

$$P[X_1 = 1, X_2 = 1 \mid X_0 = 2] = T(2, 1) \, T(1, 1) = \tfrac{1}{2} \cdot \tfrac{1}{2} = \tfrac{1}{4} \tag{3}$$

By time-homogeneity and the Markov property, the above probability is the same as, for instance,

$$P[X_{n+1} = 1, X_{n+2} = 1 \mid X_n = 2, X_{n-1} = 3, X_{n-2} = 3] \tag{4}$$

for any $n \geq 2$.

Next, suppose that initially company 1 has 1/8 of the market, company 2 has 3/8, and company 3 has 1/2. What share does each company have after one, two, three, and four days of ads? So, given $\mathbf{p}^{(0)} = (1/8, 3/8, 1/2)$, we would like to compute each of $\mathbf{p}^{(1)}$, $\mathbf{p}^{(2)}$, $\mathbf{p}^{(3)}$, and $\mathbf{p}^{(4)}$. By Theorem 1(b), or alternatively by the observation in Activity 1, we can enter the transition matrix and the vector of initial probabilities into *Mathematica* and compute:

```
T = {{1 / 2, 1 / 4, 1 / 4},
     {1 / 2, 1 / 2, 0}, {0, 1 / 4, 3 / 4}};
p0 = {1 / 8, 3 / 8, 1 / 2};
p1 = N[p0.T]
```

{0.25, 0.34375, 0.40625}

```
p2 = N[p1.T]
```

{0.296875, 0.335938, 0.367188}

```
p3 = N[p2.T]
```

{0.316406, 0.333984, 0.349609}

```
p4 = N[p3.T]
```

{0.325195, 0.333496, 0.341309}

So we see that company 1 grows quickly from its initial share of the market of $1 / 8 = 0.125$ to 0.25 after 1 day, then to over 0.29 after 2 days, and then more modest growth occurs over the next two days to about 32.5% of the market. Meanwhile, companies 2 and 3 are losing their shares, from starting values of 37.5% and 50%, respectively, to ending values of about 33.3% and 34.1%. ■

Activity 2 – As you read down the columns of the output matrix in Example 1, the market shares of the three companies seem to be stabilizing. Compute a few more of the vectors $\mathbf{p}^{(n)}$ for $n \geq 5$ to verify that this is happening.

EXAMPLE 2. Consider the Markov chain whose transition diagram is in Figure 4.4. In the closed cell above the graphics, the transition matrix was named T.

Figure 4.4 – Transition diagram of a cyclic Markov chain

The transition matrix, and its second and third powers are as follows:

```
{MatrixForm[T],
  MatrixForm[T.T], MatrixForm[T.T.T]}
```

$$\left\{ \begin{pmatrix} 0 & 1 & 0 \\ 0 & 0 & 1 \\ 1 & 0 & 0 \end{pmatrix}, \begin{pmatrix} 0 & 0 & 1 \\ 1 & 0 & 0 \\ 0 & 1 & 0 \end{pmatrix}, \begin{pmatrix} 1 & 0 & 0 \\ 0 & 1 & 0 \\ 0 & 0 & 1 \end{pmatrix} \right\}$$

Then, for example, using the second matrix we see that $P[X_2 = 2 \mid X_0 = 1] = 0$ and $P[X_2 = 3 \mid X_0 = 1] = 1$, hence the chain must be at state 3 at time 2, given that it starts at state 1. The third power of T is the identity matrix. This tells us that with certainty the chain will return to its initial state at time 3. Moreover, T^4 will equal T, since $T^4 = T^3 \cdot T$, and similarly $T^5 = T^2$, $T^6 = T^3$, etc. All of these observations make sense with the dynamics of the chain shown by the transition diagram. ∎

Computation of the short-run distributions is reduced by Theorem 1 to computation of powers of the transition matrix T. In theory, this solves the problem, but in practice the state space may be large (or infinite) and the desired time n may also be so large as to make it very costly to compute T^n. But perhaps more importantly still, it may be possible to find analytical expressions for the entries of T^n that would allow us to answer questions like the one in Activity 2, which was essentially asking for the limiting value of these matrix powers. The following ideas and results from linear algebra involving eigenvalues and diagonalization of a matrix can help.

Recall that a number λ is called an *eigenvalue* of an $m \times m$ matrix A, and a column vector $\mathbf{x}$ is an *eigenvector* for λ if

$$A \cdot \mathbf{x} = \lambda \cdot \mathbf{x} \qquad (5)$$

Suppose that A has m distinct eigenvalues $\lambda_1, \lambda_2, \ldots, \lambda_m$.   Let $\mathbf{x}_1, \mathbf{x}_2, \ldots, \mathbf{x}_m$ be eigenvectors for these eigenvalues, in the corresponding order. Define the diagonal matrix and corresponding eigenvector matrix by

$$D = \begin{pmatrix} \lambda_1 & 0 & 0 & \cdots & 0 \\ 0 & \lambda_2 & 0 & \cdots & 0 \\ \vdots & \vdots & \vdots & \ddots & \vdots \\ 0 & 0 & 0 & \cdots & \lambda_m \end{pmatrix}, \quad N = (\mathbf{x}_1, \mathbf{x}_2, \ldots, \mathbf{x}_m) \tag{6}$$

so that the columns of N are the eigenvectors, in the corresponding order. Then it can be proved that N^{-1} exists, and

$$A = N \cdot D \cdot N^{-1} \tag{7}$$

The hypothesis of distinct eigenvalues can be weakened. All that is really needed is that there exist m linearly independent eigenvectors.

This can be used to express the transition matrix in the form $T = N \cdot D \cdot N^{-1}$, in which case

$$\begin{aligned} T^n &= (NDN^{-1})(NDN^{-1})\cdots(NDN^{-1}) \\ &= ND(N^{-1}N)D(N^{-1}N)\cdots(N^{-1}N)DN^{-1} \\ &= ND^nN^{-1} \end{aligned} \tag{8}$$

The matrix ND^nN^{-1} is simple to calculate, since the diagonal matrix D raised to the nth power is the diagonal matrix whose i^{th} diagonal element is $(\lambda_i)^n$. Thus, the main work in calculating the probability distribution of X_n is in the computation of the eigenvalues and eigenvectors of T. The following facts can be helpful.

> For a transition matrix T, 1 is always an eigenvalue with eigenvector $\mathbf{1}$, by which we mean a column of 1's (see Exercise 5). $\hfill (9)$

> The trace of T, that is, the sum of the diagonal entries of T, equals the sum of all eigenvalues of T. $\hfill (10)$

> The trace of T^k is the sum of the kth powers of the eigenvalues of T. $\hfill (11)$

EXAMPLE 3. In Example 1, we computed some short-run probabilities for the Markov chain related to the frozen food market. Now we will use the diagonalization procedure to obtain a formula for T^n for arbitrary n. We need the eigenvalues λ_1, λ_2, and λ_3, and an eigenvector for each. By observation (9), we can put $\lambda_1 = 1$ and $\mathbf{x}_1 = \mathbf{1}$. As for the other eigenvalues,

(10)–(11) give us a system of equations for these two unknowns. First we compute T^2.

```
T = {{1 / 2, 1 / 4, 1 / 4},
     {1 / 2, 1 / 2, 0}, {0, 1 / 4, 3 / 4}};
{MatrixForm[T], MatrixForm[T.T]}
```

$$\left\{ \begin{pmatrix} \frac{1}{2} & \frac{1}{4} & \frac{1}{4} \\ \frac{1}{2} & \frac{1}{2} & 0 \\ 0 & \frac{1}{4} & \frac{3}{4} \end{pmatrix}, \begin{pmatrix} \frac{3}{8} & \frac{5}{16} & \frac{5}{16} \\ \frac{1}{2} & \frac{3}{8} & \frac{1}{8} \\ \frac{1}{8} & \frac{5}{16} & \frac{9}{16} \end{pmatrix} \right\}$$

Then,

$$\text{trace } T = 1/2 + 1/2 + 3/4 = 7/4 = 1 + \lambda_2 + \lambda_3$$

$$\text{trace } T^2 = 3/8 + 3/8 + 9/16 = 21/16 = 1 + (\lambda_2)^2 + (\lambda_3)^2$$

```
Solve[{1 + λ₂ + λ₃ == 7 / 4,
    1 + (λ₂)² + (λ₃)² == 21 / 16}, {λ₂, λ₃}]
```

$$\left\{ \left\{ \lambda_2 \to \frac{1}{4}, \ \lambda_3 \to \frac{1}{2} \right\}, \ \left\{ \lambda_2 \to \frac{1}{2}, \ \lambda_3 \to \frac{1}{4} \right\} \right\}$$

Alternatively, we could have asked *Mathematica* directly for the eigenvalues.

```
Eigenvalues[T]
```

$$\left\{ 1, \ \frac{1}{2}, \ \frac{1}{4} \right\}$$

Written in decreasing order, our three eigenvalues are $\lambda_1 = 1$, $\lambda_2 = 1/2$, and $\lambda_3 = 1/4$. To find the eigenvector $x_i = (x, y, z)$ for eigenvalue λ_i, the form of T implies that we must solve

$$\begin{cases} x/2 + y/4 + z/4 = \lambda_i x \\ x/2 + y/2 \qquad\quad = \lambda_i y \\ \qquad\quad y/4 + 3z/4 = \lambda_i z \end{cases}$$

You can either do this longhand, or with the *Mathematica* Solve command, or use the following Eigenvectors command. Note that the eigenvectors correspond to eigenvalues written in the same order as in the output of the Eigenvalues command.

```
Eigenvectors[T]
```

$$\{\{1, 1, 1\}, \{0, -1, 1\}, \{1, -2, 1\}\}$$

Then we can set

$$D = \begin{pmatrix} 1 & 0 & 0 \\ 0 & 1/2 & 0 \\ 0 & 0 & 1/4 \end{pmatrix}, \quad N = \begin{pmatrix} 1 & 0 & 1 \\ 1 & -1 & -2 \\ 1 & 1 & 1 \end{pmatrix}$$

and N^{-1} turns out to be:

```
D0 = {{1, 0, 0}, {0, 1 / 2, 0}, {0, 0, 1 / 4}};
N0 = {{1, 0, 1}, {1, -1, -2}, {1, 1, 1}};
NInv = Inverse[N0];
MatrixForm[NInv]
```

$$\begin{pmatrix} \frac{1}{3} & \frac{1}{3} & \frac{1}{3} \\ -1 & 0 & 1 \\ \frac{2}{3} & -\frac{1}{3} & -\frac{1}{3} \end{pmatrix}$$

From these matrices and formula (8) we find that T^n is

```
TPower[n_] := N0.MatrixPower[D0, n].NInv;
MatrixForm[TPower[n]]
```

$$\begin{pmatrix} \frac{1}{3} + \frac{1}{3}\,2^{1-2n} & \frac{1}{3} - \frac{4^{-n}}{3} & \frac{1}{3} - \frac{4^{-n}}{3} \\ \frac{1}{3} - \frac{1}{3}\,2^{2-2n} + 2^{-n} & \frac{1}{3} + \frac{1}{3}\,2^{1-2n} & \frac{1}{3} + \frac{1}{3}\,2^{1-2n} - 2^{-n} \\ \frac{1}{3} + \frac{1}{3}\,2^{1-2n} - 2^{-n} & \frac{1}{3} - \frac{4^{-n}}{3} & \frac{1}{3} + 2^{-n} - \frac{4^{-n}}{3} \end{pmatrix}$$

Our labor has returned a very powerful result. For any time n we wish, we may calculate the probability that an individual favors each company, given that at the start he favored some other company. For instance, the probability that a customer who began with company 2 favors company 1 on the sixth day is the (2, 1) entry of the preceding matrix evaluated at $n = 6$, which is 0.348633 as shown below.

```
N[MatrixForm[TPower[6]]]
```

$$\begin{pmatrix} 0.333496 & 0.333252 & 0.333252 \\ 0.348633 & 0.333496 & 0.317871 \\ 0.317871 & 0.333252 & 0.348877 \end{pmatrix}$$

When T^n is pre-multiplied by $\mathbf{p}^{(0)} = [1/8\ 3/8\ 1/2]$, one gets the unconditional distribution of X_n, which is the vector $\mathbf{p}^{(n)}$ representing the shares of the market possessed by each company at time n. For example, $\mathbf{p}^{(6)}(1)$ is the share of the market that company 1 has at time 6, which is 0.33136, the first component of the vector below:

```
N[{1 / 8, 3 / 8, 1 / 2}.TPower[6]]
```

{0.33136, 0.333344, 0.335297}

One last important observation is that as $n \longrightarrow \infty$, the matrix T^n approaches a matrix all of whose rows are $[1/3\ 1/3\ 1/3]$, because the exponent n appears with a negative coefficient in all of the exponential expressions in the matrix T^n. In terms of the applied problem, no matter what the initial state, the shares of the market possessed by the three companies are (rapidly) approaching equal shares of 1/3 apiece as time progresses. It is common that such limits exist and are independent of the initial state, though it should be pointed out that the equality of the three numbers is coincidental to this problem. We will have a more convenient way of finding limiting probabilities in Section 5. ■

Activity 3 – Find the number of days it takes for the first company to achieve a share of 0.333 of the market, using the same initial distribution $\mathbf{p}^{(0)}$ as in the example.

Exercises 4.2

1. (*Mathematica*) Consider the Markov chain whose transition diagram is below. Assume that it is certain that the chain begins in state 3.
   (a) Find the probability distribution of X_2.
   (b) For arbitrary n, find the distribution of X_n.

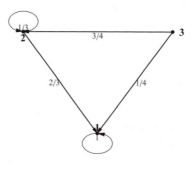

Exercise 1

2. (*Mathematica*) Consider the Markov chain with transition matrix below. Suppose the initial distribution is $\mathbf{p}^{(0)} = (1/4, 1/4, 1/4, 1/4)$. Find and interpret: (a) $T^3(3, 4)$; (b) $\mathbf{p}^{(0)} \cdot T^5(3)$; (c) $T^n(1, 1)$.

$$\begin{pmatrix} .27 & .15 & .30 & .28 \\ .06 & .54 & .22 & .18 \\ .90 & .02 & .06 & .02 \\ 1 & 0 & 0 & 0 \end{pmatrix}$$

3. (*Mathematica*) A *random walk with reflecting barriers* 0 and N is a Markov chain whose state space is $E = \{0, 1, 2, \ldots, N\}$, which, at any state strictly between 0 and N, moves next to either the state immediately to the left or immediately to the right with equal probability. If the chain is at state 0, then it is certain to be at state 1 at the next time; and if it is at state N, then it is certain to be at state $N - 1$ at the next time. For the random walk with reflecting barriers at 0 and 4, find the conditional distribution of the state at time 6, given that the initial state is each of: 0, 1, 2, 3, and 4.

4. Let (X_n) be an arbitrary two-state Markov chain with a transition matrix T whose every entry is non-zero. Find an expression for T^n, and find the limit as $n \longrightarrow \infty$ of T^n.

5. Verify condition (9).

6. (*Mathematica*) For Example 4 of Section 4.1, the flu recovery model with $N = 5$, compute $P[X_3 = 0 \mid X_0 = 5]$ and $P[X_2 = 1, X_3 = 0 \mid X_0 = 5]$.

7. A Markov chain of three states has the transition matrix below. Draw a tree diagram representing the first three transitions of the chain, in which each state has a directed edge pointing to the possible next states, weighted by the probabilities of visiting those states. Use this tree to find $P[X_3 = 1 \mid X_0 = 1]$.

$$\begin{pmatrix} 0 & 1/2 & 1/2 \\ 1/3 & 1/3 & 1/3 \\ 1/4 & 0 & 3/4 \end{pmatrix}$$

8. (*Mathematica*) A machine can either be in excellent, good, fair, or poor working condition at each time. Given that its current condition is any of the first three, it can be in either the same condition next time with probability .95, or in a condition that is one level worse with the remaining probability. When the machine reaches the poor condition, it stays there. At least how much time must elapse before a machine that begins in excellent condition reaches poor condition with probability at least 1/2?

9. A program vehicle is used by a car dealer until it reaches the end of its useful lifetime, and then is immediately replaced by a similar vehicle. It is reasonable to suppose that the successive lifetimes $Z_1, Z_2, Z_3, \ldots$ of these vehicles are i.i.d. with some discrete distribution: $p_j = P[Z = j]$, $j = 1, 2, 3, \ldots$. Assume also that if the old vehicle breaks during time period $[n, n + 1)$, then the new one is put into service at time $n + 1$. Let Y_n be the age of the vehicle in use at the start of time n. (Then Y_{n+1} is 0 if the vehicle in use at time n was replaced.)
   (a) Compute $q_i = P[Y_{n+1} = i + 1 \mid Y_n = i]$.
   (b) Find the transition matrix T of (Y_n).
   (c) Show that row i of T^2 has non-zero entries only in columns 0, 1, and $i + 2$, and find those entries in terms of the q_i's.

10. (*Mathematica*) Consider the Markov chain with the transition matrix below. Investigate the behavior of T^n for large n, and interpret it in terms of the geometry of the transition diagram.

$$T = \begin{pmatrix} 1 & 0 & 0 & 0 & 0 & 0 \\ 1/4 & 0 & 3/4 & 0 & 0 & 0 \\ 0 & 1/4 & 0 & 3/4 & 0 & 0 \\ 0 & 0 & 1/4 & 0 & 3/4 & 0 \\ 0 & 0 & 0 & 1/4 & 0 & 3/4 \\ 0 & 0 & 0 & 0 & 0 & 1 \end{pmatrix}$$

4.3 First Passage Times

The first time that a Markov chain hits a given state $j \in E$ is a random variable called a *first passage time*. Let T_j be that time. We are interested in the conditional probability distribution of T_j given the initial state i, which we denote by

$$F_k(i, j) = P[T_j = k \mid X_0 = i] \tag{1}$$

We will not be able to compute a closed form for F_k directly in most cases, but we will show how to calculate it recursively in k.

The computation is not difficult. For $k = 1$ we have

$$
\begin{aligned}
F_1(i, j) &= P[T_j = 1 \mid X_0 = i] \\
&= P[X_1 = j \mid X_0 = i] \\
&= T(i, j)
\end{aligned}
$$

where, as usual, T is the transition matrix of the Markov chain. For $k \geq 2$, the idea is that in order to hit state j for the first time at time k, the chain first visits some state $x \neq j$, then it stays away from j for exactly $k - 1$ more time units. By the law of total probability,

$$
\begin{aligned}
&F_k(i, j) \\
&= P[X_1 \neq j, \ldots, X_{k-1} \neq j, X_k = j \mid X_0 = i] \\
&= \sum_{x \in E - \{j\}} P[X_2 \neq j, \ldots, X_{k-1} \neq j, X_k = j \mid X_1 = x] \\
&\qquad \cdot P[X_1 = x \mid X_0 = i] \\
&= \sum_{x \in E - \{j\}} F_{k-1}(x, j) \cdot T(i, x)
\end{aligned}
$$

We therefore have the following theorem.

THEOREM 1. Let T be the transition matrix of a Markov chain, and let $F_k(i, j)$ be the first passage time probability defined by (1). Then,

$$
F_k(i, j) =
\begin{cases}
T(i, j) & \text{if } k = 1 \\
\sum_{x \in E - \{j\}} T(i, x) \cdot F_{k-1}(x, j) & \text{if } k \geq 2
\end{cases} \quad \blacksquare \tag{2}
$$

EXAMPLE 1. A student attending a certain college must satisfy a mathematics requirement. An entrance test, if passed, is enough. If the student does not pass on the first try, he must take a certain course. Let us suppose that the probability of passing the exam is 1/4, and that the probability of passing the course is 2/3, no matter how many times the course is taken. We may model the situation as a Markov chain with three states, as in Figure 4.5.

State 3 is the entrance state; state 2 is the state of being enrolled in the course, and state 1 represents passing the math requirement. The random variable X_n is the state at semester n, and the transition probabilities are easily deduced from the given information. We shall compute the distribution of the amount of time taken by a student entering in state 3 to reach state 1. This particular chain is simple enough that we can obtain closed formulas.

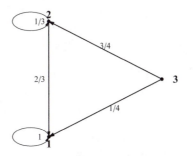

Figure 4.5 – Markov chain model for satisfaction of math requirement

In the notation of (1), this distribution is

$$F_k(3, 1) = P[T_1 = k \mid X_0 = 3], \quad k = 1, 2, 3, \ \ldots$$

First, $F_1(3, 1) = T(3, 1) = 1/4$, by Theorem 1 and the given transition probabilities. Again by the theorem,

$$
\begin{aligned}
F_2(3, 1) &= \textstyle\sum_{x \neq 1} T(3, x)\,F_1(x, 1) \\
&= T(3, 2)\,F_1(2, 1) + T(3, 3)\,F_1(3, 1) \\
&= T(3, 2)\,T(2, 1) + 0 = (3/4)\,(2/3)
\end{aligned}
$$

since $T(3, 3) = 0$. To find $F_3(3, 1)$, we will need $F_2(2, 1)$ as well, which is

$$
\begin{aligned}
F_2(2, 1) &= \textstyle\sum_{x \neq 1} T(2, x)\,F_1(x, 1) \\
&= T(2, 2)\,F_1(2, 1) + T(2, 3)\,F_1(3, 1) \\
&= (1/3)\,(2/3) + 0
\end{aligned}
$$

The computation of $F_3(3, 1)$ is similar:

$$
\begin{aligned}
F_3(3, 1) &= \textstyle\sum_{x \neq 1} T(3, x)\,F_2(x, 1) \\
&= T(3, 2)\,F_2(2, 1) + T(3, 3)\,F_2(3, 1) \\
&= (3/4)\,(1/3)\,(2/3) + 0
\end{aligned}
$$

A pattern has emerged. Since $T(3, 3) = 0$ and $T(2, 3) = 0$, we can write for general $k \geq 2$,

$$F_k(3, 1) = T(3, 2) F_{k-1}(2, 1) \quad \text{and} \quad F_k(2, 1) = T(2, 2) F_{k-1}(2, 1) \quad (3)$$

The second equation in (3) says that $F_k(2, 1)$ forms a geometric progression in k with common ratio $T(2, 2) = 1/3$ and initial term $F_1(2, 1) = T(2, 1) = 2/3$. Hence,

$$F_k(2, 1) = (1/3)^{k-1} \cdot (2/3), \quad k = 1, 2, 3, \ldots \quad (4)$$

Substitution into the left-hand equation in (3) yields

$$F_k(3, 1) = \begin{cases} (1/3)^{k-2} (2/3) (3/4) & \text{if } k = 2, 3, \ldots \\ 1/4 & \text{if } k = 1 \end{cases} \quad (5)$$

Recall that (5) gives the probability that the math requirement is passed by the kth semester.

The answer is very intuitive, given the structure of the transition diagram. Starting in state 3, a student requires only one time unit to reach state 1 if the exam is passed, and this occurs with probability 1/4. In what way can exactly k semesters ($k \geq 2$) be required to reach state 1 from state 3? First, the exam must be failed, which happens with probability 3/4. Then $k - 2$ semesters must be spent in the course. Recall that one retakes the course with probability 1/3. Finally, after these $k - 2$ unsuccessful efforts, one successful effort occurs with probability 2/3. The meaning of the product in the first line of (5) should now be clear. ∎

Activity 1 – Check that the functions of k in (4) and (5) are proper probability mass functions.

The form of equation (2) suggests an efficient recursive algorithm for computing first passage time probabilities. Consider the target state j as fixed. For each $k \geq 1$, view $\mathbf{F}_k(i, j)$ as a column vector formed as i ranges through the state space E. Then the sum in the second part of the formula is almost the product of row i of the transition matrix T with the column vector $\mathbf{F}_{k-1}(x, j)$, except that the $F_{k-1}(j, j) T(i, j)$ term is excluded. If we introduce a new matrix $\tilde{T}$ into the problem, which agrees with T except that its jth column is set to zero, then (2) becomes

$$\mathbf{F}_k(\cdot, j) = \tilde{T} \cdot \mathbf{F}_{k-1}(\cdot, j) \quad (6)$$

The first such vector F_1 should be defined as column j of the original transition matrix T, by the first part of (2). We show how to use this observation in the next example.

EXAMPLE 2. A small three-space parking lot behaves such that the sequence of values $X_1, X_2, X_3,$ ... defined as the number of parked cars at the time instants 1, 2, 3, ... forms a Markov chain. If no cars are present, it is twice as likely for there to be no cars at the next time instant as it is for there to be one car. If either one or two cars are present, at the next instant there will either be one fewer car, the same number of cars, or one more car, with equal probability. And if three cars are present, it is twice as likely for there to still be three cars at the next time instant as it is for there to be two cars. We look at the probability distribution of the time it takes for the lot to become full for the first time, starting from each of states 0, 1, and 2.

The state space $E = \{0, 1, 2, 3\}$ gives the possible number of cars in the lot. We are interested in finding $F_k(i, 3)$, $k = 1, 2, 3,$ ... for each $i = 0, 1, 2$. The transition matrix of the chain is below, and to use (6) we must zero out the last column corresponding to state $j = 3$ to obtain $\tilde{T}$. The starting vector $\mathbf{F}_1$ is the last column of T.

$$T = \begin{pmatrix} 2/3 & 1/3 & 0 & 0 \\ 1/3 & 1/3 & 1/3 & 0 \\ 0 & 1/3 & 1/3 & 1/3 \\ 0 & 0 & 1/3 & 2/3 \end{pmatrix}, \quad \tilde{T} = \begin{pmatrix} 2/3 & 1/3 & 0 & 0 \\ 1/3 & 1/3 & 1/3 & 0 \\ 0 & 1/3 & 1/3 & 0 \\ 0 & 0 & 1/3 & 0 \end{pmatrix},$$

$$\mathbf{F}_1 = \begin{pmatrix} 0 \\ 0 \\ 1/3 \\ 2/3 \end{pmatrix}$$

We now set these definitions into *Mathematica* and compute the next several vectors $\mathbf{F}_k$.

```
F1 = {0, 0, 1 / 3, 2 / 3};
Ttilde = {{2 / 3, 1 / 3, 0, 0}, {1 / 3, 1 / 3, 1 / 3, 0},
    {0, 1 / 3, 1 / 3, 0}, {0, 0, 1 / 3, 0}};
F2 = Ttilde.F1
F3 = Ttilde.F2
F4 = Ttilde.F3
F5 = Ttilde.F4
```

$$\left\{0, \frac{1}{9}, \frac{1}{9}, \frac{1}{9}\right\}$$

$$\left\{\frac{1}{27}, \frac{2}{27}, \frac{2}{27}, \frac{1}{27}\right\}$$

$$\left\{\frac{4}{81}, \frac{5}{81}, \frac{4}{81}, \frac{2}{81}\right\}$$

$$\left\{\frac{13}{243}, \frac{13}{243}, \frac{1}{27}, \frac{4}{243}\right\}$$

For instance, the first components of these vectors, 0, 1/27, 4/81, and 13/243, respectively, give the probabilities that it takes 2 time units, 3 time units, 4 time units, and 5 time units for an empty lot to first reach the state of three cars. The fact that F4(2) = 4/81 (the third element of the F4 vector) means that the probability that 4 time units are required to reach state 3 from state 2 is 4/81. Much other similar information can be gained by continuing the iterative process. ∎

Activity 2 – Use the vector approach to verify the results we found for the Markov chain of Example 1.

EXAMPLE 3. In theory, the distribution F_k characterized by Theorem 1 could be used to calculate expected values of first passage times, but in particular problems there are sometimes faster ways. Recall the flu recovery model of Example 4 of Section 4.1. Let $T = T_0$ be the time at which the healthy state 0 is reached. We would like a closed-form expression for

$$f(i) = E[T \mid X_0 = i], \qquad (7)$$

which is the expected number of steps required to reach the healthy state, beginning at state i, for $i = 1, 2, \ldots, N$. If we begin at state 0, then no more steps are required, hence $f(0) = 0$; and if we begin at state 1, then exactly one more step is required, consequently $f(1) = 1$. To find $f(i)$ for the other states, we will write $f(i)$ in terms of previous $f(j)$, $j < i$, by using a conditioning argument similar to the one used to calculate the distribution of first passage times.

Roughly, the reasoning is as follows. Begin at a state $X_0 = i \geq 2$. The time needed to reach the optimal state is one plus the time needed from the next state X_1. Condition on X_1, and uncondition, knowing that X_1 is one of the i equally likely states that are better than i. By the law of total probability, applied to the conditional probability measure $P[\cdot \mid X_0 = i]$,

$$P[T = k \mid X_0 = i] = \sum_{j=0}^{i-1} P[T = k \mid X_1 = j, X_0 = i] P[X_1 = j \mid X_0 = i] \quad (8)$$

Thus, multiplying both sides by k, summing, and interchanging order of summation yields:

$$\sum_{k=1}^{i} k \, P[T = k \mid X_0 = i]$$

$$= \sum_{k=1}^{i} k \sum_{j=0}^{i-1} P[T = k \mid X_1 = j, X_0 = i] \, P[X_1 = j \mid X_0 = i]$$

$$= \sum_{j=0}^{i-1} P[X_1 = j \mid X_0 = i] \sum_{k=1}^{i} k \, P[T = k \mid X_1 = j, X_0 = i]$$

The last formula implies

$$E[T \mid X_0 = i]$$

$$= \sum_{j=0}^{i-1} E[T \mid X_1 = j, X_0 = i] \, P[X_1 = j \mid X_0 = i]$$

$$= \sum_{j=0}^{i-1} (1 + E[T \mid X_0 = j]) \, \frac{1}{i} \tag{9}$$

$$= \frac{1}{i} \sum_{j=0}^{i-1} 1 + \frac{1}{i} \sum_{j=0}^{i-1} E[T \mid X_0 = j]$$

The second line results from the time-homogeneous Markov property and the known conditional distribution of X_1 given $X_0 = i$. We now have the recursive formula:

$$f(i) = \begin{cases} 1 + \frac{1}{i} \sum_{j=1}^{i-1} f(j) & \text{if } i = 2, 3, \ldots, N \\ 1 & \text{if } i = 1 \end{cases} \tag{10}$$

Computation of $f(2)$ gives 1, and computation of $f(3)$ gives $1 + 1/2$. We define this function recursively in *Mathematica*, and check a few more values:

```
f[1] := 1;
f[i_] := 1 + 1/i Sum[f[j], {j, 1, i - 1}];
```

```
{f[2], f[3], f[4], f[5]}
```

$$\left\{ \frac{3}{2}, \frac{11}{6}, \frac{25}{12}, \frac{137}{60} \right\}$$

Noting that $11/6 = 1 + 1/2 + 1/3$ and $25/12 = 1 + 1/2 + 1/3 + 1/4$, a reasonable guess at a closed-form expression for $f(i)$ is therefore

$$f(i) = \begin{cases} \sum_{j=1}^{i} 1/j & \text{if } i = 1, 2, 3, \ldots, N \\ 0 & \text{if } i = 0 \end{cases} \tag{11}$$

You are asked to show (11) by induction on i in Exercise 10. For instance, in a problem with six states in which the starting point is 6 students with the flu, the expected number of steps to the healthy state is

$$1 + 1/2 + 1/3 + 1/4 + 1/5 + 1/6 = 49/20$$

The expected number of steps required to reach the optimum point in this simple algorithm model is roughly a logarithmic function of the number of states. ∎

Exercises 4.3

1. Find $F_k(3, 1)$ for all $k \geq 1$ for the frozen food companies of Example 2, Section 4.1. For your convenience, the transition matrix of the chain is reproduced below.

$$T = \begin{pmatrix} 1/2 & 1/4 & 1/4 \\ 1/2 & 1/2 & 0 \\ 0 & 1/4 & 3/4 \end{pmatrix} \qquad T = \begin{array}{c} \\ 1 \\ 2 \\ 3 \\ 4 \end{array}\begin{array}{cccc} 1 & 2 & 3 & 4 \\ \begin{pmatrix} 0 & 1/2 & 1/2 & 0 \\ 1/3 & 2/3 & 0 & 0 \\ 0 & 0 & 3/4 & 1/4 \\ 0 & 0 & 1 & 0 \end{pmatrix} \end{array}$$

Exercise 1   **Exercise 2**

2. (*Mathematica*) For the Markov chain of Figure 4.2, whose transition matrix is above, compute $F_k(1, 3)$ for $k = 1, 2, 3, 4, 5$.

3. Let (X_n) be the chain with transition matrix below. Find $F_k(i, 2)$ for $i = 1, 2, 3$ and all $k = 1, 2, 3, \ldots$.

$$T = \begin{pmatrix} 1 & 0 & 0 \\ 1/2 & 1/4 & 1/4 \\ 1/3 & 3/5 & 1/15 \end{pmatrix}$$

4. (*Mathematica*) Write a *Mathematica* program to compute the vector $F_k(i, j)$ (as i ranges through the state space) given j and the transition matrix T.

5. Compute the distribution of the time of first passage of a television set from the fair state to the excellent state for the chain of Exercise 4 of Section 4.1. (Note that it is possible for this time to be $+\infty$.)

6. A judicial case can be heard at three levels: lower court (1), appellate court (2), and high court (3). State 4 in the transition diagram below represents final termination of the case. The weights in the directed graph are the probabilities that the case will move from one court to another, e.g., an appellate court case will return there with probability 1/4, or be appealed to

high court with probability 3/4. Find the probability that a case that begins in lower court is finally terminated after exactly k hearings.

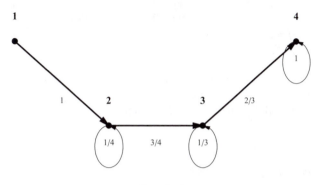

Exercise 6

7. (*Mathematica*) A game is played so that the wealth of the gambler at each play either rises by 1 with probability .51 or falls by 1 with probability .49, until the wealth either hits 0 or 8, at which point the game stops. For each interior state 1, 2, ..., 7 in the transition diagram below, find the probability that it takes k units of time to reach state 8, for values of $k = 1, 2, 3, 4, 5, 6, 7, 8, 9, 10$. Comment on whether, if the computation were continued indefinitely, these probabilities would sum to 1.

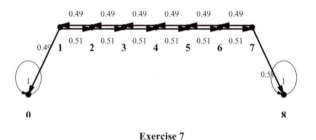

Exercise 7

8. In Exercise 6, the time of first visit from state 3 to state 4 has a geometric distribution with success parameter 2/3, so that $E[T_4 \mid X_0 = 3] = 3/2$. Find $E[T_4 \mid X_0 = 2]$ without finding the conditional distribution of T_4 given $X_0 = 2$.

9. For a general two-state Markov chain with all transition probabilities non-zero, find expressions for $F_k(1, 2)$ and $F_k(2, 1)$.

10. Verify formula (11) for the expected time to reach the healthy state in the flu model.

11. For a general three-state Markov chain in which state 1 is an absorbing state, find formulas for $F_k(i, j)$ for all pairs of states $i, j = 2, 3$. What is $F_k(1, j)$ for $j = 2, 3$? Set up, but do not attempt to solve, equations for $F_k(2, 1)$ and $F_k(3, 1)$.

12. For a cyclic Markov chain with five states, that is, a chain in which state 1 must go to state 2, state 2 must go to state 3, etc., what does formula (2) reduce to? Find all first passage time probabilities $F_k(i, j)$.

4.4 Classification of States

We now begin an examination of the long-run behavior of a Markov chain. Recall that in Section 2 of Chapter 1 we discussed what it meant for a graph to be closed. In the language of Markov chains, a set S of states is called *closed* if there does not exist a path from a state in S to a state outside of S. For example, in Figure 4.6 below, the set $\{5, 6, 7\}$ is closed, as are the sets $\{2, 3, 4\}$, $\{8\}$, and $\{1, 2, 3, 4\}$. The set $\{2, 3\}$ is not closed, since state 4 can be reached from it.

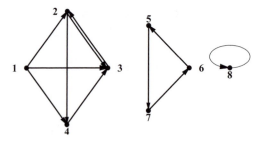

Figure 4.6 – Some closed sets of states in a Markov chain

A more important property of state spaces of Markov chains is the following.

DEFINITION 1. A set of states is *irreducible* if it is closed and contains no proper closed subsets. The Markov chain itself is called *irreducible* if its state space is irreducible.

A look back at Figure 4.6 shows that the set {1, 2, 3, 4} is not irreducible, since {2, 3, 4} is a proper closed subset. But the sets {2, 3, 4} and {5, 6, 7} are both irreducible. Single states, such as state 8 in Figure 4.6, from which no escape is possible, form their own irreducible sets, and are given the special name of *absorbing states*. The chain whose transition diagram is shown in Figure 4.7 is irreducible.

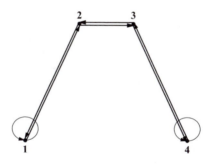

Figure 4.7 – An irreducible Markov chain

Activity 1 – Look back at the frozen food company example of Section 4.1. Is that Markov chain irreducible? Does the Markov chain that we used as a simple model for recovery from a flu epidemic (Example 4 of Section 4.1) have any closed sets of states?

To motivate the ideas that we will discuss in this section, refer again to the chain whose transition diagram is in Figure 4.6. Beginning at state 8, the probability of being again at state 8 at any later time n is 1. Hence, trivially $P[X_n = 8 \mid X_0 = 8]$ approaches 1 as $n \longrightarrow \infty$. Beginning at state 1, the chain leaves state 1 immediately and permanently, so that the probability of being at state 1 at any time $n > 0$ is 0. Again, trivially $P[X_n = 1 \mid X_0 = 1]$ approaches 0 as $n \longrightarrow \infty$. We call such a state *transient*. It is only the irreducible sets {2, 3, 4} and {5, 6, 7} that have non-trivial long-run behavior. Beginning in either of these sets, the chain visits each state in the set over and over again, and the state space is essentially limited to that set. We refer to such states as *recurrent*. One might ask what fraction of its time the chain spends in each recurrent state, in the long run. This is the main question to be answered in Section 5 on limiting probabilities. Here we will be concerned only with classifying states as either transient or recurrent.

We consider a Markov chain (X_n) with a finite state space E and transition matrix T. Recall that

$$T^n(i, j) = P[X_n = j \mid X_0 = i]$$

The finiteness of the state space removes certain technical obstacles that obscure the main issue upon first reading. The main results do extend to the countable case, however, with small alterations. The goal is to partition the state space of the Markov chain into several sets called *recurrence classes*, and one other set of transient states .

First let us give definitions for the ideas motivated in the above discussion.

DEFINITION 2. A state j is called *recurrent* if the probability that the chain eventually returns to j, given that it started there, is one. Denoting by S (or S_j) the first time $n \geq 1$ such that $X_n = j$, we say that j is *recurrent* if and only if

$$P[S < \infty \mid X_0 = j] = 1$$

A state j that is not recurrent is called *transient*. If S is as above, then j is transient if and only if either of

$$P[S < \infty \mid X_0 = j] < 1 \ \text{ or } \ P[S = \infty \mid X_0 = j] > 0$$

Note that we require a somewhat weaker condition for transience than state 1 in Figure 4.6 actually satisfies. It is only necessary to have a non-zero probability of never returning in order for a state to be transient, whereas state 1 was certain never to be visited again after the first time.

It is convenient to introduce another piece of notation. We write $i \longrightarrow j$ if there is a path from i to j in the transition diagram. In this case we say that " i reaches j" , or "j is accessible from i ". It is easy to see that

$$i \longrightarrow j \iff \exists\, n \geq 0 \ \text{ such that } T^n(i, j) > 0 \tag{1}$$

EXAMPLE 1. In Figure 4.8, we see that for example $1 \longrightarrow 2$, $2 \longrightarrow 3$, $3 \longrightarrow 2$, $1 \longrightarrow 5$ (since $1 \longrightarrow 6$ in one step and $6 \longrightarrow 5$ in one step), etc.

Several things can be seen about the chain in Figure 4.8. First, state 1 must be transient, since the probability is 3/4 that the next step is into one of the two closed sets {2, 3} or {4, 5, 6}, from which 1 can never again be visited. State 2 should be recurrent, because starting there, the chain either returns there immediately or goes to state 3. In the latter situation the probability of staying in state 3 forever is $1/4 \cdot 1/4 \cdot 1/4 \cdots = 0$, i.e., the chain must eventually return to state 2. State 6 is clearly recurrent, because from it the chain could either go to state 5 (after which it must go back to 6), or to state 4. From state 4 the chain could either return directly to 6 or pass through 5 before returning to 6. In any case, a return to state 6 is inevitable. ∎

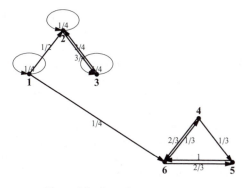

Figure 4.8 – Several recurrent states

Activity 2 – If the Markov chain of Figure 4.8 starts at state 5, is a return to 5 inevitable? Why?

The next results give a more systematic approach to the problem of classifying states as transient or recurrent. They employ the following easy lemma, whose proof we leave to the reader as Exercise 9.

LEMMA 1. If $i \longrightarrow j$ and $j \longrightarrow k$, then $i \longrightarrow k$. ∎

A second lemma will also be necessary. This lemma gives an interesting characterization of recurrence in terms of the expected number of visits to the alleged recurrent state.

LEMMA 2. A state k is recurrent iff

$$\sum_{n=1}^{\infty} T^n(k, k) = \infty \qquad (2)$$

(This sum is the expected number of return visits to k.)

Proof. Define random variables:

$$I_n = \begin{cases} 1 & \text{if } X_n = k \\ 0 & \text{otherwise} \end{cases} \tag{3}$$

Then the expected number of visits to k starting from k is

$$
\begin{aligned}
E[\textstyle\sum_{n=1}^{\infty} I_n \mid X_0 = k] &= \textstyle\sum_{n=1}^{\infty} E[I_n \mid X_0 = k] \\
&= \textstyle\sum_{n=1}^{\infty} P[X_n = k \mid X_0 = k] \\
&= \textstyle\sum_{n=1}^{\infty} T^n(k, k)
\end{aligned}
$$

If state k is recurrent, then with certainty it will be visited infinitely often, hence the number N_k of visits to k has infinite expectation. This shows the forward part of the double implication. For the converse, assume that the expected number of visits to k is infinite. An open-ended sequence of Bernoulli trials is generated, in which the ith trial is a success if, after the $(i-1)^{\text{st}}$ visit to k, state k is eventually visited again. The success probability for a trial is $f_{kk} = P[S_k < \infty \mid X_0 = k]$. The number of return visits to j is therefore the number of successes until the first failure, which has the geometric distribution with success parameter f_{kk}. The expected number of successes is $1/(1 - f_{kk})$, which has been assumed to be infinite. The only way for this to happen is for f_{kk} to be 1, i.e., k is recurrent. This proves the lemma. ∎

THEOREM 1. If j is recurrent and $j \longrightarrow k$, then $k \longrightarrow j$ and k is recurrent.

Proof. Suppose that j is recurrent and that $j \longrightarrow k$. We show first that $k \longrightarrow j$. Since there is a path from j to k, there is also a simple path from j to k, in particular, one that does not pass through j as an intermediate state. Figure 4.9 illustrates the simple path.

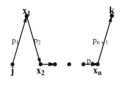

Figure 4.9 – A simple path from state j to state k

The event

$$\{X_1 = x_1, X_2 = x_2, \ldots, X_n = x_n, X_{n+1} = k, X_i \neq j \text{ for all } i > n + 1\}$$

is contained in the event $\{X_i \neq j \text{ for all } i \geq 1\}$. Therefore,

$$
\begin{aligned}
P[X_1 = x_1, X_2 = x_2, \ldots, X_n = x_n, \\
X_{n+1} = k, X_i \neq j \text{ for all } i > n + 1 \mid X_0 = j] \\
\leq P[X_i \neq j \text{ for all } i \geq 1 \mid X_0 = j]
\end{aligned}
\tag{4}
$$

The probability on the left side of (4) is

$$p_1 \cdot p_2 \cdots p_n \cdot p_{n+1} \cdot (1 - P[S_j < \infty \mid X_0 = k])$$

The probability on the right side of (4) is $1 - P[S_j < \infty \mid X_0 = j]$. But since j is recurrent, the right side of (4) is zero. Consequently the left side of (4) is also zero, which yields

$$P[S_j < \infty \mid X_0 = k] = 1$$

The latter can only be true if there is a path from k to j.

Now we show that k is recurrent. There is a path, say of length n, from j to k, hence $T^n(j, k) > 0$. Also, by the argument of the last paragraph there is a path, say of length m, from k to j, hence $T^m(k, j) > 0$. The event that the chain goes from state k to itself in some number $m + n + x$ of steps contains the event that the chain goes from k to j in m steps, and from j to itself in x steps, and from j to k in n steps. Therefore,

$$T^{m+n+x}(k, k) \geq T^m(k, j) \cdot T^x(j, j) \cdot T^n(j, k)$$

Because of this, we have

$$
\begin{aligned}
\sum_{i=1}^{\infty} T^i(k, k) &= \sum_{i=1}^{m+n} T^i(k, k) + \sum_{x=1}^{\infty} T^{m+n+x}(k, k) \\
&\geq \sum_{i=1}^{m+n} T^i(k, k) + \sum_{x=1}^{\infty} T^m(k, j) \, T^x(j, j) \, T^n(j, k) \\
&= \sum_{i=1}^{m+n} T^i(k, k) + T^m(k, j) \, T^n(j, k) \sum_{x=1}^{\infty} T^x(j, j)
\end{aligned}
$$

The right side is infinite by Lemma 2, since j is recurrent. Thus, the left side is infinite, and the same lemma implies that k is recurrent. This completes the proof of Theorem 1. ∎

The last theorem helps to classify recurrent states, because if one state is known to be recurrent, then all states that it can reach are also recurrent. The next result helps to classify transient states. You are asked for a proof in Exercise 7.

THEOREM 2. Let j be a state. If there is a state k such that $j \longrightarrow k$ but k does not reach j, then j is transient. ∎

Finally, Theorem 3 uses the previous results to generate a method for recognizing transient and recurrent states.

THEOREM 3. Fix a state j. Let C be the set of all states in the finite state space E that are reachable from j (together with j itself). If every $k \in C$ reaches j, then all states in C are recurrent and C is an irreducible set. If, on the other hand, there is some $k \in C$ that cannot reach j, then j is transient.

Proof. First, if there is $k \in C$ that cannot reach j, then j must be transient, by Theorem 2. This proves the second assertion of the theorem. Henceforth assume that every state k in C reaches j.

It is easy to see that C is closed, by construction. We claim that there must be at least one recurrent state in C. If not, then all states in C would be transient, and by Lemma 2, the expected number $E[N_k]$ of return visits to each state k in C would be finite. This would mean that for almost every outcome, and every $k \in C$, N_k itself must be finite. But this is a contradiction, since C is closed and of finite size, hence the total of the number of visits to all states $k \in C$ must be infinite.

The argument in the last paragraph allows us to suppose that there is a recurrent state, say i, in C. Let k be any other state in C. We have $i \longrightarrow j$ by assumption, and $j \longrightarrow k$ by construction of C. Thus, the recurrent state i reaches k, and Theorem 1 implies that k is recurrent.

We have already seen that C is closed. It remains only to show that C is irreducible. Let k_1 and k_2 be in C; we will show that both $k_1 \longrightarrow k_2$ and $k_2 \longrightarrow k_1$. By assumption, $k_1 \longrightarrow j$, and by construction, $j \longrightarrow k_2$. Thus, $k_1 \longrightarrow k_2$. The same argument can be used to show that $k_2 \longrightarrow k_1$. Since all states communicate with each other, C can have no proper closed subsets. The proof is complete. ∎

In the first case of Theorem 3, the irreducible set of recurrent states C is called the *recurrence class* of state j.

Activity 3 – Argue that a finite, irreducible Markov chain such as the one in Figure 4.7 must consist of a single recurrence class.

Given a state j, the Connected Components Algorithm of Chapter 1 allows us to find the set C of all states reachable from j. The last theorem gives a condition that allows us to tell if j is transient, namely if not all $k \in C$ can reach j. If j is not transient, then j, as well as all $k \in C$, are recurrent. The recurrence class C is also irreducible, so that there are no cliques of states inside C that can capture the chain forever. Since C is closed, if the chain starts in C, then C is essentially a state space of its own, for the chain can never leave C. By finding recurrence classes and transient states in this way, we can partition the state space into fundamental subspaces, and we will see in the next two sections how to use this partition to fully characterize the limiting behavior of the chain.

EXAMPLE 2. We now classify the states of the diagram in Figure 4.8. State 1 can reach all of states 2, 3, 4, 5, 6, but none of these states can reach 1, hence 1 is transient. State 2 is the next unclassified state; it reaches only state 3. Since $3 \longrightarrow 2$, the set $\{2, 3\}$ forms a recurrence class C_1. The next unclassified state is 4, which reaches 5 and 6. Both 5 and 6 reach 4, hence the set $\{4, 5, 6\}$ forms a recurrence class C_2.

Mathematica confirms our analysis, and can be used as below to assist in finding transient states and recurrence classes for larger, more complicated chains. A version of the Components function that was developed in Chapter 1 is in the package KnoxOR`StochasticProcesses`. We give it a name that is more appropriate for this context.

```
(**   ReachableSet[transmatrix,state]   **)
```

The ReachableSet command takes the transition matrix of the Markov chain, and the number of a state. It returns a list of all states reachable from the given state. In the closed cell that produced the graph in Figure 4.8, the adjacency matrix of the graph was named graph8. We call on ReachableSet for all states in the chain, and find that state 1 reaches all states, but none of them can get to 1. States 2 and 3 reach only each other, and states 4, 5, 6 only reach each other, as we noted above. ∎

```
Needs["KnoxOR`StochasticProcesses`"];
```

```
{ReachableSet[graph8, 1],
 ReachableSet[graph8, 2],
 ReachableSet[graph8, 3],
 ReachableSet[graph8, 4],
 ReachableSet[graph8, 5],
 ReachableSet[graph8, 6]}
```

```
{{1, 2, 3, 4, 5, 6}, {2, 3},
 {2, 3}, {4, 5, 6}, {4, 5, 6}, {4, 5, 6}}
```

EXAMPLE 3. Consider the Markov chain whose transition diagram is in Figure 4.10. Let us partition the state space into transient states and recurrence classes. (We suppress the transition probabilities as being irrelevant for this purpose.)

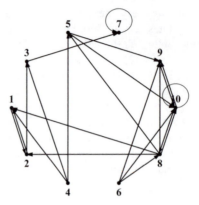

Figure 4.10 – Classifying states of a Markov chain

The reachable sets for the ten states are computed below.

```
{ReachableSet[graph10, 1],
 ReachableSet[graph10, 2],
 ReachableSet[graph10, 3],
 ReachableSet[graph10, 4],
 ReachableSet[graph10, 5],
 ReachableSet[graph10, 6],
 ReachableSet[graph10, 7],
 ReachableSet[graph10, 8],
 ReachableSet[graph10, 9],
 ReachableSet[graph10, 10]}
```

{{1, 2, 3, 7}, {1, 2, 3, 7},
 {3, 7}, {1, 2, 3, 4, 5, 7, 8, 9, 10},
 {1, 2, 3, 5, 7, 8, 9, 10}, {1, 2, 3, 5, 6, 7, 8, 9, 10},
 {7}, {1, 2, 3, 5, 7, 8, 9, 10},
 {1, 2, 3, 5, 7, 8, 9, 10}, {1, 2, 3, 5, 7, 8, 9, 10}}

Note that state 7 only reaches itself, that is, state 7 is an absorbing state. But all of the other states can reach 7, which does not reach any of them. So, we have a large set of transient states {1, 2, 3, 4, 5, 6, 8, 9, 10}, and a single recurrence class C consisting of state 7 only. ∎

Exercises 4.4

1. Find all closed sets for the court case chain of Exercise 6, Section 4.3.

2. (*Mathematica*) Find all irreducible sets of states for the chain with the transition diagram below. (The arrows represent all transitions that have non-zero probability.) Find the recurrence classes and the set of transient states.

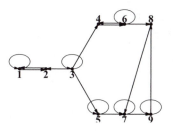

Exercise 2

3. (*Mathematica*) Calculate the first four powers of the transition matrix T for the chain whose transition diagram is below. Is the graph a regular graph in the sense of Chapter 1? Is the chain irreducible? Can anything be said about the behavior of T^n as $n \longrightarrow \infty$?

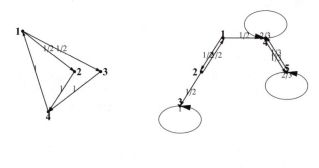

Exercise 3   **Exercise 4**

4. Classify all states in the Markov chain with transition diagram above.

5. (*Mathematica*) Find the recurrence classes and transient states of the chain whose transition matrix is below.

$$\begin{pmatrix} 1/4 & 0 & 0 & 0 & 1/2 & 1/4 & 0 & 0 \\ 0 & 0 & 0 & 1 & 0 & 0 & 0 & 0 \\ 0 & 0 & 1 & 0 & 0 & 0 & 0 & 0 \\ 0 & 1/2 & 0 & 0 & 0 & 0 & 1/2 & 0 \\ 1/3 & 0 & 0 & 0 & 2/3 & 0 & 0 & 0 \\ 1/4 & 0 & 0 & 0 & 1/4 & 1/2 & 0 & 0 \\ 0 & 1/2 & 0 & 0 & 0 & 0 & 1/2 & 0 \\ 0 & 1/4 & 0 & 1/4 & 0 & 0 & 1/4 & 1/4 \end{pmatrix}$$

Exercise 5

$$\begin{pmatrix} 0 & 1 & 0 & 0 & 0 & 0 & 0 & 0 \\ 3/4 & 0 & 1/4 & 0 & 0 & 0 & 0 & 0 \\ 0 & 0 & 0 & 1 & 0 & 0 & 0 & 0 \\ 0 & 0 & 0 & 2/3 & 1/3 & 0 & 0 & 0 \\ 0 & 0 & 1/2 & 1/2 & 0 & 0 & 0 & 0 \\ 0 & 0 & 0 & 1/3 & 0 & 0 & 1/3 & 1/3 \\ 0 & 0 & 0 & 0 & 0 & 1 & 0 & 0 \\ 0 & 0 & 0 & 0 & 0 & 0 & 0 & 1 \end{pmatrix}$$

Exercise 6

6. (*Mathematica*) Repeat Exercise 5 for the Markov chain with transition matrix above.

7. Prove Theorem 2.

8. Show that if (X_n) is a Markov chain with finite state space E such that $i \longrightarrow j$ for all states i and j, and such that there exists a state i_0 with $T(i_0, i_0) > 0$, then the chain is regular (i.e., it has a regular transition diagram, in the sense of Chapter 1).

9. Prove Lemma 1.

10. Devise an example of a Markov chain with two absorbing states 1,2 and three transient states 3,4,5 in which the transient states have self-loops, and the probability that, starting from each transient state, the chain is eventually absorbed in state 1 is 1/2 and the probability that it is eventually absorbed in state 2 is also 1/2.

11. In view of the recursive equation $\mathbf{p}^{(n)} = \mathbf{p}^{(n-1)} \cdot T$ for the short-run distributions of a Markov chain, assuming that there is such a thing as a long-run or limiting distribution represented by a vector $\mathbf{p}$, what equation should that vector satisfy? Check your hypothesis against the computations of Examples 1 and 3 in Section 4.2 on the frozen food companies.

4.5 Limiting Probabilities

Main Results

In this section we will derive the limiting probabilities

$$\lim_{n \to \infty} P[X_n = j \mid X_0 = i]$$

for a Markov chain. The notation is as usual, and the state space E is assumed to be finite in this section. Recall that we have shown that the conditional probability whose limit is being taken is the same as $T^n(i, j)$. The following is a preliminary result that clarifies the limiting behavior for transient states, shows that the limiting probabilities for recurrent states do exist, and gives an intuitive interpretation of the limiting probabilities for a recurrent state in terms of the time between visits and the long-run proportion of time spent in the state. One extra term must be introduced in order to state part (c) of the theorem. We say that a Markov chain is *regular* if its transition diagram represents a regular graph, i.e., there is some power T^n of the transition matrix which is non-zero in every component.

THEOREM 1. (a) If a state j is transient, then

$$\lim_{n \to \infty} T^n(i, j) = \lim P[X_n = j \mid X_0 = i] = 0 , \forall i \in E$$

(b) If i cannot reach j, then $\lim T^n(i, j) = 0$. This occurs in particular if i is recurrent and j is transient, or if i and j belong to different recurrence classes.

(c) If i and j belong to the same recurrence class, and the transition matrix restricted to that class is regular, then the quantity

$$\pi_j = \lim_{n \to \infty} T^n(i, j)$$

exists and does not depend on i.

(d) Under the hypothesis of part (c), the limiting probability π_j equals the reciprocal of the mean time between visits to state j.

(e) Under the hypothesis of part (c), for almost every outcome ω,

$$\pi_j = \lim_{n \to \infty} N_t(\omega)/t$$

where N_t is the number of visits by the chain to state j through time t. ∎

We will not give the proof here, since it depends on results from *renewal theory*, to be introduced in Chapter 5. In the exercise set of Section 5.3, you are led through a proof. Part (a) is intuitively obvious, since a transient state will eventually be left forever. Part (b) is also clear, since the condition that i does not reach j implies that $P[X_n = j \mid X_0 = i] = 0$ for all n. In part (d), for example, if a recurrent state has limiting probability 1/4 then we can say that the average number of time units between visits to that state is 4. And part (e) is analogous to the Strong Law of Large Numbers for independent events, in the sense that it says that, except for an exceptional set of outcomes of probability zero, if we follow the path of the chain for a very long time, it will spend a fraction of the time π_j in the recurrent state j.

Activity 1 – Make up an example of a regular Markov chain with four states.

Let us look for a moment at the structure of the transition matrix T. The states in E can be reordered so that the states in the first recurrence class C_1 appear first, then the states in the second recurrence class C_2, etc. out to class C_m, and after all of those classes is the set of transient states, denoted by S. Recall that a recurrence class is closed, therefore different classes cannot reach each other. Also, no recurrent state can reach a transient state. Then the transition matrix T can be written in block form as in Figure 4.11. Each block T_{ii} is the transition matrix of the chain restricted to class C_i. Blocks $Q_1, Q_2, \ldots, Q_m$ give transition probabilities from transient states to recurrent states in recurrence classes $1, 2, \ldots, m$, respectively. We denote by R the submatrix of T containing the transition probabilities from transient states to transient states.

$$
\begin{array}{c}
\begin{array}{cccccc}
C_1 & C_2 & \cdots & C_m & & S
\end{array} \\
\begin{array}{c} C_1 \\ C_2 \\ \vdots \\ C_m \\ \\ S \end{array}
\left(
\begin{array}{cccc|c}
T_{11} & 0 & \cdots & 0 & 0 \\
0 & T_{22} & \cdots & 0 & 0 \\
\vdots & \vdots & \ddots & \vdots & \vdots \\
0 & 0 & \cdots & T_{mm} & 0 \\
\hline
Q_1 & Q_2 & \cdots & Q_m & R
\end{array}
\right)
\end{array}
$$

Figure 4.11 – Block structure of a transition matrix

Now consider the limit as $n \longrightarrow \infty$ of T^n. By part (a) of Theorem 1, the entire S "column" of this limiting matrix is 0. By part (b) of the same theorem, the block of the limiting matrix corresponding to the C_i "row" and C_j "column" for $i \neq j$ must be 0 as well. Therefore the northwest quadrant of the limiting matrix will have the same diagonal structure as the matrix T

in Figure 4.11. The southeast quadrant of the limiting matrix will be a zero block, as opposed to the block R in the southeast corner of T.

Each recurrence class C_i corresponds to a Markov chain of its own, whose state space is the recurrence class. Just two tasks must be done in order to compute all limiting probabilities. The first is to find the limit of $T^n(i, j)$ for i and j in the same recurrence class, and the second is to find this limit for i transient and j recurrent. The latter problem is solved in Section 6. The next theorem solves the former problem by showing that limiting probabilities restricted to a given class may be found by solving a system of linear equations. The hypothesis of regularity of the restricted chain is needed, however.

THEOREM 2. Let (X_n) be a regular Markov chain with transition matrix T and finite state space E. Let π_j be as in (3), and let π be the row vector formed by the π_j as j ranges through E. Then

$$\pi = \pi \cdot T \text{ and } \pi \cdot \mathbf{1} = 1 \tag{5}$$

where $\mathbf{1}$ is the column vector of length equal to the size of the state space, all of whose entries are equal to 1.

Proof. We claim that the state space E consists of a single recurrence class. By assumption, for some m, T^m is entirely non-zero, hence there is a path of length m from every i to every j in E. By Theorem 3 of Section 4.4, E is irreducible, and each state in E is recurrent.

Thus, by part (c) of Theorem 1, the limiting matrix $\lim T^n$ exists, and every row of this limiting matrix is π. For each $i \in E$ and each $n > 0$, the sum of the entries in row i of T^n equals 1, hence:

$$1 = \lim_{n \to \infty} \sum_{j \in E} T^n(i, j) = \sum_{j \in E} \lim_{n \to \infty} T^n(i, j) = \sum_{j \in E} \pi_j$$

which shows that $\pi \cdot \mathbf{1} = 1$.

Expressing T^{n+1} as $T^n \cdot T$, we can write that, for all states i and j,

$$T^{n+1}(i, j) = \sum_{k \in E} T^n(i, k) T(k, j) \tag{6}$$

We wish to send $n \to \infty$ on both sides of (6). Because the state space is finite, the limit can be brought past the sum on the right side, to yield:

$$\begin{aligned}
\pi_j = \lim_{n \to \infty} T^{n+1}(i, j) &= \sum_{k \in E} \lim_{n \to \infty} T^n(i, k) T(k, j) \\
&= \sum_{k \in E} \pi_k \cdot T(k, j) \\
&= \pi \cdot T(j)
\end{aligned} \tag{7}$$

Therefore $\pi = \pi \cdot T$, which completes the proof. ∎

Referring again to the block form of T in Figure 4.11, each recurrence class C_i can be treated as the state space of an irreducible Markov chain with transition matrix T_{ii}. As long as T_{ii} is regular, the limiting vector π for that class can be found by solving the linear system $\pi = \pi \cdot T_{ii}$ subject to the condition that the entries of π sum to 1. We hasten to point out that the finiteness of the state space is not a necessary hypothesis in Theorem 2. It was only included to facilitate the exchange of limit and sum in (7). The same result can be shown in the infinite state case; for the more subtle proof, see Ross ([52], Thm. 4.3.3).

Activity 2 – Check that the Markov chain with transition matrix below is regular, and write out in full detail the system of equations for the limiting probabilities. If you solve for π_2 in terms of π_3 in the second equation, and then substitute that into the first equation, what do you notice?

$$\begin{pmatrix} 1/3 & 0 & 2/3 \\ 1/2 & 1/2 & 0 \\ 0 & 3/4 & 1/4 \end{pmatrix}$$

Examples

We now illustrate the application of the preceding results.

EXAMPLE 1. Return to the Markov chain of Figure 4.8 in Section 4. In the electronic version of the text, you can execute the cell below this one to reproduce the transition diagram. Recall that state 1 was the only transient state, states 2 and 3 formed a recurrence class, and states 4, 5, and 6 formed a second recurrence class.

We can block off the transition matrix as follows:

$$T = \begin{array}{c} \\ 2 \\ 3 \\ \\ 4 \\ 5 \\ 6 \\ \\ 1 \end{array} \begin{pmatrix} \begin{array}{cc} 2 & 3 \end{array} & \begin{array}{ccc} 4 & 5 & 6 \end{array} & \begin{array}{c} 1 \end{array} \\ \begin{array}{cc} 1/4 & 3/4 \\ 3/4 & 1/4 \end{array} \Big| & \begin{array}{ccc} 0 & 0 & 0 \end{array} \Big| & \begin{array}{c} 0 \\ 0 \end{array} \\ \hline \begin{array}{cc} 0 & 0 \\ 0 & 0 \\ 0 & 0 \end{array} \Big| & \begin{array}{ccc} 0 & 1/3 & 2/3 \\ 0 & 0 & 1 \\ 1/3 & 2/3 & 0 \end{array} \Big| & \begin{array}{c} 0 \\ 0 \\ 0 \end{array} \\ \hline \begin{array}{cc} 1/2 & 0 \end{array} \Big| & \begin{array}{ccc} 0 & 0 & 1/4 \end{array} \Big| & \begin{array}{c} 1/4 \end{array} \end{pmatrix}$$

The block corresponding to the set of states $\{2, 3\}$ is itself a transition matrix of a regular chain. Let $\pi = (x, y)$. The relevant system of equations for the limiting probabilities for the irreducible set $\{2, 3\}$ is

$$\pi = \pi \cdot \begin{pmatrix} 1/4 & 3/4 \\ 3/4 & 1/4 \end{pmatrix} \implies \begin{cases} x = \frac{1}{4} x + \frac{3}{4} y \\ y = \frac{3}{4} x + \frac{1}{4} y \\ x + y = 1 \end{cases}$$

The first two equations both simplify to $x = y$, hence it is easy to find the solution to the resulting system of two equations in two unknowns, which is $x = 1/2$, $y = 1/2$. The meaning of this computation is

$$\begin{aligned} \lim_{n \to \infty} P[X_n = 2 \mid X_0 = i] &= 1/2, \quad i = 2, 3 \\ \lim_{n \to \infty} P[X_n = 3 \mid X_0 = i] &= 1/2, \quad i = 2, 3 \end{aligned} \tag{8}$$

The limit of the $\{2, 3\}$ diagonal block of T^n as $n \to \infty$ is a 2×2 matrix, both of whose rows are $(1/2 \ \ 1/2)$.

Similarly, we can calculate the limit of the $\{4, 5, 6\}$ diagonal block. You are asked in Exercise 8 to check that paths of length 4 exist from each state in $\{4, 5, 6\}$ to each other state, hence the chain restricted to $\{4, 5, 6\}$ is regular. Theorem 2 can now be applied. The equations are

$$(x \ \ y \ \ z) = (x \ \ y \ \ z) \cdot \begin{pmatrix} 0 & 1/3 & 2/3 \\ 0 & 0 & 1 \\ 1/3 & 2/3 & 0 \end{pmatrix}, \ \ x + y + z = 1$$

When the system is expanded out, you can check that we get the following system, which we solve in *Mathematica*:

```
system = {x == 1/3 z,

    y == 1/3 x + 2/3 z, z == 2/3 x + y, x + y + z == 1};
Solve[system, {x, y, z}]
```

$$\left\{ \left\{ x \to \frac{3}{19}, \ y \to \frac{7}{19}, \ z \to \frac{9}{19} \right\} \right\}$$

Thus, in the block of T corresponding to $\{4, 5, 6\}$, the limiting matrix is a 3×3 matrix all of whose rows are $(3/19 \ \ 7/19 \ \ 9/19)$. Regardless of the initial state within $\{4, 5, 6\}$, the long-run probabilities that the chain occupies states 4, 5, and 6, respectively, are given by these three values. Also, for example, by Theorem 1(d), the mean time between return visits to state 5 is $1/(7/19) = 19/7$. ∎

Activity 3 – It can be shown (see Exercise 7) that the system $\pi = \pi \cdot T$ is always dependent, hence an equation can be thrown away with no loss. Use *Mathematica* to check to see that you get the same solution to the system if you discard each of the first three equations in turn.

EXAMPLE 2. We can compute the limiting distribution for the frozen food companies of Example 2 in Section 1 much more easily now than we did earlier. The transition matrix is below, and the diagram is in Figure 4.3 of Section 1, the code for which is in the closed cell beneath the transition matrix, so that you can reproduce it in the electronic text. It is easy to check that T^2 is entirely positive, hence the chain is regular.

$$T = \begin{pmatrix} 1/2 & 1/4 & 1/4 \\ 1/2 & 1/2 & 0 \\ 0 & 1/4 & 3/4 \end{pmatrix}$$

The following system must be solved:

$$(x\ y\ z) = (x\ y\ z) \cdot \begin{pmatrix} 1/2 & 1/4 & 1/4 \\ 1/2 & 1/2 & 0 \\ 0 & 1/4 & 3/4 \end{pmatrix},\quad x + y + z = 1$$

$$\implies \begin{cases} x = \frac{1}{2}x + \frac{1}{2}y \\ y = \frac{1}{4}x + \frac{1}{2}y + \frac{1}{4}z \\ z = \frac{1}{4}x + \frac{3}{4}z \\ x + y + z = 1 \end{cases}$$

The first and third equations imply that $x = y = z$, hence the last equation yields $x = y = z = 1/3$. This is precisely what was found in Section 2. ∎

EXAMPLE 3. An air conditioning system can be running at one of three speeds: high, low, or off. The table below gives the conditional probabilities that at the next minute the speed will be j given that now the speed is i, for each possible pair (i, j). Find the long-run proportion of time spent at each speed.

	next speed		
	high	low	off
current speed high	.7	.2	.1
low	.1	.8	.1
off	0	.4	.6

The table is the transition matrix of the Markov chain (X_n), where X_n is the speed at which the air conditioner is running at minute n. By Theorem 1(e), the long-run proportion of time spent at speed j is π_j, which is, by Theorem 2, the jth component of the vector $\pi = (x \ y \ z)$ satisfying the system of equations (5). You can check that the equations translate to those below. (We discard the third equation in $\pi = \pi \cdot T$ since it is expressible in terms of the other equations.)

```
system3 =
  {x == .7 x + .1 y, y == .2 x + .8 y + .4 z, x + y + z == 1};
Solve[system3, {x, y, z}]
```

$\{\{x \to 0.2, \ y \to 0.6, \ z \to 0.2\}\}$

Thus, the machine is on high 1/5 of the time, low 3/5 of the time, and off 1/5 of the time on the average, in the long run. We also observe from part (d) of Theorem 1 that the mean time between shutoffs is $1/(1/5) = 5$ minutes. If the air conditioner costs 1 cent per minute when it is on high, 1/2 cent per minute on low, and nothing when it is off, then over a 300-minute span, we expect about 60 minutes of high speed and 180 minutes of low speed, for an approximate cost of $60 \cdot 1 + 180 \cdot (1/2) = 150$ cents. ∎

Activity 4 – See what *Mathematica*'s reaction is if you ask it to solve for x, y, z in Example 3 using only the three equations you get from $\pi = \pi \cdot T$.

REMARK. Suppose that a regular chain is such that its initial distribution is the same as its limiting distribution π, i.e., $\pi_j = P[X_0 = j]$. Then

$$\mathbf{p}^{(n)} = \mathbf{p}^{(0)} \cdot T^n = \pi \cdot T^n = (\pi T) T^{n-1} = \pi T^{n-1} = \cdots = \pi T = \pi$$

where $\mathbf{p}^{(n)}$ is the distribution of X_n. The equation above says that $P[X_n = j] = \pi_j$ for all times n, if π is the initial distribution. For this reason, π is often referred to as the *stationary* or *steady-state* distribution of the chain. Similarly, the equations (5) that characterize π are called the *stationary*, or *steady-state equations*. ∎

EXAMPLE 4.   Our final example deals with the long-run distribution of inventory for a discrete-time demand model, with a particular kind of reordering policy, called an $s - S$ policy, in which the stock is restored to a level of S as soon as demands for the stocked item decrease the inventory to a level below s.   A hardware store stocks a certain type of lawn mower. When inventory level falls to 0, an order is placed immediately to a local distribution center so that by the next morning there are four lawn mowers in stock.   Total daily demands for the mower are independent random variables taking the values 0, 1, 2, and 3, respectively, with probabilities 2/3, 1/9, 1/9, and 1/9.   Unsatisfied demands are simply lost.   Let X_n be the number of mowers in stock at the beginning of day n.   Find the limiting distribution of X_n.

Denote by D_n the number of demands for the mower during day n. Then we can relate the inventory at the beginning of day $n + 1$ to the inventory at the beginning of day n by

$$X_{n+1} = \begin{cases} 4 & \text{if } D_n \geq X_n \\ X_n - D_n & \text{otherwise} \end{cases} \tag{9}$$

By the independence of the demands, it is clear that the chain (X_n) has the Markov property.   To solve the problem we find the transition matrix of the chain and solve the stationary equations.

Formula (9) indicates how to find the transition probabilities.   For instance, suppose that this morning the inventory level was 2.   Tomorrow morning, the level will be 4   if either 2 or 3 demands come in, forcing a reorder.   This occurs with probability 2/9.   Tommorow's level cannot be 3, since we either decrease level due to sales or reorder up to 4.   The inventory level tomorrow can be 2 if no demands come in, and this occurs with probability 2/3.   Finally, the next level could be 1 if exactly 1 demand arrives.   This happens with probability 1/9.   The preceding analysis justifies the second row of the transition matrix below.   You should check the other rows using similar reasoning.   Also, check to see that this is a regular Markov chain.

$$T = \begin{array}{c} \\ 1 \\ 2 \\ 3 \\ 4 \end{array} \begin{pmatrix} \begin{array}{cccc} 1 & 2 & 3 & 4 \end{array} \\ 2/3 & 0 & 0 & 1/3 \\ 1/9 & 2/3 & 0 & 2/9 \\ 1/9 & 1/9 & 2/3 & 1/9 \\ 1/9 & 1/9 & 1/9 & 2/3 \end{pmatrix}$$

Set $\pi = (x \ y \ z \ w)$. The stationary equations have the form below. Recall that the first four equations of the system $\pi = \pi T$ are dependent; therefore we may as well throw out the most complicated one, say the fourth involving w.

```
system4 = {x == 2/3 x + 1/9 y + 1/9 z + 1/9 w,

          y == 2/3 y + 1/9 z + 1/9 w,

          z == 2/3 z + 1/9 w,

          x + y + z + w == 1};
Solve[system4, {x, y, z, w}]
```

$$\left\{\left\{x \to \frac{1}{4}, \ y \to \frac{3}{16}, \ z \to \frac{9}{64}, \ w \to \frac{27}{64}\right\}\right\}$$

Thus, the limiting distribution of inventory is $\pi = (\frac{1}{4}, \frac{3}{16}, \frac{9}{64}, \frac{27}{64})$. If, for instance, there is a daily cost of \$10 per mower for storage, then the long-run average cost per day is

```
N[{1 / 4, 3 / 16, 9 / 64, 27 / 64}.{10, 20, 30, 40}]
```

    27.3438

■

Long-Run Discounted Cost

In Example 3 on the long-run behavior of air conditioners, and Example 4 on lawn mower inventory, we had a Markov chain (X_n) that produced a cost of $f(X_n)$ at time n. We found the limiting distribution π and computed the dot product $\pi \cdot \mathbf{f}$ to evaluate long-run average cost per unit time. To be more precise, as in Theorem 1(e), it can be shown that for almost every outcome ω,

$$\lim_{n \to \infty} \sum_{k=0}^{n} f(X_k(\omega))/n = \pi \cdot \mathbf{f} \tag{10}$$

where the function f is looked upon as a vector, with an entry for each member of the state space of the chain. A different criterion for evaluating long-run cost is the expected discounted cost:

$$R^\alpha \mathbf{f}(i) = E[\sum_{n=0}^{\infty} \alpha^n f(X_n) \mid X_0 = i] \tag{11}$$

where $\alpha \in (0, 1)$. The interpretation is that at time n, when the state of the chain is X_n, an absolute cost of $f(X_n)$ is charged. But relative to the value of money today (time 0), the cost is only α^n times this absolute cost, i.e., we discount by a factor of α in each time period after time 0. Then $R^\alpha \mathbf{f}$ gives, for each initial state i, the conditional expectation of the total discounted cost relative to the present value of money. In this subsection we develop a method to calculate the expected total discounted cost.

We assume that f is non-negative and that the state space is finite. By the monotone convergence theorem we may interchange expectation with summation in (11) to obtain

$$
\begin{aligned}
R^\alpha \mathbf{f}(i) &= \sum_{n=0}^{\infty} \alpha^n E[f(X_n) \mid X_0 = i] \\
&= \sum_{n=0}^{\infty} \alpha^n \sum_{j \in E} f(j) T^n(i, j) \\
&= \sum_{j \in E} \left[\sum_{n=0}^{\infty} \alpha^n T^n(i, j) \right] f(j)
\end{aligned}
\tag{12}
$$

The second line follows because row i of the matrix T^n gives the conditional distribution of X_n given $X_0 = i$, and the third line changes the order of summation, which is acceptable for series of non-negative terms.

For small state spaces, one way of computing this is to diagonalize T as in Section 2. But a faster method is possible, as we have seen in the computation of long-run probabilities. The vector $R^\alpha \mathbf{f}$ can be found by solving a system of linear equations. To see this, consider the matrix defined by

$$
R^\alpha = \sum_{n=0}^{\infty} \alpha^n T^n
\tag{13}
$$

From (12), we see that the quantity $R^\alpha \mathbf{f}$ is just the matrix product $R^\alpha \cdot \mathbf{f}$. Also,

$$
\begin{aligned}
R^\alpha &= I + \alpha T + \alpha^2 T^2 + \alpha^3 T^3 \cdots \\
&= I + \alpha T(I + \alpha T + \alpha^2 T^2 + \cdots) \\
&= I + \alpha T R^\alpha
\end{aligned}
$$

where I is the identity matrix of the appropriate size. Solving for R^α,

$$
(I - \alpha T) R^\alpha = I \implies R^\alpha = (I - \alpha T)^{-1}
\tag{14}
$$

which implies that

$$
(I - \alpha T) R^\alpha \mathbf{f} = \mathbf{f}
\tag{15}
$$

We have the option of either solving for the matrix R^α by inverting $I - \alpha T$, or solving directly for the vector $R^\alpha \mathbf{f}$ by solving the linear system in (15). The latter might save slightly in the amount of computation; but if one wants to reuse the same chain, the same discount factor α, and compare several different cost functions **f**, then it is efficient to compute the matrix R^α and just matrix multiply to get R^α **f** for all desired cost functions.

Though we have been speaking in the context of cost, the function f might just as well be a reward function, in which case R^α **f** becomes the long-run expected discounted reward.

EXAMPLE 5. A gambler has the choice of two games, each consisting of repeated plays. In the first game, he either wins \$3 or nothing on each play according to whether the result of the play is state 1 or state 2. If state 1 comes up on a given play, state 1 will be next with probability .25, or state 2 with probability .75; and if state 2 comes up on a play, then state 1 will be next with probability .15, so that as the game progresses, the outcome of the next play is conditionally independent of the past history of the game given the present state. The second game available to the gambler is similar to the first, except that he wins \$2 or nothing on each play, and a return to state 1 happens with probability 1/3 and a transition from state 2 to state 1 happens with probability .2, as shown in Figure 4.12. Which game is preferable if an absolute dollar earned on the nth play is only worth $.9^n$ in present terms?

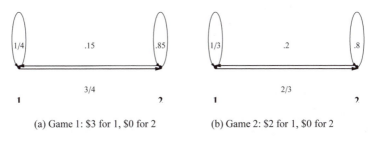

(a) Game 1: \$3 for 1, \$0 for 2   (b) Game 2: \$2 for 1, \$0 for 2

Figure 4.12 – Which is the better gamble?

There is a trade-off that we must analyze, between game 1 which pays more per visit to state 1, and game 2 which visits state 1 more often. For each game, we compute the expected long-run discounted reward with discount factor $\alpha = .9$. The Markov chain involved is defined in the same way for each game, namely X_n = state at play n. The absolute reward earned when the state is k is the kth component of the reward vector for the corresponding game. Those vectors are, respectively, $\mathbf{f}_1 = (3\ 0)'$, $\mathbf{f}_2 = (2, 0)'$. The transition matrices for the two games were defined as matrix12a and matrix12b in the closed cell that generated the graphics for Figure 4.12. For the first game, by (14) we have:

```
ident = {{1, 0}, {0, 1}};
Ralpha1 = Inverse[ident - .9 * matrix12a];
MatrixForm[Ralpha1]
```

$$\begin{pmatrix} 2.58242 & 7.41758 \\ 1.48352 & 8.51648 \end{pmatrix}$$

Hence, the expected discounted reward for game 1 is

```
Ralpha1.{3, 0}
```

{7.74725, 4.45055}

Similarly for the second game,

```
Ralpha2 = Inverse[ident - .9 * matrix12b];
MatrixForm[Ralpha2]
Ralpha1.{2, 0}
```

$$\begin{pmatrix} 3.18073 & 6.81927 \\ 2.04476 & 7.95524 \end{pmatrix}$$

{5.16484, 2.96703}

We conclude that game 1 is better, no matter whether the chain begins with state 1 or state 2. We can easily answer another question: how much should be won on a play of game 2 so that this game is equal in value to game 1, when the initial state is 1? If the winnings at state 1 for game 2 equal some unknown c, then for parity between games we require that row 1 of R_2^{α} times $(c\ 0)'$ must equal the expected discounted reward from state 1 in game 1, namely 7.74725.

```
Solve[3.18073 c == 7.74725, c]
```

{{c → 2.43568}}

Again we see the versatility gained by computing the matrix R^{α} rather than solving system (15) for R^{α} **f**. ∎

Activity 5 – In Example 5, suppose that the game 1 transition matrix is as on the left below, and the game 2 transition matrix is as on the right. Do you really need the machinery of Markov chains to compare the two games? Why or why not?

$$T_1 = \begin{pmatrix} 1/4 & 3/4 \\ 1/4 & 3/4 \end{pmatrix}, \quad T_2 = \begin{pmatrix} 1/3 & 2/3 \\ 1/3 & 2/3 \end{pmatrix}$$

Exercises 4.5

1. (*Mathematica*) For the chain of Exercise 6 of Section 4.4, whose transition matrix is reproduced below, find the limiting distribution π within each recurrence class.

$$\begin{pmatrix}
0 & 1 & 0 & 0 & 0 & 0 & 0 & 0 \\
3/4 & 0 & 1/4 & 0 & 0 & 0 & 0 & 0 \\
0 & 0 & 0 & 1 & 0 & 0 & 0 & 0 \\
0 & 0 & 0 & 2/3 & 1/3 & 0 & 0 & 0 \\
0 & 0 & 1/2 & 1/2 & 0 & 0 & 0 & 0 \\
0 & 0 & 0 & 1/3 & 0 & 0 & 1/3 & 1/3 \\
0 & 0 & 0 & 0 & 0 & 1 & 0 & 0 \\
0 & 0 & 0 & 0 & 0 & 0 & 0 & 1
\end{pmatrix}$$

Exercise 1

2. (*Mathematica*) Find the limiting distribution for the following random walk with "sticky barriers."

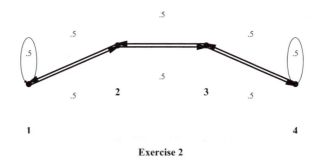

Exercise 2

3. (*Mathematica*) A company rents vans for personal moving. There are three districts from which vans can be rented, and to which they can be returned. The conditional probabilities that vans originating in each district are returned to each district are given in the table below. Find the long-run proportion of vans in each district.

$$
\begin{array}{c}
\text{to}\\
\begin{array}{cccc}
 & 1 & 2 & 3\\
1 & .1 & .5 & .4\\
\text{from} \quad 2 & .6 & .2 & .2\\
3 & .3 & .3 & .4
\end{array}
\end{array}
$$

4. For a general regular two-state Markov chain, find closed-form expressions for the limiting probabilities.

5. (*Mathematica*) For the sales representative of Exercise 1 of Section 4.1, suppose that there are weekly travel expenses of $500, $600, $700, and $800 respectively, in the four regions. Find the long-run average weekly travel expense.

6. Two drill presses are under consideration by a manufacturer. If the first press works one day, then the probability is .9 that it will also work the next. If this press does not work one day, then it will be back in service on the next day with probability .7. For the other drill press, these two conditional probabilities are .95 and .6, respectively. Presuming that the two presses cost the same, which should the manufacturer purchase?

7. Show that the system of equations $\pi = \pi \cdot T$, where T is the transition matrix of a Markov chain, must have infinitely many solutions.

8. For the chain of Example 1, find paths of length 4 from every state to every other state in the set $\{4, 5, 6\}$, and check that 4 is the shortest path length with this property.

9. For the random walk with a reflecting barrier at 0 pictured below, write the stationary equations and verify that the vector **0** is the only solution such that the sum of its components is bounded.

Exercise 9

10. A college-owned van is used until it will not run anymore, and then it is immediately replaced by a similar new one whose lifetime Z has discrete distribution $p_1 = P[Z = 1]$, $p_2 = P[Z = 2]$, ... , where the times are in months. The process continues through successive van replacements. Let (X_n) be the chain defined by X_n = remaining life of the van in use at month n. Notice that a new van has a remaining life with the distribution of Z; otherwise the remaining life of a van next month is one month less than the remaining life this month. Find the transition matrix of this Markov chain. Show that if the mean van lifetime m is finite, then the limiting distribution of X_n exists. Find that limiting distribution.

11. Let (X_n) be a finite, irreducible Markov chain with limiting distribution π, and let f be a real-valued function on its state space. We think of $f(X_n)$ as the reward earned at time n. Use the Dominated Convergence Theorem (see Appendix A) to show that

$$\lim_{m \to \infty} \frac{1}{m+1} \sum_{n=0}^{m} E[f(X_n) \mid X_0 = i] = \pi \cdot \mathbf{f}$$

i.e., the time average expected reward converges to $\pi \cdot \mathbf{f}$, independent of the initial state, as time becomes infinite.

12. A substitute teacher must choose between two school systems. In the first, the probability that he will work on the next school day given that he worked today is 2/3. The probability that he will work on the next day given that he does not work today is 1/4. The corresponding probabilities for the second school system are 3/5 and 1/5. In the first system he is paid $80 per day worked, and in the second he is paid $90 per day worked. Compare the two school systems based on: (a) long-run expected salary per day; and (b) total expected discounted salary, based on a daily discount factor of $\alpha = .95$.

13. Let (X_n) be a two-state Markov chain over which we have a degree of control, in the sense that the transition matrix is

$$T = \begin{pmatrix} .5 + \varepsilon & .5 - \varepsilon \\ .2 - 2\varepsilon & .8 + 2\varepsilon \end{pmatrix}$$

where ε may be chosen from $[-.1, .1]$. If we receive a reward of $2 when state 1 is occupied, and $1 when state 2 is occupied, and there is a discount factor $\alpha = .9$, find ε to maximize the expected total discounted reward starting at state 1.

14. A store stocks an item, for which there is a random demand D each day. We suppose that demands on successive days are i.i.d. random variables with the discrete uniform distribution on $\{0, 1, 2\}$. When the demand exceeds the stock, excess demand is lost. If there are no items left in stock at the end of the previous day, then an order for S items is placed. The order is

filled by morning. Let X_n be the number of items in stock at the beginning of day n.

(a) Write the transition matrix of the chain (X_n).

(b) Write the stationary equations and verify that the following is a solution:

$$x_k = \tfrac{2}{3} x_S \cdot (1 - (-1/2)^{S-k+1}) \quad , k = 1, 2, \ \ldots, \ S,$$

where x_k is the limit as $n \longrightarrow \infty$ of $P[X_n = k]$.

(c) Find x_S. (The information in this problem would be needed, for instance, to minimize, with respect to the reorder quantity S, the sum of long-run expected storage cost and cost due to lost sales.)

4.6 Absorption Probabilities

We have not yet solved the problem of finding the limit as $n \longrightarrow \infty$ of $T^n(i, j)$ for i transient and j recurrent. From Figure 4.8, repeated below, we see that starting from state 1 the chain may or may not ever go to the set $\{2, 3\}$, since the set $\{4, 5, 6\}$ may capture the chain first.

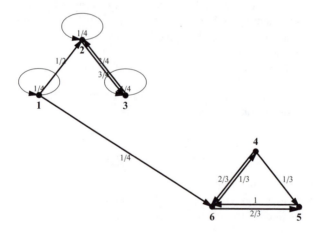

Figure 4.8 (revisited) – Limiting probabilities from transient states to recurrent states

Intuitively, we suspect that the probability that the chain ever reaches $\{2, 3\}$ starting from state 1 might be the ratio of the one-step probability of going to $\{2, 3\}$ to the total one-step probability of going to either $\{2, 3\}$ or $\{4, 5, 6\}$, i.e.,

$$\tfrac{1}{2} / (\tfrac{1}{2} + \tfrac{1}{4}) = \tfrac{2}{3}$$

Our next aim is to develop a systematic approach, which works for more complicated chains.

Activity 1 – Draw the transition diagram of the Markov chain with transition matrix below, and guess the probabilities of ever going to states 2, 3, 4, and 5 starting from state 1.

$$T = \begin{pmatrix} 1/2 & 1/4 & 1/8 & 1/16 & 1/16 \\ 0 & 1 & 0 & 0 & 0 \\ 0 & 0 & 1 & 0 & 0 \\ 0 & 0 & 0 & 1 & 0 \\ 0 & 0 & 0 & 0 & 1 \end{pmatrix}$$

Denote $S_j = \inf_{t \geq 1} \{t : X_t = j\}$, the time of first visit to state j, and

$$f_{ij} = P[S_j < \infty \mid X_0 = i] \tag{1}$$

which is the probability that the chain will ever reach j, starting from state i.

THEOREM 1. Let D be the set of transient states of a Markov chain (X_n) with finite state space E and transition matrix T. Let $i \in D$, and let j belong to a recurrence class C. Then

$$f_{ij} = \sum_{k \in D} T(i, k) f_{kj} + \sum_{k \in C} T(i, k) \tag{2}$$

Proof. The idea of the proof is to condition and uncondition on X_1. If $X_1 = i$, then at time 1 the chain could move to another state in D, to a state in C, or to a state in some other recurrence class, i.e., to the set $(E - D) - C$. From a state in $(E - D) - C$, j cannot be reached; and from a state in C, j is certain to be reached. Thus, we have

$$
\begin{aligned}
f_{ij} &= P[S < \infty \mid X_0 = i] \\
&= \sum_{k \in E} P[S < \infty \mid X_1 = k, X_0 = i]\, P[X_1 = k \mid X_0 = i] \\
&= \sum_{k \in D} P[S < \infty \mid X_1 = k, X_0 = i]\, T(i, k) \\
&\quad + \sum_{k \in C} P[S < \infty \mid X_1 = k, X_0 = i]\, T(i, k) \\
&= \sum_{k \in D} f_{kj}\, T(i, k) + \sum_{k \in C} 1 \cdot T(i, k) \quad \blacksquare
\end{aligned}
$$

We can now finish the computation of the limit of T^n. If i is transient and j is recurrent, then

$\lim_{n \to \infty} T^n(i, j)$

$= \lim_{n \to \infty} P[X_n = j \mid X_0 = i]$

$= \lim_{n \to \infty} P[X_n = j \mid S_j < \infty, X_0 = i] \, P[S_j < \infty \mid X_0 = i]$

The second factor in the line above is f_{ij}. The fact that the former is π_j requires an argument in which we condition on the state X_S at the random time $S = S_j$. To do this, an extension of the Markov property to random times is needed. This extension is called the *strong Markov property*; for a discusssion see Cinlar ([15], Cor. 5.1.26). We omit these details, and merely appeal to the reader's intuition that, if it is known that j will be visited, then in the long run, the proportion of time spent in j is unaffected by the finite amount of time spent outside of the recurrence class of j. To summarize, we have that for i transient and j recurrent,

$$\lim_{n \to \infty} T^n(i, j) = f_{ij} \pi_j \qquad (3)$$

where f_{ij} may be computed by solving the system of linear equations in (2) and π_j may be obtained from the stationary equations in the usual way.

Exercise 8 asks you to show that system (2) may be rewritten in a compact matrix form.

EXAMPLE 1. Let us now complete the determination of the limiting transition matrix for the chain of Figure 4.8. The only transient state is $i = 1$. Theorem 1 gives us

$$f_{12} = T_{11} f_{12} + T_{12} + T_{13}$$
$$\implies (1 - T_{11}) f_{12} = T_{12} + T_{13}$$
$$\implies \tfrac{3}{4} f_{12} = \tfrac{1}{2} + 0$$
$$\implies f_{12} = 2/3$$

A similar computation shows that f_{13} also equals 2/3. In fact, this is a general result: If states k and j are in the same recurrence class and state i is transient, then $f_{ij} = f_{ik}$ (see Exercise 3). Also, we have

$$f_{16} = T_{11} f_{16} + T_{14} + T_{15} + T_{16}$$
$$\implies (1 - T_{11}) f_{16} = T_{14} + T_{15} + T_{16}$$
$$\implies \tfrac{3}{4} f_{16} = 0 + 0 + 1/4$$
$$\implies f_{16} = 1/3$$

From the last line, and the observation above, we see that $f_{14} = f_{15} = f_{16} = 1/3$. In light of the fact that the probability was 2/3 that the chain would be absorbed by the other recurrence class {2, 3}, this answer is not surprising.

Combining the results of Example 1 with those of Example 1 of Section 4.5, the limit of the nth power of T is

$$
\lim_{n \to \infty} T^n = \begin{array}{c} \\ 2 \\ 3 \\ 4 \\ 5 \\ 6 \\ \\ 1 \end{array}
\begin{array}{c}
\begin{array}{cccccc} 2 & 3 & 4 & 5 & 6 & 1 \end{array} \\
\left(\begin{array}{cc|ccc|c}
1/2 & 1/2 & 0 & 0 & 0 & 0 \\
1/2 & 1/2 & 0 & 0 & 0 & 0 \\ \hline
0 & 0 & 3/13 & 1/13 & 9/13 & 0 \\
0 & 0 & 3/13 & 1/13 & 9/13 & 0 \\
0 & 0 & 3/13 & 1/13 & 9/13 & 0 \\ \hline
1/3 & 1/3 & 1/13 & 1/39 & 3/13 & 0
\end{array} \right)
\end{array}
$$

The row corresponding to state 1 was obtained by multiplying f_{1j} by π_j as suggested by (3). ∎

Activity 2 – Apply Theorem 1 to the Markov chain of Activity 1.

EXAMPLE 2. This example illustrates a class of Markov chains called *random walks*, as well as a famous problem in stochastic processes called the *gambler's ruin problem*. In a random walk, a particle moves in discrete time on an integer grid. In one time unit, the particle is allowed to move only to an adjacent grid point. Future positions are independent of the past, given the present position. In the gambler's ruin problem, the net wealth of a gambler, who either wins or loses $1 on each independent play of a game, follows a one-dimensional random walk. The gambler's ruin problem is to find the probability that the gambler goes bankrupt before some level N of wealth is reached.

In our application, at each discrete instant of time, a small business has some whole number i of millions of dollars in assets. At the next time instant the firm will gain a million with probability 5/8 or lose a million with probability 3/8. Given an initial amount of 4 million in assets, what is the probability that the business will reach 8 million before it becomes bankrupt?

If future asset levels are independent of the past given the present, then the successive asset levels of the firm can be modeled as a Markov chain with transition matrix as in the *Mathematica* command below, and transition diagram as in Figure 4.13.

```
matrix13 = {{1, 0, 0, 0, 0, 0, 0, 0, 0},
   {3 / 8, 0, 5 / 8, 0, 0, 0, 0, 0, 0},
   {0, 3 / 8, 0, 5 / 8, 0, 0, 0, 0, 0},
   {0, 0, 3 / 8, 0, 5 / 8, 0, 0, 0, 0},
   {0, 0, 0, 3 / 8, 0, 5 / 8, 0, 0, 0},
   {0, 0, 0, 0, 3 / 8, 0, 5 / 8, 0, 0},
   {0, 0, 0, 0, 0, 3 / 8, 0, 5 / 8, 0},
   {0, 0, 0, 0, 0, 0, 3 / 8, 0, 5 / 8},
   {0, 0, 0, 0, 0, 0, 0, 0, 1}};
```

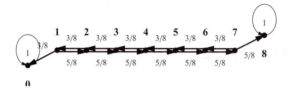

Figure 4.13 – The gambler's ruin problem

Starting from state 0 (bankruptcy) the chain will remain there throughout time, so that 0 is an absorbing state. For our purposes we can suppose that state 8 is also absorbing, since we are only interested in the motion of the chain up to the time that it hits 8. Note that from states 1, 2, ... , 7, it is only possible to step to an adjacent state. Steps to the right have probability 5/8, and steps to the left have probability 3/8. So, this chain is a random walk on $\{0, 1, 2, 3, ..., 8\}$ with absorbing barriers at 0 and 8. Defining X_n = assets at time n, we see that we have been asked for the quantity:

$$f_{48} = P[S_8 < \infty \mid X_0 = 4]$$

States 1 through 7 are clearly transient. Theorem 1 yields a system of equations in the unknowns f_{18}, f_{28}, etc., as in the *Mathematica* command below.

system13 = $\left\{ f_{18} == \frac{5}{8} f_{28}, \right.$

$f_{28} == \frac{3}{8} f_{18} + \frac{5}{8} f_{38}, \quad f_{38} == \frac{3}{8} f_{28} + \frac{5}{8} f_{48},$

$f_{48} == \frac{3}{8} f_{38} + \frac{5}{8} f_{58}, \quad f_{58} == \frac{3}{8} f_{48} + \frac{5}{8} f_{68},$

$f_{68} == \frac{3}{8} f_{58} + \frac{5}{8} f_{78}, \quad f_{78} == \frac{3}{8} f_{68} + \frac{5}{8} \left. \right\};$

Solve[system13, $\{f_{18}, f_{28}, f_{38}, f_{48}, f_{58}, f_{68}, f_{78}\}$]

$\left\{ \left\{ f_{18} \rightarrow \frac{78125}{192032}, \ f_{28} \rightarrow \frac{15625}{24004}, \ f_{38} \rightarrow \frac{153125}{192032}, \ f_{48} \rightarrow \frac{625}{706}, \right. \right.$
$\left. \left. f_{58} \rightarrow \frac{180125}{192032}, \ f_{68} \rightarrow \frac{23275}{24004}, \ f_{78} \rightarrow \frac{189845}{192032} \right\} \right\}$

N[%]

$\left\{ \left\{ f_{18.} \rightarrow 0.406833, \ f_{28.} \rightarrow 0.650933, \right. \right.$
$f_{38.} \rightarrow 0.797393, \ f_{48.} \rightarrow 0.885269,$
$\left. \left. f_{58.} \rightarrow 0.937995, \ f_{68.} \rightarrow 0.96963, \ f_{78.} \rightarrow 0.988611 \right\} \right\}$

The answer originally sought was the probability of going to state 8 from state 4, which is 625/706, but the computation has given much more information. We have computed the probability of reaching 8 million before bankruptcy starting from each of the initial states. Also, since we must either reach 8 million or bankruptcy eventually, the complementary probabilities to the ones above are the bankruptcy probabilities. For instance, the probability that the firm goes bankrupt starting with 2 million dollars is about $1 - .65 = .35$. ∎

Exercises 4.6

1. For Exercise 4 of Section 4.1, the television inspection problem, find the probability, starting from each of the states F and G, of being absorbed by each of the states P and E. If half of the sets that are made are in fair condition initially, and half are in good condition, what proportion of sets are eventually sent out for sale?

2. (*Mathematica*) For the chain with transition matrix below, find the probabilities of absorption into each of the classes {1, 2, 3} and {4, 5} starting from each of the transient states 6 and 7. Find also the limiting transition matrix.

$$\begin{pmatrix}
1/3 & 1/3 & 1/3 & 0 & 0 & 0 & 0 \\
1/2 & 0 & 1/2 & 0 & 0 & 0 & 0 \\
1/4 & 0 & 3/4 & 0 & 0 & 0 & 0 \\
0 & 0 & 0 & 1/2 & 1/2 & 0 & 0 \\
0 & 0 & 0 & 2/3 & 1/3 & 0 & 0 \\
1/8 & 1/8 & 0 & 1/4 & 1/4 & 0 & 1/4 \\
0 & 0 & 0 & 0 & 0 & 1/2 & 1/2
\end{pmatrix}$$

3. Let i, j, and k be states of a finite state Markov chain. If i is transient and j and k are in the same recurrence class C, show that $f_{ij} = f_{ik}$.

4. (*Mathematica*) Consider again Example 2, in which the firm gains or loses one million in assets at each instant. If we now let the gain probability p be general, what is the smallest p such that, starting with four million, the probability is at least 2/3 that assets will hit eight million before bankruptcy?

5. A retail clothing store has begun to issue credit cards in May. Of its card holders, 1000 have not paid the minimum payment in June. Company policy states that if an account has still not been paid after two months, card privileges will be revoked. The experience of similar stores indicates that 60% of one-month delinquent accounts pay up, and 75% of two-month delinquent accounts pay up. How many of the 1000 cards mentioned above can be expected to be revoked?

6. (*Mathematica*) Find the limit as $n \longrightarrow \infty$ of T^n, where T is the transition matrix of the Markov chain of Exercise 2 of Section 4.4. The transition matrix and diagram are shown below for your convenience; note that when multiple edges are directed out of a state, the chain is equally likely to visit any of the states to which these edges point.

$$T = \begin{pmatrix}
1/2 & 1/2 & 0 & 0 & 0 & 0 & 0 & 0 & 0 \\
1 & 0 & 0 & 0 & 0 & 0 & 0 & 0 & 0 \\
0 & 1/4 & 1/4 & 1/4 & 1/4 & 0 & 0 & 0 & 0 \\
0 & 0 & 0 & 0 & 0 & 1 & 0 & 0 & 0 \\
0 & 0 & 0 & 0 & 1/2 & 0 & 1/2 & 0 & 0 \\
0 & 0 & 0 & 1/3 & 0 & 1/3 & 0 & 1/3 & 0 \\
0 & 0 & 0 & 0 & 0 & 0 & 1/2 & 0 & 1/2 \\
0 & 0 & 0 & 0 & 0 & 0 & 1 & 0 & 0 \\
0 & 0 & 0 & 0 & 0 & 0 & 0 & 1/2 & 1/2
\end{pmatrix}$$

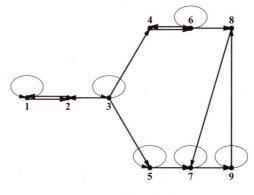

Exercise 6

7. (*Mathematica*) A graduate school offers a 5-year Ph.D. program in mathematics. Its records show a 50% attrition rate between the first and second years, 40% between the second and third, 10% between the third and fourth, 10% between the fourth and fifth, and also 10% of those who reach their fifth year fail to receive their degrees. Model this phenomenon as a Markov chain, and find the probabilities that an entering student, a beginning second-year, third-year, fourth-year, and fifth-year student will successfully complete their degree.

8. (*Mathematica*) In system (2), define a matrix $F = (f_{ij})$ to have a row for each transient state in the set D of all transient states and a column for each recurrent state in a given recurrence class C; define T_{DC} to be the portion of the transition matrix of the chain whose rows are the rows of the transient states in D and whose columns are the states in C; define T_{DD} to be the portion of the transition matrix corresponding to the transient states, and let **1** be a square matrix consisting entirely of 1's, whose size equals the number of states in C. Argue that (2) in matrix form is

$$F = T_{DD} \cdot F + T_{DC} \cdot \mathbf{1}$$

Execute this in *Mathematica* for the gambler's ruin problem of Example 2 for recurrence class $\{8\}$, and make sure that you get the same equations for f_{i8} as in that example.

9. Some elementary texts on Markov chains present the following procedure for chains with absorbing states and transient states. Let S be the submatrix of the full transition matrix T corresponding to the rows of transient states and the columns of absorbing states. Let R be the submatrix corresponding to the rows and the columns of transient states. Compute the matrix

$Q = (I - R)^{-1}$, called the *fundamental matrix* of the chain. Show that $Q(i, j)$ is the expected number of visits to transient state j if the chain begins in transient state i. (Here we suppose that $Q(i, i)$ includes the occupancy of state i at time 0.) Finally, compute the matrix $A = QS$. Show that $A(i, j)$ is the probability that the chain will be absorbed in state j given that it started in state i. (Hint: To justify the meaning of Q, consider $E[\sum_{n=0}^{\infty} I_j(X_n) \mid X_0 = i]$, where the indicator function I_j is 1 if its argument equals j and 0 otherwise. For A, rewrite formula (2) in the notation of these texts.)

10. (See Exercise 9) Let (X_n) be the Markov chain with transition matrix below. Find the expected number of visits from each transient state i to each other transient state j. Find, for each transient state i and absorbing state j, the probability that the chain will be absorbed at j given that it starts at i.

$$T = \begin{pmatrix} 1 & 0 & 0 & 0 & 0 \\ 1/2 & 1/4 & 1/4 & 0 & 0 \\ 0 & 0 & 1 & 0 & 0 \\ 0 & 1/3 & 1/3 & 0 & 1/3 \\ 0 & 1/4 & 0 & 1/2 & 1/4 \end{pmatrix}$$

Continuous Time Processes

Introduction

We turn now to stochastic processes that evolve in continuous time. In Section 1, we discuss the simplest such process, the *Poisson process*, which jumps upward by one unit in such a way that the times between jumps are i.i.d. exponential random variables. This is generalized in Section 2 to the *birth-death process*, which is allowed to jump downward by one unit as well as upward. Then a different generalization of the Poisson process is introduced in Section 3, called the *renewal process*, which again is constrained to jump upward by one unit, but whose inter-jump times are not necessarily exponentially distributed. The rudiments of one of the main applications of stochastic processes, namely *queueing theory*, are covered in Section 4. Finally, we study a process with continuous state space called the *Brownian motion* in Section 5, together with some of its applications.

5.1 Poisson Processes

Definitions and Main Results

The process that we wish to study in this section is a member of a class called *arrival counting processes*. The path of such a process for a typical outcome is depicted in Figure 5.1. The main feature is that the process proceeds consecutively through the non-negative integers, jumping by one unit at each of a sequence of random times T_1, T_2, T_3, ... For the outcome ω displayed in the figure, $T_1(\omega) = 1$, $T_2(\omega) = 2.5$, $T_3(\omega) = 3$, $T_4(\omega) = 3.8$, ... A different outcome might have jumps at different times, but the structure of the graph is the same. The function $t \longrightarrow N_t(\omega)$ for fixed ω that is graphed below is referred to as a *sample path* of the process.

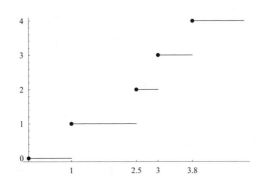

Figure 5.1 – A sample path $t \longrightarrow N_t(\omega)$ for a typical outcome ω

DEFINITION 1. A process $(N_t)_{t \in \mathbb{R}_+}$ is called an *arrival counting process* if there is an increasing sequence of random variables $T_0 = 0, T_1, T_2, T_3, \ldots$ taking values in $\mathbb{R}_+$, such that

$$N_t(\omega) = n \quad \text{if} \quad T_n(\omega) \le t < T_{n+1}(\omega)$$

The defining condition can be understood by thinking of a service facility at which the nth customer arrives at time T_n. In order for t to be between T_n and T_{n+1}, the nth customer must have arrived by time t but the $(n + 1)^{\text{st}}$ customer has not. A total of n customers have therefore arrived by time t. The random variable N_t is a counter that indicates this fact. More generally, the times T_n may be successive times of occurrence of some phenomenon, and N_t is the number of such occurrences up to and including time t. Some common applications include arrivals of planes to an airport, messages to a communication station, customers to a store, and users to a time-sharing computer system.

Activity 1 – Suppose that the first five times between successive arrivals of a certain arrival counting process are 2.1, 1.7, 3.4, 1.2, and 0.8. Sketch the path of the process.

In order to derive some results easily, we specialize to a particular kind of arrival process.

DEFINITION 2. An arrival counting process is called a *Poisson process with rate* λ if the differences:

$$S_1 = T_1 - T_0,\ S_2 = T_2 - T_1,\ S_3 = T_3 - T_2,\ \ldots$$

are independent, identically distributed exponential random variables with parameter λ.

Since we know the probability law of the inter-arrival times S_i, we suspect that we should be able to calculate many things about Poisson processes. To compute the probability distribution of N_t = number of arrivals by time t, we first derive the distribution of T_n = time of the nth arrival.

THEOREM 1. If T_n is the time of the nth arrival in a Poisson process with rate λ, then T_n has the $\Gamma(n, 1/\lambda)$ density, and furthermore,

$$P[T_n > t] = \sum_{k=0}^{n-1} e^{-\lambda t}(\lambda t)^k / k! \tag{1}$$

Proof. We can express T_n as the sum from $i = 1$ to n of i.i.d. exponential inter-arrival times S_i. The moment-generating function of T_n is therefore

$$
\begin{aligned}
E[e^{t\,T_n}] &= E[\exp(t \sum_{i=1}^{n} S_i)] \\
&= E[\exp(t\,S_1) \cdots \exp(t\,S_n)] \\
&= E[\exp(t\,S_1)] \cdots E[\exp(t\,S_n)] \\
&= E[\exp(t\,S_1)]^n = (1 - t/\lambda)^{-n}
\end{aligned}
$$

by the independence of the S_i, and the well-known formula for the moment-generating function of the exponential distribution. The last formula is the moment-generating function of the gamma distribution with parameters n and $1/\lambda$. By uniqueness of the moment-generating function, T_n must have the gamma density:

$$f(x) = \lambda^n \cdot x^{n-1} e^{-\lambda x} / (n-1)!, \quad x > 0$$

It remains to show that the integral of this density from t to ∞ equals the sum in the statement of the theorem. One application of integration by parts (with $u = x^{n-1}$, $dv = e^{-\lambda x}$) yields

$$
\begin{aligned}
&\int_t^\infty \lambda^n \cdot x^{n-1} e^{-\lambda x} / (n-1)!\, dx \\
&= \frac{\lambda^n}{(n-1)!} \left(x^{n-1} e^{-\lambda x}/\lambda + \frac{n-1}{\lambda} \int_t^\infty x^{n-2} e^{-\lambda x}\, dx \right) \\
&= \frac{\lambda^{n-1} t^{n-1} e^{-\lambda t}}{(n-1)!} + \frac{\lambda^{n-1}}{(n-2)!} \int_t^\infty x^{n-2} e^{-\lambda x}\, dx
\end{aligned}
$$

The first term on the right is the highest term in the summation of formula (1). You should verify that repeated integration by parts generates the other terms in the summation. ∎

With Theorem 1 in hand, it is easy to compute the probability law of N_t.

THEOREM 2. If (N_t) is a Poisson process with rate λ, then

$$P[N_t = n] = \frac{e^{-\lambda t}(\lambda t)^n}{n!}, \quad n = 0, 1, 2, \ldots \tag{2}$$

in other words, N_t has the Poisson distribution with parameter λt.

Proof.   The events $\{N_t < n\}$ and $\{T_n > t\}$ are the same, since both happen if and only if the nth arrival has not occurred by time t. Thus, $P[N_t < n]$ is equal to the expression in (1). Therefore,

$$
\begin{aligned}
P[N_t = n] &= P[N_t < n + 1] - P[N_t < n] \\
&= \sum_{k=0}^{n} e^{-\lambda t}(\lambda t)^k \big/ k! - \sum_{k=0}^{n-1} e^{-\lambda t}(\lambda t)^k \big/ k! \\
&= e^{-\lambda t}(\lambda t)^n \big/ n! \quad \blacksquare
\end{aligned}
$$

Activity 2 – The exponential distribution modeling the inter-arrival times is highly skewed to the right, with a lot of its probability weight near zero and a long right tail. What effect do you think that will have on the appearance of typical sample paths as in Figure 5.1?

The definition of Poisson process that we have given is a constructive definition. An arrival counting process is Poisson if it is the counting process associated with   i.i.d. exponential inter-arrival times.   It can be shown that this condition is equivalent to the following axiomatic definition.

DEFINITION 3. (Axiomatic definition of Poisson process) An arrival counting process (N_t) is a *Poisson process* if both:
    (a) For any $t, s \geq 0$, the difference $N_{t+s} - N_t$ is independent of all random variables N_u for $u \leq t$;
    (b) For any $t, s \geq 0$, the probability law of $N_{t+s} - N_t$ does not depend on t.

The first property in Definition 3 is a sort of probabilistic amnesia, which says that arrivals occurring between times t and $t + s$ are independent of the past history prior to t. This condition is usually called the *independent increments* condition. Property (b) is called *stationarity*, which says that it doesn't matter whether we begin to count arrivals at time 0 or at time t. The number of arrivals in a period of s time units (in the former case, the interval $(0, s]$ and in the latter, the interval $(t, t + s]$) is a random variable whose distribution depends only on s. Under our constructive definition, these properties can be verified by appealing to properties of the exponential distribution of the i.i.d inter-arrival times. To show the converse, that is to show that the axiomatic definition implies that the inter-arrival times are i.i.d. exponential random variables, is more difficult. We refer the reader to Ross [52] or Cinlar [15] for this.

Another interesting and useful result about Poisson processes involves the joint distribution of the arrival times given the number of arrivals in $[0, t]$. It turns out that if the number of arrivals, say n, during $[0, t]$ is known, then it is as if the n arrival times were thrown down at random on $[0, t]$. To be more precise, let $U_1, U_2, ..., U_n$ be i.i.d. uniform random variables on $[0, t]$, and let $U_{(1)}, U_{(2)}, ..., U_{(n)}$ be these same values written in increasing order (i.e., the order statistics of the sample). Then it is easy to compute (see, e.g., Hogg & Craig ([36], Sec. 4.6)) that the joint density of $U_{(1)}, U_{(2)}, ..., U_{(n)}$ is

$$f(u_1, u_2, ..., u_n) = \frac{n!}{t^n} \quad \text{if } 0 < u_1 < u_2 < ... < u_n < t \tag{3}$$

THEOREM 3. Let $T_1, T_2, T_3, ...$ be the arrival times of a Poisson process (N_t) with rate λ. Then the conditional density of $T_1, T_2, T_3, ...$ given that $N_t = n$ is the function f in formula (3).

Proof. We will not prove Theorem 3 in complete generality, but we will do the proof for $n = 1$, i.e., we show that

$$P[T_1 < s \mid N_t = 1] = s / t , \quad s \in (0, t)$$

In words, conditioned on the event $N_t = 1$, T_1 has the uniform distribution on $(0, t)$.

The conditional probability above is equal to

$$\begin{aligned} P[T_1 < s, N_t = 1] / P[N_t = 1] &= P[N_s = 1, N_t - N_s = 0] / P[N_t = 1] \\ &= P[N_s = 1] \, P[N_{t-s} = 0] / P[N_t = 1] \\ &= \frac{\left(e^{-\lambda s}(\lambda s)^1 / 1!\right)\left(e^{-\lambda(t-s)}(\lambda(t-s))^0 / 0!\right)}{e^{-\lambda t}(\lambda t)^1 / 1!} \\ &= s / t \end{aligned}$$

In the second line of the computation we have used the stationarity and the independent increments properties. (See Exercise 10 for the $n = 2$ case.) Since the last expression is the uniform c.d.f. on $(0, t)$, the proof for the $n = 1$ case is complete. ∎

Examples

With the main results in hand, let us now show some examples of Poisson process computations.

Recall that *Mathematica* has standard packages that contain the distribution objects we need in order to make computations. Discrete distributions like Poisson are contained in the first package loaded below, and continuous distributions like the exponential and gamma distributions are contained in the second.

```
Needs["Statistics`DiscreteDistributions`"];
Needs["Statistics`ContinuousDistributions`"];
```

The syntax for the three distributions we have met so far in our study of Poisson processes is below. The usages of the parameter λ for the exponential distribution and the parameter μ for the Poisson distribution are precisely as we are using it. Note that the second argument for the gamma distribution is its scale parameter β, which for us is $1/\lambda$.

```
PoissonDistribution[μ];
ExponentialDistribution[λ];
GammaDistribution[α, β];
```

Computations of probabilities can be done by using the PDF and CDF functions, contained in both the DiscreteDistributions and the ContinuousDistributions packages:

```
PDF[distribution, x];
CDF[distribution, x];
```

The PDF function returns the probability mass function value $f(x) = P[X = x]$ for a discrete random variable X that has the distribution specified in the first argument (including parameters of the distribution). In the case of continuous random variables, PDF returns the density at point x. The CDF function returns the cumulative distribution function value $F(x) = P[X \le x]$ in both the discrete and continuous cases. You might also find occasion to use the functions

```
Mean[distribution];
Variance[distribution];
```

which return the mean and variance of the given distribution. To simulate values from a given distribution, we have

```
Random[distribution];
RandomArray[distribution, n];
```

which, respectively, return a single value and a list of n values simulated from the given distribution.

EXAMPLE 1. Suppose that arrivals of buses to a stop form a Poisson process with rate λ equal to 4 per hour. Then, for instance, the probability that there are strictly more than 4 arrivals during the next hour is:

$$P[T_5 \le 1] = P[N_1 \ge 5] = 1 - P[N_1 \le 4] = 1 - \sum_{k=0}^{4} \frac{e^{-4} 4^k}{k!}$$

Note that by Theorem 2, the random variable N_1 has the Poisson(4·1) distribution. In *Mathematica* we can calculate $P[N_1 \le 4]$ either by directly requesting the c.d.f. value at 4, or by adding the terms of the probability mass function from 0 to 4, as shown below.

```
dist = PoissonDistribution[4];
N[1 - CDF[dist, 4]]
N[1 - (Sum[PDF[dist, k], {k, 0, 4}])]
```

```
0.371163
```

```
0.371163
```

Carrying the example further, what is the probability that, given that there have been exactly 4 arrivals in the first hour, there will be exactly 8 buses by the end of the third hour? This is

$$
\begin{aligned}
P[N_3 = 8 \mid N_1 = 4] &= P[N_3 - N_1 = 4 \mid N_1 = 4] \\
&= P[N_3 - N_1 = 4] \\
&= P[N_2 = 4] \\
&= \frac{e^{-4\,(2)}(4 \cdot 2)^4}{4!}
\end{aligned}
\tag{4}
$$

```
N[PDF[PoissonDistribution[8], 4]]
```

0.0572523

You are asked to verify the first line of (4) in Exercise 2. The second line of (4) is by the independent increments property, and the third line is by stationarity, since the number of arrivals between times 1 and 3 has the same distribution as the number of arrivals in the first 2 hours. Only the elapsed time matters.

Thus far, the small questions that we have asked could be answered by hand computation; *Mathematica* was just a convenience. A more sophisticated question necessitates the use of the computer. How fast must the rate of arrivals be such that the probability that at least 4 buses will come in an hour is at least 80%?

The probability of at least 4 buses in the first hour (or any hour) is $P[N_1 \geq 4] = 1 - P[N_1 \leq 3]$. Since the time period is 1 hour, the distribution of N_1 is Poisson($\lambda \cdot 1$). To require that $P[N_1 \geq 4]$ be at least .8 is to require that $P[N_1 \leq 3]$ be no more than .2. So we define a function of the parameter λ that returns the probability that $N_1 \leq 3$, and study it to find out where the probability hits .2.

```
P3orfewer[λ_] := CDF[PoissonDistribution[λ], 3];
g1 = Plot[{P3orfewer[λ], .2},
     {λ, 1, 10}, DefaultFont → {"Times", 8},
     DisplayFunction → Identity];
g2 = Plot[{P3orfewer[λ], .2}, {λ, 5.51, 5.52},
     DefaultFont → {"Times", 8},
     DisplayFunction → Identity];
Show[GraphicsArray[{g1, g2}],
   DisplayFunction → $DisplayFunction];
```

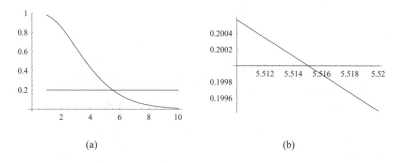

Figure 5.2 – Finding the rate λ such that $P[N_1 \le 3]$ is no more than .2

(Why does this curve decrease?) As shown in part (a) of Figure 5.2, the crossover point seems to be around $\lambda = 5.5$. Some zooming shows that to the second decimal place, $\lambda = 5.52$, as in part (b) of the figure. ∎

Activity 3 – Referring to Example 1, at least how fast must the rate λ be such that the expected number of arrivals in an hour is 4? (Yes, this is a "Who is buried in Grant's Tomb?" question, but it directs your attention to the meaning of λ, and why it is called the rate parameter of the process.) How slow must the rate be so that the probability of 2 or fewer buses in an hour is at least 50%?

EXAMPLE 2. Let us continue the bus example by calculating some expectations. First, the expected number of buses to arrive during the time period (2.5, 10] is

$$E[N_{10} - N_{2.5}] = E[N_{7.5}] = 4\,(7.5) = 30$$

by stationarity. By independent increments, this is the same, for example, as $E[N_{10} - N_{2.5} \mid N_{1.5} = 3]$.

The expected time of arrival of the fourth bus is

$$\begin{aligned} E[T_4] &= E[T_1 + (T_2 - T_1) + (T_3 - T_2) + (T_4 - T_3)] \\ &= 4\,E[T_1] = 4\,(1/4) = 1 \end{aligned}$$

since the inter-arrival times $(T_{i+1} - T_i)$ have identical exponential distributions with parameter 4, and the mean of such a distribution is 1/4.

If we know that the first bus arrived at time 2 hours (a long wait), then the expected time of arrival of the second bus is

$$E[T_2 \mid T_1 = 2] = E[T_1 + (T_2 - T_1) \mid T_1 = 2]$$
$$= E[T_1 \mid T_1 = 2] + E[T_2 - T_1 \mid T_1 = 2]$$
$$= 2 + E[T_2 - T_1] = 2 + 1/4$$

Notice that it does not matter that the first bus was late; the second bus is still expected to arrive 15 minutes later, which is the average inter-arrival time. This is another instance of the "memoryless" property of the exponential distribution.

Finally, to compute probabilities involving the arrival times T_n we can appeal to *Mathematica*. For instance, the probability that the 10th bus comes between time 5 and 6 is $P[5 \le T_n \le 6] = F(6) - F(5)$, where F is the c.d.f. of the distribution of T_n, which is $\Gamma(10, 1/4)$, by Theorem 1. We compute this below. ∎

```
F[x_] := N[CDF[GammaDistribution[10, 1/4], x]];
F[6] - F[5]
```

0.00457002

EXAMPLE 3. Suppose that dividends on an investment arrive at the times T_i of a Poisson process with parameter λ, and the worth to the investor of the ith dividend in present day terms is $f(T_i)$. Typically the function f would be a decreasing function of time, such as $f(t) = c \cdot e^{-\alpha t}$, where $\alpha \ge 0$. Let us calculate the expected present worth of dividends received during the time interval $[0, t]$. Observe that this is the same as:

$$E[\textstyle\sum_{i=1}^{N_t} f(T_i)] = E[E[\textstyle\sum_{i=1}^{N_t} f(T_i) \mid N_t]]$$

where N_t is the number of dividends by time t, whose distribution is known to be Poisson(λt). By Theorem 3, given $N_t = n$, the random variables T_i, $i = 1, \dots, n$ have the same joint distribution as the uniform order statistics on $[0, t]$, and thus $\sum f(T_i)$ has the same distribution as $\sum f(U_i)$, where the variables U_i are independent and uniformly distributed on $[0, t]$. The expected value of this sum is therefore:

$$E[\textstyle\sum_{i=1}^{n} f(U_i)] = \textstyle\sum_{i=1}^{n} E[f(U_i)] = n\, E[f(U_1)] = n \int_0^t f(u) \cdot \tfrac{1}{t}\, du$$

Hence,

$$E[\textstyle\sum_{i=1}^{N_t} f(T_i)] = E\big[N_t \cdot \tfrac{1}{t} \cdot \int_0^t f(u)\, du\big]$$
$$= \tfrac{1}{t} \cdot \int_0^t f(u)\, du \cdot E[N_t]$$
$$= \lambda \cdot \int_0^t f(u)\, du$$

since $E[N_t] = \lambda\, t$. ∎

Activity 4 – In Example 2, if dividends are constantly $50 and money discounts continuously at rate 5% (that is, $\alpha = .05$), find the present value of the dividend stream through time 10.

EXAMPLE 4. Commuters arrive to a train station at which trains are ready to load and depart. Under the current plan of the transit authority, a single train leaves at precisely T minutes after the hour, and it is assumed to be large enough to carry all waiting passengers. But the transit authority finds that it is able to increase service by providing another train, to leave at some time t earlier than T, taking with it all the commuters who have arrived to the station by that time. At what time should this new train be scheduled to leave so as to minimize the expected total waiting time of all commuters arriving in $[0,\ T]$?

We shall assume that arrivals of commuters form a Poisson process (N_t) with some rate λ. Let their times of arrival be denoted by $T_1, T_2, T_3, \ \dots\ ,$ as usual. The expected total waiting time of all commuters is the sum of the expected waits of those arriving before t plus the expected waits of those arriving in $(t,\ T]$. If $T_i < t$, then commuter i waits for $t - T_i$ minutes. Thus, the total expected wait by arrivals prior to t is

$$E[\textstyle\sum_{i=1}^{N_t} (t - T_i)] = \sum_{n=0}^{\infty} E[\textstyle\sum_{i=1}^{N_t} (t - T_i) \mid N_t = n] \cdot P[N_t = n] \qquad (5)$$

We have used the law of total probability in (5). Given $N_t = n$, the inner sum has the same distribution as the sum of n random variables $(t - U_i)$, where the U_i's are uniformly distributed on $[0,\ t]$. The conditional expectation in (5) is therefore

$$n \cdot \int_0^t (t - u) \cdot \tfrac{1}{t}\, du = n \cdot t / 2$$

after calculation. Substituting back into (5) gives us the expected total wait of commuters arriving prior to time t:

$$\textstyle\sum_{n=0}^{\infty} n \cdot (t/2) \cdot P[N_t = n] = (t/2)\, E[N_t] = \lambda\, t^2 / 2$$

The same reasoning applied to the time interval $(t,\ T]$, which has length $T - t$, shows that the total expected wait of commuters arriving after time t is $\lambda(T - t)^2 / 2$. Therefore the objective function for the problem of minimizing expected total waiting time has the form

$$f(t) = \lambda\, t^2 / 2 + \lambda(T - t)^2 / 2, \ t \in [0,\ T]$$
$$\Longrightarrow f'(t) = \lambda\, t + \lambda(T - t)\,(-1)$$

It is easy to check that the critical point is $t = T/2$, and that the absolute minimum of the objective is taken on at the critical point. The conclusion is that, regardless of the actual value of the arrival rate, as long as arrivals are Poisson, we should dispatch the first train at the halfway mark $T/2$ in order to minimize expected total waiting time. The value of the objective at the optimum point, however, does depend on λ. ∎

Activity 5 – Verify the claim in Example 4 that the expected total wait of commuters arriving after time t is $\lambda(T - t)^2/2$.

Exercises 5.1

1. (*Mathematica*) Calls arrive to a central telephone exchange according to a Poisson process with rate $\lambda = 3.6$ per minute. Let N_t be the total number of calls up to and including the tth minute. Compute
   (a) $P[N_{1.5} = 1]$
   (b) $P[N_{1.5} = 1 \mid N_{.7} = 0]$
   (c) $P[N_{.7} = 0, N_{1.2} = 0, N_{1.5} = 1]$

2. Verify the first line of (4).

3. A clock hangs precariously on a wall, falling occasionally. The clock ceases to work when it falls for the kth time. If falls occur according to a Poisson process with rate $\lambda = 2$ per week, find the probability distribution, mean, and variance of the time until the clock breaks.

4. If (N_t) is a Poisson process with rate λ, find
   (a) $E[N_{t+s} \mid N_t]$
   (b) $E[N_t \cdot N_{t+s}]$

5. Suppose that patients arriving to a doctor's office form a Poisson process with rate λ per hour. Given that there are n patients during the 8-hour day, what is the probability that there were k patients during the first 3 hours?

6. (*Mathematica*) In order for a machine to continue functioning, each of two parts must work. One replacement is available for each of the parts. A part lasts for an exponentially distributed amount of time with parameter $\lambda = .03$ before breaking, and parts behave independently of one another.
   (a) If T is the time when the machine will no longer function, find and graph as a function of t, $P[T > t]$.
   (b) Compute the probability density function of T.

(c) Find $E[T]$.

(d) Graph as a function of λ the probability that the machine lasts for at least 100 time units.

7. Suppose that cars traveling west on a two-lane highway pass a fixed point on the road at the times of a Poisson process with rate λ_1, and similarly the eastbound cars form a Poisson process with rate λ_2, independent of the first process. Find the distribution of the total number of cars passing the fixed point by time t. (Hint: Use moment-generating functions.)

8. In Example 3, suppose that f is a non-negative function such that $\int_0^\infty f(t)\,dt$ is finite. Find the expected present value of all dividends earned throughout time.

9. (*Mathematica*) A telephone customer service system can give an acceptable service level if it receives no more than two calls per minute. The supervisor of the system has a quality goal of giving acceptable service 95% of the time. If incoming calls form a Poisson process with rate λ, at most how large can λ be without failing the quality goal?

10. Let T_1 and T_2 be the first two arrival times of a Poisson process (N_t). Show that the joint conditional density of T_1 and T_2, given $N_t = 2$, is

$$f(t_1, t_2) = \begin{cases} 2/t^2 & \text{if } 0 < t_1 < t_2 < t \\ 0 & \text{otherwise} \end{cases}$$

11. Cabs arrive to drop passengers off at an airport according to a Poisson process with rate 1 per minute. A cab can contain 1, 2, or 3 passengers with probabilities 2/3, 1/6, and 1/6, respectively. The number of passengers in a cab is independent of the number in every other cab, and is also independent of the cab arrival process. Find the expected number of passengers that arrive in a 30-minute period. (To justify your answer, condition and uncondition on the number of cabs that arrive.)

12. Customers arrive to a store according to a Poisson process with rate $\lambda = 10$ per hour. On average, one fourth of all customers buy something, and their decisions are made independently of other customers and of the arrival process. Find the expected number of purchases made during a 2-hour period. (To justify your answer, condition and uncondition on the number of customers that arrive.)

13. Suppose that the number N_t of salmon that have passed a point on a river by time t forms a Poisson process with rate 2 per minute. The probability is 1/4 that a given salmon is over five pounds, and successive salmon weights are independent of one another. Show that the arrival process that

counts only the number of salmon exceeding five pounds is also Poisson, and find its rate.

14. As illustrated in the diagram below, Wagner Ct. and Schneider Dr. are parallel one-way eastbound roads, and Scott Ave. is a one-way northbound road that terminates at Wagner. Cars arriving to intersection 1 on Wagner, intersection 2 on Schneider, and intersection 2 on Scott form Poisson processes with rates 4 per minute, 6 per minute, and 3 per minute, respectively. At intersection 2, the probability that a car from Schneider turns left on Scott is 1/2 and the probability that a car from Scott turns right on Schneider is 1/3. Decisions by a car to turn are made independently of all other cars. Find the expected number of cars per unit time passing out of intersection 1, east on Wagner.

Exercise 14

15. (*Mathematica*) (a) Write a *Mathematica* command to simulate a desired number of arrival times T_1, T_2, ... of a Poisson process with a desired rate λ. (b) Write a *Mathematica* command that accepts a list of arrival times and a particular time t, and returns the value of N_t.

5.2 Birth and Death Processes

Preliminaries

We now study a different generalization of the Poisson process. The times between successive jumps are once again exponentially distributed, but the rate parameter may depend on the current state i of the process. Also, we allow the process to jump down by one unit, as well as up by one unit. A typical path is in Figure 5.3.

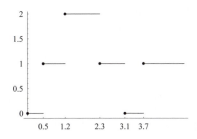

Figure 5.3 – A sample path $t \longrightarrow N_t(\omega)$ for a typical outcome ω of a birth–death process

One might choose such a model for population processes, in which a jump upward represents a "birth" or migration into the system, and a jump downward represents a "death" or migration out of the system. These processes are also used as models in queueing theory, where a birth is an arrival of a customer to the waiting line and a death is a departure from the line after service. In Section 4, we will apply the results of this section to queueing problems.

Before giving the precise definition of a birth–death process, some motivation is required. At a moment when the current population level is i, we may determine the amount of time until the next change in population, as well as whether that change is a birth or a death, by observing two independent random variables. The first, say U, is thought of as a "birth time," and is exponentially distributed with rate λ_i. The second is a "death time" W, which is also exponentially distributed with some rate μ_i. The next population change will occur at the random time $S = \min\{U, W\}$. If this minimum is U, then the next change is a birth, otherwise it is a death. To find the probability distribution of the time S until the next change in population, we compute

$$
\begin{aligned}
P[S > h] &= P[U > h, W > h] \\
&= P[U > h]\,P[W > h] \\
&= e^{-\lambda_i h} \cdot e^{-\mu_i h} \\
&= e^{-(\lambda_i + \mu_i) h}
\end{aligned}
\tag{1}
$$

Therefore the time until the next jump is exponentially distributed with rate equal to the sum of the birth and death rates. The probability that the next change is a birth is the same as the following probability:

$$P[U < W] = \int_0^\infty \left(\int_u^\infty \mu_i e^{-\mu_i w} \, dw \right) \lambda_i e^{-\lambda_i u} \, du$$

$$= \lambda_i \int_0^\infty e^{-(\lambda_i + \mu_i) u} \, du \tag{2}$$

$$= \frac{\lambda_i}{\lambda_i + \mu_i}$$

Hence, the probability that the next change is a death is

$$1 - \frac{\lambda_i}{\lambda_i + \mu_i} = \frac{\mu_i}{\lambda_i + \mu_i} \tag{3}$$

The birth and death rates therefore completely determine the probabilistic structure of the process.

Activity 1 – Is the exponential distribution special in this construction? To address this, what do you get for the probability distribution of S if you assume that U has the uniform distribution on $[0, 2/\lambda_i]$ and W has the uniform distribution on $[0, 2/\mu_i]$? What is the probability that the next change is a birth?

The discussion above should make the following definition intuitive.

DEFINITION 1. For each $i \in E = \{0, 1, 2, 3, \ldots\}$, let there be given a constant λ_i and a constant μ_i. Let $(N_t)_{t \geq 0}$ be a process with state space E, such that almost every path $t \longrightarrow N_t(\omega)$ increases or decreases by jumps of size ± 1 only. Let $T_0 = 0, T_1, T_2, \ldots$ be the sequence of jump times of (N_t), and let $X_0, X_1, X_2, \ldots$ be the sequence of states occupied by (N_t), i.e.:

$$N_t = X_n \quad \text{if} \quad T_n \leq t < T_{n+1}$$

The process (N_t) is called a *birth–death process* with *birth rates* $\{\lambda_i\}$ and *death rates* $\{\mu_j\}$ if:

(a)   For each $n \geq 1$, the conditional distribution given $X_n = j$ of $T_{n+1} - T_n$ is exponential with rate $\lambda_j + \mu_j$, and is conditionally independent of $T_1, T_2, \ldots, T_n$ and $X_1, X_2, \ldots, X_{n-1}$ given $X_n = j$;

(b) (X_n) forms a Markov chain with transition probabilities

$$P[X_{n+1} = j + 1 \mid X_n = j] = \frac{\lambda_j}{\lambda_j + \mu_j} \quad , \quad P[X_{n+1} = j - 1 \mid X_n = j] = \frac{\mu_j}{\lambda_j + \mu_j}$$

Roughly speaking, a birth–death process spends an exponential amount of time in each state before moving to an adjacent state. Only the current state affects the probability law of the waiting time until the next change, and the law of the next state visited. The parameters λ_j and μ_j are the rates at which transitions upward and downward, respectively, are made. Their sum is the rate at which any sort of transition is made. The relative size of λ_j to $\lambda_j + \mu_j$ dictates the probability that the next jump will be a birth rather than a death. Observe that in order for the formula for $P[X_{n+1} = j - 1 \mid X_n = j]$ in part (b) of the definition to make sense in the case $j = 0$, we must have $\mu_0 = 0$. That is, when the population has size 0, deaths occur at rate 0.

EXAMPLE 1. A hospital has N beds. New patients appear according to a Poisson process with rate λ, but they are turned away if the hospital is full. When j beds are occupied, we assume that the time until the next departure, either by a death or a release of a patient, is exponentially distributed with rate $\mu \cdot j$. (This is an appropriate assumption if each of the j individuals present tends to depart at the same rate μ.) Granting the independence of future arrivals and departures from the past history, it is reasonable to model the number of beds N_t occupied at time t by a birth–death process. The parameters are

$$\lambda_j = \begin{cases} \lambda & \text{if } j = 0, 1, 2, \ldots, N - 1 \\ 0 & \text{if } j \geq N \end{cases} \quad , \quad \mu_j = \mu \cdot j, \text{ if } j = 0, 1, 2, \ldots \quad (4)$$

We have set $\lambda_j = 0$ for $j \geq N$ because when all beds are occupied, no one is admitted, even if people are arriving in search of care. This fact also limits the state space to $\{0, 1, \ldots, N\}$, because in order to reach the complement of this set, the process first must go through state N and hence,

$$P[X_{n+1} = N + 1 \mid X_n = N] = \frac{\lambda_N}{\lambda_N + \mu_N} = \frac{0}{0 + N\mu} = 0$$

It follows that for $j > N$, it does not matter how μ_j is defined. ∎

To understand birth–death processes better, and to enable us to observe their properties, it is helpful to develop a command in *Mathematica* to simulate them. To characterize the sample path, we will need to produce the sequence of jump times $T_1, T_2, T_3, \ldots$ and the sequence of states visited $X_1, X_2, X_3, \ldots$ up to some fixed time t, which will be one of the input parameters. As other input parameters, we will need the initial state X_0, and functions $\lambda(i)$ and $\mu(i)$ that give the birth and death rates. We will use the strategy of the beginning of the section, simulating an inter-jump time as $S = \min\{U, W\}$, where U and W are suitable exponential random variables, and simulating the next state as the current state plus or minus 1, depending on whether U was smaller than W or not. Based on this discussion, the following code should be straightforward. A slight technicality is that the ExponentialDistribution is undefined if the parameter is 0; so we add If statements to make sure that the parameter is positive, which will return ∞ if not. The command below is also contained in the KnoxOR`StochasticProcesses` package, which is loaded automatically in the electronic version of this section when you evaluate initialization cells.

```
SimBirthDeathProcess[x0_,
   finaltime_, birthrate_, deathrate_] :=
  Module[{timelist, statelist,
    U, W, S, currtime, i},
   (* initialize the lists *)
   timelist = {};
   statelist = {x0};
   currtime = 0;
   i = 1;
   While[currtime < finaltime,
    (* while there is more time to go in
       the simulation, simulate the next
       birth and death time intervals *)
    U = If[birthrate[statelist[[i]]] > 0,
      Random[ExponentialDistribution[
        birthrate[statelist[[i]]]]], Infinity];
    W = If[deathrate[statelist[[i]]] > 0,
      Random[ExponentialDistribution[
        deathrate[statelist[[i]]]]], Infinity];
    S = Min[U, W];
    (* update the list of jump times,
      and the list of states *)
    AppendTo[timelist, currtime + S];
    If[U < W,
      AppendTo[statelist, statelist[[i]] + 1],
      AppendTo[statelist, statelist[[i]] - 1]];
    (* advance clock to the jump time just
       added, and advance subscript *)
    currtime = timelist[[i]];
    i = i + 1];
   {timelist, statelist}];
```

Here is an example in which births occur at a constant rate of 2 per unit time when the population size is at least 0, and deaths occur at a rate of 3 per unit time when the population size is at least 1.

```
λ[i_] := If[i < 0, 0, 2];
μ[i_] := If[i ≤ 0, 0, 3];
```

```
SeedRandom[456347];
SimBirthDeathProcess[0, 4, λ, μ]
```

```
{{0, 0.16583, 0.387183, 1.06746,
  1.29945, 1.52056, 1.7819, 1.81948, 1.83295,
  2.4456, 2.57291, 2.62224, 2.89913, 3.04436,
  3.12149, 3.43512, 3.57244, 4.04572},
 {0, 1, 0, 1, 2, 3, 2, 3, 4, 3, 2, 3, 4, 3, 4, 3, 2, 3}}
```

Then starting from state 0, at about time 0.166 the first birth happened, followed by a death at about time 0.387. At about time 1.07 the next birth occurrred, bringing the population size back up to 1, etc.

Activity 2 – Run the SimBirthDeathProcess command several more times, for longer time periods than just 4 units. Make note of the largest population size throughout time for each of your runs. Is it common to get very large populations? What do you think the choice of values for the constant rates λ and μ have to do with this question? Try to find values of λ and μ that make the population grow very large within 5 time units.

Kolmogorov Equations

In problems such as the hospital example above, we might well be interested in such quantities as

$$P_{ij}(t) = P[N_t = j \mid N_0 = i], \; i, j \in E \qquad (5)$$

which is the probability that j beds will be occupied at time t, if there were i patients initially. Unfortunately, these short-run probabilities are hard to calculate. But by looking at $P_{ij}(t)$ as a function of t, we can heuristically derive a system of differential equations satisfied jointly by these functions. The reasoning alone is worth seeing, for it appears often in the study of stochastic processes. But more than that, as a consequence we will derive explicit expressions for the limiting probabilities:

$$p_j = \lim_{t \to \infty} P[N_t = j \mid N_0 = i] \tag{6}$$

which turn out to be independent of the initial population i.

Referring to equation (1), $P[S > h] = e^{-(\lambda_i + \mu_i)h}$, by a Taylor expansion of the function $\exp[-(\lambda_i + \mu_i)h]$ about 0, we see that when the current population is i, the probability that there will be no change in population in the next h time units is approximately

$$1 - (\lambda_i + \mu_i)h$$

The error of approximation involves powers h^2 or higher. Similarly, the probability of a birth in the next h time units is approximately $\lambda_i h$, and that of a death is approximately $\mu_i h$. In an effort to form a differential equation for $P_{ij}(t)$, let us write an expression for $P_{ij}(t + h)$ by conditioning on the population at time t. First take $j \geq 1$. If h is small enough, then the event of having two or more births or deaths in h time units should have negligible probability (order h^2 or smaller). Ignoring such unlikely events, there are only three ways in which we could have a population of j at time $t + h$:

1. The population at time t was $j - 1$, and there was a birth during the time interval $(t, t + h]$; or

2. The population at time t was j, and there was neither a birth nor a death during $(t, t + h]$; or

3. The population at time t was $j + 1$, and there was a death during $(t, t + h]$.

Therefore, by the law of total probability, we have the following approximation, which ignores only events having probability of the order h^2:

$$P_{ij}(t + h) \approx P_{i,j-1}(t) \cdot \lambda_{j-1} h + P_{ij}(t)(1 - (\lambda_j + \mu_j)h) + P_{i,j+1}(t) \cdot \mu_{j+1} h \tag{7}$$

Subtraction of $P_{ij}(t)$ from both sides yields

$$P_{ij}(t + h) - P_{ij}(t) \approx$$
$$h(\lambda_{j-1} P_{i,j-1}(t) - (\lambda_j + \mu_j) P_{ij}(t) + \mu_{j+1} P_{i,j+1}(t))$$

Now divide both sides by h and let h approach 0. The error term in the above approximation, involving powers h^2 or higher, approaches 0. Thus we have the system of differential equations:

$$P_{ij}'(t) = \lambda_{j-1} P_{i,j-1}(t) - (\lambda_j + \mu_j) P_{ij}(t) + \mu_{j+1} P_{i,j+1}(t), \quad j \geq 1 \qquad (8)$$

Similar reasoning (see Exercise 3) for the $j = 0$ case leads to the differential equation

$$P_{i0}'(t) = -\lambda_0 P_{i0}(t) + \mu_1 P_{i1}(t) \qquad (9)$$

The equations derived heuristically in the last paragraph are called the *Kolmogorov forward equations* for birth–death processes, and can be given a rigorous foundation under regularity conditions on the birth and death rates. Reasoning again nonrigorously, it is at least plausible that if the limit as $t \to \infty$ of $P_{ij}(t)$ does exist, then this function should be stabilizing, so that its derivative approaches 0 as $t \to \infty$. Letting $t \to \infty$ on both sides of (8) and (9) gives a system of linear equations for the limiting probabilities $p_j = \lim_{t \to \infty} P_{ij}(t)$:

$$0 = -\lambda_0 p_0 + \mu_1 p_1$$
$$\qquad (10)$$
$$0 = \lambda_{j-1} p_{j-1} - (\lambda_j + \mu_j) p_j + \mu_{j+1} p_{j+1}$$

These equations can be solved recursively. We leave p_0 temporarily arbitrary, solve for the other p_j in terms of p_0, and then set $(p_0 + p_1 + p_2 + \cdots) = 1$ to find p_0. The first equation gives

$$\mu_1 p_1 = \lambda_0 p_0 \Longrightarrow p_1 = \frac{\lambda_0}{\mu_1} p_0$$

Substituting this into the next equation in (10) (i.e., the equation for the case $j = 1$) yields

$$\lambda_0 p_0 - (\lambda_1 + \mu_1) p_1 + \mu_2 p_2 = 0$$
$$\Longrightarrow \mu_2 p_2 = \lambda_1 p_1 + \mu_1 p_1 - \lambda_0 p_0 = \lambda_1 p_1$$
$$\Longrightarrow p_2 = \frac{\lambda_1 p_1}{\mu_2} = \frac{\lambda_0 \lambda_1}{\mu_1 \mu_2} p_0$$

Activity 3 – Show as above that $p_3 = \frac{\lambda_0 \lambda_1 \lambda_2}{\mu_1 \mu_2 \mu_3} p_0$.

One can show easily by induction that, since the remaining equations in system (10) have the same form as the second equation,

$$p_j = \frac{\lambda_0 \lambda_1 \cdots \lambda_{j-1}}{\mu_1 \mu_2 \cdots \mu_j} p_0 \qquad (11)$$

The condition that the sum of all of the p_j equals 1 implies that

$$1 = p_0 + p_0 \sum_{j=1}^{\infty} \frac{\lambda_0 \lambda_1 \cdots \lambda_{j-1}}{\mu_1 \mu_2 \cdots \mu_j}$$

If the series is convergent, then we have the non-zero solution:

$$p_0 = \left(1 + \sum_{j=1}^{\infty} \frac{\lambda_0 \lambda_1 \cdots \lambda_{j-1}}{\mu_1 \mu_2 \cdots \mu_j}\right)^{-1} \tag{12}$$

Our detailed plausibility argument motivates the following theorem, whose formal proof we omit.

THEOREM 1.   If the infinite series in (12) converges, then the limiting probabilities $p_j = \lim_{t \to \infty} P_{i\,j}(t)$ for a birth-death process (N_t) with birth rates $\{\lambda_i\}$ and death rates $\{\mu_i\}$ are as in (11) and (12). ■

Examples

EXAMPLE 2. We return to the hospital problem, Example 1. The long-run probability of having no patients is, by Theorem 1,

$$p_0 = \left(1 + \sum_{k=1}^{N} \frac{\lambda^k}{\mu^k k!}\right)^{-1} = \left(\sum_{k=0}^{N} \frac{\lambda^k}{\mu^k k!}\right)^{-1}$$

For $j = 1, 2, \ldots, N$, the long-run probability of having j patients is

$$p_j = \frac{\lambda^j}{\mu^j j!} \, p_0$$

Here are *Mathematica* functions that let us compute these quantities, in terms of λ, μ, and the number of beds N_0.

```
Clear[λ, μ, N0, p0];
p0[λ_, μ_, N0_] :=
   1 / (1 + NSum[λ^k / (μ^k * k!), {k, 1, N0}]);

                     λʲ
p[j_, λ_, μ_, N0_] := ――――― * p0[λ, μ, N0];
                    μʲ * j !
```

These functions allow us to easily examine the sensitivity of the distribution of the number of patients to changes in the arrival rate λ, the departure rate μ, and the number of beds N_0. For example, for $N_0 = 20$, and departure rate $\mu = 4$ per unit time, Figure 5.4 contains connected list plots of the distributions for $\lambda = 8$ and $\lambda = 15$ per unit time. The median occupancy grows only from about 2 to about 4, which is unexpectedly modest; however, it is not out of line when you remember that the overall rate of departure from the system when k beds are occupied is $k\,\mu$, not just μ, which tends to empty out the hospital.

```
Needs["Graphics`MultipleListPlot`"];
problist8 = Join[{{0, p0[8, 4, 20]}},
    Table[{j, p[j, 8, 4, 20]}, {j, 1, 20}]];
problist15 = Join[{{0, p0[15, 4, 20]}},
    Table[{j, p[j, 15, 4, 20]}, {j, 1, 20}]];
MultipleListPlot[{problist8, problist15},
    PlotJoined → True, PlotStyle →
    {RGBColor[0, 0, 0], Dashing[{.01, .01}]},
    DefaultFont → {"Times", 8}];
```

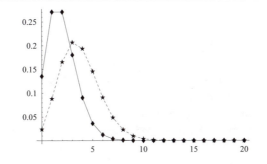

Figure 5.4 – Limiting bed occupancy probabilities, $\lambda = 8$ (solid), $\lambda = 15$ (dashed)

Most of the interesting questions about this example would have to be answered numerically as we have just done, but there are some interesting analytical results. Suppose that each patient pays at a rate of \$200 per day. Then the long-run expected revenue per day for the hospital is

$$\lim_{t\to\infty} E[200\,N_t] = 200 \cdot \lim_{t\to\infty} E[N_t]$$
$$= 200 \cdot \lim_{t\to\infty} \sum_{j=1}^{N} j \cdot P[N_t = j \mid N_0 = i]$$
$$= 200 \sum_{j=1}^{N} j \cdot p_j$$
$$= 200\, p_0 \sum_{j=1}^{N} j \cdot \frac{(\lambda/\mu)^j}{j!}$$
$$= 200\, p_0\, \frac{\lambda}{\mu} \sum_{k=0}^{N-1} \frac{(\lambda/\mu)^k}{k!}$$
$$= 200\, p_0\, \frac{\lambda}{\mu} \left(\frac{1}{p_0} - \frac{(\lambda/\mu)^N}{N!} \right)$$
$$= 200\, \frac{\lambda}{\mu} \left(1 - \frac{p_0 (\lambda/\mu)^N}{N!} \right)$$

This can be calculated for a given number of beds N and given arrival and departure rates λ and μ. It is interesting to note that this long-run expected revenue per day reaches a limit as the number of beds N approaches infinity. As $N \longrightarrow \infty$, the term $(\lambda/\mu)^N / N!$ approaches zero, thus the long-run expected revenue per day converges to $200\,\lambda/\mu$. As shown in Figure 5.5 for the case where $\lambda = 30$ and $\mu = 4$, the convergence of the expected revenue as a function of the number of beds is rather rapid; the hospital will get almost all the revenue that it will be able to get by about $N = 15$ beds. ∎

```
exprev[λ_, μ_, N0_] :=
    200 * λ/μ * ( 1 - p0[λ, μ, N0] * (λ / μ)^N0 / N0! );
exprevlist = Table[{N0, exprev[30, 4, N0]},
    {N0, 5, 22}];
ListPlot[exprevlist, PlotJoined → True,
    DefaultFont → {"Times", 8}];
```

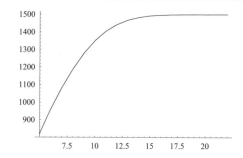

Figure 5.5 – Limiting expected revenue per day as a function of the number of beds

Activity 4 – Try several combinations of values of λ and μ for a 20-bed hospital and plot the limiting distribution of the number of patients. If μ is fixed at 4, about how large should λ be so that the peak of the distribution is around 10?

EXAMPLE 3. Suppose that at time 0, one person among a family of M individuals has a cold. Other family members contract the cold at random times T_1, T_2, ... Suppose that at any time, the distribution of the time until the next person contracts the cold is exponentially distributed, independent of the past. The rate at which the disease spreads is proportional to the product of the number infected times the number uninfected. Find the expected value and variance of the amount of time required for the whole family to contract the cold.

A typical path of the process N_t = number infected by time t is in Figure 5.6, in the case of $M = 9$ family members. (Why do you think that the path exhibits a slight S-shape, longer at the ends and more vertical in the middle?)

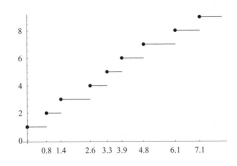

Figure 5.6 – Number infected as a function of time for a fixed outcome ω

The process proceeds deterministically through the states 1, 2, ..., M in that order, jumping at times T_i such that the intervals $T_{i+1} - T_i$ are exponential random variables. When there are i infected individuals, the uninfected population contains $M - i$ individuals, and the rate at which a jump occurs is:

$$\lambda_i = c\,i(M - i), \quad i = 0, 1, 2, \ldots, M \tag{13}$$

for some constant c. Thus, we have a special birth–death process called a *pure birth process* in which there are no deaths, and the birth rates are given by (13).

Notice that $[T_{i-1}, T_i)$ is the time interval in which exactly i individuals are infected (for convenience, we set $T_0 = 0$), for $i \geq 1$. Also, T_{M-1} is the first time that $N_t = M$, since, starting from state 1, the process must jump

$M - 1$ times to reach state M. We can compute the mean and variance of T_{M-1} by telescoping in the following way:

$$T_{M-1} = (T_{M-1} - T_{M-2}) + (T_{M-2} - T_{M-3}) + \cdots + (T_2 - T_1) + T_1 \qquad (14)$$

Because the mean of the exponential distribution is the reciprocal of the rate,

$$E[T_{M-1}] = \sum_{i=1}^{M-1} E[T_i - T_{i-1}]$$
$$= \sum_{i=1}^{M-1} \frac{1}{c\,i(M-i)} \qquad (15)$$

For example, if $c = .1$, and the family size $M = 9$, then the mean time until the whole family gets the cold is

$$\text{Sum}\left[\frac{1}{.1 * i * (9 - i)}, \{i, 1, 8\}\right]$$

6.03968

Because the times between jumps are independent, and the variance of the exponential distribution is the square of the reciprocal of the rate,

$$\text{Var}(T_{M-1}) = \sum_{i=1}^{M-1} \text{Var}(T_i - T_{i-1})$$
$$= \sum_{i=1}^{M-1} \left(\frac{1}{c\,i(M-i)}\right)^2 \qquad (16)$$

For the case $c = .1$ and $M = 9$ again, the variance of the time until they all get the cold is

$$\text{Sum}\left[\left(\frac{1}{.1 * i * (9 - i)}\right)^2, \{i, 1, 8\}\right]$$

5.26269

The previous results on limiting distributions do not apply here, because the death rates are zero. But it is intuitively obvious that as time approaches infinity, the number of infected individuals converges to M, and the limiting distribution is therefore degenerate with all of its weight on M. ∎

Exercises 5.2

1. Let (N_t) be a birth-death process with only two states, 0 and 1. Denote by λ the birth rate at state 0, and μ the death rate at state 1.

   (a) Write the Kolmogorov forward equations for $P_{00}(t)$ and $P_{11}(t)$.

   (b) Solve the equations obtained in part (a).

   (c) A pay phone may be either engaged (state 1) or unengaged (state 0). Assuming that a birth-death process is an appropriate model, find the long-run proportion of time that the phone is engaged.

2. (*Mathematica*) Write a *Mathematica* function to take the output of the SimBirthDeathProcess command and find the proportion of time that the process was in each of the states it visited.

3. Give a plausibility argument for the forward equation for P_{i0} listed in (9).

4. (*Mathematica*) In Example 2, for $\lambda = 20$ and $\mu = 5$, what number of beds N is necessary so that the hospital can come within \$5 of their highest possible long-run expected profit per day?

5. A single individual in an essentially infinite population has a disease initially. Let N_t be the cumulative number of individuals who have contracted the disease by time t; thus (N_t) is a non-decreasing process. We assume that the rate of transfer of the disease is, at every time, proportional to the number of individuals who have had the disease.

   (a) Make appropriate assumptions so that (N_t) forms a birth–death process, and find the birth and death rates.

   (b) It is possible to show inductively that the c.d.f. of the nth jump time T_n is

$$P[T_n \leq t \mid N_0 = 1] = (1 - e^{-\lambda t})^n$$

where λ is the proportionality constant mentioned above. (The idea is to condition and uncondition on the value of T_{n-1}, producing a rather messy integral that nonetheless yields to a succession of standard integration techniques.) Use this formula to show that

$$P_{1\,j}(t) = P[N_t = j \mid N_0 = 1] = e^{-\lambda t}(1 - e^{-\lambda t})^{j-1}$$

6. Write the Kolmogorov forward equations for the process in Exercise 5, and verify that $P_{1\,j}(t)$ listed in part (b) satisfies them.

7. (*Mathematica*) In Example 3, for a family of size 10, what is the smallest value of the transmission constant c such that the mean time until everyone has the cold is no more than 3 days?

8. A population begins with n individuals, who die at the times of a birth-death process with death rates μ_i, $i = 1, ..., n$. Write an expression for the mean and variance of the time until extinction of the population. If n is even, $\mu_j = 1/4$ for $j \le n/2$, and $\mu_j = 1/3$ for $j > n/2$, calculate the exact mean and variance of the extinction time.

9. A delicatessen has four service lines, each manned by one server. The food is not particularly good there, so customers who arrive when all servers are busy just decide to go to some other restaurant. Suppose that the times at which customers arrive form a Poisson process with rate 2 per minute, and the duration of a service is exponential with mean 5 minutes, independent of other services and of the arrival process. Find the long-run distribution of the number of busy servers.

10. The size of a fish population follows a birth-death process. Suppose that the birth rate when there are no fish is some positive constant λ_0 (i.e., migration to empty water is possible); otherwise the birth rate is proportional to the number of fish present. The death rate when there is only one fish is some positive constant μ_1; otherwise the death rate is proportional to the population size minus one. Find the limiting distribution of the population size.

11. An electric generator can be running at one of three speeds at any time: high, low, or off. It cannot change directly from high to off, nor from off to high. When it is on low, the probability is 2/3 that it will next go to high, and consequently 1/3 that it will shut off when the next change of state comes. The amount of time that the generator stays in each of the three states H, L, and O is exponentially distributed, with rates λ_H, λ_L, and λ_O, respectively. Model the speed as a birth-death process, and find the limiting probabilities for each state.

12. In deriving the forward equations, we conditioned on the population at time t in order to approximate $P_{ij}(t + h)$. Give a similar argument in which you condition on the population at time h instead. The resulting differential equations are called the *Kolmogorov backward equations*.

5.3 Renewal Processes

Introduction

In this section, we generalize the Poisson process by permitting inter-arrival times to have distributions other than exponential. The common distribution function of the inter-arrival times will be denoted throughout the section as F. And, in place of the phrase "arrival times" we will now use "renewal times," anticipating that these times are instants of some kind of regeneration of a process, so that the future of the process after one of these times looks probabilistically the same as the future after any other of the renewal times. As in Poisson processes, we use N_t to indicate the number of renewals through time t.

DEFINITION 1. A counting process (N_t) with renewal times $T_0 = 0, T_1, T_2, T_3, \ldots$ is called a *renewal process* if the inter-renewal times:

$$S_1 = T_1 - T_0, \ S_2 = T_2 - T_1, \ S_3 = T_3 - T_2, \ \ldots$$

are i.i.d. random variables with c.d.f. F and mean $\mu \in (0, \infty)$.

EXAMPLE 1. There are many situations in which renewal processes might reasonably be used as models. For example, the times $T_1, T_2, T_3, \ldots$ may be times of demand by customers for an item stocked in an inventory. Then N_t is the number of items requested by time t. Alternatively, T_i might represent the ith time at which a certain machine has been repaired. If each repair costs c dollars, then $c \cdot N_t$ is the total repair bill up to time t. Renewal processes have also been used to advantage in the study of queues, in which T_i is the time of arrival of the ith customer seeking service. In all of these illustrations, the main hypothesis is that the times between renewals should be i.i.d. random variables with a finite mean. ■

One interesting problem involving renewal processes is to find

$$P[N_t = n], \quad n = 0, 1, 2, \ldots \tag{1}$$

i.e., the distribution of the number of renewals up to time t. Also of interest is the mean number of renewals in $[0, t]$:

$$m(t) = E[N_t], \tag{2}$$

called the *renewal function*. Although the short-run probabilities in (1) can be characterized in terms of the inter-renewal distribution F, calculations are usually difficult. We take up this subject in the second subsection.

The long-run number of renewals per unit time:

$$\lim_{t \to \infty} N_t / t \tag{3}$$

is another item of interest. In contrast to the short-run behavior of renewal processes, the long-run behavior will be simple and intuitive. This will be examined in the third and fourth subsections. The latter concerns itself with the following problem. Let rewards (or costs), denoted by R_n, occur at the renewal times. We will compute the *long-run average reward per unit time*:

$$\lim_{t \to \infty} \left(\sum_{n=1}^{N_t} R_n / t \right) \tag{4}$$

To understand this expression, note that there are N_t renewals in $[0, t]$. At the nth renewal the reward R_n is received, and therefore the ratio is the total reward during $[0, t]$ divided by t.

Short-Run Distributions

To approach the problem of calculating short–run probabilities, recall the formula for the *convolution* of two distribution functions:

$$F * G(t) = \int_0^t G(t - s) \, d F(s) \tag{5}$$

In the case that F is continuous with density f, the notation $d F(s)$ means $f(s) \, d s$. For instance, the convolution of the exponential(1) c.d.f. F with the identity function $G(u) = u, \ u \in [0, 1]$ is

$$\int_0^t (t - s) * e^{-s} \, d s$$

$-1 + e^{-t} + t$

In the discrete case, the integral in (5) is a sum of terms $G(t - s_i) \cdot p(s_i)$ over all points $s_i \leq t$ that have positive positive probability $p(s_i)$.

The most important fact about convolutions for our purposes is that if F and G are the distribution functions of two independent random variables X and Y, then $F * G$ is the distribution function of $X + Y$. For the continuous case, we can see this from the following computation:

$$P[X + Y \le t] = \int_0^t \int_0^{t-s} f_y(u) \cdot f_x(s) \, d u \, d s$$
$$= \int_0^t \left(\int_0^{t-s} f_y(u) \, d u \right) f_x(s) \, d s$$
$$= \int_0^t G(t - s) \, f_x(s) \, d s$$

The *iterates* of a distribution function G are the repeated convolutions of G with itself:

$$G^{(1)} = G, \quad G^{(2)} = G * G, \quad G^{(3)} = G * G * G, \ldots$$

Activity 1 – Consider the cumulative distribution function $F(t)$ of the trivial distribution that puts all probability at state $t = 0$. Find the iterates $F^{(2)}$, $F^{(3)}$. If, instead, the distribution puts all of its probability at $t = 1$, find the convolution. Give an intuitive explanation of the results, and try to generalize.

Inductively, it is easy to prove that the n-fold convolution of a c.d.f. F with itself is the distribution of the random variable $X_1 + X_2 + \cdots + X_n$, where the X_i are i.i.d. with the distribution characterized by F. (Try to show this.) Thus, because inter-renewal times S_i are i.i.d. with distribution F, and because T_n is the sum of the first n of them, we have

$$P[T_n \le t] = F^{(n)}(t) \tag{6}$$

But then,

$$P[N_t = n] = P[N_t \ge n] - P[N_t \ge n + 1]$$
$$= P[T_n \le t] - P[T_{n+1} \le t]$$

which gives the following result.

THEOREM 1. If (N_t) is a renewal process with inter-renewal distribution F, then

$$P[N_t = n] = F^{(n)}(t) - F^{(n+1)}(t) \quad \blacksquare \tag{7}$$

EXAMPLE 2. Suppose that the inter-renewal times have the density

$$f(x) = \begin{cases} x e^{-x} & \text{if } x \in (0, \infty) \\ 0 & \text{otherwise} \end{cases}$$

This is a member of the $\Gamma(\alpha, \beta)$ family for which $\alpha = 2$ and $\beta = 1$, sometimes called the *Erlang* (2, 1) *density*. The convolutions can be calculated with some effort; for example,

$$F^{(2)}(t) = \int_0^t F(t - s) \, d F(s) = \int_0^t \left(\int_0^{t-s} u e^{-u} \, d u \right) s \, e^{-s} \, d s$$

Repeated integration by parts or an appeal to *Mathematica* can be used to obtain a closed formula.

F2[t_] := $\int_0^t \left(\int_0^{t-s} \mathbf{u} * \mathbf{E}^{-\mathbf{u}} \, \mathbf{du} \right) \mathbf{s} * \mathbf{E}^{-\mathbf{s}} \, \mathbf{ds};$

F2[t] // Simplify

$1 - \dfrac{1}{6} e^{-t} (6 + t (6 + t (3 + t)))$

The third iterate may be obtained by convolving $F^{(2)}$ with F as follows:

F3[t_] := $\int_0^t \mathbf{F2[t - s]} * \mathbf{s} * \mathbf{E}^{-\mathbf{s}} \, \mathbf{ds};$

F3[t] // Simplify

$\dfrac{1}{12} e^{-2t} (3 e^{2t} (17 + (-7 + t) t) -$
$\qquad 2 e^t (3 + t) (12 + t^2) + 3 (7 + t (5 + t)))$

But the larger the order of the iterate, the harder is the computation and the messier is the answer. For this example, it is much easier to use moment-generating functions to compute $F^{(n)}$. The m.g.f. of the gamma distributed inter-renewal times is $M(t) = 1/(1 - t)^2$, hence the m.g.f. of the sum T_n of n independent and identically distributed such times is the product of n identical factors of this $M(t)$. It therefore follows that the m.g.f. of T_n is $1/(1 - t)^{2n}$, which is identical to the m.g.f. of the $\Gamma(2n, 1)$ distribution. In other words, T_n has the Erlang$(2n, 1)$ distribution.

Into equation (7) we can now substitute appropriate gamma c.d.f.'s for $F^{(n)}$ and $F^{(n+1)}$:

$$P[N_t = n] = \int_0^t x^{2n-1} e^{-x}/(2n - 1)! \, d x - \int_0^t x^{2n+1} e^{-x}/(2n + 1)! \, d x$$

To simplify, integrate by parts twice on the second integral above. You should verify that the resulting expression for this integral is

$$-e^{-t} t^{2n+1}/(2n+1)! + -e^{-t} t^{2n}/(2n)! + \int_0^t x^{2n-1} e^{-x}/(2n-1)! \, dx$$

Substitution into the previous line gives us the distribution of N_t:

$$P[N_t = n] = e^{-t} t^{2n+1}/(2n+1)! + e^{-t} t^{2n}/(2n)! \quad \blacksquare \qquad (8)$$

There is a general expression for the renewal function $m(t)$, though it is usually difficult to calculate explicitly. It is listed in the next theorem.

THEOREM 2. $m(t) = E[N_t] = \sum_{n=1}^{\infty} F^{(n)}(t)$.

Proof. Define a sequence (I_n) of random variables by

$$I_n(\omega) = \begin{cases} 1 & \text{if } T_n(\omega) \le t \\ 0 & \text{otherwise} \end{cases}$$

Then $E[I_n] = P[T_n \le t]$. (See the Activity below.) Also, it is easy to see that $N_t = \sum_{n=1}^{\infty} I_n$. By the monotone convergence theorem, we may take expectation inside this infinite sum, to obtain

$$\begin{aligned} m(t) = E[N_t] &= E[\textstyle\sum_{n=1}^{\infty} I_n] \\ &= \textstyle\sum_{n=1}^{\infty} E[I_n] \\ &= \textstyle\sum_{n=1}^{\infty} P[T_n \le t] \\ &= \textstyle\sum_{n=1}^{\infty} F^{(n)}(t) \end{aligned}$$

which establishes the claim. $\blacksquare$

Activity 2 – Why is it true, in the proof of Theorem 2, that $E[I_n] = P[T_n \le t]$? Explain why it is true that $N_t = \sum_{n=1}^{\infty} I_n$.

EXAMPLE 3. Continuing Example 2, let us calculate the expected number of renewals during $[0, t]$ when the inter-renewal distribution is Erlang(2, 1). As before, since $F^{(n)}$ is the c.d.f. of the $\Gamma(2n, 1)$ distribution,

$$F^{(n)}(t) = \int_0^t x^{2n-1} e^{-x}/(2n-1)! \, dx$$

Hence, by Theorem 2,

$$
\begin{aligned}
m(t) = E[N_t] \ &= \ \textstyle\sum_{n=1}^{\infty} F^{(n)}(t) \\[2mm]
&= \ \int_0^t \left(\textstyle\sum_{n=1}^{\infty} x^{2n-1} \, e^{-x} / (2n-1)! \right) dx \\[2mm]
&= \ \frac{1}{2} \int_0^t (1 - e^{-2x}) \, dx \\[2mm]
&= \ \frac{t}{2} + \frac{e^{-2t}}{4} - \frac{1}{4}
\end{aligned}
\tag{9}
$$

In this computation, the interchange of integration and summation in the second line can be justified by the uniformity of convergence of the series for $x \in [0, t]$.   Also, the series in the second line has been computed by rewriting it as

$$
\begin{aligned}
&\frac{1}{2} e^{-x} \left(\left(1 + x + \frac{x^2}{2!} + \frac{x^3}{3!} + \cdots \right) - \left(1 + (-x) + \frac{(-x)^2}{2!} + \frac{(-x)^3}{3!} + \cdots \right) \right) \\
&= \frac{1}{2} e^{-x} (e^x - e^{-x})
\end{aligned}
$$

Incidentally, the Erlang(2, 1) distribution is the distribution of the sum of two i.i.d. exponential(1) random variables.   This kind of inter-renewal distribution might arise in the manufacturing of a piece of heavy equipment, in which there are two phases, one after the other, and each phase takes an exponential amount of time to complete. The renewal times are the successive times of completion of an entire piece, and the expression in (9) gives us the expected number of pieces made by time t. ■

Activity 3 – In Example 3, what does the ratio $m(t)/t$ converge to as $t \longrightarrow \infty$? Interpret what this limit means.

Long-Run Results

In Example 3, formula (9) allows us to find the limit of the expected number of renewals per unit time, as time becomes large:

$$
\lim_{t \to \infty} \frac{m(t)}{t} = \lim_{t \to \infty} \frac{t/2 + e^{-2t}/4 - 1/4}{t} = 1/2
$$

Notice that for the given Erlang(2, 1) distribution of inter-renewal times, the mean inter-renewal time is 2, which is the reciprocal of this long-run expected number of renewals per unit time.

We would now like to take a closer look at the limiting theory of renewal processes. The next results characterize the long-run time average behavior of N_t. First note that $N_t \longrightarrow \infty$ as $t \longrightarrow \infty$, with probability one, since

$$P[\lim_{t \to \infty} N_t < \infty] = P[\bigcup_n \{S_n = \infty\}] \leq \sum_{n=1}^{\infty} P[S_n = \infty] = 0$$

The first equation follows from the fact that the only way that the number of renewals N_t can approach a finite limit as $t \longrightarrow \infty$ is for some inter-renewal time S_n to be infinite. The inequality is by countable subadditivity. The fact that $N_t \longrightarrow \infty$ for almost every outcome is necessary for the proof of the following important theorem. Its content is that the long-run average number of renewals per unit time is the reciprocal of the average time between renewals, for all but some exceptional outcomes in a set of probability zero. The proof here follows along the lines of Ross ([52], Proposition 3.3.1).

THEOREM 3. (Renewal Law of Large Numbers) If (N_t) is a renewal process with mean inter-renewal time μ, then

$$P\left[\lim_{t \to \infty} \frac{N_t}{t} = \frac{1}{\mu}\right] = 1 \tag{10}$$

Proof. Denote by $T(N_t)$ the random time whose value for an outcome ω such that $N_t(\omega) = n$ is $T_n(\omega)$. In other words, $T(N_t)$ is the time of the N_t^{th} arrival. Similarly, let $T(N_t + 1)$ equal T_{n+1} for outcomes where $N_t = n$. It can be seen that $T(N_t)$ is the time of the last renewal prior to t, and $T(N_t + 1)$ is the time of the first renewal after t. For all outcomes, we therefore have the inequalities

$$T(N_t) \leq t < T(N_t + 1)$$

Thus,

$$\frac{N_t}{T(N_t)} \geq \frac{N_t}{t} > \frac{N_t}{T(N_t+1)} = \frac{N_t}{N_t+1} \cdot \frac{N_t+1}{T(N_t+1)} \tag{11}$$

But, by the strong law of large numbers,

$$\frac{N_t}{T(N_t)} = \left(\frac{T(N_t)}{N_t}\right)^{-1} = \left(\frac{\sum_{i=1}^{N_t} S_i}{N_t}\right)^{-1} \longrightarrow \frac{1}{\mu} \quad \text{as } t \longrightarrow \infty$$

Similarly, $(N_t + 1)/T(N_t + 1)$ approaches $1/\mu$. Because N_t converges to ∞ for almost every outcome, the ratio $N_t/(N_t + 1)$ converges to 1. Therefore, the outsides of inequality (11) force the middle to approach $1/\mu$ for almost every outcome. ∎

Surprisingly, the corresponding theorem for expectations requires more machinery; consequently we will list it without proof.

THEOREM 4. (Elementary Renewal Theorem) If (N_t) is a renewal process with mean inter-renewal time μ and renewal function $m(t) = E[N_t]$, then

$$\frac{m(t)}{t} \longrightarrow \frac{1}{\mu} \text{ as } t \longrightarrow \infty \quad \blacksquare$$

To summarize, both the expected, and the actual, number of renewals per unit time approach $1/\mu$ in the long run.

EXAMPLE 4. A computer software marketing group wants to decide how best to allocate their labor. If the supply of programmers is broken into three groups, each working on a separate phase of a project, then it is estimated that each phase requires an exponential amount of time to complete, with parameter 1/2. If the programmers are broken into two groups, each working on a separate phase, then each phase requires an exponential amount of time with parameter 1/4. We assume that successive projects are independent of one another, that the groups can work simultaneously, and that work on the next project does not commence until the previous one is complete. Which organizational structure produces the fastest output of software in the long run?

The successive completion times of software projects form a renewal process. For the first organization, a typical inter-renewal time is the maximum of the three completion times for the three work groups. Let those times be labelled X_1, X_2, and X_3. They are each exponentially distributed with parameter 1/2, and we assume that they are also independent. Because of this, the distribution function of the maximum of the X_i's is

$$
\begin{aligned}
P[S \le t] &= P[\max\{X_1, X_2, X_3\} \le t] \\
&= P[X_1 \le t]\, P[X_2 \le t]\, P[X_3 \le t] \\
&= (1 - e^{-t/2})^3
\end{aligned}
$$

Differentiation gives the probability density function of the inter-renewal times:

$$f(t) = \tfrac{3}{2}\, e^{-t/2}(1 - e^{-t/2})^2$$

And the expected inter-renewal time for the first plan is as below.

$$\mu_1 = \int_0^\infty t * \frac{3}{2} * e^{-t/2} (1 - e^{-t/2})^2 \, dt$$

$$\frac{11}{3}$$

For the second organizational plan, the model is almost the same, except that two independent groups must be finished before the project can be finished. A typical inter-renewal time is therefore the maximum of two random variables, say Y_1 and Y_2, each exponential with parameter $1/4$. Computing as in the last paragraph, we obtain the density of the inter-renewal time under plan 2:

$$g(t) = \tfrac{1}{2} e^{-t/4}(1 - e^{-t/4})$$

The mean inter-renewal time is

$$\mu_2 = \int_0^\infty t * \frac{1}{2} * e^{-t/4} (1 - e^{-t/4}) \, dt$$

$$6$$

Therefore, the long-run number of projects completed per unit time is $1/\mu_1 = 3/11$ under the first system, and $1/\mu_2 = 1/6$ under the second, by the Renewal Law of Large Numbers. The analysis shows that the first system, with three programming groups, is better under this criterion. ∎

Renewal Reward Processes

As our final renewal process model, suppose that at the renewal times $T_1, T_2, T_3, \ldots$ of a renewal process (N_t), we receive rewards (or are charged costs) $R_1, R_2, R_3, \ldots$, respectively. We assume that the sequence of pairs $(S_n, R_n)_{n=1,2,3,\ldots}$ are independent and identically distributed, where, as usual, the S_n's are inter-renewal times. Let μ be the mean inter-renewal time, and denote by r the common mean reward $r = E[R_n] < \infty$. You are asked for a proof of part (a) of the following theorem in Exercise 14. Part (b), the version for expectations, is omitted because of the extra machinery required for its proof.

THEOREM 5. (a) $P[\lim_{t \to \infty} \sum_{n=1}^{N_t} R_n / t = r/\mu] = 1$
     (b) $E[\sum_{n=1}^{N_t} R_n / t] \longrightarrow r/\mu$ as $t \to \infty$ ∎

> **Activity 4** – Try to interpret Theorem 5 intuitively, then read on to check your hypothesis.

We can interpret the result as follows. Since N_t is the number of rewards received (or costs charged) during the time interval $[0, t]$, the sum in part (a) is the total reward up to time t. For all but some exceptional outcomes of no probability, the average reward per unit time during $[0, t]$ converges to the average reward r per renewal, multiplied by the average number $1/\mu$ of renewals per unit time. Part (b) says that the expectation of the average reward per unit time reaches the same limit.

EXAMPLE 5. An office manager is considering the purchase of one of two competing duplicating machines. Machine 1 survives for a random length of time between repairs, with the following density function:

$$f_1(x) = \frac{x^{-2/3}\, e^{-(x/2)^{1/3}}}{3 \cdot 2^{1/3}} \quad \text{if } x > 0$$

(This is a case of the so-called *Weibull distribution*.) Repair costs of this machine are random variables R_1, R_2, ... , which are i.i.d. with the discrete uniform distribution on the set {50, 51, ..., 99}. The times between repairs for the second machine have the exponential distribution with parameter $1/10$, and a service agreement is available such that the cost of a repair is fixed at \$60. Which machine is preferable from the point of view of minimizing long-run average cost per unit time?

Each machine has an associated renewal process. The times of renewal are the successive times at which repair is done. For machine 1, the times between renewals have the given Weibull density f_1; and for machine 2, they have the exponential distribution, whose mean inter-renewal time is $\mu_2 = 10$. The mean inter-renewal time for machine 1 is

$$\mu 1 = \int_0^\infty x * \frac{1}{3 * 2^{1/3}} * x^{-2/3} * e^{-(x/2)^{1/3}}\, dx$$

12

Costs R_1, R_2, ... are charged at the times of repair of each machine. For machine 1, the distribution of a typical cost R is discrete uniform, and so the average cost per repair $r_1 = E[R]$ is

$$\texttt{r1 = N}\left[\sum_{k=50}^{99} \texttt{k} * \frac{1}{50}\right]$$

74.5

For machine 2, the costs are constantly $r_2 = 60$. Recall that Theorem 5 implies that as elapsed time becomes larger, the total cost incurred divided by the elapsed time converges to r/μ, so that a comparison of the two machines yields

$$r_1/\mu_1 = 74.5/12 = 6.2; \qquad r_2/\mu_2 = 60/10 = 6.0$$

Therefore machine 2 minimizes long-run time average cost. ∎

Exercises 5.3

1. If $F(x) = 1 - e^{-\lambda x}$ for $x > 0$, and zero otherwise, find a formula for the n-fold convolution $F^{(n)}(t)$.

2. Let $G(x)$ be the c.d.f. of the continuous uniform distribution on $[0, 2]$, and let $F(x)$ be the c.d.f. of the discrete uniform distribution on the set of states $\{1, 2\}$. Find $G * F(t)$ and $F * G(t)$.

3. (*Mathematica*) Suppose that the inter-renewal times of a renewal process have the (discrete) Poisson distribution with parameter 2. Note that this means it is possible for successive renewal times to be the same. Find a series expression for the distribution of the number of renewals by time t, and for the renewal function $m(t)$. Produce a connected line graph of the large portion of the probability mass function of $N_{6.5}$. Evaluate $m(1)$ explicitly.

4. The preparation of a report requires the efforts of three people, each one beginning after his predecessor finishes. When one report is finished, the next one is begun, etc. Suppose that person i among the three requires an exponentially distributed time with parameter λ_i, $i = 1, 2, 3$ to finish his portion of the report, and that reports are independent of one another. Find the long-run expected number of reports that can be finished per unit time. Is it necessary to assume that the completion times of the three workers for a given report are independent of one another?

5. The elementary renewal theorem does not follow trivially from the renewal law of large numbers, because the convergence of a sequence of random variables $X_1, X_2, X_3, \ldots$ to a constant does not necessarily imply

that their expectations converge to the same constant. Show this last statement by considering the sequence of random variables defined by

$$X_n = \begin{cases} 0 & \text{if } U > 1/n \\ n & \text{otherwise} \end{cases}$$

where U is uniformly distributed on $(0, 1)$. Show first that $P[X_n \rightarrow 0 \text{ as } n \rightarrow \infty] = 1$, and then compute $E[X_n]$.

6. (*Mathematica*) A machine receives shocks at the times T_1, T_2, T_3, ... of a renewal process, and incurs damage at a level of D_i as a result of the ith shock. The damage from each shock ebbs exponentially as time wears on. We shall suppose that the times $T_{i+1} - T_i$ between shocks are independent of the damage levels D_i, which are independent and identically distributed. Hence the total damage sustained by the machine up to time t can be written:

$$D(t) = \sum_{i=1}^{N_t} D_i \, e^{-\beta(t-T_i)}$$

Write a simulation program that takes as parameters the probability distribution of the damages, the inter-renewal distribution, the constant β, and a final time t, and returns the cumulative damage level $D(t)$ at time t. Exercise your program in the case where the inter-renewal times have the exponential(2) distribution, $\beta = .01$, the damage levels have the continuous uniform distribution on the interval $[0, 1]$. By running the simulation many times, use the average damage level $\overline{D}(t)$ in the simulated trials to estimate the expectation $E[D(t)]$. Repeat for several different times t and generate a list plot in time to get an estimate of the function that maps t to $E[D(t)]$.

7. Customers arrive to a single server according to a Poisson process with rate 3 per hour. Those customers who arrive when the server is busy are simply lost. The server requires a constant time of c to activate, then an exponential length of time with rate 2 per hour to actually perform service. Find the long-run average number of customers served per hour. How many customers are being lost per hour if the activation period is 10 minutes? 5 minutes?

8. Use Theorem 2 to find the renewal function in the case that the interrenewal distribution is deterministic with all of its weight on the point 1.5.

Exercises 9–12 lead the reader through a renewal theoretic proof of the result on convergence of Markov chains, Theorem 1 of Section 4.5.

9. A *delayed renewal process* is similar to a renewal process, except that the c.d.f. G of the first renewal time T_1 may be different from the common c.d.f. F of the inter-renewal times $T_{n+1} - T_n$, $n = 1, 2, ...$. Show that for a delayed renewal process (N_t),

$$E[N_t] = \sum_{n=1}^{\infty} G * F^{(n-1)}(t)$$

10. Given a Markov chain $(X_n)_{n \geq 0}$, argue that the times $S_1, S_2, S_3, \dots$ of successive visits to a state j, starting from a state i, form a delayed renewal process, as defined in Exercise 8.

11. There is a theorem from renewal theory called *Blackwell's Renewal Theorem* (see Ross ([52], Prop. 3.5.1) that is applicable to processes with integer-valued renewal times that are otherwise non-periodic. It implies the following for delayed renewal processes: the expected number of renewals exactly at time n converges to the reciprocal of the mean inter-renewal time as $n \longrightarrow \infty$. Prove parts (c) and (d) of Theorem 1 of Section 4.5 by presuming that the hypotheses of Blackwell's Theorem are in force, and considering the sequence of random variables $I_0, I_1, I_2, \dots$, defined by setting I_n to 1 or 0 according to whether X_n equals j or not.

12. Finally, with (I_n) as in Exercise 11 and N_t equal to the sum of the I's through t, appeal to the Renewal Law of Large Numbers to establish part (e) of Theorem 1 of Section 4.5.

13. A machine begins in good running condition, lasts for an exponential length of time with rate λ_1, then breaks down. The repair of the machine lasts for an exponential period of time with rate λ_2, after which the machine is completely repaired, lasts for another exponential length of time with rate λ_1, is repaired again, etc. If a repair costs c dollars, find the long-run cost of repairs per unit time.

14. Prove Theorem 5(a).

15. Three investments are available. The first pays fixed dividends of $100 at the times of a renewal process whose inter-renewal distribution is exponential with parameter $1/5$. The second can pay either $80, $100, or $140, each with probability $1/3$, at the times of a renewal process whose inter-renewal distribution is $\Gamma(2, 4)$. The third pays an amount $50 \cdot M_i$ at the deterministic times $i = 3, 6, 9, \dots$, where each M_i has the Poisson distribution with parameter 1, and the M_i's are mutually independent. Which investment is preferable from the point of view of maximizing the long-run expected reward per unit time?

16. (*Mathematica*) A device is currently new. We replace it with an identical new device either when it breaks or at the fixed time T, whichever comes first. The lifetime of a device has the Weibull density:

$$f(x) = \begin{cases} \frac{1}{2} x^{-1/2} e^{-x^{1/2}} & \text{if } x > 0 \\ 0 & \text{otherwise} \end{cases}$$

The cost incurred due to breakdown is twice as much as the cost due to a simple replacement of a functioning device. Show that to minimize long-run average cost per unit time, it is optimal never to replace the device. (Note that the inter-renewal time has a density on $(0, T)$ and puts positive mass on the point T. Use a sensible extension of the definition of expectation for this mixed discrete–continuous random variable.)

17. Consider a renewal process (N_t) with discrete, geometric inter-renewal distribution:

$$g(n) = (1 - p)^{n-1} p, \quad n = 1, 2, 3, \ \dots$$

(a) Find the limit as $t \longrightarrow \infty$ of N_t / t.
(b) Find explicitly the distribution of N_t. (Hint: Think about what kind of experiment gives rise to geometric times between renewal.)
(c) Find explicitly $m(t) / t$.
(d) If a reward of either \$2 or \$4, with equal probability, is earned at each renewal time, find the expected average reward in the finite time interval $[0, t]$.

18. A common stock currently sells at \$20 per share. The price changes by plus \$1 or minus \$1 (respectively with probabilities p and $1 - p$) at times $T_1, T_2, T_3, \dots$ such that the times between changes are i.i.d. $\Gamma(4, 2)$ random variables. For large time t, approximately what do we expect the price to be at that time?

5.4 Queueing Theory

Preliminaries

A vital application of stochastic processes is the study of waiting lines, or *queues*. In queueing problems, there are customers arriving to a service facility as time passes (e.g., people to a store, jobs to a central processing unit, cars to a parking lot, messages to a communication station, or planes to an airport). Customers who do not go immediately into service must wait in line. After their service is finished, they depart. One is interested in how the queue length tends to rise and fall. Also, the waiting times of individual customers are of interest.

There are several pertinent aspects of a queueing model:

1. the probabilistic arrival pattern of the customers;
2. the probability law of the time taken by a server to serve a customer;
3. the number of servers present at the service facility;
4. the size of the waiting area, if only a limited number of customers may wait;
5. the queue discipline, that is, the rule by which a new customer is selected to be served when a server becomes available;
6. the presence of one or more other queues that interact with the one of principal interest (e.g., customers may not be able to arrive to one queue until they are serviced at another).

The variations to the underlying model are many, and there are also many quantities of interest for a single model. In this brief introduction we will concentrate on single queues as opposed to networks of several queues. The queue discipline is first-in, first-out (FIFO), in which customers are served in the order in which they arrive. We present two models with multiple servers, and three with only one server. In one of our problems there will be a finite waiting room, so that arrivals who come when the waiting room is full are turned away. Either the inter-arrival times or the customer service times or both will be exponentially distributed and mutually independent. Our main goal will be to calculate the long-run distribution of queue length.

Let us look at a typical outcome of a queue length process. The table below gives the first few arrival and service times.

arrival times T_i	service times S_i
3.5	5.3
5.8	3.0
7.0	2.8
16.2	2.9

The path of the queue length process (X_t) for this outcome is sketched in Figure 5.9. The first arrival comes at time $T_1 = 3.5$ and requires 5.3 time units for service; therefore this customer departs at time 8.8. In the meantime, both the second and third arrivals have occurred, at times 5.8 and 7.0, respectively. Therefore the queue length rises to 2 at time 5.8, then to 3 at time 7.0. When the first customer leaves, the second goes into service, and the queue size is reduced to 2. Since the second service time is 3.0, customer 2 departs at time $8.8 + 3.0 = 11.8$. At this time, the third customer starts service; and since the service period takes 2.8 time units, the third customer departs at time $11.8 + 2.8 = 14.6$. The fourth arrival does not appear until time 16.2, so that the queue is empty for 1.6 time units. The queue size moves up to 1 at time 16.2, and the service process continues.

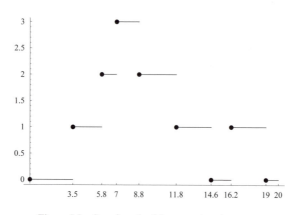

Figure 5.9 – Sample path of the queue length process

There is a standard shorthand that has arisen to specify the queueing model under study. This consists of a string of characters separated by diagonal slashes, of the form: $A/B/n/N$. In the first position is an abbreviation for the distribution of the times between successive arrivals. The symbol M is used for exponential, D for deterministic or non-random arrivals, E_k for Erlang with parameter k, and G for a general distribution, for example. The second position B represents the distribution of the service times of a single customer. The same abbreviations are used. The third position is the number of servers n, and the fourth position is the number of waiting spots N. The latter is usually left blank when there is no bound on the waiting room size. Thus, $M/M/1/6$ indicates a single server queue whose arrivals form a Poisson process, say with rate λ, such that service times are i.i.d. exponential random variables, say with parameter μ, and such that there are six waiting positions. The shorthand $M/G/2$ means a two-server queue with unlimited waiting space, such that arrivals form a Poisson process and service times have some unspecified distribution function.

Activity 1 – Think of at least two queues that you have been in recently. In the queueing shorthand, explain what the $D/M/4/5$, $M/E_k/\infty$, and $G/M/1$ models are.

Simple Poissonian Queues

EXAMPLE 1. ($M/G/\infty$) Suppose that arrivals to a service facility occur at the times $T_1, T_2, T_3, \ldots$ of a Poisson process, and upon arrival a customer is immediately served by one of an infinite number of servers. This could be an approximate model of a self-service facility such as a grocery store where shoppers select their own items, or a parking garage with a large number of spaces. Service of a customer takes a random amount of time with distribution function G, and customers are served independently of one another. We find the distribution of X_t, which denotes the number of customers in service at time t.

Normally it is difficult to calculate the short-run distribution of queue length, but here we have enough structure that it is relatively easy. The random variable N_t, defined as the total number of arrivals by time t, is Poisson with parameter λt. By the law of total probability,

$$
\begin{aligned}
P[X_t = k] &= \sum_{n=0}^{\infty} P[X_t = k \mid N_t = n] \, P[N_t = n] \\
&= \sum_{n=k}^{\infty} P[X_t = k \mid N_t = n] \cdot \frac{e^{-\lambda t}(\lambda t)^n}{n!}
\end{aligned}
\tag{1}
$$

The sum begins at k because it is impossible for the number of customers still in the system at time t to be more than the number who have arrived by time t.

Recall from Theorem 3 of Section 5.1 that, given $N_t = n$, the arrival times $T_1, T_2, \ldots, T_n$ have the joint distribution of the uniform order statistics on $[0, t]$. The event that exactly k of these arrivals are still in the system at time t is the event that exactly k of the events $\{T_i + S_i > t\}$ occur, where S_i denotes the service time of the ith customer. But this has the same probability as the event that exactly k of the events $\{U_i + S_i > t\}$ occur, where the U_i's are i.i.d. uniform random variables on $[0, t]$. By this reasoning, $P[X_t = k \mid N_t = n]$ is the probability of exactly k successes in a binomial experiment of n trials. Conditioning and unconditioning on U, the success probability per trial is

$$
\begin{aligned}
p = P[U_i + S_i > t] &= \int_0^t P[u + S > t \mid U = u] \, \tfrac{1}{t} \, du \\
&= \int_0^t P[S > t - u] \, \tfrac{1}{t} \, du \\
&= \tfrac{1}{t} \int_0^t (1 - G(t - u)) \, du \\
&= \tfrac{1}{t} \int_0^t (1 - G(x)) \, dx
\end{aligned}
\tag{2}
$$

The last line is the result of the substitution $x = t - u$. By (1),

$$P[X_t = k] = \sum_{n=k}^{\infty} \frac{n!}{k!\,(n-k)!}\, p^k (1-p)^{n-k} \cdot \frac{e^{-\lambda t}(\lambda t)^n}{n!}$$

$$= \frac{(\lambda t\, p)^k\, e^{-\lambda t}}{k!} \cdot \sum_{n=k}^{\infty} \frac{(\lambda t\,(1-p))^{n-k}}{(n-k)!}$$

$$= \frac{(\lambda t\, p)^k\, e^{-\lambda t}}{k!} \cdot e^{\lambda t(1-p)}$$

$$= \frac{(\lambda t\, p)^k\, e^{-\lambda t\, p}}{k!}$$

We have proved that the number of customers X_t in the system at time t has the Poisson distribution with parameter $\lambda t\, p$, where p is given by (2). In particular, the mean number of customers in service at time t is

$$E[X_t] = \lambda t\, p = \lambda \int_0^t (1 - G(x))\, dx \qquad (3)$$

In the case of the $M/M/\infty$ queue, where services are exponential with rate μ, the mean number in service at time t is

$$E[X_t] = \lambda \int_0^t e^{-\mu x}\, dx = \frac{\lambda}{\mu}\,(1 - e^{-\mu t})$$

As $t \longrightarrow \infty$, the mean number in service approaches λ/μ. ∎

Activity 2 – In Example 1, what is the mean number in service if the service time distribution is the continuous uniform distribution on the interval $[0, 2/\mu]$?

EXAMPLE 2. ($M/M/s$) Suppose that cars arrive to a toll station according to a Poisson process with rate λ, and that their service times are i.i.d. exponential random variables with parameter μ. There are s toll booths at the station handling the incoming traffic. Let X_t be the number of cars at the station at time t. We wish to find the limiting distribution of X_t as $t \longrightarrow \infty$.

We claim that (X_t) is a birth–death process. Consider a time such that $X_t = n$, i.e., there are n cars waiting at the toll station. Because both the inter-arrival and service time random variables have the memoryless exponential distribution, the time t may as well be time 0 and the past history of the queue is irrelevant given the current state. Regardless of how long we have waited for a change of state, there is still an exponential amount of time until the next arrival, and an exponential amount of time until the next service. The queue size can only increase by one if the arrival comes first, and decrease by one if the service happens first (if $n = 0$, then of course the queue size must increase by one). Thus, (X_t) satisfies the properties of a birth–death process.

It remains only to compute the birth and death rates. Consider the case

$1 \le n \le s$ first. Let T be the time of the next arrival, and let $S_1, S_2, ..., S_n$ be the times at which the currently busy servers 1, ..., n finish their service. Then the probability that the queue size does not change for u more time units is

$$P[T > u, S_i > u \text{ for } i = 1, ..., n] = e^{-\lambda u} \cdot (e^{-\mu u})^n = e^{-(\lambda + n\mu)u}$$

The probability that the next change in queue length is a birth is

$$P[T < S_i \text{ for } i = 1, ..., n]$$

$$= \int_0^\infty (\int_t^\infty \cdots \int_t^\infty \mu e^{-\mu s_1} \cdots \mu e^{-\mu s_n} \, ds_1 \cdots ds_n) \lambda e^{-\lambda t} \, dt$$

$$= \lambda \int_0^\infty e^{-(\lambda + n\mu)t} \, dt$$

$$= \frac{\lambda}{\lambda + n\mu}$$

The case where $n > s$ is similar, except that only the s servers can be working, so that n in the above expressions is replaced by s. When $n = 0$, it is easy to see that the birth rate is λ and the death rate is 0. In summary, the queue length process (X_t) is a birth–death process with parameters

$$\lambda_n = \lambda, \quad n = 0, 1, 2, ... \qquad \mu_n = \begin{cases} n\mu & \text{if } n = 0, 1, ..., s \\ s\mu & \text{if } n = s+1, s+2, ... \end{cases} \qquad (4)$$

We can use Theorem 1 of Section 5.2 to find the limiting probabilities. Recall that $p_0 = \lim_{t \to \infty} P[X_t = 0]$ is

$$p_0 = \left(1 + \sum_{j=1}^\infty \frac{\lambda_0 \lambda_1 \cdots \lambda_{j-1}}{\mu_1 \mu_2 \cdots \mu_j}\right)^{-1}$$

$$= \left(1 + \sum_{j=1}^s \frac{\lambda^j}{\mu^j j!} + \sum_{j=s+1}^\infty \frac{\lambda^j}{\mu^j s^{j-s} s!}\right)^{-1}$$

$$= \left(1 + \sum_{j=1}^s \frac{(\lambda/\mu)^j}{j!} + \frac{(\lambda/\mu)^s}{s!} \sum_{k=1}^\infty \left(\frac{\lambda}{\mu s}\right)^k\right)^{-1}$$

The infinite series in the last expression converges to $(\lambda/\mu s)/(1 - \lambda/\mu s)$ for $\lambda < \mu s$. Under this condition, the limiting probabilities exist and

$$p_0 = \left(\sum_{j=0}^s \frac{(\lambda/\mu)^j}{j!} + \frac{(\lambda/\mu)^s \cdot \lambda}{s! \cdot (\mu s - \lambda)}\right)^{-1}$$

after some algebraic rearrangement. Also, from the limit theorem on birth–death processes, we have

$$p_n = \lim_{t \to \infty} P[X_t = n] \;\; = \;\; \frac{\lambda_0 \, \lambda_1 \cdots \lambda_{n-1}}{\mu_1 \, \mu_2 \cdots \mu_n} \cdot p_0$$

$$= \begin{cases} \dfrac{(\lambda/\mu)^n}{n!} \cdot p_0 & \text{if } n \le s \\[2mm] \dfrac{(\lambda/\mu)^n}{s! \, s^{n-s}} \cdot p_0 & \text{if } n > s \end{cases}$$

If, for example, the arrival rate is $\lambda = 4$ cars per minute, and each booth can only serve at a rate of $\mu = 3$ cars per minute, then at least $s = 2$ booths are necessary to satisfy $\lambda < \mu s$, and thereby to save the queue from blowing up. For these numbers, we have $p_0 = 1/5$ as shown below.

$$p_0 \; = \; \left(\text{Sum}[\,(4\,/\,3)^j \,/\, j!, \; \{j, \, 0, \, 2\}] + \frac{(4\,/\,3)^2 \,*\, 4}{2\,!\, *\, (3\, *\, 2\, -\, 4)} \right)^{-1}$$

$\dfrac{1}{5}$

Therefore,

$$p_n = \begin{cases} \dfrac{1}{5} \cdot \dfrac{(4/3)^n}{n!} & \text{if } n = 1, 2 \\[3mm] \dfrac{1}{5} \cdot \dfrac{(4/3)^n}{2!\,2^{n-2}} = \dfrac{2}{5}\left(\dfrac{2}{3}\right)^n & \text{if } n = 3, 4, 5, \ldots \end{cases}$$

We show this distribution in Figure 5.10 as a connected list plot. Notice that the large preponderance of the time, there are 12 or fewer vehicles in queue.
∎

```
p[n_] := Which[n == 0, 1 / 5, 1 ≤ n ≤ 2,
    (1 / 5) * (4 / 3)ⁿ / n!, n ≥ 3, (2 / 5) * (2 / 3)ⁿ];
longrundist = Table[{n, p[n]}, {n, 0, 12}];
ListPlot[longrundist, PlotJoined → True,
    DefaultFont → {"Times", 8}];
```

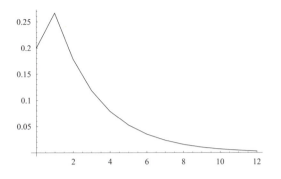

Figure 5.10 – Limiting distribution of queue length for $M/M/2$ queue with $\lambda = 4$, $\mu = 3$

Activity 3 – In Example 2, try increasing the arrival rate λ to see the effect on the long-run queue length distribution.

EXAMPLE 3. ($M/M/1/N$) Let arrivals to a barber shop form a Poisson process with rate λ, and suppose that the time required for a haircut is exponential with rate μ. There is a single barber, and there are only N chairs in the shop, including the one that the barber is using for the customer he is currently serving. When the shop is full, arrivals are turned away. We will compute the limiting distribution of the number of customers in the shop.

By the same reasoning as in the last example, we can treat the queue length process (X_t) as a birth–death process with parameters

$$\lambda_j = \begin{cases} \lambda & \text{if } j = 0, 1, \ldots, N-1 \\ 0 & \text{otherwise} \end{cases} \quad , \quad \mu_j = \begin{cases} 0 & \text{if } j = 0 \\ \mu & \text{if } j = 1, 2, 3, \cdots \end{cases}$$

Then,

$$\begin{aligned} p_0 = \lim_{t \to \infty} P[X_t = 0] &= \left(1 + \sum_{j=1}^{\infty} \frac{\lambda_0 \lambda_1 \cdots \lambda_{j-1}}{\mu_1 \mu_2 \cdots \mu_j}\right)^{-1} \\ &= \left(1 + \sum_{j=1}^{N} \frac{\lambda^j}{\mu^j}\right)^{-1} \qquad (5) \\ &= \frac{1 - \lambda/\mu}{1 - (\lambda/\mu)^{N+1}} \end{aligned}$$

The above computation is valid for $\lambda \neq \mu$; when $\lambda = \mu$ we can see directly from the second line of (5) that $p_0 = 1/(N+1)$. We also have

$$p_n = \lim_{t \to \infty} P[X_t = n] = (\lambda/\mu)^n \cdot p_0, \text{ if } n = 1, 2, \ldots, N \qquad (6)$$

Equation (6) holds in each of the cases $\lambda < \mu$, $\lambda > \mu$, and $\lambda = \mu$. (Try verifying that (5) and (6) form a valid probability mass function.) Because of the finite waiting room size, it is not necessary to impose the condition that $\lambda < \mu$ to ensure convergence of an infinite series in this problem. Also, note that in the case $\lambda = \mu$ we just have $p_n = p_0 = 1/(N + 1)$ for each $n = 1, 2, ..., N$.

In passing we note that the mean of the limiting distribution of the queue length (the limiting average number of customers in the shop) can be computed. This is

$$L = \sum_{n=0}^{N} n \, p_n = \sum_{n=1}^{N} n \, p_0 (\lambda / \mu)^n \tag{7}$$

Exercise 4 asks you to calculate that in the case $\lambda \neq \mu$,

$$L = \frac{(\lambda/\mu)\,(N(\lambda/\mu)^{N+1}-(N+1)\,(\lambda/\mu)^{N}+1)}{(1-(\lambda/\mu)^{N+1})\,(1-\lambda/\mu)} \tag{8}$$

and in the case $\lambda = \mu, L = N/2$. ∎

The ratio λ/μ has arisen several times. We denote it by ρ and call it the *traffic intensity* of the queue, since it is the ratio of the arrival rate to the service rate.

M/G/1 Queue

In this subsection we discuss a family of queues in which the service time distribution is general, and the inter-arrival times are independent and exponentially distributed. Accordingly, suppose that Poisson arrivals come in to a single server, whose service times are independent and have some unspecified distribution function G. We are interested in finding the limiting distribution as time $t \longrightarrow \infty$ of the queue length X_t. We will not be able to compute this directly, but fortunately it can be shown (see Gross and Harris [28]) to be the same as the limiting distribution of a discrete time *embedded Markov chain*, defined as follows. Let Y_n be the number of customers waiting for service just after the nth departure. For the queueing outcome depicted in Figure 5.9, the first three departure times are 8.8, 11.8, and 14.6. For this outcome, $Y_1 = 2$, since customers 2 and 3 are left behind by the departing customer $1; Y_2 = 1$, since customer 2 leaves customer 3 still standing in line; and $Y_3 = 0$, since the fourth arrival has not yet occurred when customer 3 finishes service.

Activity 4 – If the first few arrival times are 3.1, 4.5, 6.7, 8.1, and 8.7, and the service times are, respectively, 2.1, .8, 1.6, 2.4, and 1.2, sketch a graph of the queue length as a function of time and determine the first few values of the Markov chain (Y_n) embedded at the departure instants.

If Y_n is known, does Y_{n+1} depend on the past history before time n? In the case $Y_n > 0$, the number of customers after the $(n + 1)^{st}$ departure is

$$Y_{n+1} = Y_n - 1 + A \qquad (9)$$

where A is the number of arrivals during the service period of the $(n + 1)^{st}$ customer. In equation (9), 1 is subtracted because the $(n + 1)^{st}$ customer has just left. In the case $Y_n = 0$, we just have

$$Y_{n+1} = A \qquad (10)$$

Because of the memoryless property of the exponential distribution, the number of arrivals A during the service of customer $n + 1$ is completely independent of the past arrival and service stream. Thus, Y_{n+1} is conditionally independent of past Y_i's given Y_n. This indicates that the queue length chain (Y_n) embedded at departure instants is Markov.

Let us compute the transition matrix of the embedded chain (Y_n). Formulas (9) and (10) indicate that we will need to know, for each $k = 0, 1, 2, ...,$ the probability that there will be exactly k arrivals during a service. If the duration of service is known to be t, then the number of arrivals is a Poisson random variable with parameter λt. Thus, conditioning and unconditioning on the duration of service, we obtain

$$q_k = P[\text{exactly } k \text{ arrivals during a service}] = \int_0^\infty \frac{e^{-\lambda t}(\lambda t)^k}{k!} \, d\, G(t) \qquad (11)$$

where the integral with respect to the distribution function G is interpreted as $\int (\cdot) g(t) \, dt$ if services are continuously distributed with density g, and as $\sum (\cdot) g(t)$ if services are discretely distributed with probability mass function g. Recall that Y_n is the number of customers still to be served when the nth customer departs. Given $Y_n = 0$, Y_{n+1} equals k if and only if k arrivals occurred during the service of customer $n + 1$. This event occurs with probability q_k as in (11). Given that $Y_n = 1$, the event $\{Y_{n+1} = k\}$ occurs with the same probability q_k. But given $Y_n = 2$, the nth customer leaves behind the $(n + 1)^{st}$ customer plus one other, so that the event $\{Y_{n+1} = k\}$ is the same as the event that exactly $k - 1$ arrivals have occurred during the service of customer $n + 1$. By this reasoning we see that the transition matrix of (Y_n) is:

$$
T = \begin{array}{c} \\ 0 \\ 1 \\ 2 \\ 3 \\ \vdots \end{array}
\begin{array}{c} 0 \quad 1 \quad 2 \quad 3 \quad \cdots \\
\begin{pmatrix}
q_0 & q_1 & q_2 & q_3 & \cdots \\
q_0 & q_1 & q_2 & q_3 & \cdots \\
0 & q_0 & q_1 & q_2 & \cdots \\
0 & 0 & q_0 & q_1 & \cdots \\
\vdots & \vdots & \vdots & \vdots & \vdots
\end{pmatrix}
\end{array}
\tag{12}
$$

The stationary equations $\pi = \pi \cdot T$ for this transition matrix are impossible to solve explicitly. However, the limiting probabilities π_n for the embedded chain can be generated recursively. You should verify that the stationary equations have the form:

$$
\pi_i = \pi_0 q_i + \sum_{j=1}^{i+1} \pi_j q_{i-j+1}, \quad i = 0, 1, 2, \ldots
\tag{13}
$$

The next theorem shows how to calculate π_0, in order to initialize the computation, and gives a rearrangement of (13) that directly expresses π_{i+1} in terms of the previous π_i's. The proof, which involves probability generating functions, is outlined in Exercises 10–12.

THEOREM 1. Let μ denote the reciprocal of the mean of the service distribution G for the $M/G/1$ queue, let λ be the Poisson arrival rate, and let $\rho = \lambda/\mu$ be the traffic intensity. Under the condition that $\rho < 1$, the limiting distribution π of the embedded chain (Y_n) exists and satisfies

$$
\begin{aligned}
\pi_0 &= 1 - \rho \\
\pi_{i+1} q_0 &= \pi_i - \pi_0 q_i - \sum_{j=1}^{i} \pi_j q_{i-j+1}
\end{aligned}
\tag{14}
$$

EXAMPLE 4. Suppose that a manufacturer keeps a large number of machines, and that breakdowns of machines follow a Poisson process with rate three breakdowns per week. One repairman is available to service the machines, and a repair takes one day with probability 1/2, two days with probability 1/4, and three days with probability 1/4. What is the long-run probability that there will be at least three machines under repair?

Here, the queueing process is defined by $X_t =$ number of machines waiting for repair at time t. We will calculate the first few entries of π, the limiting distribution of the chain (Y_n) embedded at instants of repair. As mentioned before, the discrete chain has the same limiting distribution as the continuous process. Inter-arrival intervals of machines to the repair shop are exponential random variables with rate $\lambda = 3$ per week. The repairman is the service facility in this application. The probability mass function of the successive repair times is

$$g(x) = \begin{cases} 1/2 & \text{if } x = 1/7 \text{ week} \\ 1/4 & \text{if } x = 2/7 \text{ week} \\ 1/4 & \text{if } x = 3/7 \text{ week} \end{cases}$$

The expected value of the repair time is easily found to be 1/4 week, and consequently the repair rate is $\mu = 4$ per week. Because of this, the traffic intensity $\rho = 3/4$ satisfies the hypothesis of Theorem 1. We are asked for

$$\lim_{n \to \infty} P[Y_n \geq 3] = 1 - \pi_0 - \pi_1 - \pi_2$$

and by Theorem 1,

$$\pi_0 = 1 - \rho = 1/4$$
$$\pi_1 = \frac{1}{q_0} (\pi_0 - \pi_0 q_0)$$
$$\pi_2 = \frac{1}{q_0} (\pi_1 - \pi_0 q_1 - \pi_1 q_1)$$

To complete the solution of the problem, we must calculate q_0 and q_1. Recall that these are the probabilities of having zero and one arrival, respectively, during a service period. From (11),

$$q_k = \sum_{t=1/7,2/7,3/7} \frac{e^{-3t}(3t)^k}{k!} g(t)$$

We can set up a function to compute these as follows:

```
g[t_] := Which[t == 1 / 7,
    1 / 2, t == 2 / 7, 1 / 4, t == 3 / 7, 1 / 4];
q[k_] := NSum[ (E^-3 t * (3 t)^k) / k!  * g[t],
    {t, 1 / 7, 3 / 7, 1 / 7}];
```

Below are the q's that we need, followed by the π's, and the limiting probability of 3 or more machines under repair, which comes out to around .32.

```
{q[0], q[1]}
```

{0.500926, 0.319391}

$$\pi 1 = \frac{1}{q[0]} \left(\frac{1}{4} - \frac{1}{4} \, q[0] \right)$$

$$\pi 2 = \frac{1}{q[0]} \left(\pi 1 - \frac{1}{4} \, q[1] - \pi 1 * q[1] \right)$$

$$1 - (1 / 4 + \pi 1 + \pi 2)$$

0.249076

0.179019

0.321905

In Exercise 9 you are asked to write a *Mathematica* function that implements formula (14), with which to plot the limiting distribution. ∎

G/M/1 Queue

Up to this point we have discussed queues for which arrivals form a Poisson process. Our last example is the $G/M/1$ family of queues, where interarrival times are i.i.d. random variables with distribution function G (i.e., arrivals form a renewal process), and service times are exponential with rate μ. Let λ be the arrival rate, by which we mean the reciprocal of the mean interarrival time, and let $\rho = \lambda / \mu$ be the traffic intensity.

We again take an embedded chain approach. Let Y_n be the number of customers in the system when the nth arrival comes (excluding that n^{th} arrival). In Figure 5.9, the arrivals occur at times 3.5, 5.8, 7.0, and 16.2. The first arrival finds no customers in the queue yet, hence $Y_1 = 0$. The second customer finds the first customer ahead, so that $Y_2 = 1$. The third customer finds both of the first two still in the system, hence $Y_3 = 2$. But by the time the fourth customer has arrived, all others have been served, so that $Y_4 = 0$.

Activity 5 – For the arrivals and departures listed in Activity 4, give the first few values of the chain (Y_n) embedded at arrival instants.

To relate Y_{n+1} to Y_n, note that the $(n + 1)^{\text{st}}$ customer will arrive to find the customers that were ahead of the nth customer, plus customer n, minus the number of customers whose service has been completed since customer n arrived. Therefore,

$$Y_{n+1} = Y_n - B + 1 \tag{15}$$

where the random variable B is the number of services performed between the times of arrival of the nth and $(n + 1)^{\text{st}}$ customers. As in the $M/G/1$ queue, the Markov property of the embedded chain (Y_n) for the $G/M/1$ queue holds because of the memoryless property of the exponential service time distribution. Also, (15) suggests how to compute the transition matrix. If $Y_n = 0$, i.e., the nth customer finds no one ahead in the line, then Y_{n+1} can only be 0 or 1, according to whether or not customer n has finished service by the time that customer $n + 1$ arrives. Similarly, if $Y_n = 1$, then Y_{n+1} can only be 0, 1, or 2. It is 2 if neither customer n nor the customer in front of customer n has finished service by the time customer $n + 1$ arrives; it is 1 if the customer in front of n has been served, but customer n has not, and it is 0 if both services have been performed, perhaps with some time to spare. By the same argument used in (11) to obtain the $M/G/1$ probability of k arrivals during a service interval, we can write

$$
\begin{aligned}
q_k &= P[\text{exactly } k \text{ services during an inter–arrival interval}] \\
&= \int_0^\infty \frac{e^{-\mu s}(\mu s)^k}{k!} \, d\,G(s)
\end{aligned}
\tag{16}
$$

Thus, denoting $r_i = 1 - (q_0 + q_1 + \cdots + q_i)$, we have that the transition matrix of the chain embedded at the arrival instants of the $G/M/1$ queue is

$$
T = \begin{array}{c@{}c}
 & \begin{array}{cccccc} 0 & 1 & 2 & 3 & 4 & \cdots \end{array} \\
\begin{array}{c} 0 \\ 1 \\ 2 \\ 3 \\ \vdots \end{array} &
\left(\begin{array}{cccccc}
r_0 & q_0 & 0 & 0 & 0 & \cdots \\
r_1 & q_1 & q_0 & 0 & 0 & \cdots \\
r_2 & q_2 & q_1 & q_0 & 0 & \cdots \\
r_3 & q_3 & q_2 & q_1 & q_0 & \cdots \\
\vdots & \vdots & \vdots & \vdots & \vdots & \ddots
\end{array}\right)
\end{array}
$$

The following remarkable result is proved by solving the stationary equations.

THEOREM 2. Let (Y_n) be the Markov chain embedded at the arrival instants of the $G/M/1$ queue, and suppose that $\rho = \lambda/\mu < 1$. Then the limiting probabilities $\pi_j = \lim_{n\to\infty} P[Y_n = j]$ exist and take the form

$$
\pi_j = (1 - \beta)\,\beta^j, \quad j = 0, 1, 2, \ldots
\tag{17}
$$

where $\beta \in (0, 1)$ is a solution of the equation

$$
\beta = q_0 + q_1\,\beta + q_2\,\beta^2 + \cdots
\tag{18}
$$

Proof. The stationary equations are

$$\begin{aligned}
\pi_0 &= r_0\,\pi_0 + r_1\,\pi_1 + r_2\,\pi_2 + \cdots \\
\pi_1 &= q_0\,\pi_0 + q_1\,\pi_1 + q_2\,\pi_2 + \cdots \\
\pi_2 &= \qquad\quad\; q_0\,\pi_1 + q_2\,\pi_2 + \cdots \\
&\;\;\vdots
\end{aligned} \tag{19}$$

By the fact that $r_i = 1 - (q_0 + q_1 + q_2 + \cdots + q_i)$, it is easy to see that the top equation in (19) is just one minus the sum of all the others, and consequently it is superfluous.

We leave unproved (see Cinlar [15]) the fact that if $\rho < 1$, then the limiting distribution exists. But the limiting distribution must be unique when it does exist, so that all we need do is show that our candidate $\pi_j = (1 - \beta)\,\beta^j$ satisfies the infinite linear system (19). Note that the vector π does form a good probability mass function for β between 0 and 1.

For $k \geq 1$, the kth equation in (19) has the form

$$\pi_k = q_0\,\pi_{k-1} + q_1\,\pi_k + q_2\,\pi_{k+1} + \cdots$$

Substitute $(1 - \beta)\,\beta^i$ for each factor π_i on the right side of this equation, to get

$$\begin{aligned}
q_0(1 - \beta)\,\beta^{k-1} &+ q_1(1 - \beta)\,\beta^k + q_2(1 - \beta)\,\beta^{k+1} + \cdots \\
&= (1 - \beta)\,\beta^{k-1}(q_0 + q_1\,\beta + q_2\,\beta^2 + \cdots) \\
&= (1 - \beta)\,\beta^{k-1}\,\beta \\
&= (1 - \beta)\,\beta^k
\end{aligned}$$

The third line follows from (18). This establishes that our candidate is a solution of the stationary equations. ∎

REMARK. Unfortunately it is not true, unless the queue is $M/M/1$, that π_j is the limiting probability of j in the queue for the continuous time queue length process (X_t). Also, for both the $M/G/1$ and $G/M/1$ queues, it can be shown (see Çinlar [15]) that when $\rho \geq 1$, the limiting probabilities for all states are 0.

EXAMPLE 5. $(D/M/1)$ Suppose that at a large bakery, cakes come off of a conveyor belt non-randomly at exactly one per minute. The cakes are to be iced by a single skilled icer, who requires an exponentially distributed amount of time with mean 30 seconds to ice a cake. Find the limit as $n \longrightarrow \infty$ of the probability that the number of cakes waiting to be iced at the time of the nth cake arrival is k, for each $k = 0, 1, 2, \ldots$.

In this problem, the customers are cakes, and the inter-arrival probability mass function is

$$g(t) = \begin{cases} 1 & \text{if } t = 1 \\ 0 & \text{otherwise} \end{cases}$$

The arrival rate is $\lambda = 1$ per minute. The single server is the icer, and the service time distribution is exponential with mean $1/2$ minute, hence the rate is $\mu = 2$ per minute. Since the traffic intensity $\rho = \lambda/\mu = 1/2$, Theorem 2 can be applied. To solve for β, we will need to compute the q_j's. Referring to (16), the integral is just the discrete sum of one term, namely,

$$q_k = \frac{e^{-\mu \cdot 1}(\mu \cdot 1)^k}{k!} \cdot 1 = \frac{e^{-2} 2^k}{k!}$$

Thus, β is the solution of

$$\beta = \sum_{k=0}^{\infty} \frac{e^{-2} 2^k}{k!} \beta^k = e^{-2} \sum_{k=0}^{\infty} \frac{(2\beta)^k}{k!} = e^{-2(1-\beta)}$$

Though $\beta = 1$ is a solution, it is not the one that we want, or else all of the limiting probabilities in (17) would be 0. Other solutions are not available analytically, so we must use a numerical procedure such as FindRoot to approximate β:

```
FindRoot[β - E^-2 (1-β), {β, .5}]
```

$\{\beta \to 0.203188\}$

The value of β correct to three decimal places is $\beta = 0.203$. The limiting distribution of the chain embedded at arrival instants is therefore

$$\pi_k = \lim_{n \to \infty} P[Y_n = k] = (.797)(.203)^k, \quad k = 0, 1, 2, \ldots \quad \blacksquare$$

Exercises 5.4

1. (a) Compute the birth and death rates for the $M/M/4/6$ queue.
   (b) Find the limiting probabilities, if $\lambda = 4$ and $\mu = 2$.
   (c) What effect does doubling the service rate have on the long-run probability that the queue is full?

2. Compute the traffic intensity of a single server queue in which arrivals form a Poisson process with rate 5, and service times have the Weibull distribution: $g(t) = 2t e^{-t^2}$, if $t > 0$. Will this queue have a limiting distribution?

3. Find the limiting distribution of the $M/G/\infty$ queue.

4. Verify expressions (7) and (8) for the limiting mean queue length of the $M/M/1/N$ queue.

5. (*Mathematica*) For an $M/M/1/N$ queue with arrival rate $\lambda = 2$ and service rate $\mu = 3$, find the smallest value of N such that the limiting probability of two or fewer in the queue is less than .75.

6. (*Mathematica*) In Example 2, set $\lambda = 5.2$ and $\mu = 1.3$. At least how many toll stations should there be so that 95% of the time in the long run, there are ten or fewer vehicles in queue?

7. Suppose that the barber of Example 3 is unlucky enough to have only one employee (himself), and no waiting space other than the chair used by the customer on which he is currently working. He wishes to minimize the long-run average queue length L by decreasing the traffic intensity (i.e., by increasing the service rate μ), but counterbalancing this is an implicit cost inversely proportional to ρ. Find the traffic intensity ρ that minimizes

$$f(\rho) = L + \tfrac{1}{4\rho}$$

8. A cattle rancher is preparing to brand his cattle. A single brander is working, who can finish one steer in exactly 5 seconds. If cattle arrive to the brander according to a Poisson process with rate 10 per minute, find the limit as $t \longrightarrow \infty$ of the probability that there are 3 or fewer cattle waiting at time t.

9. (*Mathematica*) Write a *Mathematica* function that implements formula (14) for the repair time distribution of Example 4. Use it to plot the limiting distribution for states from 0 to 8.

Exercises 10–12 complete the proof of Theorem 1 on the limiting distribution of the $M/G/1$ queue, by calculating π_0.

10. Let $(p_n)_{n=0,1,2,...}$ form a discrete probability mass function on the non-negative integers. Define the *probability generating function* of this distribution as the function

$$P(z) = \sum_{n=0}^{\infty} p_n z^n = p_0 + p_1 z + p_2 z^2 + \cdots$$

Show that $P^{(n)}(0) = n! \cdot p_n$, and show that $P'(1)$ is the mean of the distribution.

11. Let $\Pi(z)$ be the probability generating function of the limiting distribution π for the $M/G/1$ queue, and let $Q(z)$ be the probability generating function of the distribution (q_n) in (11). Multiply both sides of (13) by z^i and sum from $i = 0$ to ∞ to obtain the relation

$$\Pi(z) = \pi_0 \, Q(z) \, \frac{(z-1)}{z - Q(z)}$$

12. Send $z \longrightarrow 1$ on both sides of the equation in Exercise 11, using L'Hopital's rule to evaluate the limit on the right side, to finish the proof that $\pi_0 = 1 - \rho$. (Hint: To calculate the mean number of arrivals during a service interval, condition and uncondition on the duration of the service interval.)

13. By viewing the $M/M/1$ queue as a special case of the $G/M/1$ queue, find the limiting distribution of the queue length, embedded at arrival instants. Compare this to the results of Example 2 for the $M/M/s$ queue in the special case that $s = 1$.

14. Precisely one bit of information arrives to a processor every microsecond. The processor requires an exponentially distributed amount of time with rate $\ln(4)$ per microsecond to analyze a bit and then proceed to the next bit. Find the limit as $n \longrightarrow \infty$ of the probability that two or fewer bits are waiting to be processed at the time of arrival of the nth bit.

15. (*Mathematica*) Arrivals come to a single server queue with $\exp(2.5)$ service time distribution so that only two interarrival times, 0.8 and 1.2, are possible, occurring with equal likelihood. Find the limiting distribution of the chain embedded at arrival instants.

16. Cars arrive to a state vehicle testing station according to a Poisson process with rate λ. It requires an exponential length of time with rate μ to test a car. But, cars arriving when there is at least one vehicle waiting have a probability $p \in (0, 1)$ of joining the queue, and $1 - p$ of driving away. Find the limiting distribution of the queue length.

17. Consider an $M/M/1/N$ queue whose arrival and service rates are equal. Suppose that customers who join the queue receive a reward R at the end of service, but pay at a rate of C per unit time while they are waiting. Find inequalities that characterize the optimal waiting room size N that maximizes the long-run expected profit per customer, per unit time:

$$\lambda R \cdot \lim_{t \to \infty} P[X_t < N] - C \cdot \lim_{t \to \infty} E[X_t]$$

(Hint: If a function f on the integers has a maximum at n^*, then both $f(n^*) \geq f(n^* + 1)$ and $f(n^*) \geq f(n^* - 1)$.)

5.5 Brownian Motion

Relation to Random Walks

Thus far the stochastic processes that we have studied either had a discrete time set or a discrete state space, or both. In this section we meet the most important process in the realm of continuous-time, continuous-state processes, the *Brownian motion*. It arises, however, as a limiting process of one of the simplest kind of discrete-time, discrete-state processes, the random walk. But in passing to the limit, some interesting and rather strange behaviors are introduced.

We will be able to do little more than scratch the surface of the Brownian motion process and its applications, so the goal of this section is to allow you to learn enough to give you background for future study. Because of its continuous nature, it requires heavy analytical machinery to properly derive results about this process, and in fact the true applications of Brownian motion also can only be done with some background in measure theory and partial differential equations, which we do not assume here. These applications are in such diverse areas as particle physics, the time variability of economic quantities, and the changes in populations of individuals or in levels of epidemics.

The history of the study of Brownian motion is filled with familiar names in the sciences and mathematics. An English botanist by the name of Robert Brown noted in 1827 that particles immersed in a liquid undergo continual, irregular motions. Albert Einstein in 1905 theorized that such Brownian motion was produced by countless collisions with the molecules of the surrounding liquid, and derived using physical principles a mathematical description of the motion. Other famous scientists, such as Fokker and Planck, continued to work out the physical theory of Brownian motion after this. Beginning in 1918, and continuing for years after, the great mathematician Norbert Wiener gave a mathematical formulation of the Brownian motion process and derived many of its properties. In his honor, Brownian motion is also sometimes referred to as the *Wiener process*. But not only scientists were interested in this stochastic process. As early as 1900, in his Ph.D. dissertation, the French economist Louis Bachelier used the process to model the motion of stock market prices. Its use in economic problems grew slowly at first, but then very rapidly later, fueled by the work of Black and Scholes [7] on the valuation of options in the 1970's, as well as the work of other mathematical economists such as Merton [43] on portfolio and capital market theory.

```
Needs["KnoxOR`StochasticProcesses`"];
```

To motivate Brownian motion and its paths, consider a symmetric random walk starting at a point x_0 in which a very short step up or down by an amount of Δx is taken at time intervals separated by a short amount of time Δt. So that a non-trivial limiting process of the kind we want exists, it turns out that a connection should be made between the state increment Δx and the time increment Δt. We take for the moment $\Delta x = \sqrt{\Delta t}$. To see what such a process looks like, let us simulate it.

There is a function in the KnoxOR`StochasticProcesses` package that plots the path of a simulated discretized Brownian motion as follows. The initial state is x0, the parameter deltat is as described above, and numpoints is the number of time points in the simulation.

```
(***  PlotSimulateBrownianMotion[
   x0_,deltat_,numpoints_]  ***)
```

It is educational for you to see how the function is written. First, we need a utility function that takes the argument Δt and returns the size of the step, $\pm \sqrt{\Delta t}$ each with probability 1/2. This is routine and is shown below. (It is also in the StochasticProcesses package.)

```
StepSize[deltat_] := If[Random[] < 1/2,
   Sqrt[deltat], -Sqrt[deltat]];
```

Now to simulate the full random walk, starting with a list containing only the initial state x_0, we continually append the next state, which is the most recent state plus the random step size. The function below does this, constructs the corresponding series of time points $\Delta t, 2\Delta t, 3\Delta t, ...$, and ListPlots the resulting (time, state) pairs. This version of the program joins neighboring points with a line segment, essentially forming a continuous-time process with continuous state space by linear interpolation.

```
PlotSimulateBrownianMotion[
    x0_ , deltat_ , numpoints_] := Module[
    {statelist, timepoints}, statelist = {x0};
    Do[AppendTo[statelist, Last[statelist] +
        StepSize[deltat]], {numpoints}];
    timepoints = Table[n * deltat,
        {n, 0, numpoints}];
    ListPlot[Transpose[{timepoints, statelist}],
     PlotJoined → True]];
```

Here is a simulation for initial state 0, $\Delta t = 0.001$, and 1000 time points, so that the terminal time in the simulation is 1. In the electronic text, you should rerun the command many times to get an idea of the behavior of this random walk.

```
PlotSimulateBrownianMotion[0, .001, 1000];
```

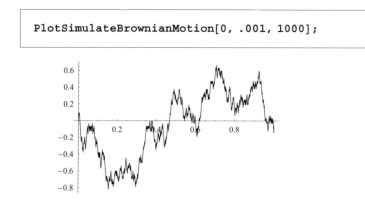

Figure 5.11 – Approximate sample path of the standard Brownian motion process

Activity 1 – Modify the program to return the state of the random walk at the final time; then write a program that generates a list of 100 such random walk final states. To make your command run faster, use $\Delta t = 0.01$, and 100 time points. Use the command below, contained in KnoxOR`StochasticProcesses`, to plot a histogram of the data with five rectangles to study the empirical distribution of the final state.

```
(**  Histogram[datalist,numrectangles]  **)
```

Definition and Properties of Standard Brownian Motion

The random walk construction above suggests several properties that the limiting process should have. First, the state $X_{n\Delta t}$ at time $n\Delta t$ consists of the sum of independent, identically distributed step sizes Y_i, $i = 1, \ldots, n$, where each Y_i has mean and variance

$$
\begin{aligned}
\mu_y &= \tfrac{1}{2}\sqrt{\Delta t} + \tfrac{1}{2}\left(-\sqrt{\Delta t}\right) = 0, \\
\sigma_y^2 &= E[Y^2] = \tfrac{1}{2}\Delta t + \tfrac{1}{2}\Delta t = \Delta t
\end{aligned}
\tag{1}
$$

The sum of the steps $X_{n\Delta t} = \sum_{i=1}^{n} Y_i$ therefore has mean 0 and variance $n\Delta t$, so the variance equals the time subscript. Because $X_{n\Delta t}$ is an independent sum of random variables, we expect it to have an approximate normal distribution, and in the limit as $n \longrightarrow \infty$ and $\Delta t \longrightarrow 0$ in such a way that $n\Delta t = t$, a constant time, the distribution of X_t ought to be normal with mean 0 and variance t.

There are further properties to expect of the limiting random walk. If the random walk is currently in a state $X_{n\Delta t}$, then the change in state over the next, say $m\Delta t$ units of time, depends only on the next m steps $Y_{n+1}, Y_{n+2}, \ldots, Y_{n+m}$. Specifically,

$$
X_{n\Delta t + m\Delta t} - X_{n\Delta t} = \sum_{i=n+1}^{n+m} Y_i
\tag{2}
$$

The probability distribution of this change in state depends only on m, since the Y_i's are independent and identically distributed, not on the current state $X_{n\Delta t}$ or the current time $n\Delta t$. So changes in state should be independent of the past, and moreover the distribution of the change in state should depend only on the amount of time that elapses between the change.

Having made these observations, we have motivation to define the following process.

DEFINITION 1. A stochastic process $(X_t)_{t \geq 0}$ is called a *standard Brownian motion* (with initial state 0) if

    (a) $X_0 = 0$ and for all outcomes ω except possibly some in a set of probability zero, the function $t \mapsto X_t(\omega)$ is continuous at every t;

    (b) For all $t > 0$, X_t is normally distributed with mean 0 and variance t;

    (c) For all $t, s > 0$, the distribution of $X_{t+s} - X_t$ does not depend on t;

    (d) For all $t, s > 0$ $X_{t+s} - X_t$ is independent of X_r for all $r \leq t$.

Property (c) is called the *stationarity property*. The distribution of the change in value of the Brownian motion between times t and $t+s$ is the same for all t; consequently, choosing $t = 0$, the distribution of $X_{t+s} - X_t$ is the same as that of $X_s - X_0 = X_s$. In light of assumption (b), it follows that $X_{t+s} - X_t \sim N(0, s)$ for all s, t. Property (d) is called the *independent incre-ments property*, which implies that the changes in value $X_{t+s} - X_t$ and $X_{r+u} - X_r$ of the Brownian motion on disjoint time intervals $[t, t + s]$ and $[r, r + u]$ are independent. We saw these properties earlier in the context of Poisson processes.

Activity 2 – What changes should be made to properties (b), (c), and (d) of Definition 1 if we allow standard Brownian motion to begin at a state x_0 other than 0?

EXAMPLE 1. The definition of Brownian motion permits us to make a number of elementary probabilistic computations. Suppose that the process $(X_t)_{t \geq 0}$ is a standard Brownian motion as in Definition 1. Let us compute:
  (a) $P[X_{2.3} \leq 1]$;
  (b) $P[X_{4.1} - X_{1.8} \leq 1 \mid X_{0.9} \geq 0]$;
  (c) $P[X_{2.3} \leq 1 \mid X_{0.9} = .5]$; and
  (d) $P[X_{t+s} \leq y \mid X_t = x]$.

For part (a), condition (b) of the definition yields directly that $X_{2.3} \sim N(0, 2.3)$, so the desired probability is the c.d.f. value below:

CDF[NormalDistribution[0, $\sqrt{2.3}$], 1]

0.745174

To answer part (b), note that the time interval $(1.8, 4.1]$ is disjoint from the time interval $[0, 0.9]$, and so by the independent increments property, the difference $X_{4.1} - X_{1.8}$ is independent of the event $X_9 \geq 0$. In addition, by stationarity, $X_{4.1} - X_{1.8}$ has the same distribution as $X_{2.3} - X_0 = X_{2.3}$. Hence,

$$P[X_{4.1} - X_{1.8} \leq 1 \mid X_{0.9} \geq 0] = P[X_{4.1} - X_{1.8} \leq 1]$$
$$= P[X_{2.3} \leq 1] = 0.745174$$

In part (c), the random variable $X_{2.3}$ is not independent of $X_{0.9}$, so we must be more clever. Adding and subtracting $X_{0.9}$ on both sides of the inequality gives

$$
\begin{aligned}
P[X_{2.3} \leq 1 \mid X_{0.9} = .5] &= P[X_{2.3} - X_{0.9} + X_{0.9} \leq 1 \mid X_{0.9} = .5] \\
&= P[X_{2.3} - X_{0.9} \leq .5 \mid X_{0.9} = .5] \\
&= P[X_{2.3} - X_{0.9} \leq .5] \\
&= P[X_{1.4} \leq .5]
\end{aligned}
\tag{3}
$$

The third line follows from the independent increments condition, and the fourth line is by stationarity. The last probability on the right is

$$
\texttt{CDF}\big[\texttt{NormalDistribution}\big[\texttt{0, } \sqrt{1.4}\,\big]\texttt{, .5}\big]
$$

0.663698

Part (d) is simply a generalization of part (c). Following the steps of display (3) for general starting time t and time increment s,

$$
\begin{aligned}
P[X_{t+s} \leq y \mid X_t = x] &= P[X_{t+s} - X_t + X_t \leq y \mid X_t = x] \\
&= P[X_{t+s} - X_t \leq y - x \mid X_t = x] \\
&= P[X_{t+s} - X_t \leq y - x] \\
&= P[X_s \leq y - x]
\end{aligned}
\tag{4}
$$

(Be sure that you can justify each line of this derivation.) The last probability is a function of the time increment s, the starting state x, and y, which we will denote by $P(s, x, y)$ and call the *transition c.d.f.* of the standard Brownian motion. Its derivative with respect to y, calculated below, is called the *transition density* of the process:

$$
\begin{aligned}
p(s, x, y) = \tfrac{d}{dy} P(s, x, y) &= \tfrac{d}{dy} P[X_s \leq y - x] \\
&= \tfrac{d}{dy} \int_{-\infty}^{y-x} \frac{1}{\sqrt{2\pi s}} \, e^{-\frac{u^2}{2s}} \, du \\
&= \frac{1}{\sqrt{2\pi s}} \, e^{-\frac{(y-x)^2}{2s}}
\end{aligned}
\tag{5}
$$

We can interpret the result of this last computation by saying that given $X_t = x$, the conditional distribution of the state at time $t + s$ is $N(x, s)$, which is consistent with the defining properties of the standard Brownian motion. ∎

One of the most prominent of the peculiar properties of Brownian motion mentioned at the start of the section is this: If $B > 0$ is a point in the state space of the Brownian motion then the first hitting time of B is finite with probability 1, but has infinite expectation. By the symmetry of the Brownian motion, it is clear that the same property holds for $B < 0$. Let us now show this property.

Denote $T_B = \inf \{s > 0, X_s = B\}$, the first hitting time of $B > 0$. Because the event $\{X_t \geq B\}$ is contained in the event $\{T_B \leq t\}$, we can write

$$P[X_t \geq B] = P[X_t \geq B, T_B \leq t] = P[X_t \geq B \mid T_B \leq t] \, P[T_B \leq t]$$

It can be shown rigorously that since $X_{T_B} = B$, the conditional probability on the right of the last string of equations is the same as $P[X_t - X_{T_B} \geq 0]$, but to do so we really need a stonger version of the Markov property applied at the random time T_B rather than at a deterministic time, so we will omit this detail. But by stationarity, the increment $X_t - X_{T_B}$ would have a normal distribution symmetric about 0, so this probability is exactly 1/2. Therefore,

$$P[X_t \geq B] = \tfrac{1}{2} \, P[T_B \leq t]$$

$$\Longrightarrow P[T_B \leq t] = 2 \, P[X_t \geq B] = 2 \int_B^\infty \frac{1}{\sqrt{2\pi t}} \, e^{-\frac{u^2}{2t}} \, du \qquad (6)$$

To show that T_B is finite with probability 1 means to show that $P[T_B < \infty] = \lim_{t \to \infty} P[T_B \leq t] = 1$. To do this, make the substitution $z = u / \sqrt{t}, \, dz = du / \sqrt{t}$ to obtain

$$
\begin{aligned}
\lim_{t \to \infty} P[T_B \leq t] &= \lim_{t \to \infty} 2 \int_B^\infty \frac{1}{\sqrt{2\pi t}} \, e^{-\frac{u^2}{2t}} \, du \\
&= \lim_{t \to \infty} 2 \int_{B/\sqrt{t}}^\infty \frac{1}{\sqrt{2\pi}} \, e^{-\frac{z^2}{2}} \, dz \\
&= 2 \int_0^\infty \frac{1}{\sqrt{2\pi}} \, e^{-\frac{z^2}{2}} \, dz = 1
\end{aligned}
$$

The last line is true because the integrand is the standard normal density, which is symmetric about 0, so the integral gives half of the total area under that density. We have therefore proved that T_B is finite with probability 1, that is, no matter how distant is the target state B from the initial state of 0, the standard Brownian motion is certain to hit it eventually. But let us furthermore show that the expected value of T_B is infinite, which means that for a significantly probable set of outcomes, it may take the process very long to reach B.

To do this, recall the standard result from the theory of non-negative, continuous random variables that $E[X] = \int_0^\infty P[X > x] \, dx$. Applying this to $X = T_B$ using the expressions derived above, we can compute that:

$$E[T_B] = \int_0^\infty P[T_B > t]\,dt$$

$$= \int_0^\infty (1 - P[T_B \le t])\,dt$$

$$= \int_0^\infty \left(1 - 2\int_{B/\sqrt{t}}^\infty \frac{1}{\sqrt{2\pi}}\,e^{-\frac{z^2}{2}}\,dz\right)dt$$

$$= \int_0^\infty \int_{-B/\sqrt{t}}^{B/\sqrt{t}} \frac{1}{\sqrt{2\pi}}\,e^{-\frac{z^2}{2}}\,dz\,dt$$

$$= \int_0^\infty 2\int_0^{B/\sqrt{t}} \frac{1}{\sqrt{2\pi}}\,e^{-\frac{z^2}{2}}\,dz\,dt$$

In the last integral on the right, change the order of integration. Since $z \le B/\sqrt{t}$, it must be that $\sqrt{t} \le B/z \Longrightarrow t \le B^2/z^2$. Thus,

$$E[T_B] = \int_0^\infty \frac{2}{\sqrt{2\pi}}\,e^{-\frac{z^2}{2}}\left(\int_0^{B^2/z^2} dt\right)dz$$

$$= \frac{2B^2}{\sqrt{2\pi}} \int_0^\infty \frac{1}{z^2}\,e^{-\frac{z^2}{2}}\,dz$$

By comparison with the integral of $1/z^2$ near the left endpoint of 0, the last integral fails to converge. Thus, $E[T_B] = \infty$ as desired.

Activity 3 – In the computation of the distribution of the hitting time T_B above, try to give a rough argument that:
$$P[X_t \ge B \mid T_B \le t] = P[X_t - X_{T_B} \ge 0]$$

EXAMPLE 2. We know that Brownian motion is the continuous analogue to the discrete random walk. Let us try to solve a problem about Brownian motion that we asked earlier about a random walk: the gambler's ruin problem. Specifically, suppose that a standard Brownian motion starting at 0 is to be stopped either when it has reached a positive number M, or a negative number $-N$, whichever comes first. What is the probability that it will reach M before $-N$?

As often happens, it is helpful to generalize a bit and solve the more general problem first, then apply it in the specific case. We will try to solve for the quantity

$$f(x) = P[X_t \text{ reaches } M \text{ before } -N \mid X_0 = x] \tag{7}$$

for all $x \in [-N, M]$. We will do so by using an approach similar to the development of the Kolmogorov equations for birth–death processes, that is, by deriving a differential equation for f.

It is clear that $f(-N) = 0$ and $f(M) = 1$, so consider $x \in (-N, M)$. Now formula (6) and L'Hopital's Rule can be used to show that the probability that standard Brownian motion hits a point at a fixed non-zero distance from its starting point within a time h is of the order $o(h)$ (see Exercise 14). In light of this, if we condition and uncondition on the state of the Brownian motion at a time h near zero, we may safely ignore the unlikely event that X_t has already hit M or $-N$ by time h starting from x. Thus, we may write:

$$\begin{aligned} f(x) &= E[P[X_t \text{ reaches } M \text{ before } -N \mid X_h, X_0] \mid X_0 = x] \\ &= E[f(X_h) + o(h) \mid X_0 = x] \end{aligned}$$

Expanding f in a Taylor series of order 2 about the point x yields

$$f(x) = E[f(x) + f'(x)(X_h - x) + \tfrac{1}{2} f''(x)(X_h - x)^2 + \text{ terms in } (X_h - x)^3 \text{ and higher} + o(h) \mid X_0 = x] \tag{8}$$

Given $X_0 = x$, the moments $E[(X_h - x)^n]$ equal zero for odd n, since the distribution of X_h is normal with mean x; in particular, it is symmetric about x. For n even and $n \geq 4$, we have by symmetry,

$$\begin{aligned} E[(X_h - x)^n \mid X_0 = x] &= 2 \int_x^\infty (y - x)^n \frac{1}{\sqrt{2\pi h}} e^{\frac{-(y-x)^2}{2h}} \, dy \\ &= 2 h^{n/2} \int_0^\infty z^n \frac{1}{\sqrt{2\pi}} e^{-z^2/2} \, dz \end{aligned}$$

after substituting $z = (y - x)/\sqrt{h}$ (check this yourself). Dividing by h, the limit is zero when $n \geq 4$; and when $n = 2$, $E[(X_h - x)^2 \mid X_0 = x] = 2 h(\tfrac{1}{2}) = h$, since the integral is half of the integral representing the variance of the $N(0, 1)$ distribution. Putting these facts together with formula (8) gives us

$$\begin{aligned} f(x) &= f(x) + f'(x) E[(X_h - x) \mid X_0 = x] + \\ &\quad \tfrac{1}{2} f''(x) E[(X_h - x)^2 \mid X_0 = x] + o(h) \\ \Longrightarrow 0 &= f'(x) \cdot 0 + \tfrac{1}{2} f''(x) \cdot h + o(h) \\ \Longrightarrow f''(x) &= 0 \end{aligned} \tag{9}$$

upon dividing both sides by h and letting $h \to 0$.

   Therefore f must be a linear function; $f(x) = m x + b$, where also $f(-N) = 0$ and $f(M) = 1$. It is easy to check that the boundary conditions imply that

$$f(x) = \tfrac{x+N}{M+N}$$

Hence the quantity that we originally sought is $f(0) = N / (M + N)$. ∎

Brownian Motion with Drift

We now generalize the standard Brownian motion to allow the process to experience a trend or "drift" with time. Here is the definition.

DEFINITION 2. A stochastic process $(X_t)_{t \in \mathbb{R}_+}$ is called a *Brownian motion with drift rate* μ, *variance rate* σ^2, and *initial state* x iff the process defined by $Y_t = \frac{X_t - \mu t - x}{\sigma}$ is a standard Brownian motion with initial state 0.

Several properties follow from Definition 2 and the definition of standard Brownian motion:

(a) $X_0 = x$ and for all outcomes ω except possibly some in a set of probability zero, the function $t \mapsto X_t(\omega)$ is continuous at every t;

(b) For all $t > 0$, X_t is normally distributed with mean $x + \mu t$ and variance $\sigma^2 t$;

(c) For all $t, s > 0$, the distribution of $X_{t+s} - X_t$ does not depend on t;

(d) For all $t, s > 0$, $X_{t+s} - X_t$ is independent of X_r for all $r \le t$.

Activity 4 – Verify properties (a)–(d) above.

The non-standard Brownian motion can also be viewed as the limiting process of a random walk as the time step and the state step approach 0. We just need to begin with a non-symmetric random walk. For simplicity let the random walk start at 0. Let the independent step size random variables Y_i have the distribution

$$Y_i = \begin{cases} \sigma \sqrt{\Delta t} & \text{with probability } p \\ -\sigma \sqrt{\Delta t} & \text{with probability } 1 - p \end{cases}$$

where the probability p of moving to the right is to be chosen presently in order to make the drift rate equal to μ. Then,

$$E[Y_i] = p \sigma \sqrt{\Delta t} + (1 - p)\left(-\sigma \sqrt{\Delta t}\right) = \sigma \sqrt{\Delta t} \, (2p - 1)$$

$$\text{Var}(Y_i) = p\left(\sigma \sqrt{\Delta t}\right)^2 + (1 - p)\left(-\sigma \sqrt{\Delta t}\right)^2 - \left(\sigma \sqrt{\Delta t} \, (2p - 1)\right)^2$$

$$= \sigma^2 \Delta t (1 - (2p - 1)^2)$$

The sum of the steps $X_{n\Delta t} = \sum_{i=1}^{n} Y_i$ therefore has mean $n\sigma\sqrt{\Delta t}\,(2p-1)$ and variance $n\sigma^2 \Delta t(1-(2p-1)^2)$. So if we let $p = \frac{1}{2}\left(1 + \mu\sqrt{\Delta t}\,/\sigma\right)$, then $E[X_{n\Delta t}] = \mu(n\Delta t)$; that is, the drift rate μ times the time subscript. Also, $\text{Var}(X_{n\Delta t}) = \sigma^2(n\Delta t)(1-\mu^2 \Delta t/\sigma^2)$; that is, the variance rate σ^2 times the time subscript times an expression approaching 1 as $\Delta t \to 0$. Property (b) is motivated, and properties (c) and (d) are inherited by the limiting process from the independence of the steps, as with standard Brownian motion.

EXAMPLE 3. An epidemic in its rapidly spreading initial stages is such that the number of units of population that are infected can be modeled by $I_t = e^{X_t}$, where (X_t) is a Brownian motion with drift rate 2, variance rate 1, and initial state .5. Find (a) $P[I_3 \geq 2]$; (b) $E[I_t]$; (c) $\text{Var}(I_t)$.

To compute the probability in part (a), first note that $X_t \sim N(.5 + 2t, t)$. Then,

$$
\begin{aligned}
P[I_3 \geq 2] &= P[e^{X_3} \geq 2] \\
&= P[X_3 \geq \log(2)] \\
&= 1 - P[X_3 < \log(2)]
\end{aligned}
$$

The numerical value is computed below, using the fact that $X_3 \sim N(.5 + 2\cdot 3, 3)$.

```
N[1 -
    CDF[NormalDistribution[.5 + 2 * 3, √3], Log[2]]]
```

0.9996

For part (b), we also need to use the distribution of X_t, and we need to recall that the moment-generating function of a $N(\mu, \sigma^2)$ random variable Y is $M(s) = E[e^{sY}] = e^{\mu s + \sigma^2 s^2/2}$. Then,

$$
\begin{aligned}
E[I_t] &= E[e^{X_t}] \\
&= M_{X_t}(1) \\
&= \exp((.5 + 2t)\cdot 1 + \tfrac{1}{2}t\cdot 1^2) \\
&= \exp(.5 + 2.5t)
\end{aligned}
$$

For the variance in part (c), we can use the computational formula $\text{Var}(I_t) = E[I_t^2] - (E[I_t])^2$. The second moment of I_t can be found similarly to the mean:

$$E[I_t^2] = E[e^{2X_t}]$$
$$= M_{X_t}(2)$$
$$= \exp((.5 + 2t) \cdot 2 + \tfrac{1}{2} t \cdot 2^2)$$
$$= \exp(1 + 6t)$$

The variance is therefore

$$\text{Var}(I_t) = \exp(1 + 6t) - \exp(2(.5 + 2.5t)) = \exp(1 + 6t) - \exp(1 + 5t). \quad \blacksquare$$

EXAMPLE 4. This example concerns the optimal balance of assets in a portfolio of one risky and one non-risky asset. Before stating the problem in a way that permits solution, we need to review a couple of economic concepts.

If the price of a deterministic asset at time t is denoted by $p_0(t)$, and the asset is experiencing exponential growth at rate r, as is often assumed, then $p_0(t) = p_0 e^{rt}$, where p_0 is the initial price of the asset. For an investment period beginning at time 0 and ending at time t, the rate of return on the asset per dollar invested is the difference between the final value and the initial value, divided by the initial value:

$$\text{rate of return} = \frac{p_0(t) - p_0}{p_0} = \frac{p_0 e^{rt} - p_0}{p_0} = e^{rt} - 1$$

Activity 5 – The instantaneous rate of return on an asset is the limit as $t \to 0$ of the rate of return on time interval $[0, t]$ divided by t. What is that limit for the deterministic asset?

A reasonable way to model a risky asset similar to the deterministic asset is such that its price behaves as $P_1(t) = p_1 e^{Y_t}$, where (Y_t) is a Brownian motion with drift rate μ, variance rate σ^2, and initial state 0. Then $Y_t \sim N(\mu t, \sigma^2 t)$ so that the expected value of the random variable in the exponent of $P_1(t)$ is a constant times t. We normally assume that risky assets grow on average faster than non-risky ones, hence $\mu > r$. Now if at time 0 an investor has a total wealth of W_0, which he chooses to apportion by buying s_0 shares of the deterministic asset and s_1 shares of the risky asset, then

$$s_0 p_0 + s_1 p_1 = W_0$$

Hence the proportions of initial wealth devoted to the deterministic and risky assets, respectively, are

$$\frac{s_0 p_0}{W_0} \equiv w_0, \quad \frac{s_1 p_1}{W_0} \equiv w_1$$

Note that $w_1 = 1 - w_0$. The final wealth at time t is

$$\begin{aligned}
W_1 &= s_0 \, p_0(t) + s_1 \, P_1(t) \\
&= s_0 \, p_0 \, e^{rt} + s_1 \, p_1 \, e^{Y_t} \\
&= w_0 \, W_0 \, e^{rt} + (1 - w_0) \, W_0 \, e^{Y_t}
\end{aligned}$$

hence the rate of return on the whole portfolio of assets, that is, the final wealth minus the initial wealth divided by the initial wealth, is

$$R = \frac{W_1 - W_0}{W_0} = \frac{w_0 \, W_0 \, e^{rt} + (1-w_0) \, W_0 \, e^{Y_t} - W_0}{W_0} = w_0 \, e^{rt} + (1 - w_0) \, e^{Y_t} - 1 \qquad (10)$$

The problem is to decide what proportion of initial wealth w_0 to invest in the deterministic asset, and consequently what proportion $w_1 = 1 - w_0$ to invest in the risky asset. So we seem to have a single variable optimization problem, with variable w_0, but we have not yet decided on a criterion for optimization. It does not make sense to optimize rate of return, because that is random and therefore not wholly subject to our control. We could optimize the expected rate of return, but that would pay no attention to the undesirability of the variation in the return on the risky asset. Reasonable investors would attempt to weigh that risk negatively in their decision. A commonly used optimization criterion is

$$\text{maximize } E[R] - a \cdot \text{Var}(R) \qquad (11)$$

where a is a constant called the *risk aversion*. The greater is the value of a, the more the investor considers the variance of the rate of return to be undesirable. Note that in formula (10) the 1 that is subtracted on the right will neither change the variance of R nor will it alter the value of w_0 at which the maximum occurs, so we will drop it.

To obtain an explicit expression for the objective function in (11), we again use the formula for the moment-generating function of a $N(\mu, \sigma^2)$ random variable: $M(s) = e^{\mu s + \sigma^2 s^2/2}$. Then,

$$\begin{aligned}
E[R] &= E[w_0 \, e^{rt} + (1 - w_0) \, e^{Y_t}] \\
&= w_0 \, e^{rt} + (1 - w_0) \, E[e^{Y_t}] \\
&= w_0 \, e^{rt} + (1 - w_0) \, e^{\mu t + \sigma^2 t/2}
\end{aligned}$$

and

$$
\begin{aligned}
\mathrm{Var}(R) &= \mathrm{Var}(w_0\, e^{rt} + (1 - w_0)\, e^{Y_t}) \\
&= (1 - w_0)^2\, \mathrm{Var}(e^{Y_t}) \\
&= (1 - w_0)^2 \left(E[e^{2\,Y_t}] - (E[e^{Y_t}])^2 \right) \\
&= (1 - w_0)^2 \left(e^{2\mu t + 2\sigma^2 t} - e^{2\mu t + \sigma^2 t} \right)
\end{aligned}
$$

Therefore we want to maximize the following with respect to w_0:

$$
f(w_0) = w_0\, e^{rt} + (1 - w_0)\, e^{\mu t + \sigma^2 t/2} - a(1 - w_0)^2 \left(e^{2\mu t + 2\sigma^2 t} - e^{2\mu t + \sigma^2 t} \right) \quad (12)
$$

For concreteness, take the final time as $t = 1$, and let $r = .05$, $\mu = .06$, $\sigma = .03$. We set up f as a function of both w_0 and a in order to see how the optimal solution changes with the risk aversion.

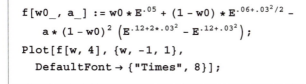

```
f[w0_, a_] := w0 * E^.05 + (1 - w0) * E^.06+.03^2/2 -
    a * (1 - w0)^2 (E^.12+2*.03^2 - E^.12+.03^2);
Plot[f[w, 4], {w, -1, 1},
    DefaultFont -> {"Times", 8}];
```

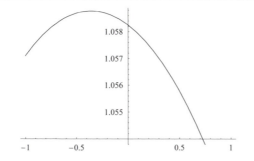

Figure 5.12 – Optimizing a portfolio with risk aversion 4

Figure 5.12 shows the case where the risk aversion constant $a = 4$. An unexpected result shows up, that hints at economic considerations. The optimal value of the objective function occurs near $w_0 = -.5$. It appears as if, given that it is possible to hold a negative or *short position* (i.e., borrow cash from the non-risky asset to buy more of the risky asset) in the non-risky asset, that we should do so in order to optimize the objective. Specifically, an amount $w_0\, W_0$ should be borrowed against the non-risky asset at time zero in order to buy shares at a total value of $(1 + w_0)\, W_0$ in the risky asset. The total value of the portfolio at time zero is then $-w_0\, W_0 + (1 + w_0)\, W_0 = W_0$. If w_0 is constrained economically to be non-negative, then $w_0 = 0$ is the optimal value.

Let us use a surface graph to see how the optimal value of w_0 changes with the risk aversion a.

```
Plot3D[f[w, a], {w, -1, 1},
   {a, 4, 12}, AxesLabel → {"w₀", "a", " "},
   DefaultFont → {"Times", 8}];
```

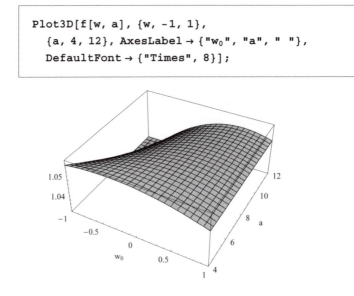

Figure 5.13 – Dependence of optimal portfolio on risk aversion

You can see from the graph in Figure 5.13 that as a moves up between 4 and 12, the peak on the curve is found at higher and higher values of w_0, which makes intuitive sense because the more risk averse the investor is, the more of his wealth he should choose to devote to the non-risky asset. In Exercise 9 you are asked to show this observation in general. ∎

Exercises 5.5

1. Let (X_t) be a standard Brownian motion. Compute (a) $P[X_{5.4} > 2]$; (b) $P[X_{3.1} - X_{2.1} < 1 \mid X_5 = 1.6]$; (c) the joint density of X_2, X_3, X_4.

2. Let (X_t) be a Brownian motion with initial state 0, drift rate 2, and variance rate 4. Compute (a) $P[X_{8.2} - X_{4.5} \le 6]$; (b) $P[X_1 \le 3, X_2 > 4]$.

3. (*Mathematica*) Write a program to simulate standard Brownian motion many times, and to return the proportion of the replications in which the process hits M before $-N$, and compare your empirical results to the analytical result in Example 2 for several choices of M and N.

4. Extend the result of Example 2 for the probability $f(x)$ of hitting M before $-N$ starting at x to the case of Brownian motion with drift μ and

variance rate 1. (Hint: Begin by following the steps of that example, but look harder at the expected powers of $X_h - x$.)

5. (*Mathematica*) Write a simulator for non-standard Brownian motion with parameters μ and σ analogous to the PlotSimulateBrownianMotion command in the section, which was designed for standard Brownian motion. (Hint: See the discussion subsequent to Definition 2.)

6. Consider as in Example 4 a risky asset whose price behaves as $P_1(t) = p_1 e^{X_t}$, where (X_t) is a Brownian motion with drift rate μ, variance rate σ^2, and initial state 0. Suppose that at time 0 an investor purchases an option to buy a share of this asset at time T at a fixed price K if it is profitable to do so. If the market price of the asset at T exceeds this K, the investor can purchase the share at K and immediately resell it at a profit; otherwise the option is worthless. Find an expression for the expected value of this option.

7. Let (X_t) be a standard Brownian motion with initial state 0, and for a point $M > 0$ let T_M be the first time that the Brownian motion achieves a value of at least M. Let $Y_t = \max_{u \leq t} X_u$. The process (Y_t) is called the *maximum process* for X_t. By relating the maximum process to hitting times T_M, show that the c.d.f. of Y_t is $G(y) = 1 - 2 \int_y^\infty \frac{1}{\sqrt{2\pi t}} e^{-x^2/2t} \, dx$.

8. (*Mathematica*) Suppose that you have a parcel of land for sale, and you receive offers of X_t at each time $t \in [0, T)$, where (X_t) is a standard Brownian motion with initial state x. At time T the game runs out and all offers are withdrawn. Consider a policy that exercises the option at the first time the offer exceeds a value y, and does not accept an offer otherwise. Find an expression for the expected profit under such a policy as a function of y. If $T = 10$, $x = 100$, use *Mathematica* to find the optimal value of y. (Hint: Express the event of earning a positive profit in terms of the maximum over times in $[0, T]$ of the Brownian motion X_t, and use the result of Exercise 7 regarding the distribution of the maximum.)

9. Use formula (12) to solve explicitly for the optimal value of w_0 in terms of a, and show that the optimal portion of wealth invested in the non-risky asset increases as a increases.

10. In the portfolio problem with the same choice of parameters $t = 1$, $r = .05$, $\mu = .06$, $\sigma = .03$, how large should a be so that it is optimal to keep at least half of the initial wealth in the non-risky asset?

11. A *geometric Brownian motion* is a process (Y_t) such that its log forms a standard Brownian motion. For such a process with initial state 1, find the probability density function of Y_t and compute $P[2.3 \leq Y_1 \leq 5.6]$.

12.   If (X_t) is a standard Brownian motion, then the process $Y_t = M - |M - X_t|$ is called a *Brownian motion with reflecting barrier at M*. Explain the terminology, and find the c.d.f. and density function of Y_t.

13. In advanced courses in stochastic processes, it is possible to define a *stochastic integral with respect to a standard Brownian motion*

$$\int_a^b X_s \, dW_s$$

where (X_s) is a stochastic process and (W_s) is the Brownian motion, by a limit-taking process. These are used heavily in the area of mathematical finance. The building blocks are the following simple versions of stochastic integrals.   Let x_s be a deterministic step process, with jumps at times $t_1, t_2, \ldots, t_{n-1}$ and values

$$x_s = \begin{cases} x_0 & \text{if } a \le s < t_1 \\ x_1 & \text{if } t_1 \le s < t_2 \\ \vdots & \vdots \\ x_{n-1} & \text{if } t_{n-1} \le s \le b \end{cases}$$

Let the stochastic integral $\int_a^b x_s \, dW_s$ be defined as (the Riemann-Stieltjes integral)

$$\sum_{i=0}^{n-1} x_i(W_{t_{i+1}} - W_{t_i})$$

where t_0 is taken to be a and t_n is b. Note that $\int_a^b x_s \, dW_s$ is a random variable. Find its mean and variance.

14. Show that the probability that standard Brownian motion hits a point at a fixed distance $\epsilon > 0$ from its starting point within a time interval of length h is of the order $o(h)$. (Hint: Express the question in terms of the hitting time T_ϵ, and use formula (6) and L'Hopital's Rule.)

6

Dynamic Programming

Introduction

This chapter will introduce the reader to concepts, examples, and techniques of *stochastic dynamic programming*. Roughly described, we are to control a system that is moving from state to state as time progresses. As a result of the sequence of controlling actions that we take, the motion of the system is influenced, and a sequence of rewards (or costs) is accumulated. What actions should be taken in order to maximize total reward (or minimize total cost)? There are diverse applications of this general problem, including re-ordering in inventory problems, production control problems, gambling models, fishery harvesting models, and financial models, especially the control of stock portfolios.

In the first section we describe the problem more precisely, first in the context of the simpler *deterministic dynamic programming* model, then in the stochastic case. Section 2 illustrates the dynamic programming technique that is used to solve problems in which control is exerted over finite time. Then, Sections 3 and 4 extend the problem to control over infinite time, with a discount factor incorporated into each time period. A related problem, called the *optimal stopping problem*, is introduced in Section 5. In this problem, reward is only earned at some final time which is subject to the choice of the controller, and we will find that many ideas from other areas of this text are united in a common setting. Finally, Section 6 gives further applications of stochastic dynamic programming that are less straightforward than the examples used in the earlier sections, featuring problems of inventory control and optimal control of stock portfolios.

6.1 The Markovian Decision Model

In this chapter we will primarily focus on stochastic dynamic programming problems. But before we include the influence of randomness in these problems, we will begin with an example of a deterministic dynamic programming problem to gain a better understanding of the structure and components of the model. This example will also introduce us to the method of solving finite horizon problems that starts at the end of the problem and works backward. We explore this method in the stochastic context in Section 2.

Deterministic Dynamic Programming

EXAMPLE 1. One of the simplest dynamic programming problems is to find the shortest path from a starting point to a destination. To place the problem into context, suppose that a guide is giving a tour of a college to a prospective student (see Figure 6.1). The guide wants to show the student five key attractions on campus, including the starting point A and the destination K, in 50 minutes. It is advantageous to make the travel time from A to K as short as possible, in order that the prospective student can spend as much of the 50 minutes as possible at the attractions. The other possible attractions that can be visited are labeled B, C, D, F, G, H, I, and J. They happen to be laid out in levels, so that, for instance, starting from A, you have the option of visiting B, C, or D next.

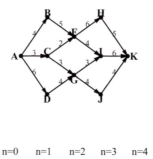

n=0   n=1   n=2   n=3   n=4

Figure 6.1 – Possible routes on a campus tour

The edge weights indicate the number of minutes it takes to walk from the attraction on the left of the arrow to the attraction on the right of the arrow. An admissible campus tour is a path from the starting point A to the destination K that follows existing arrows. Thus, the amount of time spent walking during the tour is equal to the total weight of the path taken. We want to minimize this sum among all admissible tours, so as to find the optimal path that spends the least amount of time walking.

A useful mathematical model for the problem is a sequence of *states* x_0, x_1, x_2, x_3, and x_4 where x_n represents the state of a system at time n. The times here correspond to the levels of the graph in Figure 6.1 as shown, from left to right, and the state x_n is the attraction the student is looking at in level n. In particular, our tour constrains x_0 to be A and x_4 to be K. The set E of all states consists of all the attractions, $E = \{A, B, C, D, F, G, H, I, J, K\}$.

Another useful element of the model is a sequence of *action functions*

$u_0, u_1, u_2,$ ..., where u_n depends on the current attraction and gives the next attraction chosen for the tour. An action function is therefore a function of the state: $u_n(i)$ is the action taken when the state is i and the time is n. The graph tells us not only what actions are possible at each state, but also what the next state will be given the current state and the action we take.

As you can see from the graph, for each current state, there are a limited number of possible actions to take. For example, at state C you can only move to state F or state G. In many problems, not all actions are possible for every current state. Rather, when the current state is i, there is a given subset A_i of *admissible actions*. This imposes a condition on the action functions described below:

$$u_n(i) \in A_i, \text{ for each } n = 0, 1, 2, \text{ ... and each } i \in E.$$

In addition to describing the permissible actions, the graph also shows the cost structure of the problem. There is a cost of taking an action when we are at a state, namely the amount of time it takes to go from current attraction to the next attraction. In general, costs can be described by a function, called the *cost function* $r(i, a)$, giving the cost when the state is i and the action is a. The actual cost incurred at time n can then be denoted by: $R_n = r(x_n, u_n(x_n))$. For example, if we were at attraction B and decided to go to attraction F, then we would have $n = 1$, $x_1 = B$, $u_1(x_1) = F$, and $R_1 = r(B, F) = 5$. ∎

In summary, the building blocks of a mathematical model for the shortest path problem, as well as many other deterministic dynamic programming problems, are:

1. the space E of possible states the system can be in;
2. the sets A_i of possible actions that can be taken when the state is i;
3. a means for determining the next state given the current state and the action taken;
4. the cost function $r(i, a)$.

The goal in such problems will be to find the sequence of action functions (u_n) that yield smallest total cost.

Though we will go on to solve the campus tour problem in Section 2, a word or two in the way of preview is in order here. This particular example has relatively few possible tours, and with patience we could itemize them all, compute the travel time that each one takes, and pick the shortest. If there were many more attractions at each level, or many more levels, this brute-force approach could quickly become unmanageable. So we will look for something more systematic, and also amenable to solution by computer. What makes the problem complicated is that there are several levels to go through. If only levels 3 and 4 existed in the graph of Figure 6.1, for example, then there would have been no decisions to make at all. The shortest (and only) paths from attractions H, I, and J, respectively, to K are the single edges (H, K), (I, K), and (J, K) requiring 5, 6, and 4 minutes, respectively. If we back up to level 2 and ask what is the shortest path from attraction F to K, it seems intuitively reasonable that we can find it by minimizing among the neighbors of F (H and I) the total of the edge cost from F to the neighbor plus the shortest path cost from the neighbor to K, which is known from the previous stage of the computation. We would choose between path F, H, K with a total cost of $6 + 5 = 11$ and path F, I, K with a total cost of $4 + 6 = 10$; hence the latter is the shortest path from F to K. We can continue to back up to level 1, and then to level 0 in the same way to ultimately find the shortest path from A to K, which was our original goal. The complete solution will be given in the next section; the following activity is a stepping stone.

Activity 1 – Use the approach suggested in the last paragraph to find the shortest path from G to K, and use that to find the shortest path from C to K.

Stochastic Dynamic Programming: The Finite Horizon Problem

With an understanding of the concepts of states, actions, and costs in hand, we can move on to the problem of stochastic dynamic programming. There is a large amount of structure and notation in stochastic dynamic programming problems; consequently we will use this section to set the notation and discuss the main features of the problem, without yet trying to solve the problem. Once again we will be working with a sequence $X_0, X_1, X_2, \dots$, where X_n represents the state of a system at time n. These X_n's are random variables taking values in the state space E, which we will once again assume to be finite. We also have a sequence of actions $U_0, U_1, U_2, \dots$, where U_n is interpreted as the action taken by the controller at time n. These actions are also random variables, because of their relationship with the X_n's, which take values in a set A, called the *action set*. The sequence of pairs $(X_n, U_n)_{n \geq 0}$ will be called a *Markov decision process* if, for each $n \geq 0$, the probability distribution of X_{n+1} is completely determined by the previous state-action pair (X_n, U_n), in particular, it is not dependent on past states or actions prior to time n. (Note that this does not necessarily mean that the chain (X_n) itself is Markov, for we have not yet restricted to the case where action U_n is conditionally independent of the past prior to time n, given X_n. We will do this shortly when we discuss admissible policies.)

The basic difference between deterministic dynamic programming and stochastic dynamic programming is that in a deterministic problem, the action taken at a current state completely determines what the next state will be; while in a probabilistic problem, the action taken at a current state alters the probability law of the next state of the process, but the next state is still a random variable.

> **Activity 2** – Consider the following situation. You are playing a game in which you have \$5. If you use strategy A, you could win \$2 with probability 1/4, or stay the same with probability 3/4. If instead you use strategy B, you could win \$5 with probability 1/3, or lose \$2 with probability 2/3. Formulate this as a single-stage stochastic dynamic programming problem. Which action should you take, and on what basis do you make that decision?

To describe the probability law of the chain of states, we assume that there is, for each possible action $a \in A$, a transition matrix T_a such that:

$$T_a(i, j) = P[X_{n+1} = j \mid X_n = i, U_n = a] . \tag{1}$$

The notations $T_a(i, j)$ and $T(i, j; a)$ will be used interchangably for the probability that the next state will be j, given that the current state is i and the action taken is a.

For example, if the two following transition matrices describe a problem

$$T_1 = \begin{pmatrix} \frac{1}{3} & \frac{2}{3} \\ \frac{3}{4} & \frac{1}{4} \end{pmatrix} \text{ and } T_2 = \begin{pmatrix} \frac{1}{5} & \frac{4}{5} \\ \frac{1}{4} & \frac{3}{4} \end{pmatrix}$$

then, when action 1 is taken, the process behaves as on the left of Figure 6.2; and when action 2 is taken, the process obeys the transition diagram on the right.

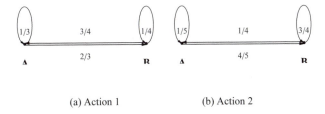

| (a) Action 1 | (b) Action 2 |

Figure 6.2 – Transition diagrams for two actions in a Markov decision process

For the most general kind of policy, the action U_n taken at time n could depend on the entire previous history of states and actions $X_0, U_0, X_1, U_1, \ldots, X_{n-1}, U_{n-1}, X_n$. In fact, U_n may even be randomized, in the sense that for a given history, we may flip a coin to determine whether to take one action or another. But for the problems that we will consider, it will be enough to take actions of a simpler kind, called *feedback actions*. That is, we assume that the action taken at each time n is a deterministic function of the state at time n:

$$U_n = u_n(X_n), \quad \text{for each } n = 0, 1, 2, \ldots \tag{2}$$

where u_n is a function mapping the state space E to the action space A. In many problems, not all actions are practical for every current state. So, we suppose that when the current state is i, there is a given subset A_i of permissible actions. This imposes a condition on the action functions described in (2):

$$u_n(i) \in A_i, \quad \text{for each } n = 0, 1, 2, \ldots \text{ and each } i \in E. \tag{3}$$

DEFINITION 1. An *admissible (feedback) policy* is a sequence $\mathbf{u} = (u_0, u_1, u_2, \ldots)$ of functions from E to A such that (3) holds. Such a policy prescribes that action $u_n(X_n)$ be taken if the state of the system at time n is X_n. An admissible policy $\mathbf{u}$ is called *stationary* if all of its component functions u_i are the same.

REMARK. Under an admissible feedback policy, the chain (X_n) has the Markov property (i.e., independence of past and future given present), but it is not time-homogeneous (i.e., the transition probabilities are not independent of time) unless the policy is stationary. The reader can check that we have, for instance,

(a) $P[X_{n+1} = j \mid X_n = i] = P[X_{n+1} = j \mid X_n = i, U_n = u_n(i)]$
$\qquad = T(i, j; u_n(i))$

(b) $P[X_1 = j_1, X_2 = j_2, \ldots, X_n = j_n \mid X_0 = i]$
$\qquad = T(i, j_1; u_0(i)) \cdots T(j_{n-1}, j_n; u_{n-1}(j_{n-1}))$ (4)

(c) $P[X_n = j_n \mid X_0 = i]$
$\qquad = \sum_{j_1, \ldots, j_{n-1}} T(i, j_1; u_0(i)) \cdots T(j_{n-1}, j_n; u_{n-1}(j_{n-1}))$

For the Markov decision process of Figure 6.2, for example, suppose that a policy uses action function $u_0(A) = 2$, $u_0(B) = 1$ at time 0, and action function $u_1(A) = 1$, $u_1(B) = 2$ at time 1. To form the transition matrix of the controlled process for time 0, we glue together the first row of T_2 and the second row of T_1, since at state A action 2 is taken by u_0, and at state B action 1 is taken. Similarly, to form the transition matrix for time 1, because of the way that u_1 is defined, glue together the first row of T_1 and the second row of T_2. The two transition matrices are then

$$T_{u_0} = \begin{pmatrix} 1/5 & 4/5 \\ 3/4 & 1/4 \end{pmatrix}, \quad T_{u_1} = \begin{pmatrix} 1/3 & 2/3 \\ 1/4 & 3/4 \end{pmatrix}$$

Under this policy for example, following (4)(b),

$$P[X_1 = B, X_2 = A \mid X_0 = A] = (4/5)(1/4) = 1/5$$

When more than one policy $\mathbf{u}$ is under study, it may be necessary to exhibit the dependence of the probability distributions in (4) (a), (b), and (c) on the policy. We do this by subscripting: $P_{\mathbf{u}}$. Similarly, the associated expectation operator under a policy is written $E_{\mathbf{u}}$.

At a time when the state is x and the action taken is a, there will be a current reward (or cost) $r(x, a)$. In the case that the state space E and the action space A are both finite, this reward function is necessarily bounded. Then, under an admissible feedback policy $u = (u_0, u_1, u_2, \ldots)$, there is a sequence of reward random variables $R_0, R_1, R_2, \ldots$, where

$$R_n = r(X_n, u_n(X_n))$$

The expected value of the reward received at time n, given the initial state, is

$$E_{\mathbf{u}}[R_n \mid X_0 = i] = \sum_j r(j, u_n(j)) P_{\mathbf{u}}[X_n = j \mid X_0 = i] \qquad (5)$$

The probability inside the summation is given by (4)(c).

Activity 3 – Consider again the Markov decision process illustrated by Figure 6.2 and the policy with action functions u_0 and u_1 described above. If the reward function is $r(A, 1) = 4$, $r(B, 1) = 3$, $r(A, 2) = 2$, $r(B, 2) = 5$, find the expected reward $E_u[R_1 \mid X_0 = A]$.

Let us review for a moment. The building blocks of the Markov decision model are the following:

(1) State space E;
(2) Sets A_i of actions that can be taken when the state is i;
(3) Transition matrices $T_a(i, j)$ for each action a;
(4) Reward function $r(i, a)$.

The system begins in state X_0. The controller takes an admissible action $u_0(X_0)$. As a result, a reward $r(X_0, u_0(X_0))$ is earned, and the system moves to a new state X_1 according to transition probabilities $T(X_0, j; u_0(X_0))$. The controller then takes another admissible action $u_1(X_1)$, dependent only on the current state X_1. Another reward $r(X_1, u_1(X_1))$ is earned, and the system moves to a state X_2 according to the transition probabilities $T(X_1, k; u_1(X_1))$. The system continues to move in this way.

The Markov decision problem is to find a policy, consisting of the action functions $u_0, u_1, u_2, \ldots$ to maximize the expected total reward (or minimize expected total cost). The following definition carefully sets down the *finite horizon problem*, in which control is only exerted up to a finite time T, and a terminal reward $R(X_T)$ is earned at time T.

DEFINITION 2. Let $R(x)$ be a non-negative valued function on E, and let T be a positive integer. The function

$$V(i, \mathbf{u}) = E_{\mathbf{u}}[\sum_{n=0}^{T-1} r(X_n, u_n(X_n)) + R(X_T) \mid X_0 = i\,]$$

is called the *value function* of policy $\mathbf{u}$ for the finite horizon problem with time horizon T. The *optimal value function* for this problem is

$$V(i) = \max_{\mathbf{u}} V(i, \mathbf{u}) \quad \text{(minimum for a minimum cost problem)}$$

A policy $\mathbf{u}^*$ is *optimal* if $V(i, \mathbf{u}^*) = V(i)$ for all initial states i.

In this section and the following one, we will consider optimal policies in Markov decision models for the finite horizon problem in the definition above. Later, in Sections 6.3 and 6.4, we will focus on the infinite horizon problem with a discounted reward, in which control is exerted forever.

A remark on the magnitude of the problem is in order. For the finite horizon problem, a policy is a sequence $(u_0, u_1, \ldots, u_{T-1})$ of functions from the finite set E to the finite set A (possibly with some admissibility restrictions). If the elements of E are labeled $1, 2, \ldots, N$, then each component function u_i can be thought of as an N-tuple:

$$u_i = (a_1, a_2, \ldots, a_N),$$

where a_j is the action that u_i prescribes if the state is j. There are $n(A_j)$ possible such actions, so that there are $n(A_1) \cdot n(A_2) \cdots n(A_N)$ action functions u_i, by the fundamental counting principle. A sequence of T such functions u_i makes up a policy, hence there are

$$[n(A_1) \cdot n(A_2) \cdots n(A_N)]^T$$

possible policies. As the action set, the state space, or the time horizon grows, the number of policies grows rapidly. Even if we restrict to only the class of stationary policies $\mathbf{u} = (u)$, by the above reasoning there are about

$[n(A)]^{n(E)}$ of those, a potentially very large number. Thus we need more efficient methods of locating optimal policies than simple brute-force evaluation of all policies.

Examples

Thus far, we have discussed the Markov decision problem in the abstract. A "system" is in "state" i, an "action" a is taken that determines both a "reward" $r(i, a)$ and the probability $T(i, j; a)$ that the system next will move to state j. What concrete experiments can be so formulated? To close this section we present two models. Other examples will be studied later in the chapter, and still more are contained in the exercises.

EXAMPLE 2. The defense department has ordered exactly one rocket booster from a contractor. Quality standards are strict, so that the chance that a single manufactured rocket will be acceptable is just 1/4. The plan is to use no more than 3 production runs. Either 0, 1, 2, or 3 rockets are to be made on each run, and are checked after the run is complete. No more runs will occur after a satisfactory rocket has finally been made. If a production run makes any rockets, there is a fixed set–up cost $C_1 = 10$ and a production cost of $C_2 = 5$ per rocket. There is a penalty cost of $C_3 = 64$ if a successful rocket has not been made by the third production run. What production strategy minimizes the expected total cost of the process?

The "times" are represented by the production runs, so $n \in \{0, 1, 2, 3\}$, with time 0 representing the initial time before manufacturing begins. At time 3, a terminal penalty may or may not be incurred. The action at time n will be defined as the number of rockets made in production run n, but as it often happens in dynamic programming, it is not so easy to see how the "state" of the system should be defined. Since the cost at time n is only to be dependent on the action, and the terminal cost depends only on whether a successful rocket has been made, it would seem that we need to keep track of whether or not a good rocket has been made yet. Accordingly, denote

$$X_n = \text{\# of good rockets (0 or 1) still to be made at the beginning of} \atop \text{the } n^{\text{th}} \text{ production run;}} \tag{6}$$

$$U_n = \text{\# rockets made in production run } n. \tag{7}$$

The state space is clearly $E = \{0, 1\}$. State 0 means that there are no more rockets to make, consequently only action 0, the action of making no rockets, is permissible at state 0. Likewise, for state 1, the state where 1 good rocket still needs to be made, $A_1 = \{0, 1, 2, 3\}$. The cost function for state–action pairs (x, a) for each time period (prior to the terminal time) is

$$r(x, a) = \begin{cases} 10 + 5\,a & \text{if } x = 1 \text{ and } a > 0 \\ 0 & \text{otherwise} \end{cases} \tag{8}$$

and the terminal cost function at time 3 is

$$R(x) = \begin{cases} 64 & \text{if } x = 1 \\ 0 & \text{otherwise} \end{cases} \tag{9}$$

Among N rockets, the probability that all are unacceptable is $(3/4)^N$. Thus, a run that makes N rockets will produce an acceptable rocket with probability $1 - (3/4)^N$. Because of this, it is easy to see that the transition matrices T_a are as follows:

$$T_0 = \begin{matrix} & 0 & 1 \\ 0 & \\ 1 & \end{matrix}\begin{pmatrix} 1 & 0 \\ 0 & 1 \end{pmatrix} \qquad T_1 = \begin{matrix} & 0 & 1 \\ 0 & \\ 1 & \end{matrix}\begin{pmatrix} - & - \\ 1/4 & 3/4 \end{pmatrix}$$

$$T_2 = \begin{matrix} & 0 & 1 \\ 0 & \\ 1 & \end{matrix}\begin{pmatrix} - & - \\ 7/16 & 9/16 \end{pmatrix} \qquad T_3 = \begin{matrix} & 0 & 1 \\ 0 & \\ 1 & \end{matrix}\begin{pmatrix} - & - \\ 37/64 & 27/64 \end{pmatrix}$$

A policy for this production problem is a sequence $\mathbf{u} = (u_0, u_1, u_2)$, where $u_n(x)$ is the number of rockets to be made on production run n, if x good rockets ($x = 0$ or 1) remain to be made. Since we will obviously make no more rockets after a good one has been made, $u_n(0)$ is constrained to be 0. We wish to minimize over $\mathbf{u}$:

$$V(x, \mathbf{u}) = E_{\mathbf{u}}[\textstyle\sum_{n=0}^{2} r(X_n, u_n(X_n)) + R(X_3) \mid X_0 = x] \tag{10}$$

for $x = 0, 1$, and we are of course most interested in the case $x = 1$. We will solve this problem in the next section. ∎

EXAMPLE 3. A certain type of shellfish is harvested by fishermen as years pass. The population can be at six levels, 0 population units up to 5 units. In the absence of harvesting, the population would form a Markov chain with the transition diagram in Figure 6.3.

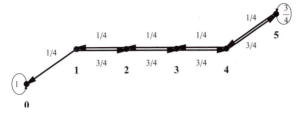

Figure 6.3 – Transition diagram for shellfish population change

Fishermen can only harvest as many population units of fish in a year as there are in existence at the start of the year, and then a "natural" change occurs due to randomness described by the transition diagram above. The population at the start of a year will be the population at the start of the previous year minus the amount harvested in the previous year plus the natural change. To clarify the behavior near state 0, suppose that if i units are harvested in a year in which the population level started at i units, the population will remain extinct with certainty for all following years. The government fishing bureau wants to determine how many fish to allow to be harvested over the next 3 years. There is a net benefit of h monetary units per population unit harvested, and a terminal benefit of R monetary units for each population unit of fish remaining at the end of the 3 year fishing plan. Model this problem as a Markov decision problem, including a description of the state space, the action sets, the transition matrices, the reward function, and the optimal value function.

First, we need to determine what the state space is. The problem explains that the population of fish can change over time and determines the terminal reward. Thus, a reasonable choice is to let X_n be the population level at time n, so that the state space is $E = \{0, 1, 2, 3, 4, 5\}$.

The actions can be defined to be the number of units of fish to harvest. The maximum that can be harvested is the population at the beginning of the year. Thus, the action sets are as follows:

$$A_0 = \{0\}, \quad A_1 = \{0, 1\}, \quad A_2 = \{0, 1, 2\}, \quad A_3 = \{0, 1, 2, 3\},$$
$$A_4 = \{0, 1, 2, 3, 4\}, \quad A_5 = \{0, 1, 2, 3, 4, 5\}$$

The dynamics of the fish population that are described in the problem allow us to write the transition matrices T_a that describe how the chain moves given that action a is taken (see below). For example, consider action 2 and the corresponding matrix T_2; the states cannot be 0 or 1 (it is not feasible to harvest more units than there are). If the initial state is 2 and 2 units are harvested, the population becomes extinct, that is the next state is 0, with probability 1. If the initial state is 3, then after the 2 units are harvested, leaving 1 unit, the population can go back up to 2 with probability 3/4, or down to 0 with probability 1/4. This explains the third and fourth rows of T_2, and the rest should be self evident.

$$
T_0 = \begin{pmatrix}
1 & 0 & 0 & 0 & 0 & 0 \\
1/4 & 0 & 3/4 & 0 & 0 & 0 \\
0 & 1/4 & 0 & 3/4 & 0 & 0 \\
0 & 0 & 1/4 & 0 & 3/4 & 0 \\
0 & 0 & 0 & 1/4 & 0 & 3/4 \\
0 & 0 & 0 & 0 & 1/4 & 3/4
\end{pmatrix},
$$

$$
T_1 = \begin{pmatrix}
- & - & - & - & - & - \\
1 & 0 & 0 & 0 & 0 & 0 \\
1/4 & 0 & 3/4 & 0 & 0 & 0 \\
0 & 1/4 & 0 & 3/4 & 0 & 0 \\
0 & 0 & 1/4 & 0 & 3/4 & 0 \\
0 & 0 & 0 & 1/4 & 0 & 3/4
\end{pmatrix},
$$

$$
T_2 = \begin{pmatrix}
- & - & - & - & - & - \\
- & - & - & - & - & - \\
1 & 0 & 0 & 0 & 0 & 0 \\
1/4 & 0 & 3/4 & 0 & 0 & 0 \\
0 & 1/4 & 0 & 3/4 & 0 & 0 \\
0 & 0 & 1/4 & 0 & 3/4 & 0
\end{pmatrix},
$$

$$
T_3 = \begin{pmatrix}
- & - & - & - & - & - \\
- & - & - & - & - & - \\
- & - & - & - & - & - \\
1 & 0 & 0 & 0 & 0 & 0 \\
1/4 & 0 & 3/4 & 0 & 0 & 0 \\
0 & 1/4 & 0 & 3/4 & 0 & 0
\end{pmatrix}
$$

$$
T_4 = \begin{pmatrix}
- & - & - & - & - & - \\
- & - & - & - & - & - \\
- & - & - & - & - & - \\
- & - & - & - & - & - \\
1 & 0 & 0 & 0 & 0 & 0 \\
1/4 & 0 & 3/4 & 0 & 0 & 0
\end{pmatrix}, \quad
T_5 = \begin{pmatrix}
- & - & - & - & - & - \\
- & - & - & - & - & - \\
- & - & - & - & - & - \\
- & - & - & - & - & - \\
- & - & - & - & - & - \\
1 & 0 & 0 & 0 & 0 & 0
\end{pmatrix}
$$

The reward for harvesting a units is ah, by the conditions stated in the problem. We could of course characterize the reward function by the formula $r(x, a) = ah$ for $x \geq a$, but for computational solution, it is also convenient to characterize the reward as the matrix below, where $r(x, a)$ is the entry in row x and column a of this matrix.

$$
r = x \begin{array}{c}
 \\ 0 \\ 1 \\ 2 \\ 3 \\ 4 \\ 5
\end{array}
\begin{pmatrix}
0 & 1 & 2 & 3 & 4 & 5 \\
0 & - & - & - & - & - \\
0 & h & - & - & - & - \\
0 & h & 2h & - & - & - \\
0 & h & 2h & 3h & - & - \\
0 & h & 2h & 3h & 4h & - \\
0 & h & 2h & 3h & 4h & 5h
\end{pmatrix}
$$

At time $T = 3$, a terminal reward may be incurred, given by $R(x) = R\,x$. This terminal reward can also be thought of as a vector:

$$R = \begin{pmatrix} 0 \\ R \\ 2R \\ 3R \\ 4R \\ 5R \end{pmatrix}$$

We wish to find a policy $u = (u_0, u_1, u_2)$ to maxmize

$$V(x, \mathbf{u}) = E_{\mathbf{u}}[\textstyle\sum_{n=0}^{2} r(X_n, u_n(X_n)) + R(X_3) \mid X_0 = x]$$

for every initial state x. ∎

Activity 4 – Verify the entries of the other transition matrices T_a for the fish harvesting example.

Exercises 6.1

1. Consider the Markov chain with the transition diagram below. Suppose that there are two possible actions, labelled 0 and 1. Under action 0, the chain moves according to this transition diagram, and under action 1, the chain moves with certainty to state 2. Let $\mathbf{u}$ be the stationary policy with action function defined by:

$$u(i) = \begin{cases} 0 & \text{if } i = 1, 2 \\ 1 & \text{if } i = 3, 4 \end{cases}$$

If, for a certain experimental outcome ω, we observe $X_0(\omega) = 1$, $X_2(\omega) = 3$, $X_3(\omega) = 2$, what are the first four actions taken under this policy? Again for this outcome, if the reward function is $r(i, a) = i - a$, what are the first four rewards?

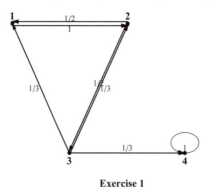

Exercise 1

2. Prove formula (4)(b).

3. If the state space of a Markov decision process has size 4, the action space has size 3, and all actions are admissible at all states, then how many stationary policies are there? How many admissible feedback policies are there for a finite horizon problem with terminal time $T = 5$?

4. Below is a directed graph seen earlier in the book as Figure 1.26. Find the shortest paths in the graph from vertices 7, 8, and 9 to vertex 10, and use these to find the shortest paths from vertices 5 and 6 to 10. (Note that this graph is not as simply partitioned into levels as Figure 6.1 was.)

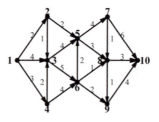

Exercise 4

5. An advertising agency will conduct a campaign for a new soft drink. The agency follows the share of the market possessed by the soft drink, in increments of 5%, from month to month. Each month that advertising continues, there is a cost of c dollars to the drink manufacturer. But there is a reward of r dollars to the manufacturer for each 5% of the market pos-

sessed by their product. If no advertising is used in any given month, the share of the market will not change. If there is advertising, the share of the market will either increase by 5% (with probability 1/2) or stay the same (also with probability 1/2), until a maximum of 30% is reached, where it will stay forever. Formulate the problem of finding the optimal advertising policy as a Markov decision problem, including a description of the state space, the admissible action sets, the transition matrices, the reward function, and the optimal value function.

6. Consider the Markov decision process illustrated by Figure 6.2. Suppose that the time horizon is $T = 3$, the terminal reward function is $R(x) = 0$; $x = A, B$, and the per period reward function is $r(A, 1) = 4$, $r(B, 1) = 3$, $r(A, 2) = 2$, $r(B, 2) = 5$. For the policy $\mathbf{u}$ that always uses action 1 at time 0, action 2 at time 1, and action 1 at time 2, compute $V(A, \mathbf{u})$ and $V(B, \mathbf{u})$.

7. How many stationary policies are there in the problem of Exercise 5?

8. How many admissible feedback policies are there in the fish harvesting problem of Example 3?

9. For the Markov decision process of Exercise 1, find the transition matrix corresponding to the stationary policy $\mathbf{u}$, and calculate $E_{\mathbf{u}}[\sum_{n=0}^{2} R_n \mid X_0 = 1]$.

10. For the Markov decision process of Exercise 1, calculate the expectation $E_{\mathbf{u}}[\sum_{n=0}^{2} R_n \mid X_0 = 3]$ for the non–stationary policy for which $u_0(i) = 0$ for all i, $u_1(i) = 1$ for all i, and $u_2(i) = 0$ for all i.

11. At a small cellular phone company, servers must spend a half hour discussing options with each possible customer who comes in. During any half hour period, either 0, 1, or 2 customers will come in, with probabilities 1/2, 1/4, and 1/4 respectively. A total of three servers can be summoned to work if necessary. Customers who are being served will not leave before their service is complete. If there are 6 or more customers in the store, including those in service, all of those not currently being served will leave without being served, otherwise if there are 5 or fewer customers in the store all of those who are not being served will stay for the next half hour period. The company can control how many servers are on the floor, but they pay a price of s dollars per half hour per server to keep them there. They earn a profit of p dollars for each customer who stays and gets served. The store is open from 9:00 am to 4:00 pm. Customers left over at closing time leave without service or profit to the company. Formulate this problem as a Markov decision problem, describing the state and action spaces, the transition matrices, and the per period and terminal reward functions.

12. The population of a country can be approximately modelled so that it has a value of either 0 units, 1 unit, 2, 3, 4, or 5 units. The population undergoes a natural change from one time period to the next, increasing by one unit with probability p and decreasing one unit with probability $1 - p$. At the boundaries, when the population is 5, it stays the same at the next time period with probability p or goes down to 4 with probability $1 - p$, and at 0 it stays at 0 with probability 1. In any time period there is a net benefit to the economy of the country of 10 units per unit of population. Immigration is possible; at any time a number of units of population can be admitted to bring the net population after the natural change to 5 or less. But there is a resettlement cost of 8 units per unit of immigrant population admitted. Assume that new immigrants in a time period do not contribute to the net benefit until the next time period. Let the immigration control be done for 6 time periods, and let the economic benefit at the end also be 10 units per unit of population. Formulate this as a Markov decision problem.

6.2 The Finite Horizon Problem

Let us now turn to the solution of finite horizon dynamic programming problems. Recall that for deterministic dynamic programming problems with finite time horizon, we have a sequence of states $x_0, x_1, x_2, \ldots x_T$, where x_n represents the state of a system at time n. The state space E is finite. A policy $\mathbf{u} = (u_0, u_1, u_2 \ldots, u_{T-1})$ induces a chain of actions $u_0(x_0)$, $u_1(x_1), \ldots, u_{T-1}(x_{T-1})$. Some deterministic mechanism is in place to generate the next state x_{n+1} from the previous state x_n and action $u_n(x_n)$. For times $n = 0, \ldots, T - 1$ there is a cost (or reward) $r(x_n, u_n(x_n))$. There may or may not be a terminal cost (or reward) $R(x_T)$ at time T. We are to minimize total cost subject to the choice of actions.

To gain an understanding of how to solve a deterministic dynamic programming problem, we will start by solving the shortest tour problem outlined in the previous section.

EXAMPLE 1. The diagram for the shortest campus tour problem is repeated as Figure 6.4 for your convenience. Remember, the state space is the set of vertices of the graph, the actions are the next attractions to be visited, and the costs, indicated as edge weights, are functions of the current state and the action taken at that state. There is no terminal cost at the final level $T = 4$, since that is our destination.

One last idea is important to understand before going on to solving the problem: the optimal value function. Let $V_n(i)$ be the optimal value function starting from level n and state i, which in our context means the cost of the shortest path from the current attraction i at level n to the destination. We are really interested in finding $V_0(A)$, but we can do so by finding the other $V_n(i)$, working backwards from the right side of the graph.

```
Needs["KnoxOR`Graphs`"]
```

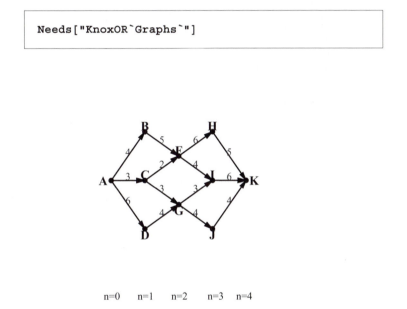

n=0   n=1   n=2   n=3   n=4

Figure 6.4 – Possible routes on a campus tour

First, we will determine the shortest path to use during the last stage of the tour from each possible current state at time $n = 3$. Since there is only one possible path from these attractions to the destination, we can see immediately the shortest one: the only one. Hence,

$$V_3(H) = 5 \qquad V_3(I) = 6 \qquad V_3(J) = 4$$

Next, we step backwards one time unit and determine the shortest path to the destination from each possible attraction at $n = 2$. It is rather simple to find $V_2(F)$, which is the length of the optimal path from F to the destination, and $V_2(G)$, the length of the optimal path from G to the destination. The shortest path from attraction F must start with the edge from attraction F to H or from F to I. From H or I, the path must follow the shortest path from that attraction to the destination given by $V_3(H)$ or $V_3(I)$, respectively.

$$
\begin{aligned}
V_2(F) &= \min\{r(F, H) + V_3(H),\ r(F, I) + V_3(I)\} \\
&= \min\{6 + 5,\ 4 + 6\} = 10 \text{ at } F \longrightarrow I
\end{aligned}
$$

Thus, we have found that the shortest path from F to the destination is the path $F \longrightarrow I \longrightarrow K$, which has cost 10. Similarly,

$$
\begin{aligned}
V_2(G) &= \min\{r(G, I) + V_3(I),\ r(G, J) + V_3(J)\} \\
&= \min\{3 + 6,\ 4 + 4\} = 8 \text{ at } G \longrightarrow J
\end{aligned}
$$

Hence the shortest path from G to the destination is $G \longrightarrow J \longrightarrow K$, with cost 8.

Now, we step backward one more time unit and determine the shortest path to the destination from each possible attraction at $n = 1$ using the information: $V_2(F) = 10$ and $V_2(G) = 8$. The shortest path from attraction B must start with the edge from attraction B to F. From F, the path must follow the shortest path from that attraction to the destination given by $V_2(F)$. From vertex C, the two choices for the next attraction are F and G; and from vertex D, the next attraction must be G. The appropriate computations are as follows:

$$
\begin{aligned}
V_1(B) &= \min\{r(B, F) + V_2(F)\} = 5 + 10 = 15 \text{ at } B \longrightarrow F \\
V_1(C) &= \min\{r(C, F) + V_2(F),\ r(C, G) + V_2(G)\} \\
&= \min\{2 + 10,\ 3 + 8\} = 11 \quad \text{at } C \longrightarrow G \\
V_1(D) &= \min\{r(D, G) + V_2(G)\} = 4 + 8 = 12 \text{ at } D \longrightarrow G
\end{aligned}
$$

The optimal path $B \longrightarrow F \longrightarrow I \longrightarrow K$ from B to the destination has cost 15, the optimal path $C \longrightarrow G \longrightarrow J \longrightarrow K$ from C to the destination has cost 11, and the optimal path $D \longrightarrow G \longrightarrow J \longrightarrow K$ from D to the destination has cost 12.

Finally, from A there are three choices: to use the edge from A to B, then the optimal path from B to K, or to use the edge from A to C or from A to D and then the corresponding shortest paths:

$$
\begin{aligned}
V_0(A) &= \min\{r(A, B) + V_1(B),\ r(A, C) + V_1(C),\ r(A, D) + V_1(D)\} \\
&= \min\{4 + 15,\ 3 + 11,\ 6 + 12\} \\
&= \min\{19,\ 14,\ 18\} = 14 \text{ at } A \longrightarrow C
\end{aligned}
$$

In words, the path that spends the least amount of time walking and allows for the most time to look at the attractions is $A \longrightarrow C \longrightarrow G \longrightarrow J \longrightarrow K$ and this path takes 14 minutes to walk, leaving the student a total of 36 minutes to look at the five attractions. ∎

Activity 1 – At the beginning of the last example, we stated that there is no terminal cost function. Could you take a slightly different modeling perspective and recast the problem so that there is a terminal cost? (Hint: Does the time horizon have to be $T = 4$?)

Let us review how we solved this finite deterministic dynamic programming problem. We started at the destination and worked backwards step by step from the terminal time T to find the optimal value functions V_n evaluated at each state, for times $n \leq T - 1$, using the following equation:

$$
\begin{aligned}
V_{n-1}(i) &= \min_{a \in A_i} \{r(i, a) + V_n(j)\} \\
&= \min_{\text{available actions}} \{\text{current cost for next action} \\
&\qquad + \text{smallest total future cost resulting from that action}\}
\end{aligned} \tag{1}
$$

In formula (1), the state j is the next state to be visited if the current state is i and the action taken is a. Thus, we worked backwards, at each time n computing the optimal value function for a problem starting at time n and ending at T, until we found the optimal value function for the complete problem starting at $n = 0$ and ending at time T.

Now we can move on to the stochastic version. Like the deterministic problem, we start at the end and work backwards, but the problem is slightly more complicated due to the random nature of the dynamics that govern the motion of the controlled process.

Dynamic Programming Algorithm, Stochastic Case

Once again, let $X_0, X_1, \ldots, X_T$ be the chain of states for a Markov decision process with a finite time horizon T and finite state space E. A policy $\mathbf{u} = (u_0, u_2, \ldots, u_{T-1})$ induces a chain of actions $U_0, U_1, \ldots, U_{T-1}$ by $U_n = u_n(X_n)$. For times $n = 0, \ldots, T - 1$ there is a reward $R_n = r(X_n, U_n)$. A terminal reward $R(X_T)$ is earned at time T. The probabilistic motion of the chain is described by the one-step transition matrices T_a. Recall that $T(i, j; a) = T_a(i, j)$ is the conditional probability, under action a, that the next state will be j, given that the current state is i. Then,

$$
P_{\mathbf{u}}[X_{n+1} = j \mid X_n = i] = T(i, j; u_n(i))
$$

For a maximum problem, we wish to devise an algorithm to find a policy $\mathbf{u}$ to maximize for all starting states i the expected total reward

$$
V(i, \mathbf{u}) = E_{\mathbf{u}}[\textstyle\sum_{n=0}^{T-1} r(X_n, u_n(X_n)) + R(X_T) \mid X_0 = i]
$$

The strategy will be to find u_{T-1} first, then work backwards step by step to u_0. At each time m we must compute the optimal value function for a problem that starts at time m, and ends at T. So, let us define, for a policy $\mathbf{u}$ and a time $m \leq T - 1$:

$$V_m (i, \mathbf{u}) = E_{\mathbf{u}}[\sum_{n=m}^{T-1} r(X_n, u_n(X_n)) + R(X_T) \mid X_m = i] \qquad (2)$$

and the related time m optimal value function:

$$V_m(i) = \max_{\mathbf{u}} V_m (i, \mathbf{u}) \qquad (3)$$

The maximum is taken over all admissible policies. Notice that for time $m = T - 1$,

$$
\begin{aligned}
V_{T-1} (i, \mathbf{u}) &= r(i, u_{T-1}(i)) + E_{\mathbf{u}}[R(X_T) \mid X_{T-1} = i] \\
&= r(i, u_{T-1}(i)) + \sum_{j \in E} R(j)\, T(i, j; u_{T-1}(i))
\end{aligned}
\qquad (4)
$$

At time $T - 1$ we have a simple decision to make. Knowing that the state of the system is i, we must pick exactly one more action $a = u_{T-1}(i)$ to maximize the immediate reward $r(i, a)$, plus the expected reward $\sum_j R(j)\, T(i, j; a)$ to be earned at the terminal time. It is now clear that

$$V_{T-1}(i) = \max_{a \in A_i} \{r(i, a) + \sum_j R(j)\, T(i, j; a)\} \qquad (5)$$

which enables us to initiate the backwards programming process, since r, R, $T(i, j; a)$, and the action sets A_i are all known quantities.

The key idea of dynamic programming, usually called the *Principle of Optimality*, is this: an optimal policy is also optimal from each time m onward. This means that if we have the optimal policy $(u_m, u_{m+1}, \ldots, u_{T-1})$ from time m onward, then we can find the optimal policy from time $m - 1$ onward by adjoining an optimal action function u_{m-1} to the front of this policy, chosen at state i so as to maximize the sum of the immediate reward $r(i, u_{m-1}(i))$ plus the expected reward from time m onward under the optimal time m policy. We step back in time until time 0 is reached, at which point the optimal policy is completely determined. The following theorem makes this idea more precise.

THEOREM 1. Let V_m, $m = 0, \ldots, T - 1$, be the sequence of optimal value functions defined in (3) for the finite horizon Markov decision problem with finite state and action spaces. Also, let V_T equal the terminal reward function R. Then,

$$V_{m-1}(i) = \max_{a \in A_i} \{r(i, a) + \sum_j V_m(j)\, T(i, j; a)\}, \quad m = 1, \ldots, T \qquad (6)$$

For $m = 1, \ldots, T$, if we define a policy $\mathbf{u}^*$ such that $u_{m-1}^*(i)$ is the maximizing action in expression (6) for each time m and state i, then $\mathbf{u}^*$ is optimal.

Proof. The technique will be to show inductively on m, proceeding backward from time T, that (6) holds for all m, and in so doing, show that $\mathbf{u}^*$ is optimal from time m onward. In particular, for $m = 0$, $\mathbf{u}^*$ is optimal from time 0 onward, and V_0 is the optimal value function for the entire problem.

Since there are no more actions to be taken at time T, the anchoring step is trivial:

$$V_T(i) = V_T(i, \mathbf{u}^*) = R(i) \, , \, \forall \, i \in E$$

To proceed with the proof, we will need the following computation. For times $m = 1, \ldots, T$ and an arbitrary admissible policy $\mathbf{u} = (u_0, \ldots, u_{T-1})$, we have that

$$
\begin{aligned}
V_{m-1} &(i, \mathbf{u}) \\
&= E_\mathbf{u}[\textstyle\sum_{n=m-1}^{T-1} r(X_n, u_n(X_n)) + R(X_T) \mid X_{m-1} = i] \\
&= E_\mathbf{u}[E_\mathbf{u}[\textstyle\sum_{n=m-1}^{T-1} r(X_n, u_n(X_n)) + R(X_T) \mid X_{m-1} = i, X_m] \mid X_{m-1} = i] \\
&= E_\mathbf{u}[r(i, u_{m-1}(i)) \\
&\quad + E_\mathbf{u}[\textstyle\sum_{n=m}^{T-1} r(X_n, u_n(X_n)) + R(X_T) \mid X_{m-1} = i, X_m] \mid X_{m-1}] \\
&= E_\mathbf{u}[r(i, u_{m-1}(i)) + E_\mathbf{u}[\textstyle\sum_{n=m}^{T-1} r(X_n, u_n(X_n)) + R(X_T) \mid X_m] \mid X_{m-1} = i] \\
&= E_\mathbf{u}[r(i, u_{m-1}(i)) + V_m(X_m, \mathbf{u}) \mid X_{m-1} = i] \\
&= r(i, u_{m-1}(i)) + \textstyle\sum_{j \in E} V_m(j, \mathbf{u}) T(i, j; u_{m-1}(i))
\end{aligned}
\tag{7}
$$

Now we suppose that $\mathbf{u}^*$ is optimal from time m onward, i.e., $V_m(i, \mathbf{u}^*) = V_m(i)$ for all $i \in E$, and that the value functions $V_T, \ldots, V_m$ satisfy the dynamic programming equation (6). We must extend this to time $m - 1$; that is, we must show (6) itself and also show that $\mathbf{u}^*$ is optimal from time $m - 1$ onward.

Return to the last line of (7). For all policies $\mathbf{u}$ and all states j, $V_m(j, \mathbf{u}) \leq V_m(j)$. We can take the maximum over all admissible policies $\mathbf{u}$ on the right side of (7) to obtain the inequality

$$V_{m-1}(i, \mathbf{u}) \leq r(i, u_{m-1}(i)) + \textstyle\sum_{j \in E} V_m(j) T(i, j; u_{m-1}(i))$$

But the right side of the last inequality is smaller than or equal to the maximum over all possible actions of a sum of similar form:

$$V_{m-1}(i, \mathbf{u}) \leq \max_{a \in A_i} \{r(i, a) + \textstyle\sum_{j \in E} V_m(j) T(i, j; a)\}$$

For $\mathbf{u} = \mathbf{u}^*$, the first inequality is an equality by the induction hypothesis, and the second inequality is also an equality by (7) and the construction of the time $m - 1$ component of $\mathbf{u}^*$. Thus,

$$V_{m-1}(i, \mathbf{u}^*) = \max_{a \in A_i} \{r(i, a) + \sum_{j \in E} V_m(j) T(i, j; a)\} \geq V_{m-1}(i, \mathbf{u}),$$

$\forall\, \mathbf{u}$

This implies that $\mathbf{u}^*$ is optimal from time $m - 1$ onward, i.e., $V_{m-1}(i, \mathbf{u}^*) = V_{m-1}(i)$ for all states i, and that V_{m-1} satisfies (6). ∎

Activity 2 – Explain why each line of the computation in derivation (7) is true.

Formula (6) is called the *dynamic programming (DP) equation.* In the case of cost minimization, the procedure is the same, but the maxima are replaced by minima. The theorem gives rise to the following algorithm, which requires knowledge of the state space E, the admissible action sets A_i, the transition matrices T_a, the reward functions r and R, and the terminal time T. Since the state space is finite, we can consider the terminal reward R and the time m optimal value function V_m as column vectors, with an entry for each state. Note that the sum $\sum_{j \in E} V_m(j) T(i, j; a)$ in the DP equation is just the ith row of the matrix T_a dotted with the column vector V_m, or what is the same thing, the ith row of the matrix product $T_a \cdot V_m$.

ALGORITHM. (Finite horizon Markov decision problem)

    (1) Initialize column vector V_T by $V_T(i) = R(i)$ for each state i;
    (2) For $m = T$ down to 1, do (3)–(4):
      (3) For each $i \in E$, do (a)–(c);
        (a) Find $a_i \in A_i$ to maximize $r(i, a) + T_a \cdot V_m(i)$
        (b) Let $u_{m-1}(i) = a_i$
        (c) Let $V_{m-1}(i) = r(i, a_i) + T_{a_i} \cdot V_m(i)$
    (4) Output $V_{m-1}(i)$ and $u_{m-1}(i)$ for each i.

To paraphrase the algorithm, after first determining the parameters of the problem, we set $V_T = R$, in order to initialize the computation. For each time m, working backward from T, we do the following. In step (3)(a)–(c), we find the optimal actions for time $m - 1$ for each state, and at the same time compute the next value function V_{m-1}, using the DP equation and the known, current value function V_m. We continue to step back one time period at a time until we reach time 0, when all optimal actions have been determined.

Examples

We will now use the dynamic programming algorithm to find the optimal solutions to the rocket production problem and the fish harvesting problem. You should refer now to Examples 2 and 3 of Section 1 for the notation and problem conditions.

EXAMPLE 2. In Example 2 of Section 6.1, we were to decide how many rockets to make at each of the times 0, 1, and 2. The per-period and terminal cost functions were as follows:

$$r(x, a) = \begin{cases} 10 + 5a & \text{if } x = 1 \text{ and } a > 0 \\ 0 & \text{otherwise} \end{cases}$$

$$R(x) = \begin{cases} 64 & \text{if } x = 1 \\ 0 & \text{otherwise} \end{cases}$$

These result from the assumptions that 10 is the set-up cost for a manufacturing run, 5 is the cost per rocket manufactured, and 64 is the penalty if the manufacturer does not fulfill the contract. There are two states, 0 and 1, which indicate the number of good rockets remaining to be produced. At state 0, only action 0 is permissible; and at state 1, the permissible actions are $A_1 = \{0, 1, 2, 3\}$, meaning that on a production run we can choose to make between 0 and 3 rockets if we have not yet made a good rocket. The transition matrices T_a are reproduced below for your convenience:

$$T_0 = \begin{matrix} 0 \\ 1 \end{matrix} \begin{pmatrix} \overset{0}{1} & \overset{1}{0} \\ 0 & 1 \end{pmatrix} \qquad T_1 = \begin{matrix} 0 \\ 1 \end{matrix} \begin{pmatrix} \overset{0}{-} & \overset{1}{-} \\ 1/4 & 3/4 \end{pmatrix}$$

$$T_2 = \begin{matrix} 0 \\ 1 \end{matrix} \begin{pmatrix} \overset{0}{-} & \overset{1}{-} \\ 7/16 & 9/16 \end{pmatrix} \qquad T_3 = \begin{matrix} 0 \\ 1 \end{matrix} \begin{pmatrix} \overset{0}{-} & \overset{1}{-} \\ 37/64 & 27/64 \end{pmatrix}$$

The terminal time is $T = 3$. In the algorithm, the maxima are replaced by minima.

Since $V_3(0) = 0$, and for $m = 1, 2, 3$,

$$V_{m-1}(0) = r(0, 0) + V_m(0) \cdot T_0(0, 0) = 0 + V_m(0),$$

it is clear that $V_m(0) = 0$ for all m, and the optimal (and only) action is $a = 0$. Thus, we confine our attention to the computation of $V_m(1)$ for each m. According to the dynamic programming algorithm, we initialize $V_3 = R = [0 \ 64]^t$. Then we calculate

$$V_2(1) = \min_{a \in \{0,1,2,3\}} \{r(1, a) + V_3(1) \cdot T_a(1, 1)\}$$

$$= \min \{0 + 64, \ 10 + 5 + 64 \left(\tfrac{3}{4}\right), \ 10 + 10 + 64 \left(\tfrac{9}{16}\right),$$

$$10 + 15 + 64 \left(\tfrac{27}{64}\right)\} \tag{8}$$

$$= \min \{64, 63, 56, 52^*\}$$

$$= 52 \qquad \qquad \text{at action 3}$$

To proceed backward to V_1, we have

$$V_1(1) = \min_{a \in \{0,1,2,3\}} \{r(1, a) + V_2(1) \cdot T_a(1, 1)\}$$

$$= \min \{0 + 52, \ 10 + 5 + 52 \left(\tfrac{3}{4}\right), \ 10 + 10 + 52 \left(\tfrac{9}{16}\right),$$

$$10 + 15 + 52 \left(\tfrac{27}{64}\right)\} \tag{9}$$

$$= \min \{52, 54, 49.25, 46.9375^*\}$$

$$= 46.9375 \qquad \qquad \text{at action 3}$$

V_0 is computed in the same way. You should check that the result is

$$V_0(1) = 44.801758, \quad \text{taken on at action 3.} \tag{10}$$

Combining these facts, we see that $\mathbf{u} = (u_0, u_1, u_2)$ is optimal, where for each n,

$$u_n(1) = 3 \quad \text{and} \quad u_n(0) = 0. \tag{11}$$

In words, we should make three rockets on each production run until an acceptable rocket is made. ■

Activity 3 – Intuitively, what is it about the problem parameters in Example 2 that gave us an optimal policy that always makes three rockets? Speculate on how the nature of the solution would change if the problem parameters change. In Exercises 1 and 2 you will be asked to do some computations of this kind.

EXAMPLE 3. In Example 3 of Section 6.1, the problem was to determine how many units of fish should be harvested at each time 0, 1, and 2 in order to maximize the reward received. We will complete that problem now and show how the use of *Mathematica* simplifies some of the computations. Let us fix the values $h = 5$ and $R = 50$ for the parameters of the problem. Recall that h is the amount paid for each population unit harvested, and R is the terminal benefit for each population unit remaining after the third year. Then we can write the two reward functions as:

```
r = {{0, -∞, -∞, -∞, -∞, -∞},
     {0, 5, -∞, -∞, -∞, -∞}, {0, 5, 10, -∞, -∞, -∞},
     {0, 5, 10, 15, -∞, -∞}, {0, 5, 10, 15, 20, -∞},
     {0, 5, 10, 15, 20, 25}};
R = {0, 50, 100, 150, 200, 250};
{MatrixForm[r], MatrixForm[R]}
```

$$
\left\{
\begin{pmatrix}
0 & -\infty & -\infty & -\infty & -\infty & -\infty \\
0 & 5 & -\infty & -\infty & -\infty & -\infty \\
0 & 5 & 10 & -\infty & -\infty & -\infty \\
0 & 5 & 10 & 15 & -\infty & -\infty \\
0 & 5 & 10 & 15 & 20 & -\infty \\
0 & 5 & 10 & 15 & 20 & 25
\end{pmatrix}
,
\begin{pmatrix}
0 \\
50 \\
100 \\
150 \\
200 \\
250
\end{pmatrix}
\right\}
$$

We are using $-\infty$ to encode infeasible combinations of state and action. Since we are maximizing total reward, it will never come out that one of these infeasible combinations could be optimal. (What would you do for a minimum cost problem?) Recall that the state space is $E = \{0, 1, 2, 3, 4, 5\}$, which is the set of all possible population levels. Permissible actions are limited by the fact that no more fish than are present at the beginning of the year may be harvested. Thus, the action sets are $A_i = \{0, \ldots, i\}$. The transition matrices, as defined in the original example, are as in the output cell below.

```
T0 = {{1, 0, 0, 0, 0, 0},
   {1 / 4, 0, 3 / 4, 0, 0, 0}, {0, 1 / 4, 0, 3 / 4, 0, 0},
   {0, 0, 1 / 4, 0, 3 / 4, 0}, {0, 0, 0, 1 / 4, 0, 3 / 4},
   {0, 0, 0, 0, 1 / 4, 3 / 4}};
T1 = {{0, 0, 0, 0, 0, 0}, {1, 0, 0, 0, 0, 0},
   {1 / 4, 0, 3 / 4, 0, 0, 0},
   {0, 1 / 4, 0, 3 / 4, 0, 0}, {0, 0, 1 / 4, 0, 3 / 4, 0},
   {0, 0, 0, 1 / 4, 0, 3 / 4}};
T2 = {{0, 0, 0, 0, 0, 0}, {0, 0, 0, 0, 0, 0},
   {1, 0, 0, 0, 0, 0}, {1 / 4, 0, 3 / 4, 0, 0, 0}, {0,
   1 / 4, 0, 3 / 4, 0, 0}, {0, 0, 1 / 4, 0, 3 / 4, 0}};
T3 = {{0, 0, 0, 0, 0, 0}, {0, 0, 0, 0, 0, 0},
   {0, 0, 0, 0, 0, 0}, {1, 0, 0, 0, 0, 0}, {1 / 4,
   0, 3 / 4, 0, 0, 0}, {0, 1 / 4, 0, 3 / 4, 0, 0}};
T4 = {{0, 0, 0, 0, 0, 0}, {0, 0, 0, 0, 0, 0},
   {0, 0, 0, 0, 0, 0}, {0, 0, 0, 0, 0, 0},
   {1, 0, 0, 0, 0, 0}, {1 / 4, 0, 3 / 4, 0, 0, 0}};
T5 = {{0, 0, 0, 0, 0, 0}, {0, 0, 0, 0, 0, 0},
   {0, 0, 0, 0, 0, 0}, {0, 0, 0, 0, 0, 0},
   {0, 0, 0, 0, 0, 0}, {1, 0, 0, 0, 0, 0}};
AllTa = {T0, T1, T2, T3, T4, T5};
{MatrixForm[T0], MatrixForm[T1], MatrixForm[T2],
 MatrixForm[T3], MatrixForm[T4], MatrixForm[T5]}
```

$$\left\{ \begin{pmatrix} 1 & 0 & 0 & 0 & 0 & 0 \\ \frac{1}{4} & 0 & \frac{3}{4} & 0 & 0 & 0 \\ 0 & \frac{1}{4} & 0 & \frac{3}{4} & 0 & 0 \\ 0 & 0 & \frac{1}{4} & 0 & \frac{3}{4} & 0 \\ 0 & 0 & 0 & \frac{1}{4} & 0 & \frac{3}{4} \\ 0 & 0 & 0 & 0 & \frac{1}{4} & \frac{3}{4} \end{pmatrix}, \begin{pmatrix} 0 & 0 & 0 & 0 & 0 & 0 \\ 1 & 0 & 0 & 0 & 0 & 0 \\ \frac{1}{4} & 0 & \frac{3}{4} & 0 & 0 & 0 \\ 0 & \frac{1}{4} & 0 & \frac{3}{4} & 0 & 0 \\ 0 & 0 & \frac{1}{4} & 0 & \frac{3}{4} & 0 \\ 0 & 0 & 0 & \frac{1}{4} & 0 & \frac{3}{4} \end{pmatrix}, \right.$$

$$\begin{pmatrix} 0 & 0 & 0 & 0 & 0 & 0 \\ 0 & 0 & 0 & 0 & 0 & 0 \\ 1 & 0 & 0 & 0 & 0 & 0 \\ \frac{1}{4} & 0 & \frac{3}{4} & 0 & 0 & 0 \\ 0 & \frac{1}{4} & 0 & \frac{3}{4} & 0 & 0 \\ 0 & 0 & \frac{1}{4} & 0 & \frac{3}{4} & 0 \end{pmatrix}, \begin{pmatrix} 0 & 0 & 0 & 0 & 0 & 0 \\ 0 & 0 & 0 & 0 & 0 & 0 \\ 0 & 0 & 0 & 0 & 0 & 0 \\ 1 & 0 & 0 & 0 & 0 & 0 \\ \frac{1}{4} & 0 & \frac{3}{4} & 0 & 0 & 0 \\ 0 & \frac{1}{4} & 0 & \frac{3}{4} & 0 & 0 \end{pmatrix},$$

$$\left. \begin{pmatrix} 0 & 0 & 0 & 0 & 0 & 0 \\ 0 & 0 & 0 & 0 & 0 & 0 \\ 0 & 0 & 0 & 0 & 0 & 0 \\ 0 & 0 & 0 & 0 & 0 & 0 \\ 1 & 0 & 0 & 0 & 0 & 0 \\ \frac{1}{4} & 0 & \frac{3}{4} & 0 & 0 & 0 \end{pmatrix}, \begin{pmatrix} 0 & 0 & 0 & 0 & 0 & 0 \\ 0 & 0 & 0 & 0 & 0 & 0 \\ 0 & 0 & 0 & 0 & 0 & 0 \\ 0 & 0 & 0 & 0 & 0 & 0 \\ 0 & 0 & 0 & 0 & 0 & 0 \\ 1 & 0 & 0 & 0 & 0 & 0 \end{pmatrix} \right\}$$

Here we are setting row i of transition matrix T_a to be completely zero to encode infeasible state-action combinations (i, a). The sum $\sum_j V_m(j) T(i, j; a)$ in (6) is then computable in *Mathematica* as the ith row of the matrix product $T_a \cdot V_m$, and the answer will be zero when action a is not permissble for state i. In this case, when $r(i, a)$ is added, the result of $-\infty$ will remind us that action a is not only suboptimal, but cannot be considered for that state.

The following command will allow us to use the dynamic programming equation (6) to solve the problem easily. It is contained in the KnoxOR`DynamicProgramming` package, but we also show its code. The input parameters are the list of transition matrices, called TransMats; the per-period reward function r in matrix form, called RewardMatrix; and the current value function V_m, called Val. You should compare to step 3(a) of the DP algorithm above; for each state i, we make a list for each action a of the quantities $r(i, a) + T_a \cdot V_m(i)$, suitably converted to *Mathematica* syntax.

```
Needs["KnoxOR`DynamicProgramming`"]
```

```
(** DPEquation[TransMats,RewardMatrix,Val] **)
```

```
DPEquation[TransMats_ , RewardMatrix_ , Val_] :=
  Table[RewardMatrix[[i, a]] +
    (TransMats[[a]].Val)[[i]],
   {i, 1, Length[TransMats[[1]]]},
   {a, 1, Length[TransMats]}]
```

To initialize the computation at time 3, we must call DPEquation with Val = *R* the terminal reward vector. The sublists in the output correspond to states 0, 1, ... , 5 respectively, and within each sublist the entries correspond to actions 0, 1, ... , 5.

```
time2list = DPEquation[AllTa, r, R]
```

$$\{\{0, -\infty, -\infty, -\infty, -\infty, -\infty\},$$
$$\{75, 5, -\infty, -\infty, -\infty, -\infty\}, \{125, 80, 10, -\infty, -\infty, -\infty\},$$
$$\{175, 130, 85, 15, -\infty, -\infty\},$$
$$\{225, 180, 135, 90, 20, -\infty\},$$
$$\{\frac{475}{2}, 230, 185, 140, 95, 25\}\}$$

The next command picks out the maximum elements in each sublist, and forms them into the time 2 optimal value function V_2 for the next step. The actions that achive the maximum values are 0 in each case.

```
V2 = Table[Max[time2list[[i]]],
     {i, 1, Length[AllTa[[1]]]}]
```

$$\{0, 75, 125, 175, 225, \frac{475}{2}\}$$

Thus the time 2 optimal action function is $u_2 = \{0, 0, 0, 0, 0, 0\}$. Here is the analogous computation for time 1.

```
time1list = DPEquation[AllTa, r, V2]
V1 = Table[Max[time1list[[i]]],
   {i, 1, Length[AllTa[[1]]]}]
```

$$\left\{ \{0, -\infty, -\infty, -\infty, -\infty, -\infty\}, \left\{ \frac{375}{4}, 5, -\infty, -\infty, -\infty, -\infty \right\}, \right.$$

$$\left\{ 150, \frac{395}{4}, 10, -\infty, -\infty, -\infty \right\},$$

$$\left\{ 200, 155, \frac{415}{4}, 15, -\infty, -\infty \right\},$$

$$\left\{ \frac{1775}{8}, 205, 160, \frac{435}{4}, 20, -\infty \right\},$$

$$\left. \left\{ \frac{1875}{8}, \frac{1815}{8}, 210, 165, \frac{455}{4}, 25 \right\} \right\}$$

$$\left\{ 0, \frac{375}{4}, 150, 200, \frac{1775}{8}, \frac{1875}{8} \right\}$$

Again, the time 1 optimal action function is $u_1 = \{0, 0, 0, 0, 0, 0\}$. Finally, here is the result for time 0.

```
time0list = DPEquation[AllTa, r, V1]
V0 = Table[Max[time0list[[i]]],
   {i, 1, Length[AllTa[[1]]]}]
```

$$\left\{ \{0, -\infty, -\infty, -\infty, -\infty, -\infty\}, \left\{ \frac{225}{2}, 5, -\infty, -\infty, -\infty, -\infty \right\}, \right.$$

$$\left\{ \frac{2775}{16}, \frac{235}{2}, 10, -\infty, -\infty, -\infty \right\},$$

$$\left\{ \frac{6525}{32}, \frac{2855}{16}, \frac{245}{2}, 15, -\infty, -\infty \right\},$$

$$\left\{ \frac{7225}{32}, \frac{6685}{32}, \frac{2935}{16}, \frac{255}{2}, 20, -\infty \right\},$$

$$\left. \left\{ \frac{925}{4}, \frac{7385}{32}, \frac{6845}{32}, \frac{3015}{16}, \frac{265}{2}, 25 \right\} \right\}$$

$$\left\{ 0, \frac{225}{2}, \frac{2775}{16}, \frac{6525}{32}, \frac{7225}{32}, \frac{925}{4} \right\}$$

Comparing the maxima in the second output to the lists in the first, we again see that $u_0 = (0, 0, 0, 0, 0, 0)$. Thus, for all times it is optimal not to harvest any units of fish for this choice of parameters. It must be the case that the relative size of the terminal reward R to the per period reward r was so great that there is a lot of incentive to wait and let the fish population grow. Do the activity below to follow up on this result. ∎

Activity 4 – Change the parameter R from 50 to 20 in Example 3 and recompute the optimal policy. Does it change? If not, try reducing the value of R gradually until you find a value for which it becomes optimal to harvest under some conditions.

Exercises 6.2

1. Working by hand (rather than using the DPEquation command), find the optimal policy for the rocket production problem, Example 2, if the cost function $r(x, a)$ is changed to

$$r(x, a) = \begin{cases} 16 + 8\,a & \text{if } x = 1 \text{ and } a > 0 \\ 0 & \text{otherwise} \end{cases}$$

2. (*Mathematica*) (a) Use the DPEquation command to confirm the computations in Example 2.
(b) Repeat the solution of Example 2 holding r as it is, but with values of the terminal cost function of (i) 50; (ii) 45; (iii) 40 if a successful rocket is not made.

3. (*Mathematica*) In Section 1 we introduced an example with two states and two actions in which the reward function was $r(A, 1) = 4$, $r(B, 1) = 3$, $r(A, 2) = 2$, $r(B, 2) = 5$ and the transition matrices were as below. For a finite horizon stochastic dynamic programming problem with time horizon $T = 6$ and terminal reward $R(A) = 3$, $R(B) = 5$, find the optimal policy.

$$T_1 = \begin{pmatrix} 1/3 & 2/3 \\ 3/4 & 1/4 \end{pmatrix} \text{ and } T_2 = \begin{pmatrix} 1/5 & 4/5 \\ 1/4 & 3/4 \end{pmatrix}$$

4. (*Mathematica*) A house has a simple thermostat that can be set at 1 to turn the furnace on, and 0 to turn it off. Potential changes in setting take effect every 10 minutes. If the thermostat is on 0, in a 10-minute period the room temperature will either stay the same or go down by a degree, with equal probability. If it is on 1, the room temperature will go up by a degree with certainty in the next 10-minute period. There is an energy cost of 1 cent for each 10-minute period during which the furnace is on. There are also discomfort costs of 1.5 cents per degree for each 10-minute period for each degree of room temperature difference between the current temperature and the ideal temperature of 68. Assume that room temperature must be kept at all times between 65 and 71, and that the thermostat must turn on when the temperature is 65, and must turn off when it is 71. How should the thermostat be programmed to operate? Formulate the problem as a Markov decision problem, write out the DP equation, and solve it using *Mathematica* for

a time horizon of 50 minutes, using a terminal cost that penalizes differences between final temperature and 68 as described above.

5. (*Mathematica*) In Exercise 1 of Section 6.1, suppose that the single period reward function is $r(i, a) = i - a$, and at the terminal time $T = 4$, a final reward $R(X_4) = X_4$ is received. Find the optimal policy.

6. In the fishery example of Example 3, assume that there are only two fishing seasons under study, and that the net benefit per unit of fish harvested is 5 and the net benefit per unit remaining at the end is 10. Find the optimal policy by hand, that is, without using the DPEquation command.

7. (*Mathematica*) Using the original problem parameters of Example 3, find the smallest time horizon T such that it is beneficial to harvest fish at some time prior to that horizon.

8. For the two-state, two-action Markov decision process with transition matrices and per period reward function as below, consider the finite horizon problem with time horizon $T = 4$ and terminal reward $R(1) = 2$, $R(2) = 1$. Find the optimal policy.

$$T_1 = \begin{pmatrix} 1/2 & 1/2 \\ 2/3 & 1/3 \end{pmatrix}, T_2 = \begin{pmatrix} 1/4 & 3/4 \\ 1/3 & 2/3 \end{pmatrix}$$

$$r(1, a) = \begin{cases} 5 & \text{if } a = 1 \\ 4 & \text{if } a = 2 \end{cases}, r(2, a) = \begin{cases} 2 & \text{if } a = 1 \\ 3 & \text{if } a = 2 \end{cases}$$

9. Let us presume that the dynamic programming equation (6) still holds when the state and action spaces are not finite, for the purposes of the following problem. An owner of a baseball team can spend any proportion $p \in [0, 1]$ of his currrent assets on free agents. He estimates that the team will come through and return him twice the amount that he spent with probability w, but the team will fail and he will lose what he spent with probability $l = 1 - w$. The owner plans to keep the team for T years before selling out. His goal is to maximize the expected value of the logarithm of his wealth when he sells the team.
   (a) Model this problem as a Markov decision problem, including a description of the state and action spaces, a formula for the transition probabilities $T(x, y; a)$, and the single period and terminal reward functions.
   (b) Write the dynamic programming equation for the problem.
   (c) If $T = 3$ and $w > 1/2$, show that the optimal action at each time 0, 1, and 2 is to bet a proportion $a = 2w - 1$ of the current wealth.

10. A person has \$4000 available initially for investment in two risky ventures A and B. Venture A will return nothing in a time period with probability 2/3, and will return \$3000 per thousand invested with probability 1/3. Venture B will return either \$1000 or \$2000 per thousand invested, each with probability 1/2. Investment amounts are in units of a thousand dollars, and the person may risk as much as \$2000 per time period. Find the strategy that maximizes the expected value of the square of the terminal wealth at the end of 2 time periods.

11. The following is a deterministic dynamic programming problem. A company is planning a marketing strategy for a new product. There are three phases of the plan: (1) an introductory low price; (2) a subsequent intensive advertising campaign in newspapers and magazines; and (3) a follow-up ad campaign on radio. A total of \$4 million, which can be spent in \$1 million blocks, is available. After each phase, it is possible for the product to have one of the following shares of the market:

$$5\% \qquad 10\% \qquad 15\% \qquad 20\% \qquad 25\%$$

In the initial phase, allotments of \$0–\$4 million result in these five percentages, respectively. The following table shows the changes in market share that will result between phases 1 and 2, and between phases 2 and 3, under the five possible investments.

Amount invested (\$millions)	Old share → new share
0	5 % → 5 %, 10 % → 5 %, 15 % → 5 %, 20 % → 10 %, 25 % → 15 %
1	5 % → 5 %, 10 % → 5 %, 15 % → 10 %, 20 % → 15 %, 25 % → 20 %
2	Old = new
3	5 % → 10 %, 10 % → 15 %, 15 % → 20 %, 20 % → 25 %, 25 % → 25 %
4	5 % → 15 %, 10 % → 20 %, 15 % → 25 %, 20 % → 25 %, 25 % → 25 %

Find the amount of money to be allocated in each phase in order to maximize the share of the market at the end of the plan (there is no single period reward r).

12. Solve Exercise 12 of Section 6.1 on immigration if the time horizon is $T = 4$ and the probability of population increase is $p = 1/2$.

6.3 The Discounted Reward Problem

Method of Successive Approximations

To this point we have studied only the problem of maximization (or minimization) over a finite time horizon. Now we examine the problem of maximizing the *infinite horizon discounted reward*:

> **DEFINITION 1.** Let $\alpha \in (0, 1)$. The *value function* of policy $\mathbf{u} = (u_0, u_1, u_2, \ldots)$ for the *infinite horizon discounted problem* with *discount factor* α is
>
> $$W(i, \mathbf{u}) = E_{\mathbf{u}}[\textstyle\sum_{n=0}^{\infty} \alpha^n r(X_n, u_n(X_n)) \mid X_0 = i]$$
>
> and the *optimal value function* for this problem is $W(i) = \max_{\mathbf{u}} W(i, \mathbf{u})$ (minimum for a minimum cost problem).

In this problem, control is exerted forever, but the present value of a reward of d absolute dollars earned at time n is only $\alpha^n d$. Since the state space is finite, the reward function r is bounded, and therefore the expected total discounted reward is also bounded (see Exercise 2).

> **Activity 1** – For the infinite horizon discounted problem there are infinitely many policies; but even if we restrict to only the class of stationary policies $\mathbf{u} = (u, u, u, \ldots)$, there are potentially a very large number of them. At most how many?

To motivate the dynamic programming equation that is the focus of our investigation, consider a policy $\mathbf{u} = (u_0, u_1, u_2, \ldots)$. In the following, we will denote by $\mathbf{u}^1$ the policy $(u_1, u_2, \ldots)$ obtained from $\mathbf{u}$ by truncating the first action function u_0. In the infinite series in $W(i, \mathbf{u})$, split the time zero reward away from the sum, factor out α from what remains, and then condition and un-condition on X_1. We obtain the following expression:

$W(i, \mathbf{u})$

$= E[r(i, u_0(i)) + \alpha \sum_{n=1}^{\infty} \alpha^{n-1} r(X_n, u_n(X_n)) \mid X_0 = i]$

$= E[r(i, u_0(i))$

$\qquad + \alpha E[\sum_{n=1}^{\infty} \alpha^{n-1} r(X_n, u_n(X_n)) \mid X_1, X_0 = i] \mid X_0 = i]$ $\qquad$ (1)

$= E[r(i, u_0(i)) + \alpha W(X_1, \mathbf{u}^1) \mid X_0 = i]$

$= r(i, u_0(i)) + \alpha \sum_{j \in E} W(j, \mathbf{u}^1) T(i, j; u_0(i))$

If $\mathbf{u}$ is a stationary policy with action function u, then the changes that result in formula (1) are that $u_0 = u$ and $\mathbf{u}^1 = \mathbf{u}$. In order for such a stationary policy to be optimal, it must choose the best action $a = u_0(i) = u(i)$ in the above equation, and its value $W(i, \mathbf{u})$ will equal the optimal value function $W(i)$ for all states i. These remarks should help to motivate the following theorem.

THEOREM 1. The optimal value function W satisfies the equation:

$$W(i) = \max_{a \in A_i} \{ r(i, a) + \alpha \sum_{j \in E} W(j) T(i, j; a) \}$$ $\qquad$ (2)

If $\mathbf{u}^*$ is defined as the stationary policy such that the action $u^*(i)$ taken at state i maximizes the right side of (2) for every i, then $\mathbf{u}^*$ is optimal.

Proof. We give a proof that assumes the existence of an optimal policy $\mathbf{v}^* = (v_0, v_1, v_2, ...)$, a fact that we will not prove. The existence can be shown (see Derman ([17], Lemma 3.5) by proving that $W(i, \mathbf{u})$ is a continuous function of $\mathbf{u}$, and that the set of admissible policies is compact.

Let $\mathbf{v}^* = (v_0, v_1, ...)$ be an optimal policy. Then $W(i, \mathbf{v}^*) = W(i)$ for all i. As in the theorem statement, let $u^*(i)$ be the maximizing action for state i. We consider a sequence of policies $\mathbf{u}^1, \mathbf{u}^2, \mathbf{u}^3$ defined by

$$\mathbf{u}^n = (u^*, u^*, ..., u^*, v_0, v_1, ...),$$

that is, $\mathbf{u}^n$ uses the actions $u^*(i)$ up through time $n - 1$, and follows the optimal policy thereafter. Notice that

$W(i, \mathbf{u}^n) =$
$\qquad E[\sum_{k=0}^{n-1} \alpha^k r(X_k, u^*(X_k)) + \sum_{k=n}^{\infty} \alpha^k r(X_k, v_{k-n}(X_k)) \mid X_0 = i]$

Thus, it is clear that $W(i, \mathbf{u}^n) \to W(i, \mathbf{u}^*)$ as $n \to \infty$ (since the two values differ only in the tail sum, which is bounded by α^n times a constant).

Consider $\mathbf{u}^1 = (u^*, v_0, v_1, ...)$. By (1), the choice of u^* and the optimality of $\mathbf{v}^*$, we have

$$W(i, \mathbf{u}^1) = r(i, u^*(i)) + \alpha \sum_{j \in E} W(j, \mathbf{v}^*) T(i, j; u^*(i))$$
$$\geq r(i, v_0(i)) + \alpha \sum_{j \in E} W(j, \mathbf{v}^*) T(i, j; v_0(i))$$
$$= W(i, \mathbf{v}^*) = W(i)$$

Thus, $\mathbf{u}^1$ is at least as good as $\mathbf{v}^*$. Now consider the policy $\mathbf{u}^2 = (u^*, u^*, v_0, v_1, ...)$. Iterating (1), that is, replacing $W(j, \mathbf{u}^1)$ in that formula by the immediate reward r plus α times the expected future reward gives

$$W(i, \mathbf{u}^2) = r(i, u^*(i)) + \alpha \sum_{j \in E} T(i, j; u^*(i))$$
$$\times [r(j, u^*(j)) + \alpha \sum_{k \in E} W(k, \mathbf{v}^*) T(j, k; u^*(j))]$$
$$\geq r(i, u^*(i)) + \alpha \sum_{j \in E} T(i, j; u^*(i))$$
$$[r(j, v_0(j)) + \alpha \sum_{k \in E} W(k, \mathbf{v}^*) T(j, k; v_0(j))]$$
$$= r(i, u^*(i)) + \alpha \sum_{j \in E} W(j) T(i, j; u^*(i))$$
$$\geq r(i, v_0(i)) + \alpha \sum_{j \in E} W(j) T(i, j; v_0(i))$$
$$= W(i, \mathbf{v}^*) = W(i)$$

Thus, the policy $\mathbf{u}^2$ is at least as good as $\mathbf{v}^*$.

One can obviously repeat the process to obtain

$$W(i, \mathbf{u}^n) \geq W(i)$$

Therefore, in the limit as $n \to \infty$, $W(i, \mathbf{u}^*) \geq W(i)$. Since W is the optimal value function, the reverse inequality is obvious; consequently, $\mathbf{u}^*$ is optimal and (2) follows from (1) and the choice of $u^*(i)$. ∎

Equation (2) is called the *dynamic programming equation* for the discounted problem. Intuitively, it says that if the initial state is i, then the optimal action maximizes the sum of the immediate reward plus the discount factor times the expected total reward under an optimal policy from time one onward. For the problem of cost minimization, the maximum is simply replaced by a minimum.

At first glance, the infinite horizon problem appears to be solved, since we have characterized the optimal policy. Unfortunately, in order to find the optimal actions in (2), we must know the optimal value function W. Unlike the finite horizon problem, there is no terminal time from which we can slowly step back until the optimal value function is reached. But there is a way of approximating W to any desired accuracy by a sequence of functions. We will describe this method next. In the next section a different way of finding the optimal policy is given, in which we begin with an arbitrary policy and successively improve it until, after finitely many steps, we reach optimality.

The next theorem gives us the so-called *method of successive approxima-tions*.

THEOREM 2. For each $j \in E$, let $w_0(j)$ be an arbitrary real number, i.e., let $\mathbf{w}_0$ be an arbitrary column vector. Define a sequence of vectors $\mathbf{w}_1, \mathbf{w}_2, \mathbf{w}_3, \cdots$ by

$$\mathbf{w}_{n+1}(i) = \max_{a \in A_i} \{r(i, a) + \alpha \, T_a \cdot \mathbf{w}_n(i)\} \tag{3}$$

Then the sequence $(\mathbf{w}_n)$ converges to the optimal value function W (again, replace max by min in the case of cost minimization).

Proof. We will use the functional notation w_n instead of the boldface vector notation. We would first like to establish the inequality

$$\max_{i \in E} |w_{n+1}(i) - W(i)| \leq \alpha(\max_{i \in E} |w_n(i) - W(i)|) \tag{4}$$

Let a^* be the maximizing action in (3) for state i. Since a^* may not be the maximizer in the DP equation (2), we have the inequality

$$
\begin{aligned}
w_{n+1}(i) - W(i) &\leq [r(i, a^*) + \alpha \sum_{j \in E} w_n(j) \, T(i, j; a^*)] \\
&\quad -[r(i, a^*) + \alpha \sum_{j \in E} W(j) \, T(i, j; a^*)] \\
&= \alpha \sum_{j \in E} (w_n(j) - W(j)) \, T(i, j; a^*) \\
&\leq \alpha(\max_{k \in E} \{|w_n(k) - W(k)|\})
\end{aligned} \tag{5}
$$

The last line occurs because $(w_n(j) - W(j))$ is no larger than the stated maximum, which is constant as far as j is concerned. The remaining sum $\sum T(i, j; a^*)$ is the sum of all the entries in the ith row of a transition matrix, which is 1.

By considering, instead of a^*, the maximizers $u^*(i)$ in the DP equation (2), one can show in a similar way (see Exercise 4) that

$$W(i) - w_{n+1}(i) \leq \alpha(\max_{k \in E} \{|w_n(k) - W(k)|\}) \tag{6}$$

Since both of the inequalities (5) and (6) are true for all $i \in E$, (4) is true.

Iterating (4), we obtain

$$\max_{i \in E} \{|w_{n+1}(i) - W(i)|\} \leq \alpha^{n+1} \max_{i \in E} \{|w_0(i) - W(i)|\} \tag{7}$$

Since the state space is finite, since the maximum on the right side is some non-negative real constant, and since $\alpha \in (0, 1)$, the right side of (7) forces the left side to zero, which means that $w_n(i) \to W(i)$ for each $i \in E$. ∎

REMARK. The convergence of the sequence (w_n) to W shows the uniqueness of the solution to the dynamic programming equation, which is something that we will use implicitly in the examples. For, if W_0 is another solution to (2), consider the sequence of functions generated in (3) by using W_0 as the initial function. Since W_0 is a solution of the DP equation, it is easy to see that $W_0 = w_1 = w_2 = \cdots$, hence the limit W of the sequence must equal W_0.

The two theorems together give us a rough procedure for finding the optimal policy. Beginning with an arbitrary function w_0 on the state space, we can generate a sequence of functions w_1, w_2, ... by (3), which approach the optimal value function W. At some n, we decide that we have a good enough approximation. This decision may be made on the basis of the stabilization of the functions w_n, or the stabilization of the optimal actions in (3). For the function w_n at which we stop, the optimal actions a_i are found for each state i from (3). One forms a stationary policy $\mathbf{u}$ from these actions and computes its value function $W(i, \mathbf{u})$. If this value satisfies the DP equation (2), then the policy is optimal. If not, then one can return to the method of successive approximations to find a closer approximator for the optimal value function, together with its corresponding optimal actions. To solve for the value function of $\mathbf{u}$, we use (1) to obtain a system of linear equations for the unknowns $x_i = W(i, \mathbf{u})$:

$$(I - \alpha\, T_{\mathbf{u}})\, \mathbf{x} = \mathbf{r_u} \tag{8}$$

where $T_{\mathbf{u}}(i, j) = T(i, j; u(i))$ and $\mathbf{r_u} = r(i, u(i))$. Note that this is just the system derived in Chapter 4, Section 5 for the long-run discounted cost (or reward).

The structure of the DP equation (3) for the method of successive approximations is identical to the DP equation for the finite horizon problem, with the single exception that the discount factor α is a coefficient of the second term. This means that the DPEquation command of the previous section can be modified to produce the next function w_{n+1} given the current function w_n, and also it can be used to check to see whether a current value function $W(i, \mathbf{u})$ satisfies the DP equation (2), which indicates that policy $\mathbf{u}$ is optimal. The command DiscountedDPEquation contained in the KnoxOR`-DynamicProgramming` package takes the list of transition matrices Trans-Mats, the reward function RewardMatrix, the current approximating value function Val (i.e., w_n), and the discount factor α, and returns the list (for each state i) of sublists (for each action a) on the right side of formulas (2) or (3), the optimum values of which form next value function approximator w_{n+1}.

```
Needs["KnoxOR`DynamicProgramming`"]
```

```
(*** DiscountedDPEquation[
     TransMats,RewardMatrix,Val,α] ***)
```

Activity 2 – Try without referring to Section 2 to write the code for the DiscountedDPEquation command. Compare your version to the code in the closed cell above this Activity.

The method for using DiscountedDPEquation is to start with an arbitrary initial function (vector) $\mathbf{w}_0$, iteratively compute a few of the next $\mathbf{w}_i$, noting whether the optimal actions a_i all remain the same from one time to the other. When they do, pause to compute the value function of the current policy $\mathbf{u}$ associated with the actions a_i. Form the transition matrix $T_\mathbf{u}$ and reward vector $\mathbf{r}_\mathbf{u}$ for this policy, where $T_\mathbf{u}(i, j) = T(i, j; u(i))$ and $\mathbf{r}_\mathbf{u} = r(i, u(i))$. Solve the linear system $(I - \alpha T_\mathbf{u}) \mathbf{x} = \mathbf{r}_\mathbf{u}$; the solution vector $\mathbf{x}$ is the value of the policy $W_\mathbf{u}$. Use $W_\mathbf{u}$ as the Val argument in Discounted-DPEquation command, and check whether the maxima are identical to the values of $W_\mathbf{u}$; and if so, $\mathbf{u}$ is an optimal policy. If not, resume the successive approximations until the optimal actions restabilize differently and check for optimality as in the previous sentence. Continue to do this until the optimal policy is found. (Make sure that you can explain why Theorems 1 and 2 justify this approach.) We will use the standard *Mathematica* command LinearSolve to do the necessary equation solving.

```
? LinearSolve
```

```
LinearSolve[m, b] finds an x which solves
   the matrix equation m.x==b. LinearSolve[m]
   generates a LinearSolveFunction[ ... ] which
   can be applied repeatedly to different b. More...
```

Examples

The application of the method of successive approximations is illustrated by the following example.

EXAMPLE 1. Consider a two-state, two-action problem in which the states are labeled 1, 2 and the actions are also labeled 1, 2. Let the two transition matrices for these actions be

$$T_1 = \begin{matrix} 1 \\ 2 \end{matrix} \begin{pmatrix} 1 & 2 \\ 1/2 & 1/2 \\ 2/3 & 1/3 \end{pmatrix}, \quad T_2 = \begin{matrix} 1 \\ 2 \end{matrix} \begin{pmatrix} 1 & 2 \\ 1/4 & 3/4 \\ 1/3 & 2/3 \end{pmatrix}$$

and suppose that the reward function is

$$r(1, a) = \begin{cases} 5 & \text{if } a = 1 \\ 4 & \text{if } a = 2 \end{cases}, \quad r(2, a) = \begin{cases} 2 & \text{if } a = 1 \\ 3 & \text{if } a = 2 \end{cases}$$

Let the discount factor α be .9. We will need the *Mathematica* definitions below.

```
matT1 = {{1 / 2, 1 / 2}, {2 / 3, 1 / 3}};
matT2 = {{1 / 4, 3 / 4}, {1 / 3, 2 / 3}};
AllTas = {matT1, matT2};
r = {{5, 4}, {2, 3}};
α = .9;
```

Start the sequence with $w_0(1) = w_0(2) = 0$. Then $\mathbf{w}_1$ is generated as follows:

```
w0 = {0, 0};
list1 = DiscountedDPEquation[AllTas, r, w0, α]
w1 = Table[Max[list1[[i]]],
    {i, 1, Length[AllTas[[1]]]}]
```

{{5, 4}, {2, 3}}

{5, 3}

The optimal actions are $a_1 = 1$, $a_2 = 2$. Let us compute $\mathbf{w}_2$ and $\mathbf{w}_3$.

```
list2 = DiscountedDPEquation[AllTas, r, w1, α]
w2 = Table[Max[list2[[i]]],
    {i, 1, Length[AllTas[[1]]]}]
```

{{8.6, 7.15}, {5.9, 6.3}}

{8.6, 6.3}

```
list3 = DiscountedDPEquation[AllTas, r, w2, α]
w3 = Table[Max[list3[[i]]],
   {i, 1, Length[AllTas[[1]]]}]
```

{{11.705, 10.1875}, {9.05, 9.36}}

{11.705, 9.36}

In each case the optimal actions are $a_1 = 1$, $a_2 = 2$. It is time to stop and check the stationary policy defined by $u(1) = 1$, $u(2) = 2$ for optimality.

When the system is in state 1, this policy takes action 1; and when the system is in state 2, action 2 is taken. Thus, the chain of states $X_0, X_1, X_2, \dots$ is a time-homogeneous Markov chain with transition matrix:

$$T = T_\mathbf{u} = \begin{matrix} 1 \\ 2 \end{matrix} \begin{pmatrix} \overset{1}{1/2} & \overset{2}{1/2} \\ 1/3 & 2/3 \end{pmatrix}$$

We have a reward function

$$\mathbf{r_u} = [\, r(1, 1) \quad r(2, 2)\,]' = [\, 5 \quad 3\,]'$$

The value of the policy is

```
Tu = {{1 / 2, 1 / 2}, {1 / 3, 2 / 3}};
ru = {5, 3};
Ident = {{1, 0}, {0, 1}};
Wu = LinearSolve[Ident - α * Tu, ru]
```

{39.4118, 37.0588}

Checking the DP equation (2),

```
ulist = DiscountedDPEquation[AllTas, r, Wu, α]
Table[Max[ulist[[i]]],
   {i, 1, Length[AllTas[[1]]]}]
```

{{39.4118, 37.8824}, {36.7647, 37.0588}}

{39.4118, 37.0588}

Since $W_{\mathbf{u}}$ matches the maximum values on the right side of (2), policy $\mathbf{u}$ is optimal. Though it has turned out here that at each state i one acts in order to receive the best immediate reward $r(i, a)$, it was not altogether obvious at the outset that this had to be the case. For instance, from state 2 we receive 3 monetary units if we take action 2; but if this action is taken, there is a relatively low probability that the chain next will go to the comparatively high reward state 1. ∎

> **Activity 3** – If you compare the results for $\mathbf{w}_3$ in Example 1 to the final result for $W_{\mathbf{u}}$, you see that $\mathbf{w}_3$ was not very close at all to the optimal value function. Nevertheless, we quickly located the optimal policy itself by interrupting the successive approximations and checking for optimality of the current policy. Try computing the next several $\mathbf{w}_n$ to see whether they tend slowly or quickly to $W_{\mathbf{u}}$. (See also Exercise 5.)

The following simple model of machine repair is an example in which we can actually solve the dynamic programming equation without resorting to successive approximations.

EXAMPLE 2. A machine can be in one of three conditions: like new, mildly deteriorated, or badly deteriorated. At each time, our options are to do nothing to the machine, to attempt a repair, or to replace the machine with another that is like new. To simplify matters, we will assume that we never interfere with a machine that is like new, and we must replace a badly deteriorated machine. There is a known repair cost and a known replacement cost. In addition, there are costs due to production of inferior items by the machine, when it is not in best possible condition. The transition probabilities under our various possible actions are known. Find the repair schedule that will minimize expected total discounted cost for a given discount factor α.

Let us set down some notation first. The state space and the associated costs for inferior production are

States	Cost (bad output)
0 (like new)	0
1 (deteriorated)	C_1
2 (badly deteriorated)	C_2

Writing 0 for the action of doing nothing, 1 for repairing, and 2 for replacing, we have that the action sets are

$$A_0 = \{0\}, \quad A_1 = \{0, 1, 2\}, \quad A_2 = \{2\}$$

Since only one action is allowed at each of states 0 and 2, the problem is simply to decide what to do when the machine is in the mildly deteriorated state. This simple structure will permit us to compute the optimal value function directly from the dynamic programming equation. Suppose that

$$C_3 = \text{cost of repairing a deteriorated machine}$$
$$C_4 = \text{cost of a new machine}$$

Therefore the cost $c(i, a)$ when the state is i and the action is a can be written

$$c(i, a) = \begin{cases} 0 & \text{if } i = 0 \\ C_1 & \text{if } i = 1, a = 0 \\ C_1 + C_3 & \text{if } i = 1, a = 1 \\ C_1 + C_4 & \text{if } i = 1, a = 2 \\ C_2 + C_4 & \text{if } i = 2 \end{cases}$$

Suppose that the transition matrices for the three actions are

$$T_0 = \begin{matrix} 0 \\ 1 \\ 2 \end{matrix}\begin{pmatrix} 3/4 & 1/4 & 0 \\ 0 & 7/8 & 1/8 \\ - & - & - \end{pmatrix}, \quad T_1 = \begin{matrix} 0 \\ 1 \\ 2 \end{matrix}\begin{pmatrix} - & - & - \\ 1/2 & 1/2 & 0 \\ - & - & - \end{pmatrix},$$

$$T_2 = \begin{matrix} 0 \\ 1 \\ 2 \end{matrix}\begin{pmatrix} - & - & - \\ 1 & 0 & 0 \\ 1 & 0 & 0 \end{pmatrix}$$

To say that $T_1(1, 0) = 1/2$, for example, says that if we choose to repair a mildly deteriorated machine, the chance is only 50% that the repair will be successful. To say that $T_0(0, 1) = 1/4$ means that a good machine will deteriorate with probability 1/4 if no maintenance is done.

Equation (2) can now be written for each of the states. Using *Mathematica*, we obtain the expressions inside the minimum.

```
T0 = {{3 / 4, 1 / 4, 0}, {0, 7 / 8, 1 / 8}, {0, 0, 0}};
T1 = {{0, 0, 0}, {1 / 2, 1 / 2, 0}, {0, 0, 0}};
T2 = {{0, 0, 0}, {1, 0, 0}, {1, 0, 0}};
c = {{0, ∞, ∞},
    {C1, C1 + C3, C1 + C4}, {∞, ∞, C2 + C4}};
W = {W0, W1, W2};
AllTas = {T0, T1, T2};
ulist = DiscountedDPEquation[AllTas, c, W, α]
```

$$\left\{\left\{\left(\frac{3\ W0}{4} + \frac{W1}{4}\right)\alpha,\ \infty,\ \infty\right\},\right.$$
$$\left\{C1 + \left(\frac{7\ W1}{8} + \frac{W2}{8}\right)\alpha,\ C1 + C3 + \left(\frac{W0}{2} + \frac{W1}{2}\right)\alpha,\right.$$
$$\left.C1 + C4 + W0\ \alpha\right\},\ \left\{\infty,\ \infty,\ C2 + C4 + W0\ \alpha\right\}\right\}$$

(Why did we define the cost function c in this way?) We have in full form the following equations:

$$W(0) = \alpha[(3/4)\,W(0) + (1/4)\,W(1)]$$
$$W(1) = \min\{C_1 + \alpha[(7/8)\,W(1) + (1/8)\,W(2)],$$
$$\qquad\qquad C_1 + C_3 + \alpha[(1/2)\,W(0) + (1/2)\,W(1)], \qquad\qquad (9)$$
$$\qquad\qquad C_1 + C_4 + \alpha\,W(0)\}$$
$$W(2) = C_2 + C_4 + \alpha\,W(0)$$

The first and third equations allow us to solve for $W(0)$ and $W(2)$ in terms of $W(1)$:

$$W(0) = [\alpha/(4 - 3\,\alpha)]\,W(1),$$
$$W(2) = C_2 + C_4 + [\alpha^2/(4 - 3\,\alpha)]\,W(1)$$

These may be substituted into equation (9) for $W(1)$ to give

$$W(1) = \min\{C_1 + \alpha(C_2 + C_4)/8 + (7\,\alpha/8 + \alpha^3/[8\,(4 - 3\,\alpha)])\,W(1),$$
$$\qquad C_1 + C_3 + (\alpha/2 + \alpha^2/[2\,(4 - 3\,\alpha)])\,W(1), \qquad\qquad (10)$$
$$\qquad C_1 + C_4 + [\alpha^2/(4 - 3\,\alpha)]\,W(1)\}$$

We now take some specific numbers in order to obtain a numerical solution. Let

$$\alpha = 0.9, \quad C_1 = 1, \quad C_2 = 2, \quad C_3 = 6, \quad C_4 = 10$$

The three expressions inside the minimum simplify as follows:

```
action0value = 1 + .9 (2 + 10) / 8 +
   (7 (.9) / 8 + (.9)³ / (8 (4 - 3 (.9)))) W1
action1value = 1 + 6 +
   (.9 / 2 + (.9)² / (2 (4 - 3 (.9)))) W1
action2value = 1 + 10 + ((.9)² / (4 - 3 (.9))) W1
```

2.35 + 0.857596 W1

7 + 0.761538 W1

11 + 0.623077 W1

Now $W(1)$ must equal one of these three expressions. This gives us three linear equations and three corresponding solutions that are candidates for the true $W(1)$. These turn out to be

```
{Solve[x == 2.35 + .857596 x, x],
  Solve[x == 7 + .761538 x, x],
  Solve[x == 11 + .623077 x, x]}
```

$\{\{\{x \to 16.5023\}\}, \{\{x \to 29.3548\}\}, \{\{x \to 29.1837\}\}\}$

If action 0 is optimal for state 1, then when $W(1) = 16.5023$ is substituted into the expression to be minimized in the DP equation, action 0 should produce the smallest number among the three. Upon substituting, we find that the three numbers are

```
{2.35 + 0.857596 (16.5023),
  7 + 0.761538 (16.5023), 11 + 0.623077 (16.5023)}
```

$\{16.5023, 19.5671, 21.2822\}$

Because 16.5023 agrees with the minimum of these which does occur for action 0, action 0 is indeed optimal, and the dynamic programming equation is satisfied if $W(1) = 16.5023$. (To see what goes wrong when a non-optimal action is picked, do the activity following this example.) This means that the optimal policy is to wait until it is not functioning at all and replace it then. For this choice of constants, the repair and replacement costs are apparently too large in comparison with the costs of inferior production to attempt any maintenance. To see how sensitive the optimal policy is to the costs, the reader can do Exercise 8, in which the cost structure is more favorable to maintenance. Exercise 9 looks at the sensitivity of the optimal policy to changes in the discount factor α. ∎

Activity 4 – In Example 2, one of our candidate values for $W(1)$ was 29.3501, corresponding to the case where action 1 is optimal at state 1. By substituting into the minimum expressions in the DP equation, check to see that this value cannot be the correct $W(1)$.

Exercises 6.3

1. Argue that for fixed i, the maximum in the optimal value function $W(i) = \max_{\mathbf{u}} W(i, \mathbf{u})$ among only all stationary policies must be assumed by some policy. Does your argument extend to the case where the supremum is taken over all admissible policies?

2. Show that if the reward function r of a Markov decision problem is bounded in absolute value by a constant c, then for any policy $\mathbf{u}$, the infinite horizon discounted value function of $\mathbf{u}$ with discount factor α is bounded in absolute value by $c/(1 - \alpha)$.

3. Write an expression similar to (1) relating the value of a policy $\mathbf{u} = (u_0, u_1, u_2, u_3, \ldots)$ to that of $\mathbf{u}^2 = (u_2, u_3, \ldots)$.

4. Prove inequality (6) as suggested in the proof of Theorem 2.

5. (*Mathematica*) For the two-state, two-action problem (Example 1), write a *Mathematica* program to compute the sequence of functions generated by the method of successive approximations, until a termination condition is achieved, which stops the computation after the successive approximating functions differ by no more than a desired tolerance. The program should output the number of iterations necessary to terminate and the final function w_n. Run the program for the initial function $w_0(1) = w_0(2) = 0$ and for the initial function $w_0(1) = 40$, $w_0(2) = 30$, and find the number of iterations necessary to make the successive approximations differ by no more than .01.

6. Using the same problem parameters as in Example 1 and the initial function $w_0(1) = w_0(2) = 0$, estimate analytically how large n must be so that w_n is within .1 of the optimal value function W. (Hint: Use (7) and Exercise 2.)

7. (*Mathematica*) Redo Example 1, changing $T_1(2, 1)$ to 7/8 and $r(1, 1)$ to 8.

8. Redo Example 2, changing the costs to $C_1 = 4$, $C_2 = 6$, $C_3 = 3$, $C_4 = 5$. Keep α set at .9.

9. (a) Redo Example 2, changing α to .5.

    (b) Redo Example 2, changing α to .95.

    (c) What happens to the solution $W(1)$ of the dynamic programming equation as $\alpha \longrightarrow 1$? Why is this result intuitively obvious?

10. (*Mathematica*) Let us expand the model of Example 2. Suppose now that the machine can be in "like new" condition (state 0), "badly deteriorated" condition (state 4), or one of three intermediate states of deterioration, labeled 1, 2, 3, in order of increasing severity. Again suppose that we do not act if the machine is in state 0, we must replace the machine if it is in state 4, and for the intermediate states we have the option of doing nothing, attempting a repair, or replacing the machine. Suppose that if we do nothing, the machine stays in its current state with probability 3/4 and reduces to the next lower state with probability 1/4 (except that if it is badly deteriorated, it stays in that state). If the repair option is chosen, the machine goes to the next higher level with probability 1/2, or stays the same with probability 1/2. Replacement always restores the state to "like new." Suppose that the costs of poor output for the five states are 0, 3, 6, 9, and 12; the repair cost is 4; and the replacement cost is 10. Let the discount factor α be .95. Find the optimal policy.

11. (a) Recall the advertising problem (Exercise 5 of Section 6.1). Considering the problem as an infinite horizon discounted reward problem with discount factor $\alpha = .9$, write the DP equation.

(b) Find the optimal value function and the optimal action at the state where the soft drink has the maximum 30% of the market, as a function of the problem parameters r and c.

(c) Let $r = 10$, $c = 2$, and find the optimal value function and the optimal policy for the advertising problem, without resorting to successive approximations.

12. (*Mathematica*) For the advertising problem (Exercise 5 of Section 6.1) viewed as an infinite horizon problem with discount factor $\alpha = .9$, use the parameters in Exercise 11(c) to find the value of the policy that never advertises. Use this value as the initial function in the method of successive approximations and compute w_1, w_2, and w_3.

13. (*Mathematica*) A reservoir holds 3 units of water. We will control the chain defined by X_n = # units of water in the reservoir at the beginning of month n, by deciding how much water to release from the reservoir at the beginning of the month. Each unit of water released produces a monetary benefit of 1 unit, due to the production of power and irrigation. But if the

reservoir is dry, there is a loss of 2 monetary units due to the need to purchase power from another source. During a month, rainfall produces either 0 units or 1 unit of inflow to the reservoir, each with probability 1/2. Any rainfall occurring when the reservoir is already full is simply lost. Use the method of successive approximations to find the optimal value function and optimal water release policy for the infinite horizon Markov decision problem with discount factor $\alpha = .95$.

6.4 Policy Improvement

Main Theorem and Policy Improvement Algorithm

Consider again the discounted reward problem with discount factor α, expressed by Definition 1 of the last section. Recall that there exists an optimal stationary policy. When $\mathbf{u} = (u)$ is a stationary policy, the chain of states (X_n) is Markov, with transition matrix defined by

$$T_{\mathbf{u}}(i, j) = P[X_{n+1} = j \mid X_n = i, U_n = u(i)] = T(i, j; u(i)) \tag{1}$$

As we saw in Chapter 4, there is a system of linear equations for the value $W(i, \mathbf{u}) = W_{\mathbf{u}}(i)$ of a stationary policy. Defining $r_{\mathbf{u}}(i) = r(i, u(i))$, this system can be written in matrix form as

$$(I - \alpha\, T_{\mathbf{u}})\, \mathbf{W_u} = \mathbf{r_u} \tag{2}$$

where I is the identity matrix of the appropriate size, and we view $W_{\mathbf{u}}$ and $r_{\mathbf{u}}$ as column vectors.

Since there are usually many policies, we need an efficient way of searching through policies to find an optimal one. In the last section, we discussed the method of successive approximations, which can be used to approximate the optimal value function W, from which the optimal policy can be found. But there are problems with this method. Nothing guarantees that the sequence (w_n) of approximating functions converges to W at some finite step n. For a given reward function, it is possible to find n large enough to ensure that w_n is very close to W, but the policy we find from w_n may not be optimal. As in Example 1 of Section 6.3, the value function of this policy must be computed, and checked with the dynamic programming equation below:

$$W(i) = \max_{a \in A_i} \{ r(i, a) + \alpha \sum_{j \in E} W(j)\, T(i, j\,; a) \} \tag{3}$$

If the policy is still not optimal, then our only recourse would be to continue to generate functions w_n by successive approximations, in hopes that later we may locate the optimal policy.

Thus, the method of successive approximations does not give us a perfectly satisfactory algorithm, and we now look for something else. The algorithm studied in this section is called the *policy improvement algorithm*. Instead of creating a sequence of better and better approximate value functions, we form a sequence of better and better policies. Below is the key result.

THEOREM 1. Let $\mathbf{u} = (u)$ be a stationary policy, whose value function is $W(i, \mathbf{u}) = W_{\mathbf{u}}(i)$. Define a new stationary policy $\mathbf{v} = (v)$, such that for each $i \in E$, $v(i)$ achieves the maximum in

$$\max_{a \in A_i} \{r(i, a) + \alpha \sum_{j \in E} T(i, j; a) W_{\mathbf{u}}(j)\} \qquad (4)$$

Then for each $i \in E$,

$$W_{\mathbf{v}}(i) \geq W_{\mathbf{u}}(i)$$

and if equality holds for all i, then $\mathbf{u}$ and $\mathbf{v}$ are optimal policies. (As always, the maxima are replaced by minima in the case of cost minimization problems.)

Proof. By choice of $v(i)$ and formula (2),

$$r(i, v(i)) + \alpha \sum_{j \in E} T(i, j; v(i)) W_{\mathbf{u}}(j) \geq$$
$$r(i, u(i)) + \alpha \sum_{j \in E} T(i, j; u(i)) W_{\mathbf{u}}(j) = W_{\mathbf{u}}(i)$$

The left side of this inequality is the value of the non-stationary policy $\mathbf{v}^1$ that uses $\mathbf{v}$ for one period, and uses $\mathbf{u}$ in every period thereafter. Thus, the inequality can be restated as

$$W(i, \mathbf{v}^1) \geq W(i, \mathbf{u}) \text{ for each } i \in E$$

For $n = 1, 2, 3, \ldots$, let the policy $\mathbf{v}^n$ use action function v up through time $n - 1$, and u thereafter. We have just shown that $\mathbf{v}^1$ is better than $\mathbf{u}$, and we will now prove inductively that $\mathbf{v}^n$ is better than $\mathbf{u}$ for all n. Assuming that this is true for a given $\mathbf{v}^n$, it suffices to prove that $\mathbf{v}^{n+1}$ is a better policy than $\mathbf{v}^n$ in order to complete the argument.

The policies $\mathbf{v}^n$ and $\mathbf{v}^{n+1}$ both use v through time n. Thereafter, $\mathbf{v}^{n+1}$ uses the non-stationary policy $\mathbf{v}^1$ introduced above, whereas $\mathbf{v}^n$ uses $\mathbf{u}$. Therefore,

$$W(i, \mathbf{v}^{n+1}) =$$
$$E_v[\sum_{k=0}^{n} \alpha^k r(X_k, v(X_k)) \mid X_0 = i] + E_v[\alpha^{n+1} W(X_{n+1}, \mathbf{v}^1) \mid X_0 = i]$$

and

$$W(i, \mathbf{v}^n) =$$
$$E_v[\textstyle\sum_{k=0}^{n} \alpha^k \, r(X_k, v(X_k)) \mid X_0 = i] + E_v[\alpha^{n+1} \, W(X_{n+1}, \mathbf{u}) \mid X_0 = i]$$

The difference between the values of the two policies is

$$W(i, \mathbf{v}^{n+1}) - W(i, \mathbf{v}^n) = \alpha^{n+1} \, E_v[W(X_{n+1}, \mathbf{v}^1) - W(X_{n+1}, \mathbf{u}) \mid X_0 = i]$$

Since $\mathbf{v}^1$ is better than $\mathbf{u}$, and expectation is a monotonic operator, this difference exceeds 0, as desired.

As in the proof of Theorem 2 of Section 6.3, $W(i, \mathbf{v}^n) \to W(i, \mathbf{v})$ as $n \to \infty$. Since each member of this sequence is bounded below by $W(i, \mathbf{u})$, we must have that $W(i, \mathbf{v}) \geq W(i, \mathbf{u})$, which establishes the first claim of Theorem 1.

Now suppose the value of the new policy $\mathbf{v}$ equals the value of the old policy $\mathbf{u}$. Then, by the choice of $\mathbf{v}$, we have that for each $i \in E$,

$$
\begin{aligned}
W_{\mathbf{v}}(i) &= r(i, v(i)) + \alpha \textstyle\sum_{j \in E} T(i, j; v(i)) \, W_{\mathbf{v}}(j) \\
&= r(i, v(i)) + \alpha \textstyle\sum_{j \in E} T(i, j; v(i)) \, W_{\mathbf{u}}(j) \\
&= \max_{a \in A_i} \{ r(i, a) + \alpha \textstyle\sum_{j \in E} T(i, j; a) \, W_{\mathbf{u}}(j) \} \\
&= \max_{a \in A_i} \{ r(i, a) + \alpha \textstyle\sum_{j \in E} T(i, j; a) \, W_{\mathbf{v}}(j) \}
\end{aligned}
\tag{5}
$$

Since $W_{\mathbf{v}}$ satisfies the dynamic programming equation, $W_{\mathbf{v}}$ must equal the optimal value function, i.e., $\mathbf{v}$ is optimal. Since the value of $\mathbf{u}$ was the same as the value of $\mathbf{v}$, the old policy $\mathbf{u}$ is also optimal. ∎

Thus, starting with a policy, we may compute its value by (2). Theorem 1 gives us a new policy that is at least as good. If there is a state i such that the new value for i is strictly better than the old, then find the value of the new policy, and improve once again using Theorem 1. Since there are finitely many stationary policies, we can only strictly improve the policy finitely many times. At some stage we will find a new policy that has the same value as the old. The theorem proves that this policy is optimal. This discussion motivates the following algorithm.

ALGORITHM. (Policy improvement for discounted Markov decision problem)

   (1) Pick an initial policy $\mathbf{u} = (u)$.

   (2) Repeat steps (3)–(6) until done:

      (3) Let $T_{\mathbf{u}}(i, j) = T(i, j; u(i))$ and let $\mathbf{r_u}(i) = r(i, u(i))$ for $i, j \in E$.

      (4) Find the solution $\mathbf{W}$ to $(I - \alpha\, T_{\mathbf{u}})\, \mathbf{W} = \mathbf{r_u}$.

      (5) For each $i \in E$, find the action $a_i = v(i) \in A_i$ to maximize (minimize for costs)

$$r(i, a) + \alpha \sum_{j \in E} T(i, j; a)\, \mathbf{W}(j)$$

      (6) If $v(i) = u(i)$ for each i, then $\mathbf{u}$ and $\mathbf{v}$ are optimal and the algorithm is done, otherwise let $\mathbf{u} = \mathbf{v}$ to set up the next pass through loop (3)–(6).

Activity 1 – What happens in the policy improvement algorithm if the initial policy happens to be optimal? Must there be just one optimal policy?

Step 5 in the algorithm suggests that once again it is possible to make use of a version of the DiscountedDPEquation command to carry out the algorithm. Actually, the KnoxOR`DynamicProgramming` package contains a streamlined function (see Exercise 5) that essentially does steps (3)–(5) all at once.

```
Needs["KnoxOR`DynamicProgramming`"]
```

```
(***  PolicyImprovementOneStep[
     TransMats,RewardMatrix,α,policy] ***)
```

The command called PolicyImprovementOneStep takes the list of transition matrices, the reward function in matrix form, the discount factor, and a current policy represented as a list $\{u(1), u(2), ...\}$ and outputs a list of sublists like DiscountedDPEquation from which the next policy can be obtained. It assumes that the state space is of the form $\{1, 2, ..., n\}$ and the action space is of the form $\{1, 2, ..., a\}$. We illustrate its use in the next example.

Examples

EXAMPLE 1. To illustrate the policy improvement approach, consider a three-state, three-action problem with $\alpha = .9$, and transition matrices and reward matrix as defined below.

```
Clear[T1, T2, T3, AllTas, r, α];
α = 9 / 10;
T1 = {{3 / 10, 2 / 10, 5 / 10},
     {4 / 10, 1 / 10, 5 / 10}, {6 / 10, 2 / 10, 2 / 10}};
T2 = {{0, 5 / 10, 5 / 10}, {5 / 10, 0, 5 / 10},
     {5 / 10, 5 / 10, 0}};
T3 = {{0, 1, 0}, {0, 0, 1}, {1, 0, 0}};
AllTas = {T1, T2, T3};
{MatrixForm[T1], MatrixForm[T2], MatrixForm[T3]}
```

$$\left\{ \begin{pmatrix} \frac{3}{10} & \frac{1}{5} & \frac{1}{2} \\ \frac{2}{5} & \frac{1}{10} & \frac{1}{2} \\ \frac{3}{5} & \frac{1}{5} & \frac{1}{5} \end{pmatrix}, \begin{pmatrix} 0 & \frac{1}{2} & \frac{1}{2} \\ \frac{1}{2} & 0 & \frac{1}{2} \\ \frac{1}{2} & \frac{1}{2} & 0 \end{pmatrix}, \begin{pmatrix} 0 & 1 & 0 \\ 0 & 0 & 1 \\ 1 & 0 & 0 \end{pmatrix} \right\}$$

```
r = {{3, 3, 1}, {4, 5, 2}, {2, 3, 5}};
MatrixForm[r]
```

$$\begin{pmatrix} 3 & 3 & 1 \\ 4 & 5 & 2 \\ 2 & 3 & 5 \end{pmatrix}$$

Let us begin the policy improvement algorithm with the stationary policy $\mathbf{u} = (u)$ that takes action 1 at state 1, action 2 at state 2, and action 3 at state 3. The transition matrix and reward vector for this policy are

```
{MatrixForm[{{3 / 10, 2 / 10, 5 / 10}, {1 / 2, 0, 1 / 2},
     {1, 0, 0}}], MatrixForm[{3, 5, 5}]}
```

$$\left\{ \begin{pmatrix} \frac{3}{10} & \frac{1}{5} & \frac{1}{2} \\ \frac{1}{2} & 0 & \frac{1}{2} \\ 1 & 0 & 0 \end{pmatrix}, \begin{pmatrix} 3 \\ 5 \\ 5 \end{pmatrix} \right\}$$

(Make sure you see where these came from.) The command PolicyImprovementOneStep finds these, solves the linear equations (2) to find $W_{\mathbf{u}}$, and outputs a list whose i^{th} element is the sublist, one for each action, of the quantities in braces in the DP equation (4).

```
u = {1, 2, 3};
N[PolicyImprovementOneStep[AllTas, r, α, u]]
```

```
{{38.3109, 38.7686, 37.0053},
 {39.1584, 40.0058, 37.5319},
 {36.9953, 38.2425, 39.4798}}
```

The output shows us that the maximizing actions are now action 2 at state 1, action 2 at state 2, and action 3 at state 3. With this as our next policy, we repeat the computation:

```
u = {2, 2, 3};
N[PolicyImprovementOneStep[AllTas, r, α, u]]
```

```
{{40.1151, 40.4875, 38.6801},
 {40.991, 41.8668, 39.2949},
 {38.8583, 40.0595, 41.4388}}
```

We see that the maximizing actions are still 2, 2, and 3, respectively, at states 1, 2, and 3, and so the stationary policy with action function $u(1) = 2$, $u(2) = 2$, $u(3) = 3$ is optimal. Notice that the actions taken at each state maximize the immediate reward earned. Here the shortsighted policy turns out to be optimal.

The other important thing to notice is that we can easily check the result; examining the sublist maxima in the output above, we have $W_u(1) = 40.4875$, $W_u(2) = 41.8668$, $W_u(3) = 41.4388$. The computation below verifies that this W_u is the solution of the linear system (2), and that these values agree respectively with the maxima for states 1, 2, and 3 found from the sublists in the second part of the output generated by the Discounted-DPEquation command. This means that this particular W_u satisfies the dynamic programming equation (3), hence this **u** is optimal. ∎

```
Clear[Tu, ru, Ident, Wu];
Tu = {{0, 1 / 2, 1 / 2}, {1 / 2, 0, 1 / 2}, {1, 0, 0}};
ru = {3, 5, 5};
Ident = IdentityMatrix[3];
Wu = N[LinearSolve[Ident - α * Tu, ru]]
N[DiscountedDPEquation[AllTas, r, Wu, α]]
```

{40.4875, 41.8668, 41.4388}

{{40.1151, 40.4875, 38.6801},
 {40.991, 41.8668, 39.2949},
 {38.8583, 40.0595, 41.4388}}

Activity 2 – Redo the problem in Example 1 starting from the policy that takes action 3 at state 1, action 2 at state 2, and action 1 at state 3. Are more or fewer iterations required than were required above? What do you predict the answer would be before performing the calculation?

EXAMPLE 2. A small rental van operator must service returned vans before returning them to the pool. One van is returned in a day with probability $p = .6$, otherwise no vans are returned. The operator can either service all of the vans that might be waiting on a particular day, at a fixed cost of $c = \$300$ (since he has a contract with a local handyman), or service none of them, except that the contract says that the repairman will do no more than five at a time, so that when the fifth van comes the service must be done. Each van waiting for service on a particular day entails a cost for lost opportunity of $l = \$100$. Using a discount factor of .95, model the problem as an infinite horizon discounted Markov decision problem, and find the optimal servicing policy.

The states and actions are rather easy to recognize: let $X_n = $ # vans waiting for service at the end of day n, and let action U_n be the number of vans serviced at the end of day n. We will suppose that we know whether another van has arrived in a particular day before we make the decision to service all of them or none, and that the service is complete within a day, so that those vehicles are back in the pool and do not contribute to X_{n+1}. The state space is then $E = \{0, 1, 2, 3, 4, 5\}$, and by the conditions of the problem, the action space is $A = \{0, 1, 2, 3, 4, 5\}$ and the admissible action sets are $A_i = \{0, i\}$, $i = 0, ..., 4$ due to the "all or nothing" nature of the decision. The action set for state 5 is $A_5 = \{5\}$. The probabilistic dynamics of the system are summarized by the equation:

$$X_{n+1} = \begin{cases} X_n - U_n + 1 & \text{with probability } p = .6 \\ X_n - U_n & \text{with probability } 1 - p = .4 \end{cases} \qquad (6)$$

This indicates that on the next day, we have the vans waiting from the previous day, less those that were serviced, plus either one or zero, depending on whether a new van has come in. Admissibility also implies that U_n can only be either X_n or 0 for $X_n \in \{0, 1, 2, 3, 4\}$ and $U_n = X_n$ for $X_n = 5$. From these observations we can create the transition matrices T_a:

```
Clear[T0, T1, T2, T3, T4, T5, AllTas, α, c, Tu, cu];
T0 = {{.4, .6, 0, 0, 0, 0}, {0, .4, .6, 0, 0, 0},
      {0, 0, .4, .6, 0, 0}, {0, 0, 0, .4, .6, 0},
      {0, 0, 0, 0, .4, .6}, {0, 0, 0, 0, 0, 0}};
T1 = {{0, 0, 0, 0, 0, 0}, {.4, .6, 0, 0, 0, 0},
      {0, 0, 0, 0, 0, 0}, {0, 0, 0, 0, 0, 0},
      {0, 0, 0, 0, 0, 0}, {0, 0, 0, 0, 0, 0}};
T2 = {{0, 0, 0, 0, 0, 0}, {0, 0, 0, 0, 0, 0},
      {.4, .6, 0, 0, 0, 0}, {0, 0, 0, 0, 0, 0},
      {0, 0, 0, 0, 0, 0}, {0, 0, 0, 0, 0, 0}};
T3 = {{0, 0, 0, 0, 0, 0}, {0, 0, 0, 0, 0, 0},
      {0, 0, 0, 0, 0, 0}, {.4, .6, 0, 0, 0, 0},
      {0, 0, 0, 0, 0, 0}, {0, 0, 0, 0, 0, 0}};
T4 = {{0, 0, 0, 0, 0, 0}, {0, 0, 0, 0, 0, 0},
      {0, 0, 0, 0, 0, 0}, {0, 0, 0, 0, 0, 0},
      {.4, .6, 0, 0, 0, 0}, {0, 0, 0, 0, 0, 0}};
T5 = {{0, 0, 0, 0, 0, 0}, {0, 0, 0, 0, 0, 0},
      {0, 0, 0, 0, 0, 0}, {0, 0, 0, 0, 0, 0},
      {0, 0, 0, 0, 0, 0}, {.4, .6, 0, 0, 0, 0}};
AllTas = {T0, T1, T2, T3, T4, T5};
α = .95;
{MatrixForm[T0], MatrixForm[T1], MatrixForm[T2],
 MatrixForm[T3], MatrixForm[T4], MatrixForm[T5]}
```

$$\left\{ \begin{pmatrix} 0.4 & 0.6 & 0 & 0 & 0 & 0 \\ 0 & 0.4 & 0.6 & 0 & 0 & 0 \\ 0 & 0 & 0.4 & 0.6 & 0 & 0 \\ 0 & 0 & 0 & 0.4 & 0.6 & 0 \\ 0 & 0 & 0 & 0 & 0.4 & 0.6 \\ 0 & 0 & 0 & 0 & 0 & 0 \end{pmatrix}, \right.$$

$$\begin{pmatrix} 0 & 0 & 0 & 0 & 0 & 0 \\ 0.4 & 0.6 & 0 & 0 & 0 & 0 \\ 0 & 0 & 0 & 0 & 0 & 0 \\ 0 & 0 & 0 & 0 & 0 & 0 \\ 0 & 0 & 0 & 0 & 0 & 0 \\ 0 & 0 & 0 & 0 & 0 & 0 \end{pmatrix},$$

$$\begin{pmatrix} 0 & 0 & 0 & 0 & 0 & 0 \\ 0 & 0 & 0 & 0 & 0 & 0 \\ 0.4 & 0.6 & 0 & 0 & 0 & 0 \\ 0 & 0 & 0 & 0 & 0 & 0 \\ 0 & 0 & 0 & 0 & 0 & 0 \\ 0 & 0 & 0 & 0 & 0 & 0 \end{pmatrix}, \begin{pmatrix} 0 & 0 & 0 & 0 & 0 & 0 \\ 0 & 0 & 0 & 0 & 0 & 0 \\ 0 & 0 & 0 & 0 & 0 & 0 \\ 0.4 & 0.6 & 0 & 0 & 0 & 0 \\ 0 & 0 & 0 & 0 & 0 & 0 \\ 0 & 0 & 0 & 0 & 0 & 0 \end{pmatrix},$$

$$\begin{pmatrix} 0 & 0 & 0 & 0 & 0 & 0 \\ 0 & 0 & 0 & 0 & 0 & 0 \\ 0 & 0 & 0 & 0 & 0 & 0 \\ 0 & 0 & 0 & 0 & 0 & 0 \\ 0.4 & 0.6 & 0 & 0 & 0 & 0 \\ 0 & 0 & 0 & 0 & 0 & 0 \end{pmatrix}, \left. \begin{pmatrix} 0 & 0 & 0 & 0 & 0 & 0 \\ 0 & 0 & 0 & 0 & 0 & 0 \\ 0 & 0 & 0 & 0 & 0 & 0 \\ 0 & 0 & 0 & 0 & 0 & 0 \\ 0 & 0 & 0 & 0 & 0 & 0 \\ 0.4 & 0.6 & 0 & 0 & 0 & 0 \end{pmatrix} \right\}$$

The cost structure has two components: servicing cost and lost opportunity cost. The following per-period cost function captures the problem assumptions:

$$c(i, a) = \begin{cases} 300 & \text{if } a = i, a \neq 0 \\ 100\,i & \text{if } a = 0 \end{cases} \tag{7}$$

This means that the cost matrix is as follows:

```
c = {{0, ∞, ∞, ∞, ∞, ∞},
    {100, 300, ∞, ∞, ∞, ∞}, {200, ∞, 300, ∞, ∞, ∞},
    {300, ∞, ∞, 300, ∞, ∞}, {400, ∞, ∞, ∞, 300, ∞},
    {∞, ∞, ∞, ∞, ∞, 300}}; MatrixForm[c]
```

$$\begin{pmatrix} 0 & \infty & \infty & \infty & \infty & \infty \\ 100 & 300 & \infty & \infty & \infty & \infty \\ 200 & \infty & 300 & \infty & \infty & \infty \\ 300 & \infty & \infty & 300 & \infty & \infty \\ 400 & \infty & \infty & \infty & 300 & \infty \\ \infty & \infty & \infty & \infty & \infty & 300 \end{pmatrix}$$

The discount factor was given to be $\alpha = .95$, and we are interested in minimizing among stationary policies, for each initial state i, the policy value $W_{\mathbf{u}}(i) = E_u[\sum_{k=0}^{\infty} \alpha^k c(X_k, u(X_k)) \mid X_0 = i]$.

Let us try a policy of the "threshhold" type, that is, service all of the vans that are waiting if and only if the number waiting is at least some threshhold value m. If we choose $m = 3$ to start, then the action function determining the stationary policy is

$$u(0) = 0, \ u(1) = 0, \ u(2) = 0, \ u(3) = 3, \ u(4) = 4, \ u(5) = 5$$

Remember, though, that because *Mathematica* indexes lists beginning at 1, we should treat our states as $\{1, 2, 3, 4, 5, 6\}$ and our actions as $\{1, 2, 3, 4, 5, 6\}$ as well. So we can define the policy as below, and make a first attempt at policy improvement.

```
u = {1, 1, 1, 4, 5, 6};
N[PolicyImprovementOneStep[AllTas, c, α, u]]
```

{{2933.62, ∞, ∞, ∞, ∞, ∞},
 {3190.96, 3233.62, ∞, ∞, ∞, ∞},
 {3295.42, ∞, 3233.62, ∞, ∞, ∞},
 {3371.94, ∞, ∞, 3233.62, ∞, ∞},
 {3471.94, ∞, ∞, ∞, 3233.62, ∞},
 {∞, ∞, ∞, ∞, ∞, 3233.62}}

We seem to have come close to the optimal policy. It is only at state $i = 2$ that the value $W_u(2) = 3295.42$ is not equal to the minimum element of the list for $i = 2$, $\{3295.42, \infty, 3233.62, \infty, \infty, \infty\}$. The optimal action there is $a = 2$, so we switch to the threshhold policy $\mathbf{v} = (v)$ that services the vans when 2 or more are there. Explicitly, we have $v(0) = 0$, $v(1) = 0$, $v(2) = 2$, $v(3) = 3$, $v(4) = 4$, $v(5) = 5$. We can reuse earlier commands in edited form to recalculate the policy value W_v and the DP equation lists.

```
v = {1, 1, 3, 4, 5, 6};
N[PolicyImprovementOneStep[AllTas, c, α, v]]
```

```
{{2596.13, ∞, ∞, ∞, ∞, ∞},
 {2823.87, 2896.13, ∞, ∞, ∞, ∞},
 {2951.33, ∞, 2896.13, ∞, ∞, ∞},
 {3051.33, ∞, ∞, 2896.13, ∞, ∞},
 {3151.33, ∞, ∞, ∞, 2896.13, ∞},
 {∞, ∞, ∞, ∞, ∞, 2896.13}}
```

Since the optimal actions in the last output above are again 0, 0, 2, 3, 4, 5, policy v is optimal by Theorem 1. ∎

Activity 3 – Write out the dynamic programming equation for the problem in Example 2. Notice that the optimal value function takes the same value at states 2, 3, 4, and 5. Why intuitively do you think that happens?

REMARK. It is very interesting to note that there is yet another attack on the infinite horizon discounted Markov decision problem, which uses linear programming. It turns out that the optimal value function is the smallest function $V = V(i)$ satisfying:

$$V(i) \geq \max_{a \in A_i} \{r(i, a) + \alpha \sum_{j \in E} T(i, j; a) V(j)\}$$

for each $i \in E$. Therefore one can find it by solving the linear programming problem:

$$
\begin{aligned}
&\text{minimize: } \sum_{j \in E} V(i) \\
&\text{subject to: } V(i) \geq r(i, a) + \alpha \sum_{j \in E} T(i, j; a) V(j) \qquad (8) \\
&\qquad \text{for all } a \in A_i \text{ and all } i \in E
\end{aligned}
$$

For more information on linear programming approaches to dynamic programming problems, the reader may refer to Ross [53] or Derman [17]. The reader is asked to show the above result in Exercise 12. We will see this idea again when we study optimal stopping problems in the next section.

Exercises 6.4

1. (*Mathematica*) Solve Example 1 of Section 6.3 by policy improvement, starting with the policy that takes action 1 at both states.

2. (*Mathematica*) Redo Example 2 on van servicing with problem parameters: (a) $c = 400$, $l = 300$; (b) $c = 500$, $l = 200$.

3. (*Mathematica*) Solve the advertising problem (Exercise 5 of Section 6.1) viewed as an infinite horizon problem with $\alpha = .9$, $r = 10$, and $c = 2$ by policy improvement, starting with the policy that never advertises.

4. (*Mathematica*) Solve the reservoir problem (Exercise 13 of Section 6.3) by policy improvement, starting with the policy that releases all water present.

5. (*Mathematica*) Write your own version of the command PolicyImprovementOneStep as described in the section.

6. (*Mathematica*) We have a machine that is in one of five possible conditions at each time. State 1 is the best condition, etc. down to state 5, which is the worst condition. We can replace a machine with one that is in condition 1 at any time. If we do not replace, the machine in operation will stay in its current condition with probability 3/4, or go to the next worst condition with probability 1/4 at the next instant of time. If the machine is currently in the worst condition 5, it will stay there until it is replaced. There are costs for inferior production in each period, dependent on the machine's condition:

$$C_1 = 0, \quad C_2 = 2, \quad C_3 = 4, \quad C_4 = 6, \quad C_5 = 8.$$

A new machine costs a constant $C_6 = 12$ monetary units. What replacement strategy should be adopted in order to minimize expected total discounted cost, with a discount factor $\alpha = 9/10$? For your solution, use policy improvement with the initial policy of replacing the machine whatever is its condition.

7. (*Mathematica*) Redo the machine replacement example Exercise 6 by policy improvement, beginning with the initial policy of replacing the current machine if it is in condition 4 or worse.

8. Consider the machine replacement example Exercise 6, in which all costs C_i, $i = 1, ..., 6$ are replaced by $b \cdot C_i$, where b is a positive constant. Show that the policy that replaces the current machine if its condition is 3 or worse is still optimal for this new problem.

9. Create an infinite horizon problem in which the optimal policy is not unique.

10. Consider again Example 3 of Section 6.2 on fishery planning, but in an infinite horizon context with discount factor $\alpha = .99$. Assume the same per-period reward function r and transition matrices T_a are in effect. Solve

for the optimal harvesting policy starting with the initial policy of harvesting nothing when the population is 0 or 1, and 1 unit otherwise.

11. Does the optimal policy in the fishery example (see Exercise 10) change if the reward function $r(i, a)$ is changed to $10\,a$? If not, can you provide an intuitive reason for it?

12. Consider an infinite horizon Markov decision problem with the usual notation, and let V be a function on the state space satisfying the linear programming problem (8). Adjoin to E an absorbing state Δ of no reward, and adjoin to the action space an action $a(\Delta)$ such that $r(i, a(\Delta)) = V(i)$ and $T(i, \Delta; a(\Delta)) = 1$. In other words, under action $a(\Delta)$ the chain proceeds immediately to absorbing state Δ, entailing a reward $V(i)$ for that move but no reward thereafter.

(a) For the new problem, use (8) to show that the optimal policy is to take action $a(\Delta)$ immediately.

(b) Deduce from (a) that V exceeds the optimal value function W. Conclude that W satisfies the LP problem (8).

13. Use the linear programming formulation (8) in the remark at the end of the section to solve for the optimal value function in Example 1. Once the optimal value function is in hand, discuss how you would obtain the optimal policy from it.

6.5 Optimal Stopping of a Markov Chain

Dynamic Programming Approach

The problem that we will discuss in this section shows beautifully the unity of the four areas of Operations Research that have been studied in this book: graph theory, linear programming, stochastic processes, and dynamic programming. To illustrate the problem, consider the six-state Markov chain whose transition diagram is in Figure 6.4. Graph-theoretic considerations show us easily that states 1–5 are transient, since there is a path from each of these states to the absorbing state 6.

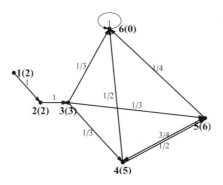

Figure 6.4 – A Markov chain with rewards

Suppose that at any time we can stop the chain and collect a reward dependent on the state occupied at that time. These potential rewards are shown in parentheses in the diagram beside the state numbers. At what time should we collect the reward in order to maximize the expected payoff? We would like to wait long enough to hit one of the large payoffs (5 and 6 monetary units at states 4 and 5, respectively), but to wait carries a risk that the chain will be absorbed by state 6, which yields no reward.

The optimal time to stop and collect the reward will turn out to be the time of first entry to a set of states characterized by the optimal value function of a Markov decision problem. Moreover, we will be able to solve for that value function using linear programming techniques. In this way, the optimal stopping problem ties together the apparently distantly related topics of this book.

The Markov chain of Figure 6.4 has enough special structure that we can use our intuition to find the optimal stopping time. First, examine states 1 and 2. Beginning from these states, state 3 will be reached with certainty. Since state 3 has a higher reward than states 1 and 2, it is clear that we should not stop at states 1 and 2. The decision is also clear at states 5 and 6. State 5 possesses the best possible reward, hence we should stop at state 5. There is no decision to be made at the absorbing state 6; we must stop and collect a reward of 0.

The interesting behavior is at states 3 and 4. Consider state 4 first. If the chain is at state 4, we can either stop immediately, for a certain reward of 5, or permit the chain to make one more jump. In the latter case, the chain either moves to state 5 to produce a reward of 6, with probability 1/2, or the chain moves to state 6 for no reward, with probability 1/2. The expected reward if we do not stop at state 4 is therefore

$$(1/2)\,V(5) + (1/2)\,V(6) = (1/2)\,6 + (1/2)\,0 = 3,$$

where we have written $V(i)$ for the value starting at state i under the optimal policy. Since this expected reward is smaller than the immediate reward of 5 units that could be earned by stopping, it is optimal to stop at state 4. Hence, $V(4) = 5$.

Knowing the optimal value starting at states 4, 5, and 6, we can now decide what to do when the chain is at state 3. If we stop immediately, we can collect a reward of 3. Let us compare this to what we expect to get using an optimal policy, if we do not stop at state 3. Since the chain will jump to state 4, state 5, or state 6 with equal probability, the expected earnings under the policy of not stopping at state 3 are

$$(1/3)\,V(4) + (1/3)\,V(5) + (1/3)\,V(6).$$
$$= (1/3)\,5 + (1/3)\,6 + 0 = 11/3$$

This exceeds the immediate reward at state 3, thus it is optimal not to stop at state 3. Common sense has led us to the optimal stopping policy, which is that we should stop at the first time T^* such that the chain is in the set:

$$A^* = \{4,\ 5,\ 6\}.$$

The main thing that you should notice about this example is that decisions are made on the basis of finding the larger of the immediate reward $f(i)$ and the expected (optimal) value of waiting one more time unit:

$$\sum_{j \in E} T(i,\ j) \cdot V(j).$$

Viewed as a column vector, the optimal value function therefore satisfies

$$V(i) = \max\{f(i),\ T \cdot \mathbf{V}(i)\}. \tag{1}$$

We stop at states i such that this maximum is $f(i)$, i.e., the optimal stopping time is the time of first entry to the set

$$A^* = \{i \in E :\ V(i) = f(i)\}. \tag{2}$$

Our advantage in the simple example above was that we had states i for which the value $V(i)$ was "obvious," and we could use these states to compute V for the less obvious states.

Activity 1 – In the example above, at least how large would the reward at state 3 have to be so that it is optimal to stop there? By how much can the reward at state 4 be reduced without changing the optimal stopping policy?

Equation (1) looks like a dynamic programming equation, and indeed we can obtain it by formulating the optimal stopping problem as a problem of Markov decision theory. Let X_0, X_1, X_2, ... be the chain of states. Suppose that the state space E is finite and contains an absorbing state Δ of no reward. For each state i, let the action space be $A = A_i = \{0, 1\}$, where action 1 is to stop and collect the reward, and action 0 is to continue. A stationary policy for the problem is determined by a function $u : E \rightarrow A$. If $u(i) = 1$, then we stop at state i; and if $u(i) = 0$, then we let the chain continue. There are rewards $f(i)$ given for each state $i \in E$, that determine the reward function for the Markov decision process:

$$r(i, a) = \begin{cases} 0 & \text{if } a = 0 \text{ or } i = \Delta \\ f(i) & \text{if } a = 1 \text{ and } i \neq \Delta \end{cases}$$

To construct the transition matrices, note that if we choose action 0, then the probability law of the next state is the same as that of the uncontrolled Markov chain. Thus,

$$T_0(i, j) = T(i, j) , \quad \forall \ i, j \in E$$

If we choose action 1, then no more rewards are to be earned. A convenient way of reflecting this fact is to assume that under action 1, the next state is certain to be Δ, i.e.:

$$T_1(i, j) = \begin{cases} 1 & \text{if } j = \Delta \\ 0 & \text{otherwise} \end{cases}$$

for all $i \in E$. The optimal stopping problem is to find a stationary policy $\mathbf{u} = (u)$, where u is a function from the state space E to the set $A = \{0, 1\}$ of actions, to achieve the maximum value in the following expression, for all $i \in E$:

$$V(i) = \max_u V(i, u) = \max_u E_u[\textstyle\sum_{n=0}^{\infty} r(X_n, u(X_n)) \mid X_0 = i]]] \tag{3}$$

States i for which $u(i) = 1$ are those at which we stop the chain. This model defines an infinite horizon undiscounted Markov decision problem. The value $V(i, u)$ of any stationary policy $\mathbf{u} = (u)$ is still finite because only one term in the infinite series will actually be non-zero.

Let us proceed, at least formally, as if the results of Section 3 still apply, with $\alpha = 1$. If this is so, then the DP equation (Formula (2) of Section 3) becomes:

$$V(i) = \max_{a \in \{0,1\}} \{r(i, a) + \alpha \sum_{j \in E} V(j) T(i, j; a)\}$$

$$= \max \{r(i, 0) + \sum_{j \in E} V(j) T(i, j; 0),$$

$$r(i, 1) + \sum_{j \in E} V(j) T(i, j; 1)\} \qquad (4)$$

$$= \max \{0 + \sum_{j \in E} V(j) T(i, j), \quad f(i) + V(\Delta) \cdot 1\}$$

$$= \max \{T \cdot \mathbf{V}(i), f(i)\}$$

for all $i \neq \Delta$. Clearly $V(\Delta) = 0$. Thus, formula (1) arises as a result of a Markov decision process formulation, granting the truth of the dynamic programming equation in this new setting.

EXAMPLE 1. Let us see how expression (1) might be used directly to solve an optimal stopping problem. Suppose that you have a piece of real estate up for sale. Each day, an offer of some amount comes in. Assume that there are n possible offers, $f_1 \leq f_2 \leq \ldots \leq f_n$, and independent of the previous history of offers, you receive an offer of level f_i with probability p_i. But, there is a positive probability p_0 that another offer will never come. Find a strategy that tells you when to accept an offer, so as to maximize the expected value of the offer accepted.

To model this as an optimal stopping problem, let X_k be the value of the offer received on day k. The state space is $E = \{0, f_1, f_2, \ldots, f_n\}$, where we have included 0 to represent the state of not receiving an offer. By the stated conditions, (X_k) is almost a completely memoryless Markov chain, with the exception that once state 0 is reached, the chain stays there forever. The transition matrix is

$$T = \begin{array}{c} \\ 0 \\ f_1 \\ f_2 \\ \vdots \\ f_n \end{array} \begin{array}{cccccc} 0 & f_1 & f_2 & f_3 & \cdots & f_n \\ \left(\begin{array}{cccccc} 1 & 0 & 0 & 0 & \cdots & 0 \\ p_0 & p_1 & p_2 & p_3 & \cdots & p_n \\ p_0 & p_1 & p_2 & p_3 & \cdots & p_n \\ \vdots & \vdots & \vdots & \vdots & \vdots & \vdots \\ p_0 & p_1 & p_2 & p_3 & \cdots & p_n \end{array}\right) \end{array}$$

Since $V(0)$ is clearly 0, to characterize the optimal stopping policy we must solve for n variables $y_i = V(i)$, $i = 1, 2, \ldots, n$. The structure of the transition matrix gives a very special form to equation (1):

$$V(i) = \max \{f(i), T \cdot \mathbf{V}(i)\}$$

$$\implies y_i = \max \{f_i, \sum_{j=1}^{n} p_j y_j = \mathbf{p} \cdot \mathbf{y}\}$$

So we must have both of the following:

$$y_i \geq \mathbf{p} \cdot \mathbf{y}, \quad y_i \geq f_i \qquad i = 1, 2, \ldots, n$$

Each component y_i of the optimal solution vector $\mathbf{y}$ exceeds the corresponding f_i, and exceeds the constant $\mathbf{p} \cdot \mathbf{y}$. One of these inequalities must be an equality. Since the offers f_i increase as i increases, there must be some cutoff level i^*, such that the first set of constraints $y_i \geq \mathbf{p} \cdot \mathbf{y}$ are binding for $i < i^*$, and the second set $y_i \geq f_i$ are binding for $i \geq i^*$. For $i \geq i^*$, the optimal value function agrees with the reward function; thus the optimal policy accepts offers of f_{i^*} or higher. Hence the value function satisfies

$$y_i = V(i) = \begin{cases} \mathbf{p} \cdot \mathbf{y} & \text{for } i < i^* \\ f_i & \text{for } i \geq i^* \end{cases} \tag{5}$$

As a numerical example, suppose that the possible offers are 10, 20, 30, 40, and 50 thousand dollars, occurring with probabilities 1/16, 1/8, 1/8, 1/2, and 1/8, respectively. Thus, the probability of not receiving an offer is $p_0 = 1/16$. Since there are few possibilities here, we can attempt by trial and error to arrive at the offer that maximizes expected returns. Let us start by calculating the value of the policy that accepts offer 2 (20,000) or more. For $i \geq 2$, $y_i = f_i = 10000\,i$. Clearly $y_0 = 0$. For y_1, we can solve the linear equation $y_1 = \mathbf{p} \cdot \mathbf{y}$ in the first part of (5), where $\mathbf{p}$ is the vector of probabilities and $\mathbf{y}$ is the vector of y_i's. Then, initializing variables and solving as below we find $y_1 = 104,000/3 \approx 34666.7$.

```
f = {f0, f1, f2, f3, f4, f5} =
    {0, 10000, 20000, 30000, 40000, 50000};
y = {y0, y1, y2, y3, y4, y5} =
    {0, y1, f2, f3, f4, f5};
p = {p0, p1, p2, p3, p4, p5} =
    {1 / 16, 1 / 16, 1 / 8, 1 / 8, 1 / 2, 1 / 8};
T = {{1, 0, 0, 0, 0, 0}, p, p, p, p, p};
Solve[y1 == p.y, y1]
```

$$\left\{ \left\{ y1 \to \frac{104000}{3} \right\} \right\}$$

```
y1 = 104000 / 3;
{{f0, (T.y)[[1]]}, {f1, (T.y)[[2]]},
 {f2, (T.y)[[3]]}, {f3, (T.y)[[4]]},
 {f4, (T.y)[[5]]}, {f5, (T.y)[[6]]}}
```

$$\left\{\{0, 0\}, \left\{10000, \frac{104000}{3}\right\},\right.$$
$$\left\{20000, \frac{104000}{3}\right\}, \left\{30000, \frac{104000}{3}\right\},$$
$$\left.\left\{40000, \frac{104000}{3}\right\}, \left\{50000, \frac{104000}{3}\right\}\right\}$$

The maxima are $\{0, \frac{104000}{3}, \frac{104000}{3}, \frac{104000}{3}, 40000, 50000\}$, whereas the value of our current policy is $\{0, \frac{104000}{3}, 20000, 30000, 40000, 50000\}$. This mismatch indicates that the current policy is not optimal. Similarly we can find that the policy of accepting offers 3 (30,000) or more is unsuitable (see Activity 2 below).

Now we evaluate a policy that accepts an offer of 40,000 or more. Then $y_0 = 0$, $y_4 = 40000$, $y_5 = 50000$, and y_1, y_2, y_3 can be solved for using the equations $y_1 = y_2 = y_3 = \mathbf{p} \cdot \mathbf{V}$ from formula (5). We obtain $y_1 = y_2 = y_3 = 420000/11 \approx 38181.81$, as below.

```
Clear[y, y0, y1, y2, y3, y4, y5];
y =
   {y0, y1, y2, y3, y4, y5} = {0, y1, y2, y3, f4, f5};
Solve[{y1 == p.y, y2 == p.y, y3 == p.y}, {y1, y2, y3}]
```

$$\left\{\left\{y1 \to \frac{420000}{11}, y2 \to \frac{420000}{11}, y3 \to \frac{420000}{11}\right\}\right\}$$

Again we compute for each state the two expressions inside the maximum in (1):

```
y1 = 420000 / 11; y2 = 420000 / 11; y3 = 420000 / 11;
{{f0, (T.y)[[1]]}, {f1, (T.y)[[2]]},
 {f2, (T.y)[[3]]}, {f3, (T.y)[[4]]},
 {f4, (T.y)[[5]]}, {f5, (T.y)[[6]]}}
```

$$\left\{ \{0, 0\}, \left\{10000, \frac{420000}{11}\right\}, \right.$$

$$\left\{20000, \frac{420000}{11}\right\}, \left\{30000, \frac{420000}{11}\right\},$$

$$\left. \left\{40000, \frac{420000}{11}\right\}, \left\{50000, \frac{420000}{11}\right\}\right\}$$

The maxima are $\{0, 420000/11, 420000/11, 420000/11, 40000, 50000\}$ which agree with the values of the current policy. Therefore the policy of accepting an offer of 40000 or more is optimal. ∎

Activity 2 – Check in Example 1 that the policy that accepts an offer of 30,000 or more is not optimal.

Linear Programming Approach

We cannot count on special structure similar to Example 1 all the time. We now search for a computationally useful characterization of the optimal value function and the optimal policy. The characterization of the optimal stopping policy has already been suggested earlier in the section, and it is repeated in Theorem 1 below. To find the policy explicitly requires the calculation of the value function. A method for doing this will arise from Theorem 2. Regrettably, the proofs of these two theorems would take us rather far afield. They require a careful definition of the notion of *stopping time*, and a study of general properties of stopping times, as well as functions on Markov chains called *excessive functions* (see the Remark below Theorem 2). Consequently, we omit the proofs. For a thorough development, see Cinlar ([15], Section 7.3).

THEOREM 1. Let A^* be the set:

$$A^* = \{i \in E : V(i) = f(i)\} \tag{6}$$

where V is the optimal value function. Define a stationary policy by $u^*(i) = 1$ iff $i \in A^*$. Then u^* is optimal. In other words, the optimal time to stop is the time at which the chain (X_n) first visits A^*. ∎

Theorem 1 reduces the problem to that of computing the optimal value function V. Since f is known, it is easy to find those states i at which V and f agree, and these are the stopping states.

From (1), we see that V is a function g from E to $\mathbb{R}$ satisfying the properties:

$$\text{(a) } g(i) \geq T \cdot g(i)$$
$$\text{(b) } g(i) \geq f(i) \tag{7}$$

(We continue to view functions on E as column vectors.) We have the following.

THEOREM 2. The optimal value function V satisfies (7a) and (7b); and if g is another function on E satisfying (7a) and (7b), then $g(i) \geq V(i)$ for all $i \in E$. ∎

REMARK. A function $g : E \rightarrow \mathbb{R}^+$ is called *excessive* if it satisfies (7a). Thus, the theroem may be restated as: The value function of the optimal stopping problem is the minimal excessive function dominating the reward function. There is an interesting intuitive interpretation of excessive functions based on the fact that

$$\begin{aligned} g(i) \geq T \cdot g(i) &= \sum_{j \in E} T(i, j) g(j) \\ &= \sum_{j \in E} P[X_1 = j \mid X_0 = i] g(j) \\ &= E[g(X_1) \mid X_0 = i] \end{aligned} \tag{8}$$

One can iterate the inequality to show that also

$$g(i) \geq E[g(X_n) \mid X_0 = i] \tag{9}$$

for all $n > 0$. View g as a reward function. If g is excessive, then for each starting state i, the reward $g(i)$ that can be collected immediately exceeds the expected reward that can be collected at any later instant of time. In fact, (9) forms the basis for the proof of Theorem 2 because we can replace g by f inside expectation, by the monotonicity of expectation and the fact that $g \geq f$, to obtain

$$g(i) \geq E[f(X_n) \mid X_0 = i] \quad \forall\, n > 0$$

That is, g exceeds the expected reward we could collect at any fixed time n. From this it is not too difficult to show that g must exceed the optimal value function V of the problem.

   Theorem 2 enables us to compute V as the optimal solution of a linear program. View the optimal value function V as a column vector $\mathbf{y}^* = (y_1^*, \ldots, y_m^*)'$, where m is the size of E. Since V is the minimal excessive function dominating f, for any other vector $\mathbf{y} = (y_1, \ldots, y_m)'$ satisfying $y_i \geq (T \cdot y)(i)$ and $y_i \geq f(i)$ for all i, we have $y_i^* \leq y_i$ for all i. Since $\mathbf{y}^*$ is the smallest such vector in every component, in particular the sum of its components is the smallest among all such vectors. This means that $\mathbf{y}^*$ is an optimal solution of the linear program:

minimize $\sum_{i=1}^{m} y_i$

subject to: $y_i \geq \sum_{j=1}^{m} T(i, j) y_j$ $i = 1, 2, \ldots, m$
$$y_i \geq f(i)$$

To summarize, in order to find the optimal stopping policy, solve the linear program (10) for the values $y_i^* = V(i)$ of the optimal value function. Then compare these values to $f(i)$. The optimal policy is to stop at those states for which $f(i) = V(i)$, and continue otherwise.

One final remark is that we can often simplify the computation, because the value function is constant over any recurrence class of the Markov chain (X_n), and equals the maximum reward $f(i)$ for states i in the class. To see this, note that if C is a recurrence class and the chain begins in C, then the chain will reach every state in C. Thus, we can wait until the state i^*, whose reward is maximal in C, is reached. We stop there to receive a reward of $f(i^*)$. This reduces the problem to finding $V(i)$ for transient states i.

Activity 3 – Suppose that a Markov chain has a transient state labeled state 1, and two recurrence classes C_1 and C_2. The rewards for the three states in C_1 are 6, 5, and 3, and the rewards for the four states in C_2 are 2, 8, and 5. State 1 has reward 6, and from state 1, it is equally likely to go back to state 1, or into class C_1 or into class C_2 at the next move. What is the optimal stopping policy?

EXAMPLE 2. Let X_n be the state at time n of a game, in which the gambler can stop at any time and collect the reward that accrues to the current state. Suppose that there are two "bust" states (states 1 and 5 in the transition diagram of Figure 6.5) and three active game states (2, 3, and 4). If the potential rewards are as indicated, at what states should the gambler quit the game in order to maximize his expected reward?

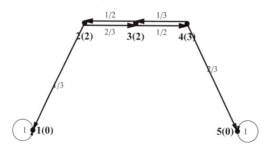

Figure 6.5 – Markov chain of Example 2

The transition matrix and reward vector are below:

$$T = \begin{pmatrix} 1 & 0 & 0 & 0 & 0 \\ 1/3 & 0 & 2/3 & 0 & 0 \\ 0 & 1/2 & 0 & 1/2 & 0 \\ 0 & 0 & 1/3 & 0 & 2/3 \\ 0 & 0 & 0 & 0 & 1 \end{pmatrix} \quad f = \begin{pmatrix} 0 \\ 2 \\ 2 \\ 3 \\ 0 \end{pmatrix}$$

There are two absorbing states, namely 1 and 5. Clearly,

$$V(1) = 0, \quad V(5) = 0$$

States 1 and 5 belong to the set A^* of states where the gambler should (in fact, must) stop. Write y_i for $V(i)$, $i = 1, 2, \ldots, 5$. Since $y_1 = y_5 = 0$, we have, from (10), that the vector $y = (0, y_2, y_3, y_4, 0)$ is the optimal solution to the three-variable linear program:

$$\text{minimize: } y_2 + y_3 + y_4$$

$$\text{subject to: } \begin{aligned} y_2 &\geq (2/3) y_3 \\ y_3 &\geq (1/2) y_2 + (1/2) y_4 \\ y_4 &\geq (1/3) y_3 \\ y_2 &\geq 2 \\ y_3 &\geq 2 \\ y_4 &\geq 3 \end{aligned} \tag{11}$$

By suitable algebraic rearrangement, this problem can be written in standard minimum form, and then dualized. You can check that the dual standard maximum problem is

$$\text{maximize: } 2 x_4 + 2 x_5 + 3 x_6$$

subject to:

$$\begin{aligned} x_1 &- \tfrac{1}{2} x_2 & & + x_4 & & \leq 1 \\ -\tfrac{2}{3} x_1 &+ x_2 &- \tfrac{1}{3} x_3 & & + x_5 & & \leq 1 \\ &- \tfrac{1}{2} x_2 &+ x_3 & & + x_6 &\leq 1 \\ & & x_i \geq 0 \; \forall \, i \end{aligned} \tag{12}$$

This can be solved by the simplex algorithm. We use the Dictionary command to produce the final system of equations. In the electronic version of the text you can reproduce the intermediate steps.

```
Needs["KnoxOR`LinearProgramming`"]
```

```
Clear[f];
system2 = {x₁ - (1 / 2) x₂ + x₄ + x₇ == 1,
    - (2 / 3) x₁ + x₂ - (1 / 3) x₃ + x₅ + x₈ == 1,
    - (1 / 2) x₂ + x₃ + x₆ + x₉ == 1, f == 2 x₄ + 2 x₅ + 3 x₆};
Dictionary[system2, {x₇, x₈, x₉, f},
  {x₁, x₂, x₃, x₄, x₅, x₆}]
```

$$x_7 = 1 - 1\ x_1 + \tfrac{1}{2}\ x_2 + 0\ x_3 - 1\ x_4 + 0\ x_5 + 0\ x_6$$

$$x_8 = 1 + \tfrac{2}{3}\ x_1 - 1\ x_2 + \tfrac{1}{3}\ x_3 + 0\ x_4 - 1\ x_5 + 0\ x_6$$

$$x_9 = 1 + 0\ x_1 + \tfrac{1}{2}\ x_2 - 1\ x_3 + 0\ x_4 + 0\ x_5 - 1\ x_6$$

$$f\ = 0 + 0\ x_1 + 0\ x_2 + 0\ x_3 + 2\ x_4 + 2\ x_5 + 3\ x_6$$

```
Dictionary[system2,
  {x₇, x₈, x₆, f}, {x₁, x₂, x₃, x₄, x₅, x₉}]
```

```
Dictionary[system2,
  {x₇, x₅, x₆, f}, {x₁, x₂, x₃, x₄, x₈, x₉}]
```

```
Dictionary[system2,
  {x₄, x₅, x₆, f}, {x₁, x₂, x₃, x₇, x₈, x₉}]
```

```
Dictionary[system2,
  {x₄, x₂, x₆, f}, {x₁, x₃, x₅, x₇, x₈, x₉}]
```

$$x_4 = \tfrac{3}{2} - \tfrac{2}{3}\ x_1 + \tfrac{1}{6}\ x_3 - \tfrac{1}{2}\ x_5 - 1\ x_7 - \tfrac{1}{2}\ x_8 + 0\ x_9$$

$$x_2 = 1 + \tfrac{2}{3}\ x_1 + \tfrac{1}{3}\ x_3 - 1\ x_5 + 0\ x_7 - 1\ x_8 + 0\ x_9$$

$$x_6 = \tfrac{3}{2} + \tfrac{1}{3}\ x_1 - \tfrac{5}{6}\ x_3 - \tfrac{1}{2}\ x_5 + 0\ x_7 - \tfrac{1}{2}\ x_8 - 1\ x_9$$

$$f\ = \tfrac{15}{2} - \tfrac{1}{3}\ x_1 - \tfrac{13}{6}\ x_3 - \tfrac{1}{2}\ x_5 - 2\ x_7 - \tfrac{5}{2}\ x_8 - 3\ x_9$$

Here, x_7, x_8, x_9 are the slack variables for the maximum problem. We find:

$$V^t = (0, 2, 5/2, 3, 0)$$

since, as you will recall, the optimal solutions y_2, y_3, y_4 to the minimum problem are the negatives of the slack coefficients in the objective row of the final system of the dual maximum problem. The stopping set is the set of states for which V and f have the same value, hence the optimal stopping time is the time of first entry to the set:

$$A^* = \{1, 2, 4, 5\} . \quad \blacksquare$$

Exercises 6.5

1. (*Mathematica*) For the chain with transition matrix below, the rewards for states 1–5 are, respectively, 1, 0, 5, 2, and 3. Draw the transition diagram, and use your intuition to guess at the optimal stopping policy. Then solve the linear program associated with the value function of the problem to verify that your solution is correct.

$$T = \begin{pmatrix} 1/2 & 1/2 & 0 & 0 & 0 \\ 1/3 & 0 & 1/3 & 1/3 & 0 \\ 2/3 & 1/3 & 0 & 0 & 0 \\ 0 & 0 & 0 & 0 & 1 \\ 0 & 0 & 0 & 1 & 0 \end{pmatrix}$$

2. For the chain with the transition diagram below, the number of the state is equal to its reward. Find the optimal stopping time.

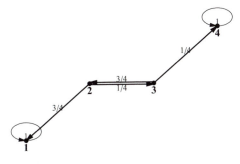

Exercise 2

3. Consider the Markov chain of Figure 4.8, whose transition matrix is reproduced below. Suppose that the reward function is $f(i) = 7 - i$, $i = 1, \ldots, 6$ Find intuitively the optimal stopping time.

$$T = \begin{array}{c@{}c} & \begin{array}{cccccc} 2 & 3 & 4 & 5 & 6 & 1 \end{array} \\ \begin{array}{c} 2 \\ 3 \\ \\ 4 \\ 5 \\ 6 \\ \\ 1 \end{array} & \left(\begin{array}{cc|cccc|c} 1/4 & 3/4 & 0 & 0 & 0 & 0 \\ 3/4 & 1/4 & 0 & 0 & 0 & 0 \\ \hline 0 & 0 & 0 & 1/3 & 2/3 & 0 \\ 0 & 0 & 0 & 0 & 1 & 0 \\ 0 & 0 & 1/3 & 2/3 & 0 & 0 \\ \hline 1/2 & 0 & 0 & 0 & 1/4 & 1/4 \end{array}\right) \end{array}$$

4. Let (X_n) be a Markov chain with the transition matrix below. Show that the constant function $f = 2$ is excessive (i.e., $f \geq T \cdot f$). More generally, show that for an arbitrary Markov chain with finite state space, the constant function $f = c$ is excessive.

$$T = \begin{pmatrix} 1/2 & 1/2 & 0 \\ 1/3 & 1/3 & 1/3 \\ 1/4 & 0 & 3/4 \end{pmatrix}$$

5. A shady character has a sports betting operation. Each month he either makes one more monetary unit, or else he is closed down by the police and all profits are confiscated. The latter occurs with probability 1/8. His desired profit level is 6 monetary units; if that is reached he will abandon the operation and retire to Florida. But he realizes that it may be better for him to quit early, since he risks the loss of all money.

(a) Formulate the problem as an optimal stopping problem.

(b) Write the system of inequalities $V \geq T \cdot V$ for the value function V of the problem.

(c) Use expressions (1) and (2) to find the optimal stopping policy.

6. (*Mathematica*) You have a contract called an *option* to purchase a share of stock when you desire, at the fixed price of 3 monetary units. The day-to-day price of the stock follows a Markov chain with the transition diagram below. If the option is exercised when the price is s, then the stock can be immediately resold at a profit of $s - 3$. You do not have to exercise the option at all if it is not beneficial, but note that there is a chance that the company will go bankrupt before the option is exercised if you wait too long, in which case it will be worthless. When should you exercise the option?

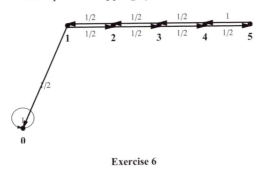

Exercise 6

7. On a television game show, a contestant is offered a sequence of prizes, which are independent and identically distributed random variables taking possible values $1000, $2000, and $3000 with probabilities 1/2, 1/8, and 1/8, respectively. The contestant may at any time choose to accept a prize, at which point the game is over. The game show host may, at some point, choose to offer no more prizes, and the contestant departs with nothing. This happens with probability 1/4. When should the contestant accept a prize?

8. Suppose, in the optimal stopping problem, that there is a discount factor of $\alpha \in (0, 1)$ per period. That is, the reward collected when the game is stopped at time S is only $\alpha^S f(X_S)$ in present-day terms.

(a) Use a dynamic programming argument to show that the optimal value function satisfies

$$V(i) = \max \{f(i), \alpha \cdot T \cdot V(i)\},$$

where T is the transition matrix of the Markov chain.

(b) Let us presume that it is still true that the optimal stopping time is the time at which the chain first enters the set $A^* = \{i \in E : f(i) = V(i)\}$. Find the optimal stopping time for the chain whose transition diagram is below, if the reward function is $f(i) = i$, and the discount factor is $\alpha = .5$.

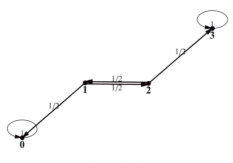

Exercise 8

9. Suppose that the reward function *f* itself in an optimal stopping problem is excessive. Show that the optimal strategy is to stop immediately.

10. (*Mathematica*) Can you develop a policy improvement approach to the optimal stopping problem? Try it on the problem of Figure 6.4.

11. (*Mathematica*) In the game show "Who Wants To Be A Millionaire?," a contestant is given a sequence of multiple choice questions with four alternative answers. The contestant can choose to keep the amount of money that he or she has currently earned and bow out of the game, or to gamble on answering the next question correctly. Each question answered correctly doubles the contestant's winnings; but with the first question that is answered incorrectly, the contestant loses the game and goes away with nothing. The game stops if the contestant reaches an amount in excess of $1 million. The questions become harder as the game proceeds; suppose that the chance of answering the ith question correctly is $(3/2)/(i+1)$. If the contestant starts with $1000, when should he quit the game?

6.6 Extended Applications

In this section we present two examples that are larger and more complicated than any we have done so far: (1) a model for the valuation of an investment object called an *American call option*, and (2) an inventory management problem. There are two purposes: first, you will see other applications of dynamic programming in which the initial modeling is more subtle, and second, the main ideas covered in this chapter will be reviewed.

American Option Problem

EXAMPLE 1. Suppose that there is a risky investment opportunity (such as a common stock) whose market price changes with time in a probabilistic way. The transition diagram of the price process is drawn in Figure 6.6.

Figure 6.6 – Markov chain for risky asset motion

The state labels in the diagram have been denoted generally as x_1, x_2, x_3, etc., but we actually assume that the states are connected to one another as follows. If at time n, the price is X, then at time $n + 1$, the price is either $(1 + u)X$ or $(1 + d)X$, where u and d represent the percentage increase and decrease, respectively, in the price of the risky asset (which we may refer to occasionally as *rates of return*, or *interest rates*). Hence if the state at time 0 is x_1, then the possible states at time 1, which are states x_2 and x_3, respectively, represent $(1 + u)x_1$ and $(1 + d)x_1$; states x_4, x_5, and x_6 represent $(1 + u)^2 x_1$, $(1 + u)(1 + d)x_1$, and $(1 + d)^2 x_1$; etc. The "up" probability is q at each time, and the "down" probability is $1 - q$.

Activity 1 – Consider as a numerical example the case where at time 0, the share price is $x_1 = \$55$. Return rates are $u = .04$ and $d = -.02$. Assume that the probability q is exactly $1/2$. Draw a tree diagram representing the possible motions of the stock price. What is the expected price at time 3?

A *call option* on a risky asset such as our stock is a contract, tradeable in the marketplace, which permits the owner of the option to purchase a share of the stock for a prespecified price E if it is in the option owner's interest. An option of so-called *American type* allows the owner to exercise the option at any time up through a fixed termination time T. For example, if the option contract specifies a termination time of three months, the exercise price is \$60 per share, and the current price is \$55, it is not in the owner's interest to exercise the option immediately because he can purchase a share in the open market for a lower price than the option permits. But if at any time in the next three months the stock price rises to \$63, for example, then the option holder can exercise the option to buy a share at \$60, then immediately resell the share in the open market for a profit of \$3. Because the option holder can benefit in this way, the option has some value of its own, and the option can be traded on the market just like the stock on which it is based. Intuitively, the option value ought to depend on the price of the stock at the current time, how much time remains before the termination time, and perhaps some other parameters of the stock price process or the economic market as a whole. We would like to find that option value.

Activity 2 – Would you expect the option value to increase or to decrease as a function of (a) stock price; (b) time remaining until termination?

Though it is not obvious, we can consider the problem as a dynamic programming problem with a finite time horizon equal to the termination time T of the option. At each (discrete) time up to time T, the investor has a decision to make as to whether to exercise the option or not. Let action 1 represent the action of exercising, and action 2 the action of not exercising the option. The state space of the system is the set of prices that are attainable by the stock on which the option is based; but as we have done before, we can introduce a special death state Δ of no reward to which the process goes if the option is exercised; otherwise the process follows the probability law of the stock price as shown on the tree of Figure 6.6. When action 1 is taken, there is a reward of max $\{X - E, 0\}$ earned by the investor by exercising the option at price E and reselling at current price X in the open market. We will also assume a time discounting factor of α using which future money values are translated to present value, as in Sections 3 and 4. The key observation is that the value of the option to the investor equals the maximum expected discounted value that the investor can possibly receive by using an optimal policy in the DP problem. This DP problem is closely related to the optimal stopping problem of Section 6.5, except for the discount factor and the fact that the termination time prevents the investor from waiting indefinitely to act.

A brief note on the economics of the situation is in order. Assume that there is a non-risky asset available whose value increases deterministically by a factor of $1 + r$. Then it can be shown that in order to avoid *arbitrage* in the market, i.e., the possibility that investors can achieve riskless profits by balancing portfolios of stocks and options on those stocks, the probability q must be related to the parameters of the model by $q = (r - d)/(u - d)$. (See, for example, Baxter and Rennie [4].)

With this introduction, it should be clear that we ought to define the parameters of the DP problem as follows:

$$T_1(x, y) = \begin{cases} 1 & \text{if } y = \Delta \\ 0 & \text{otherwise} \end{cases} \qquad T_2(x, y) = \begin{cases} q & \text{if } y = (1 + u)\,x \\ 1 - q & \text{if } y = (1 + d)\,x \end{cases}$$

$$r(x, 1) = \max(x - E, 0) \ \forall \ x \in \mathbb{R};$$
$$r(x, 2) = 0 \ \forall \ x \in \mathbb{R};$$
$$r(\Delta, a) = 0 \ \text{for } a = 1, 2$$

$$R(\Delta) = 0; \ R(x) = \max(x - E, 0) \ \forall \ x \in \mathbb{R};$$

The value of the option at time 0 solves the DP problem:

$$V_0(x) = \max_{\mathbf{u}} E_{\mathbf{u}}\left[\sum_{n=0}^{T-1} \alpha^k r(X_k, u_k(X_k)) + \alpha^T R(X_T)\right]$$

We can solve the problem of valuing the option as in Section 6.2 by the backward programming algorithm beginning at the final time T with the function $V_T = R$ and working back to time 0 using the DP equation:

$$
\begin{aligned}
V_{n-1}(x) &= \max_{a \in \{1,2\}} \{r(x, a) + \alpha \cdot E_{\mathbf{u}}[V_n(X_n)]\} \\
&= \max \{\max(x - E, 0), 0 + \alpha(q \cdot V_n((1 + u)\,x) \qquad (1) \\
&\quad + (1 - q) \cdot V_n((1 + d)\,x))\}
\end{aligned}
$$

In the process of executing the algorithm, we also derive the option value at all times n between 0 and T by finding the value functions V_n.

EXAMPLE 1. Suppose that at time 0, the share price is \$10, interest rates are $u = .04$ and $d = -.02$, and $q = .5$ (which would follow from the anti-arbitrage principle if the riskless rate was $r = .01$). Let the termination time be $T = 3$, let the discount factor be $\alpha = .9$, and let the exercise price be $E = 10.50$. Setting the initial state to be x_1, the possible values x_2 and x_3 of the stock price at time 1 are

```
x₁ = 10;
{x₂, x₃} = { (1 + .04) 10, (1 - .02) 10}
```

{10.4, 9.8}

The possible values at time 2 are

```
{x₄, x₅, x₆} =
  {(1 + .04)² 10, (1 + .04) (1 - .02) 10, (1 - .02)² 10}
```

{10.816, 10.192, 9.604}

And the possible values at time 3 are

```
{x₇, x₈, x₉, x₁₀} =
  {(1 + .04)³ 10, (1 + .04)² (1 - .02) 10,
   (1 + .04) (1 - .02)² 10, (1 - .02)³ 10}
```

{11.2486, 10.5997, 9.98816, 9.41192}

The tree representing the possible motions of the stock for this particular set of parameters is below.

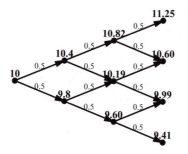

Figure 6.7 – Risky asset motion, $x_1 = \$10$, $u = .04$, $d = -.02$

The dynamic programming algorithm initializes

$$V_3(x) = R(x) = \max(x - 10.50, 0), \ \text{for } x = x_7, x_8, x_9, x_{10}$$

and the DP equation (1) becomes

$$V_{n-1}(x)$$
$$= \max\{\max(x - 10.50, 0), .45 \cdot V_n(1.04\,x) + .45 \cdot V_n(.98\,x)\} \tag{2}$$

The values of V_3 at the states $x = x_7, x_8, x_9, x_{10}$ are computed below:

```
Clear[V3, V2, V1, V0];
V3[x_] := Max[x - 10.50, 0];
{V3[x₇], V3[x₈], V3[x₉], V3[x₁₀]}
```

```
{0.74864, 0.09968, 0, 0}
```

This gives us the obvious result that the call option is only valuable at the expiration time $T = 3$ if the stock price is at one of its two highest values, $x_7 = 11.25$ or $x_8 = 10.60$, in which case the profit to the option holder is the stock price minus the exercise price of 10.50. The option is not exercised, and consequently it expires worthless, if the stock price is $x_9 = 9.99$ or $x_{10} = 9.41$.

We use *Mathematica* to simplify the expression inside the maximum in (2) to find V_2. So that we can see the maximizing action, we compute the list of two values in the DP equation, then find the maximum of the two. Beginning with x_4, note that the two states that it leads to are x_7 and x_8; hence we can find $V_2(x_4)$ as follows:

```
V2x4list =
 {Max[x₄ - 10.50, 0], .45 V3[x₇] + .45 V3[x₈]}
V2[x₄] = Max[V2x4list]
```

```
{0.316, 0.381744}
```

```
0.381744
```

Since the second action achieves the maximum, it follows that the option should not be exercised at state x_4. Repeating for x_5 below, we see that it is again optimal not to exercise.

```
V2x5list =
  {Max[x₅ - 10.50, 0], .45 V3[x₈] + .45 V3[x₉]}
V2[x₅] = Max[V2x5list]
```

{0, 0.044856}

0.044856

The next computation shows that the option is without value at x_6. (Look at the tree of Figure 6.7 to decide why this is true.)

```
V2x6list =
  {Max[x₆ - 10.50, 0], .45 V3[x₉] + .45 V3[x₁₀]}
V2[x₆] = Max[V2x6list]
```

{0, 0}

0

We proceed to V_1, working along the same lines.

```
V1x2list =
  {Max[x₂ - 10.50, 0], .45 V2[x₄] + .45 V2[x₅]}
V1[x₂] = Max[V1x2list]
```

{0, 0.19197}

0.19197

```
V1x3list =
  {Max[x₃ - 10.50, 0], .45 V2[x₅] + .45 V2[x₆]}
V1[x₃] = Max[V1x3list]
```

{0, 0.0201852}

0.0201852

At both x_2 and x_3, it is optimal not to exercise. Finally, at the initial state x_1 we have

```
V0x1list =
  {Max[x₁ - 10.50, 0], .45 V1[x₂] + .45 V1[x₃]}
V0[x₁] = Max[V0x1list]
```

{0, 0.0954698}

0.0954698

Thus, the value of the option when there are three time periods remaining and the initial price is $10 is just $V_0(x_1) = \$.10$, rounded to the nearest cent. ∎

> **Activity 3** – In Example 1, redo the American call option valuation problem assuming that the expiration time is $T = 4$. Is there any path of the stock price process under which the option is exercised before the termination time?

Inventory Problem

EXAMPLE 2. A retail store stocks a certain dishwasher. We are interested in controlling the number of dishwashers in stock over an indefinite period of weeks. During a given week, a total demand for either 0, 1, or 2 dishwashers occurs, each with probability 1/3. After the store closes on Friday night, it is possible to place an order with the manufacturer for more dishwashers. The order will be filled by the following Monday morning when the store reopens. The management has decided that the store will never keep more than two dishwashers on hand. Demands for a dishwasher that occur when no dishwashers are in stock are lost. Each dishwasher is sold at retail price r and can be ordered from the manufacturer at wholesale price $m < r$. There is a weekly storage cost, which for simplicity we assume is s dollars for each dishwasher in stock on Monday morning, regardless of whether any are sold during the week. If there is a discount factor of $\alpha = .99$, how should the store reorder dishwashers on each Friday night in order to maximize expected total discounted profit?

A reasonable first guess at how to model the problem as a Markovian decision problem is to say that the "state" of the system is the inventory level, and the "actions" are orders for more dishwashers. In each weekly time period there is a profit, which is the sales revenue minus storage and reorder costs. Let us adopt the following notation:

Y_n = # dishwashers in stock at beginning of week n, $n = 0, 1, 2, \ldots$;

D_n = # dishwashers demanded during week n, $n = 0, 1, 2, \ldots$;

U_n = # dishwashers ordered at end of week n, $n = 0, 1, 2, \ldots$;

R_n = net profit during week n (including reorder cost, if any).

Note that if the demand D_n in week n is less than or equal to the stock level Y_n, then D_n dishwashers are sold, otherwise Y_n are sold. In other words, the smaller of these two numbers gives the number of dishwashers sold. The sequence of operations is shown in Figure 6.8.

order U_0	order U_1	order U_2	
week 0	week 1	week 2	week 3
$M -- F$	$M -- F$	$M -- F$	$M -- F$
Y_0(demand D_0)	Y_1(demand D_1)	Y_2(demand D_2)	Y_3(demand D_3)

Figure 6.8 – Sequence of operations in the inventory problem

Consider the net profit during week n, for $n = 0, 1, 2, \ldots$. We sell $\min\{D_n, Y_n\}$ dishwashers at r dollars apiece, we store Y_n dishwashers at a cost of s dollars each, and we order U_n dishwashers from the manufacturer at a cost of m dollars apiece. Hence, we can write

$$R_n = r \cdot \min\{D_n, Y_n\} - s \cdot Y_n - m \cdot U_n, \quad n = 0, 1, 2, \ldots . \qquad (3)$$

Notice also that the amount that can be ordered after week n depends on both Y_n and D_n. For example, if there are 2 dishwashers at the start of the week, and none are demanded during the week, the rules say that no dishwashers can be ordered. This observation indicates that the inventory level Y_n is an insufficient description of the "state" of the system. The demand D_n should be recorded as well. A typical state is of the form $\mathbf{i} = (y, d)$, where y is the inventory level at the start of the week and d is the weekly demand. Under the problem conditions, both y and d can take on the values 0, 1, and 2, and the set of possible actions a (a = # dishwashers ordered from the manufacturer) is also $\{0, 1, 2\}$, with some restrictions dependent on the current state. You should have no difficulty checking the sets A_i of admissible actions displayed in Figure 6.9 for the states $i \in E$.

states	(0, 0)	(0, 1)	(0, 2)	(1, 0)	(1, 1)	(1, 2)	(2, 0)	(2, 1)	(2, 2)
actions	0, 1, 2	0, 1, 2	0, 1, 2	0, 1	0, 1, 2	0, 1, 2	0	0, 1	0, 1, 2

Figure 6.9 – Admissible action sets for inventory problem

By (3), the per-period reward function r_0 is

$$r_0(\mathbf{i}, a) = r_0((y, d), a) = r \cdot \min\{d, y\} - s \cdot y - m \cdot a \qquad (4)$$

Recall that r is the retail price per dishwasher, the minimum in the two expressions above is the number of dishwashers sold during the week, s is the storage cost per dishwasher, m is the reorder cost per dishwasher, and a is the number of dishwashers ordered. We use the notation r_0, since r is being used for the retail price.

The transition matrices for the three possible actions 0, 1, and 2 are below. You should check all of the entries; here we explain just a few. Consider the $\mathbf{i} = (1, 0)$ row of T_0 $(\mathbf{i}, \mathbf{j})$. We are given that this week we began with 1 dishwasher, none were demanded, and none were reordered. Then we must start next week with 1 dishwasher. Demands next week for 0, 1, or 2 dishwashers have equal probability 1/3, so that the (1, 0) row of T_0 clearly has 1/3 in each of the columns in which the first component is 1, and has zeros elsewhere in the row. This is the thought process behind all of these computations. As another example, consider the (2, 1) row of T_1. Here, we are given to have started the week with 2 dishwashers, of which 1 was demanded, and 1 was ordered at the end of the week. Thus, it is certain that we start next week with 2 dishwashers, and demands of 0, 1, or 2 will occur with probability 1/3 each. Note, for instance that the (1, 0) row of the matrix T_2 is omitted, because if we begin a week with 1 dishwasher and none are sold during the week, we are not permitted to order two more under the conditions of this problem.

		(0, 0)	(0, 1)	(0, 2)	(1, 0)	(1, 1)	(1, 2)	(2, 0)	(2, 1)	(2, 2)
	(0, 0)	1/3	1/3	1/3	0	0	0	0	0	0
	(0, 1)	1/3	1/3	1/3	0	0	0	0	0	0
	(0, 2)	1/3	1/3	1/3	0	0	0	0	0	0
$T_0 =$	(1, 0)	0	0	0	1/3	1/3	1/3	0	0	0
	(1, 1)	1/3	1/3	1/3	0	0	0	0	0	0
	(1, 2)	1/3	1/3	1/3	0	0	0	0	0	0
	(2, 0)	0	0	0	0	0	0	1/3	1/3	1/3
	(2, 1)	0	0	0	1/3	1/3	1/3	0	0	0
	(2, 2)	1/3	1/3	1/3	0	0	0	0	0	0

		(0, 0)	(0, 1)	(0, 2)	(1, 0)	(1, 1)	(1, 2)	(2, 0)	(2, 1)	(2, 2)
	(0, 0)	0	0	0	1/3	1/3	1/3	0	0	0
	(0, 1)	0	0	0	1/3	1/3	1/3	0	0	0
	(0, 2)	0	0	0	1/3	1/3	1/3	0	0	0
$T_1 =$	(1, 0)	0	0	0	0	0	0	1/3	1/3	1/3
	(1, 1)	0	0	0	1/3	1/3	1/3	0	0	0
	(1, 2)	0	0	0	1/3	1/3	1/3	0	0	0
	(2, 0)	–	–	–	–	–	–	–	–	–
	(2, 1)	0	0	0	0	0	0	1/3	1/3	1/3
	(2, 2)	0	0	0	1/3	1/3	1/3	0	0	0

	(0, 0)	(0, 1)	(0, 2)	(1, 0)	(1, 1)	(1, 2)	(2, 0)	(2, 1)	(2, 2)
(0, 0)	0	0	0	0	0	0	1/3	1/3	1/3
(0, 1)	0	0	0	0	0	0	1/3	1/3	1/3
(0, 2)	0	0	0	0	0	0	1/3	1/3	1/3
(1, 0)	–	–	–	–	–	–	–	–	–
(1, 1)	0	0	0	0	0	0	1/3	1/3	1/3
(1, 2)	0	0	0	0	0	0	1/3	1/3	1/3
(2, 0)	–	–	–	–	–	–	–	–	–
(2, 1)	–	–	–	–	–	–	–	–	–
(2, 2)	0	0	0	0	0	0	1/3	1/3	1/3

$T_2 =$ (applies to the matrix above)

Mathematica definitions of the matrices T_0, T_1, and T_2, and the list of all three denoted by AllTa are in the closed cell below this paragraph.

Activity 4 – Why is the (2, 0) row of T_1 deleted? Justify the (2, 2) row of T_2. Why are the (0, 0), (0, 1), and (0, 2) rows of each matrix identical?

A stationary policy for this problem of inventory control is a function u, where $u(\mathbf{i}) = u(y, d)$ = number of dishwashers ordered at the end of each week n, if week n started with y dishwashers and during week n, d dishwashers were demanded. We wish to find a stationary policy to maximize

$$W(\mathbf{i}, u) = E_u[\textstyle\sum_{n=0}^{\infty} \alpha^n r_0(Y_n, D_n, u_n(Y_n, D_n)) \mid (Y_0, D_0) = \mathbf{i}]$$

for every initial state $\mathbf{i}$.

Let us now fix values for the coefficients of the reward function. Let the storage cost s be 1 unit, and suppose that there is a 100% mark-up on the price of the dishwashers, i.e., the wholesale price $m = (1/2)\,r$, where r is the retail price. Then we have

$$r_0(\mathbf{i}, a) = r_0((y, d), a) = r \cdot \min\{d, y\} - y - \tfrac{1}{2}\, r \cdot a, \tag{5}$$

We will leave r as a parameter of the problem, and do two examples with different values of r. Here is the reward function in tabular form in *Mathematica*.

```
Needs["KnoxOR`DynamicProgramming`"];
```

```
Clear[r];
r0 = {{0, - (1 / 2) r, -r}, {0, - (1 / 2) r, -r},
    {0, - (1 / 2) r, -r}, {-1, -1 - (1 / 2) r, -∞},
    {r - 1, r - 1 - (1 / 2) r, r - 1 - r},
    {r - 1, r - 1 - (1 / 2) r, r - 1 - r},
    {-2, -∞, -∞}, {r - 2, r - 2 - (1 / 2) r, -∞},
    {2 r - 2, 2 r - 2 - (1 / 2) r, 2 r - 2 - r}};
tabheads = {{"(0,0)", "(0,1)", "(0,2)", "(1,0)",
    "(1,1)", "(1,2)", "(2,0)", "(2,1)",
    "(2,2)"}, {"a=0", "a=1", "a=2"}};
TableForm[r0, TableHeadings → tabheads]
```

	a=0	a=1	a=2
(0,0)	0	$-\frac{r}{2}$	$-r$
(0,1)	0	$-\frac{r}{2}$	$-r$
(0,2)	0	$-\frac{r}{2}$	$-r$
(1,0)	-1	$-1 - \frac{r}{2}$	$-\infty$
(1,1)	$-1 + r$	$-1 + \frac{r}{2}$	-1
(1,2)	$-1 + r$	$-1 + \frac{r}{2}$	-1
(2,0)	-2	$-\infty$	$-\infty$
(2,1)	$-2 + r$	$-2 + \frac{r}{2}$	$-\infty$
(2,2)	$-2 + 2 r$	$-2 + \frac{3 r}{2}$	$-2 + r$

Let us try to use the method of successive approximations to approximate the optimal value function and find the optimal policy. From Section 6.3, the DP equation is

$$\mathbf{w}_{n+1}(i) = \max_{a \in A_i} \{r(i, a) + \alpha\, T_a \cdot \mathbf{w}_n(i)\} \tag{6}$$

First, for $r = 10$, and an initial function w_0 that is identically zero, the DiscountedDPEquation command can be used to generate for each state the list of values from which to choose the maximum.

```
Clear[α, w0, reward];
α = .99; w0 = {0, 0, 0, 0, 0, 0, 0, 0, 0};
reward = r0 /. r → 10;
```

```
wlist =
  DiscountedDPEquation[AllTa, reward, w0, α];
TableForm[N[wlist], TableAlignments → Center,
  TableHeadings → tabheads]
w1 = Table[Max[wlist[[i]]],
  {i, 1, Length[AllTa[[1]]]}]
```

	a=0	a=1	a=2
(0,0)	0.	-5.	-10.
(0,1)	0.	-5.	-10.
(0,2)	0.	-5.	-10.
(1,0)	-1.	-6.	-∞
(1,1)	9.	4.	-1.
(1,2)	9.	4.	-1.
(2,0)	-2.	-∞	-∞
(2,1)	8.	3.	-∞
(2,2)	18.	13.	8.

{0, 0, 0, -1, 9, 9, -2, 8, 18}

Comparing the maximum values reported in the second list to the first column of the table, we see that action 0 for each state maximizes the quantity on the right of the DP equation. We repeat the process with w_1.

```
wlist =
  DiscountedDPEquation[AllTa, reward, w1, α];
TableForm[N[wlist], TableAlignments → Center,
  TableHeadings → tabheads]
w2 = Table[Max[wlist[[i]]],
  {i, 1, Length[AllTa[[1]]]}]
```

	a=0	a=1	a=2
(0,0)	0.	0.61	-2.08
(0,1)	0.	0.61	-2.08
(0,2)	0.	0.61	-2.08
(1,0)	4.61	1.92	-∞
(1,1)	9.	9.61	6.92
(1,2)	9.	9.61	6.92
(2,0)	5.92	-∞	-∞
(2,1)	13.61	10.92	-∞
(2,2)	18.	18.61	15.92

```
{0.61, 0.61, 0.61, 4.61,
  9.61, 9.61, 5.92, 13.61, 18.61}
```

Since action 0 is no longer the maximizing action for each state, stabilization has yet to occur, and we should try another step. The current maximizing actions are $a = 1, 1, 1, 0, 1, 1, 0, 0, 1$, respectively, for the nine states.

```
wlist =
   DiscountedDPEquation[AllTa, reward, w2, α];
TableForm[N[wlist], TableAlignments → Center,
 TableHeadings → tabheads]
w3 = Table[Max[wlist[[i]]],
   {i, 1, Length[AllTa[[1]]]}]
```

	a=0	a=1	a=2
(0,0)	0.6039	2.8639	2.5862
(0,1)	0.6039	2.8639	2.5862
(0,2)	0.6039	2.8639	2.5862
(1,0)	6.8639	6.5862	$-\infty$
(1,1)	9.6039	11.8639	11.5862
(1,2)	9.6039	11.8639	11.5862
(2,0)	10.5862	$-\infty$	$-\infty$
(2,1)	15.8639	15.5862	$-\infty$
(2,2)	18.6039	20.8639	20.5862

```
{2.8639, 2.8639, 2.8639, 6.8639,
  11.8639, 11.8639, 10.5862, 15.8639, 20.8639}
```

The maximizing actions are $a = 1, 1, 1, 0, 1, 1, 0, 0, 1$, which are the same as those in the previous step, but let us try one more time.

```
wlist =
   DiscountedDPEquation[AllTa, reward, w3, α];
TableForm[N[wlist], TableAlignments → Center,
 TableHeadings → tabheads]
w4 = Table[Max[wlist[[i]]],
   {i, 1, Length[AllTa[[1]]]}]
```

	a=0	a=1	a=2
(0,0)	2.83526	5.09526	5.61362
(0,1)	2.83526	5.09526	5.61362
(0,2)	2.83526	5.09526	5.61362
(1,0)	9.09526	9.61362	$-\infty$
(1,1)	11.8353	14.0953	14.6136
(1,2)	11.8353	14.0953	14.6136
(2,0)	13.6136	$-\infty$	$-\infty$
(2,1)	18.0953	18.6136	$-\infty$
(2,2)	20.8353	23.0953	23.6136

{5.61362, 5.61362, 5.61362, 9.61362,
   14.6136, 14.6136, 13.6136, 18.6136, 23.6136}

We hit the policy below that always orders to restore the inventory to level 2, defined in the offset formula below:

$$u_0(0, 0) = u_0(0, 1) = u_0(0, 2) = 2;$$
$$u_0(1, 0) = 1; \quad u_0(1, 1) = u_0(1, 2) = 2;$$
$$u_0(2, 0) = 0; \quad u_0(2, 1) = 1; \quad u_0(2, 2) = 2$$

This differs from the previous policies, but let us pause here to check for optimality of this policy. What we must do is to construct the transition matrix T_u of the chain under this policy and the reward vector r_u. These are below, and you should check to see that the definitions are correct.

```
Tu = {{0, 0, 0, 0, 0, 0, 1 / 3, 1 / 3, 1 / 3},
    {0, 0, 0, 0, 0, 0, 1 / 3, 1 / 3, 1 / 3},
    {0, 0, 0, 0, 0, 0, 1 / 3, 1 / 3, 1 / 3},
    {0, 0, 0, 0, 0, 0, 1 / 3, 1 / 3, 1 / 3},
    {0, 0, 0, 0, 0, 0, 1 / 3, 1 / 3, 1 / 3},
    {0, 0, 0, 0, 0, 0, 1 / 3, 1 / 3, 1 / 3},
    {0, 0, 0, 0, 0, 0, 1 / 3, 1 / 3, 1 / 3},
    {0, 0, 0, 0, 0, 0, 1 / 3, 1 / 3, 1 / 3},
    {0, 0, 0, 0, 0, 0, 1 / 3, 1 / 3, 1 / 3}};
ru = {-10, -10, -10, -6, -1, -1, -2, 3, 8};
Ident = IdentityMatrix[9];
Wu = LinearSolve[Ident - α * Tu, ru]
```

{287., 287., 287., 291., 296., 296., 295., 300., 305.}

To check whether the current policy value function W_u satisfies the DP equation, we use it as the third argument of the DiscountedDPEquation command.

```
wlist =
    DiscountedDPEquation[AllTa, reward, Wu, α];
TableForm[N[wlist], TableAlignments → Center,
    TableHeadings → tabheads]
Table[Max[wlist[[i]]],
    {i, 1, Length[AllTa[[1]]]}]
```

	a=0	a=1	a=2
(0,0)	284.13	286.39	287.
(0,1)	284.13	286.39	287.
(0,2)	284.13	286.39	287.
(1,0)	290.39	291.	$-\infty$
(1,1)	293.13	295.39	296.
(1,2)	293.13	295.39	296.
(2,0)	295.	$-\infty$	$-\infty$
(2,1)	299.39	300.	$-\infty$
(2,2)	302.13	304.39	305.

{287., 287., 287., 291., 296., 296., 295., 300., 305.}

Our function W_u does satisfy the DP equation, so for the value $r = 10$ it is optimal to restore the inventory to its maximum possible level of 2 at the start of every week. You will observe that the same solution can be found more easily using the policy improvement method in Exercise 11. ■

Activity 5 – We are about to do some parametric analysis of the inventory problem, which is continued in the exercises. Try to answer the following before reading on: In what way do you think the parameter r (the dishwasher retail price) might be changed in order that the optimal policy no longer restores the inventory level to 2 at the start of the week?

In the activity above you should have concluded that the balance between the profit on dishwashers and the cost of storing them determines the nature of the solution. If we reduce the value of r in comparison with s, we might expect that the optimal inventory level might decrease. Let us try the value $r = 2$ and proceed through the same computations as in the example.

```
Clear[α, w0, reward, r];
α = .99; w0 = {0, 0, 0, 0, 0, 0, 0, 0, 0};
reward = r0 /. r → 2;
```

```
wlist =
  DiscountedDPEquation[AllTa, reward, w0, α];
TableForm[N[wlist], TableAlignments → Center,
 TableHeadings → tabheads]
w1 = Table[Max[wlist[[i]]],
   {i, 1, Length[AllTa[[1]]]}]
```

	a=0	a=1	a=2
(0,0)	0.	-1.	-2.
(0,1)	0.	-1.	-2.
(0,2)	0.	-1.	-2.
(1,0)	-1.	-2.	-∞
(1,1)	1.	0.	-1.
(1,2)	1.	0.	-1.
(2,0)	-2.	-∞	-∞
(2,1)	0.	-1.	-∞
(2,2)	2.	1.	0.

{0, 0, 0, -1, 1, 1, -2, 0, 2}

The first policy that achieves the maximum in the DP equation is the policy that never orders. We see that in the next step of successive approximations this policy does not change.

```
wlist =
  DiscountedDPEquation[AllTa, reward, w1, α];
TableForm[N[wlist], TableAlignments → Center,
 TableHeadings → tabheads]
w2 = Table[Max[wlist[[i]]],
   {i, 1, Length[AllTa[[1]]]}]
```

	a=0	a=1	a=2
(0,0)	0.	-0.67	-2.
(0,1)	0.	-0.67	-2.
(0,2)	0.	-0.67	-2.
(1,0)	-0.67	-2.	-∞
(1,1)	1.	0.33	-1.
(1,2)	1.	0.33	-1.
(2,0)	-2.	-∞	-∞
(2,1)	0.33	-1.	-∞
(2,2)	2.	1.33	0.

{0, 0, 0, -0.67, 1, 1, -2, 0.33, 2}

Let us check to see whether this policy is optimal. Its value is computed below.

```
Ident = IdentityMatrix[9];
ru = {0, 0, 0, -1, 1, 1, -2, 0, 2};
Wu = LinearSolve[Ident - α * T₀, ru]
```

$\{2.22045 \times 10^{-16}, 2.22045 \times 10^{-16}, 2.28773 \times 10^{-16},$
$-0.507463, 1., 1., -1.75741, 0.492537, 2.\}$

```
wlist =
   DiscountedDPEquation[AllTa, reward, Wu, α];
TableForm[N[wlist], TableAlignments → Center,
  TableHeadings → tabheads]
Table[Max[wlist[[i]]],
  {i, 1, Length[AllTa[[1]]]}]
```

	a=0	a=1	a=2
(0,0)	2.22045×10^{-16}	-0.507463	-1.75741
(0,1)	2.22045×10^{-16}	-0.507463	-1.75741
(0,2)	2.22045×10^{-16}	-0.507463	-1.75741
(1,0)	-0.507463	-1.75741	-∞
(1,1)	1.	0.492537	-0.75740
(1,2)	1.	0.492537	-0.75740
(2,0)	-1.75741	-∞	-∞
(2,1)	0.492537	-0.757407	-∞
(2,2)	2.	1.49254	0.242593

$$\{2.22045 \times 10^{-16}, \ 2.22045 \times 10^{-16}, \ 2.22045 \times 10^{-16},$$
$$-0.507463, \ 1., \ 1., \ -1.75741, \ 0.492537, \ 2.\}$$

So we have swung over to the other extreme for our optimal policy, in which, for $r = 2$ it is best not to order any replacement dishwashers. (Why intuitively it is clear that this should be the optimal policy?) Exercises 8 and 9 ask you to find values of r that yield intermediate policies.

Conclusion

Let us review briefly the highlights of this chapter. In the first section we introduced the model. A controller is to make a sequence of decisions based upon observation of the system. Those decisions change the course of the system and incur rewards. The goal is to find the optimal plan, which determines what the decision will be at each time, and for each possible state. In Section 2 we introduced the backtracking procedure for finding the optimal value function at any time for the problem in which control is exerted over a finite time interval. Though the backward programming idea had been known for some time, it was only in the 1950's and 1960's that Bellman and others formalized the procedure in the modern way, and began to fully exploit it.

Section 3 began the study of the infinite horizon discounted problem, which continued in Section 4. In those sections we used a dynamic programming equation to generate two approximation procedures. The first, the method of successive approximations, constructs a sequence of functions approaching the optimal value function. This method is not guaranteed to converge in finitely many steps; hence another method, the policy improvement algorithm, due to Howard [37], was introduced. Instead of approximating the value function, the idea was to approximate the optimal policy by a sequence of strictly improving policies. Here it is required to solve for the value function of each policy in the sequence, and this can be computationally burdensome, but *Mathematica* provides assistance.

Section 5, on finding an optimal time to stop a Markov chain and accept a reward, primarily followed Cinlar [15]. There are obvious applications of this problem to games of chance, and also to problems such as the acceptance of the best offer for an item up for sale. We saw in this section that it is possible to solve for the optimal value function by the methods of linear programming. The optimal stopping problem therefore serves to unify the diverse elements of this text: graph theory (transition diagrams), linear programming, stochastic processes, and dynamic programming. There are linear programming approaches to the computation of the optimal value function for other dynamic programming problems as well. Both Ross [53] and Derman [17] have details. Finally, we presented two extended examples of the ideas and methods of this chapter in Section 6 on option valuation and inventory control. Here, more than anywhere else in our brief introduction, one sees the intricacies of the modeling process.

Our formulation of the Markovian decision problem was somewhat less general than that of some sources. In selecting actions, no dependence on the past history of the system was allowed, and the only sense in which actions were random was in their dependence on the random variables representing the state of the system. By contrast, it would be possible to consider randomized policies, in which we observe the state, then perform some randomized procedure based on the state to determine the action. The rewards were also non-random functions of state and action. The extra generality was dispensed with for the sake of clarity in this introduction to the subject. The reader may see Derman [17] for a discussion of more general problems. From the standpoint of the finite horizon and infinite horizon discounted problems, our restricted view of policy suffices, for it can be shown that the optimum among the broader class of policies occurs among policies of the special form studied in this chapter. Much of our material was patterned after Ross [53], who discusses the finite and infinite horizon discounted problems, as well as some others. Hillier and Lieberman [31] devote two chapters to dynamic programming and Markov decision processes. Additional applied examples can be found in that source. Bellman [5], and Dreyfus and Law [21] are other good references.

Such extensions and generalizations are numerous. Also, as with stochastic processes, serious technical issues arise when one moves to the case of countably or uncountably infinite state or action spaces, or to a continuous time parameter. The latter typically involves partial differential equations. But even in the context of a discrete time parameter and finite action set, questions about existence of optimal policies arise when the state space is countable. As a simple example, consider the two transition diagrams in Figure 6.10. The state space is taken to be {0, 1, 2, ...}, and there are only two actions available, a and b. The first diagram is the transition diagram of the chain of states under action a, the second, action b.

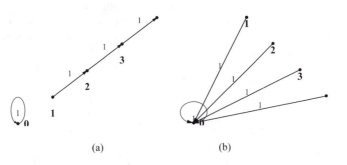

(a) (b)

Figure 6.10 – Transition diagram of a Markov chain under (a) action a; (b) action b

Suppose that the reward function is

$$r(i, a) = 0 \text{ for all } i; \qquad r(0, b) = 0, \quad r(i, b) = 1 - 1/i \text{ for } i \geq 1$$

Under action a, the chain drifts to the right deterministically and no reward can be collected. Under action b, the chain immediately goes to the absorbing state 0; an immediate reward $1 - 1/i$ can be collected if this action is taken at state i, but no further rewards are possible. The supremum of all possible rewards is 1, but no stopping policy can have a value of identically 1. Moreover, the DP equation is

$$\begin{aligned} V(i) &= \max\{r(i, a) + T_a V(i), r(i, b) + T_b V(i)\} \\ &= \max\{V(i + 1), 1 - 1/i\} \\ &= V(i + 1) = 1, \text{ at action } a \end{aligned}$$

The policy indicated by the DP equation is to take action a at every state, and the value of this policy is 0. Not only does there not exist an optimal policy, but the DP equation gives us a very bad policy. In general, there are topological and analytical subtleties in the question of existence of optimal policies when either the state or action space is not finite, and also in the case of continuous time parameter. And, the problem of actually calculating the solution in the infinite case is not insignificant.

An optimization criterion that we have not discussed is the maximization of long-run average reward:

$$\lim_{n \to \infty} E_u[\textstyle\sum_{k=0}^{n} r(X_k, u_k(X_k)) \mid X_0 = i] / (n + 1)$$

Does there exist an optimal policy at all? If so, is it stationary, as in the discounted problem? In general, the answer is no to both questions (for counterexamples, see Ross ([53], Sec. 5.1)). But under some regularity conditions, there is a dynamic programming equation that characterizes an optimal stationary policy. Furthermore, there is a way of translating the problem to a discounted problem with the same optimal policy, so that one could use the methods developed here to solve the latter problem.

To summarize, the methods of dynamic programming come into play in applications in which there is a sequence of interrelated decisions to be made. We have seen examples in inventory, fisheries, production, and option valuation among others in this chapter, and the imaginative reader can think of other applications as well. The breadth of the examples is impressive, and the methodologies used to analyze them are rather simple, yet powerful, and are often algorithmic in nature. In fact, one could make the same statements about the other subject areas in this book. We have tried to show not only the useful examples, but also some of the interesting mathematics of operations research, and how the pieces of the puzzle make an aesthetically appealing picture. This picture is by no means complete. New problems in need of study arise frequently in business, government, engineering and many other areas, and it is to be hoped that you will choose to investigate some of them.

Exercises 6.6

1. (*Mathematica*) In the American option valuation problem, suppose that at time 0, the share price is $x_1 = \$20$. Return rates are $u = .07$ and $d = 0$. Use the remark in the section to compute the probability q if there is a non-risky asset whose rate of return is $r = .02$. For an option that expires in 4 time units and has exercise price \$23, draw a tree diagram representing the possible motions of the stock price. If the discount factor is $\alpha = .995$, find the value of the option at each node of the tree, in particular at time 0.

2. (*Mathematica*) For the scenario in Exercise 1, redo the problem for values $u = .08$ and $u = .09$, and comment on how the solution depends on u.

3. (*Mathematica*) For the scenario in Exercise 1, redo the problem for values of the exercise price $E = \$22$ and $E = \$24$, and comment on how the solution depends on E.

4. (*Mathematica*) For the scenario in Exercise 1, redo the problem for values of the discount factor $\alpha = .95$ and $\alpha = .9$, and comment on how the solution depends on α.

5. (*Mathematica*) A *put option* is an option that allows its owner to sell, rather than buy, an asset at an agreed-upon exercise price E by a certain expiration time T. This option is valuable to the holder if the asset value falls below E, because then the option holder can buy the asset in the market at the lower price, then sell it at a price of E, thus making a profit. Redo the analysis of Example 1 to give the value of an American put option. Use the same parameter values, except that the initial price of the asset should be taken as \$11.

6. For the dishwasher inventory problem, Example 2, relabel the state space such that states $(0, 0)$, $(0, 1)$, and $(0, 2)$ are lumped together as, say, state 0^*, and states $(1, 1)$ and $(1, 2)$ are lumped together as state 1^*. Write the new transition matrices T_a, and write dynamic programming equations similar to (6) for the redesigned problem. Expand the equations out in full.

7. (*Mathematica*) Redo the inventory problem, Example 2, for (a) parameter value $r = 4$; (b) parameter value $r = 6$.

8. (*Mathematica*) In the inventory problem, Example 2, try to find the smallest value of $r > 2$ that you can for which for some state it is optimal to

reorder some positive number of dishwashers. Obtain the r only to the nearest .1.

9. (*Mathematica*) In the inventory problem, Example 2, try to find the largest value of $r < 10$ that you can for which for some state it is optimal to reorder less than the maximum possible number of dishwashers. Obtain the r only to the nearest .1.

10. (*Mathematica*) Suppose in the inventory problem of Example 2 that we are only interested in controlling the system during a particular ad campaign that ends after the third week. No additional order is placed at the end of week 3. The other problem conditions are the same, but we use no discount factor. Solve the problem in finite time horizon.

11. (*Mathematica*) Use the method of policy improvement to find the optimal policy in the dishwasher inventory example of Example 2, beginning with the policy that orders only enough to restore the inventory level to 1 when necessary, ordering nothing when the inventory level is 2.

12. (*Mathematica*) Consider an inventory problem as in Example 2, but with finite time horizon $T = 2$ and no discount factor. Let r remain as a parameter and work through the backward programming method for times 1 and 0 to find the optimal action functions, which will depend on the interval of values to which r belongs. (Hence you should be able to make statements of the form: "For state **i**, at time n, action a is optimal for r belonging to a set; otherwise action b is optimal" for each of the two times and nine states.)

Exercises 13–17 lead you through another financial model. A risky asset moves in the manner of Figure 6.6. We follow the progress of an investor who owns a portfolio of some shares in this asset as well as a checking account (0% interest) containing some amount of money at time 0. The investor can decide to either invest in more shares of the same asset or to sell some number of the shares he holds currently. The investor is to decide how to change the portfolio and how to consume money as time progresses, in order to maximize the expected total consumption. In the finite time horizon investment problem, all wealth is consumed at the terminal time T and only then. (In a similar infinite horizon discounted problem, the investor maximizes the expected total discounted consumption over all periods.) To begin modeling this situation, let the information contained in the state of the process be given by

x = current price of risky asset;
r = current number of shares of risky asset;
y = current checking account balance.

Therefore, a state $\mathbf{i}$ is a three-dimensional vector: $\mathbf{i} = (x, r, y)$. The associated stochastic process has three component processes: X_n, R_n, and Y_n, indicating the share price and quantity held of the risky asset and the checking account balance at time n. At a time such that the state of the process is (x, r, y), the total wealth of the investor is $r \cdot x + y$. We will suppose that as time progresses, the investor can shift money from the risky asset to the checking account and vice versa.

13. Describe the possible actions of the investor and the transition probabilities under those actions. What are the per-period and terminal reward functions for the finite horizon problem? Describe the conditions for a policy to be feasible. Write an expression for the optimal value function.

14. Write in general the DP equation for the problem formulated in Exercise 13. Suppose that at time 0, the share price $x_0 = \$5$. The investor holds exactly 10 shares initially and the checking account starting balance is $y = \$50$. Interest rates are $u = .08$ and $d = -.05$. The investor wants to maximize wealth at the end of 3 years assuming that $q = .5$. Write the DP equation using these parameter values and draw a tree diagram representing the possible paths of the risky asset.

15. (*Mathematica*) To solve the problem, it is convenient at this stage to impose a continuous approximation to the actual problem: we suppose that the investor can buy and sell fractional shares of stock so that all real values of a between the constrained bounds in Exercise 13 are possible. This could be the case if one was able to work through a consortium of investors that allows joint ownership of a share of stock. Use *Mathematica* to simplify the expression inside the DP equation for V_2 in terms of V_3. What action leads to the maximum? Find V_2. Continue to find V_1, V_0, and the optimal policy. Given the nature of the problem parameters, is the optimal policy intuitively obvious?

16. (*Mathematica*) Redo the investment problem with $q = .4$, $u = .05$, $d = -.04$, $T = 4$, and explain the result intuitively.

17. For the investment problem, show that as long as the expected change in stock price from one time to the next is positive, the optimal strategy is to invest all wealth in the stock.

Appendix A
Probability Review

Introduction

This appendix is not meant to take the place of a course in probability theory, but rather to refresh your memory of notions such as sample spaces, random variables, expectation, conditional distributions, and moment-generating functions. In addition, some limit theorems are listed at the end of the appendix that are often omitted in elementary treatments. We will try to present only those definitions and results that are used directly in Chapters 4–6. Consequently, some standard features of introductory courses, such as Chebyshev's inequality, the Central Limit Theorem, and elementary combinatorics, are omitted. Good references for the material in this appendix are: Devore [18], Hastings [30], and Hogg & Tanis [35], and at a somewhat higher level: Feller [22], Hastings [29], Hogg & Craig [36], and Parzen [48].

A.1 Definitions and Properties

Probability theory is concerned with *random experiments*, the possible results of which are called *outcomes*. We leave these two terms undefined, except to say that a random experiment is a phenomenon whose result cannot be predicted exactly before the phenomenon is observed. An outcome is one possible, indecomposable result of the experiment. By indecomposable, we mean that an outcome cannot be broken down into a combination of other outcomes.

Perhaps the simplest experiment is the flip of a coin, in which the outcomes are head (H) and tail (T). Another experiment, in which there are uncountably infinitely many outcomes, is the observation of the time until the next comet comes into view in the night sky. The set of outcomes is the set of all real-valued times $t \geq 0$. The experiment is random, since we cannot predict exactly when the next comet will appear. The reader should note that when we say random, we do not mean that nothing whatsoever is known about the experiment, but only that the exact outcome is not known in advance. In the coin flip experiment, for example, we believe that if a coin is uniformly weighted and fairly flipped, then both outcomes should have equal chance of occurring.

DEFINITION 1. The *sample space* Ω of a random experiment is the set of all possible outcomes ω of the experiment.

Consider the experiment of picking two balls in succession, without replacement, from a box that contains ten balls numbered 1–10. Typical outcomes are $(1, 5)$ and $(6, 3)$, and we can describe the sample space as

$$\Omega = \{\omega = (i, j) : i \neq j \text{ and } i, j \in \{1, 2, \ldots, 10\}\} \tag{1}$$

It is apparent that there are 90 outcomes in Ω, since there are ten possibilities for the first component i, and for each of these, nine possibilities for the second component j.

The sample space for the comet experiment is clearly

$$\Omega = \{\omega : \omega \in [0, \infty)\} \tag{2}$$

Another experiment with an uncountably large sample space is the experiment of watching a machine and noting its times of failure. A typical outcome is depicted in Figure A.1.

1.5   2   2.5   3   time

Figure A.1 – Outcome representing successive failure times

The information that is recorded is the sequence of times $t_1, t_2, t_3, \ldots$ at which the failures occur, from which comes the set theoretic description of Ω:

$$\Omega = \{\omega = (t_1, t_2, t_3, \ldots) : t_i \in [0, \infty) \text{ and } t_i \leq t_{i+1} \text{ for all } i\} \tag{3}$$

DEFINITION 2. An *event* is a subset of the sample space.

Since we have defined events as sets, they satisfy all of the usual set theoretic properties. In the experiment of picking two numbered balls from ten, the event described in English by "the first ball is 1" consists of the outcomes $\{(1, 2), (1, 3), \ldots, (1, 10)\}$. In the comet experiment, the event described by "the arrival time is less than 4" is the interval $[0, 4)$ on the real line. Although we do not discuss it in detail here, we must be careful of what subsets can be called events in the case where the sample space is not countable in size. In this case there may be difficulties in measuring the size of the set, which is roughly what must be done to give it a probability. When the real line is the sample space, the sets that we will use are expressible as countable disjoint unions of intervals, which have enough structure to avoid such technical problems. An advanced course in measure theoretic probability treats more complicated events.

If the precise outcome of a random experiment cannot be known in advance of the performance of the experiment, then the next best thing is to have a probability measure that tells us how likely each event is. Under some conditions it is possible to constructively define a probability measure. Almost trivially, the assumption of a fair coin suggests that we define for the coin flip experiment:

$$P[\{H\}] = 1/2, \quad P[\{T\}] = 1/2, \quad P[\emptyset] = 0, \quad P[\Omega] = 1 \qquad (4)$$

These four sets, $\{H\}$, $\{T\}$, $\emptyset$, Ω, are the only events for this experiment. To say that the empty set has probability zero means that it is impossible for nothing to happen (e.g., the coin cannot land on its side); and to say that the probability of the whole sample space is one means that some outcome is certain to happen.

Less trivially, for the sample space of (1), suppose that balls are drawn randomly, meaning that no preference is given to any pair (i, j). Then each of the 90 outcomes should be given $1/90^{th}$ of the total probability of the sample space, which we take to be $100\% = 1$. Hence $P[(i, j)] = 1/90$ for each outcome $\omega = (i, j)$. Also, probability should be an additive function, so that, for example, the event that the first ball is a 1 should be given probability equal to the total of the probabilities of outcomes satisfying the event, namely $9/90 = 1/10$.

For an arbitrary finite sample space one can constructively define a probability measure by assigning non-negative probabilities $P[\omega]$ to each outcome $\omega \in \Omega$ in such a way that the sum of all of the outcome probabilities is one. Then define probabilities of events by

$$P[E] = \sum_{\omega \in \Omega} P[\omega] \qquad (5)$$

The same procedure works for countably infinite sample spaces; but when the sample space is uncountable, we cannot assign non-zero probability to every outcome and still have finite total probability. As we shall see shortly, one way of defining the probability of an event E in such a case is to integrate a suitable function over the set E.

Following is a non-constructive, axiomatic definition of probability.

DEFINITION 3. A *probability measure P* on the sample space Ω of a random experiment is a function from the events of Ω to the real numbers such that:
  (a) $P[\Omega] = 1$;
  (b) If E is an event, then $P[E] \geq 0$;
  (c) If A and B are disjoint events, then
$$P[A \cup B] = P[A] + P[B].$$

It can be quickly checked that the probability defined in (5) satisfies the three axioms. We remark that by induction, Axiom (c) of Definition 3 extends to unions of finitely many pairwise disjoint events, and in fact with a little more effort it extends to countable disjoint unions.

Several useful consequences of the axioms are easy to prove.

$$P[\emptyset] = 0 \tag{6}$$

$$\text{For any event } E, \ 0 \leq P[E] \leq 1 \tag{7}$$

$$\text{For any event } E, \ P[E^c] = 1 - P[E], \text{ where } E^c \text{ denotes the} \\ \text{complement of } E \text{ in } \Omega \tag{8}$$

$$\text{If } A \subset B, \text{ then } P[A] \leq P[B] \text{ and } P[B \cap A^c] = P[B] - P[A] \tag{9}$$

For any two events A and B,

$$P[A \cup B] = P[A] + P[B] - P[A \cap B] \tag{10}$$

Property (9) is easy to see from the Venn diagram in Figure A.2(a) and the additivity property of probability on disjoint sets. The Venn diagram in Figure A.2(b) indicates why (10) is true. If $P[A]$ and $P[B]$ are simply added, then the probability associated with the overlap $P[A \cap B]$ has contributed twice; so that to obtain the correct $P[A \cup B]$, we must subtract this intersection probability. Notice also that (10) implies that $P[A \cup B] \leq P[A] + P[B]$. This result extends to the case of countable unions, and is referred to as *countable sub-additivity*:

$$P[\bigcup_{i=1}^{\infty} A_i] \leq \sum_{i=1}^{\infty} P[A_i] \tag{11}$$

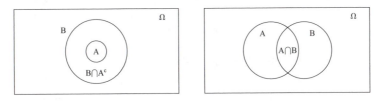

(a) Subset rule   (b) General rule for unions

Figure A.2 – Two rules of probability

A.2 Random Variables and Their Distributions

Often in applications of probability we are interested in a numerical-valued function of outcomes. For example, let X_1 be the number of the first ball drawn in the experiment of (1). Then $X_1(6, 1) = 6$, $X_1(3, 8) = 3$, and in general, $X_1(i, j) = i$. In the experiment of observing the machine breakdown times, a random variable of interest is T_2, the time of the second breakdown. Then $T_2(1, 3.5, 6.2, \ldots) = 3.5$, for example.

DEFINITION 4. A *random variable* X is a real-valued function whose domain is the sample space Ω of a random experiment. The set of values E that X can take on is called its *state space*.

Thus, as in the examples described above, a random variable is able to take an outcome of an experiment and assign a number to it.

Another simple example of a random variable is the following, in which we again consider the flip of a single coin: X is defined to be 0 for the outcome T, and X is 1 for the outcome H. The state space is $\{0, 1\}$; and if p is the probability of head, then we can write

$$P[X = 1] = p; \quad P[X = 0] = 1 - p \tag{12}$$

This is a listing of what is called the *probability distribution* of X. Incidentally, we have used a well-accepted shorthand in (12). Recall that probability measures act on subsets of the sample space. In the first part of (12), we mean

$$P[X = 1] = P[\{\omega \in \Omega : X(\omega) = 1\}] = P[\{H\}] \tag{13}$$

The short form on the left side is preferable in clarity and brevity to the long, albeit correct, expression in the middle of (13). We make the agreement that "the event $X = 1$" means the set of outcomes in the sample space for which X

has the value 1. But a random variable should never be confused with a numerical constant. The former is a function; the latter is a value taken on by the function.

We will not adopt the most general way of characterizing probability distributions of random variables, but will instead restrict to random variables belonging to one of two special classes.

DEFINITION 5. (a) X is said to be a *discrete* random variable with *probability mass function* (p.m.f.) $p(x)$ if X takes on at most countably many values, and

$$P[X = x] = p(x)$$

(b) X is said to be a *continuous* random variable with *probability density function* (p.d.f.) $f(x)$ if, for every subset A of the state space E for which the integral is defined,

$$P[X \in A] = \int_A f(x)\,dx$$

That is, the probability that X takes a value in the set A is the area beneath the graph of the density function f corresponding to A, as in Figure A.3.

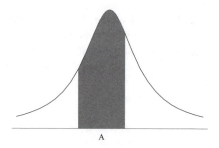

A

Figure A.3 – A density function

In order for the axioms of probability to be satisfied, a probability mass function p must be non-negative, and

$$\sum_{x \in E} p(x) = 1$$

where E is the state space. Similarly, a p.d.f. f must be non-negative and

$$\int_E f(x)\,dx = 1$$

Also, for a continuous random variable with p.d.f. f,

$$P[X = x] = \int_x^x f(t)\,dt = 0$$

EXAMPLE 1. (a) Recall the experiment of picking two numbered balls, and recall the random variable X_1, which was the number of the first ball picked. If outcomes are equally likely, then it is easy to see that the probability mass function of X_1 is

$$p(x) = P[X_1 = x] = \begin{cases} 1/10 & \text{if } x = 1, 2, 3, \ldots, 10 \\ 0 & \text{otherwise} \end{cases} \tag{14}$$

This distribution is called the *discrete uniform distribution*. Note that the sum of the values $p(x)$ for $x = 1, \ldots, 10$ is indeed 1, as it should be. Using this mass function, we can compute, for instance,

$$P[X \geq 5] = \sum_{x=5}^{10} p(x) = 6/10$$

(b) A similar *continuous uniform density* with state space $[a, b]$ can be defined by

$$f(x) = \begin{cases} 1/(b-a) & \text{if } x \in [a, b] \\ 0 & \text{otherwise} \end{cases} \tag{15}$$

This is the density function of the random variable X, which represents a randomly selected point in the interval $[a, b]$. Notice that, as desired for a density, the integral of f over the entire interval $[a, b]$ is 1. For the uniform distribution on $[0, 1]$, we can compute, for instance

$$P[X \leq 1/3] = \int_0^{1/3} 1\,dx = 1/3$$

In general, for the continuous uniform distribution, the probability that the associated random variable falls into a set is the length of that set divided by the length of the entire state space. ■

Special Discrete Distributions

Several discrete mass functions come up often in stochastic processes and operations research.

1. *Binomial distribution.* Let an experiment consist of repeated trials, in which each trial results in either a "success" or a "failure." Success occurs with probability p, and failure occurs with probability $1 - p$. Under the assumption that the trials are independent of one another (to be discussed

later), it can be shown by elementary combinatorics that the probability mass function of the random variable X = number of successes among n trials is

$$p(k) = P[X = k] = \begin{cases} \frac{n!}{k!\,(n-k)!}\, p^k(1-p)^{n-k} & \text{if } k = 0, 1, \ldots, n \\ 0 & \text{otherwise} \end{cases} \tag{16}$$

This distribution is the *binomial distribution with parameters n and p.*

2. *Geometric distribution.*   Let T be the trial on which the first success occurs for a binomial experiment as described above.   Then the probability mass function of T is

$$p(n) = P[T = n] = \begin{cases} p(1-p)^{n-1} & \text{if } n = 1, 2, 3, \ldots \\ 0 & \text{otherwise} \end{cases} \tag{17}$$

This can be seen from the observation that in order for the nth trial to be the one on which the first success occurs, we must see a string of $n-1$ failures, each happening with probability $1-p$, and then a single success, with probability p.   The distribution characterized by the mass function in (17) is called the *geometric distribution with parameter p.*

3. *Poisson distribution.*   A random variable X is said to have the *Poisson distribution with parameter* λ if

$$g(k) = P[X = k] = \begin{cases} \frac{e^{-\lambda} \lambda^k}{k!} & \text{if } k = 0, 1, 2, 3, \ldots \\ 0 & \text{otherwise} \end{cases} \tag{18}$$

This arises later as the distribution of the number of occurrences of a certain phenomenon during a fixed time interval.

Mathematica knows about the main discrete distributions, including the ones that we have introduced here. They are referred to by their names and the values of their parameters as follows.

```
BinomialDistribution[n, p]
GeometricDistribution[p]
PoissonDistribution[λ]
```

They can be used as arguments to four main commands, contained in the standard package Statistics`DiscreteDistributions`.

```
Random[distribution]
RandomArray[distribution, n]
PDF[distribution, x]
CDF[distribution, x]
```

The Random command simulates a single random observation from the given distribution. Similarly, RandomArray simulates a list of n such observations. The PDF function gives the value of the probability mass function of the distribution at the given x. For instance, the probability that a Poisson(2) distributed random variable X takes the value 1 is computed as follows. (We must load the package first.)

```
Needs["Statistics`DiscreteDistributions`"];
PDF[PoissonDistribution[2], 1]
N[%]
```

$$\frac{2}{e^2}$$

```
0.270671
```

The CDF function works like the PDF function, except that it returns values of the cumulative distribution function of the distribution, described in a later subsection.

Special Continuous Distributions

The continuous distributions that are most common in stochastic processes are instances of the *gamma density*, which depends on two constant parameters α and β. The defining formula for the $\Gamma(\alpha, \beta)$ density is

$$f(x) = \begin{cases} \frac{1}{\Gamma(\alpha)\beta^\alpha} x^{\alpha-1} e^{-x/\beta} & \text{if } x > 0 \\ 0 & \text{otherwise} \end{cases} \tag{19}$$

In this expression, the gamma function $\Gamma(\alpha)$ is given by

$$\Gamma(\alpha) = \int_0^\infty y^{\alpha-1} e^{-y} \, dy \tag{20}$$

One can show, using integration by parts, that if n is an integer, then $\Gamma(n) = (n-1)!$. We will see that gamma densities are appropriate for random variables T that are times of occurrence of certain phenomena. An important special case of the gamma density is the one for which $\alpha = 1$ and $\beta = 1/\lambda$, where λ is a constant called the rate. The resulting density is the *exponential density*:

$$f(x) = \begin{cases} \lambda e^{-\lambda x} & \text{if } x > 0 \\ 0 & \text{otherwise} \end{cases} \tag{21}$$

This density is often suitable for random variables that are defined as times between successive arrivals of customers to a service facility.

The well-known *normal distribution* will be involved when we look at Brownian motion processes. The formula for the normal density function, with parameters μ and σ^2 state space $(-\infty, \infty)$, is below:

$$f(x) = \frac{1}{\sqrt{2\pi\sigma^2}} \, e^{-(x-\mu)^2/2\sigma^2} \tag{22}$$

Its graph is similar to the density of Figure A.3, symmetric about μ and with spread determined by the magnitude of σ^2. The parameter μ is called the *mean*, and σ is the *standard deviation* of the distribution. The square σ^2 of the standard deviation is called the *variance*.

Again the Random, RandomArray, PDF, and CDF functions are available to apply to continuous distributions after you load the standard package Statistics`ContinuousDistributions`. The names of the distributions above are

```
GammaDistribution[α, β]
ExponentialDistribution[λ]
NormalDistribution[μ, σ]
```

Notice that the second argument in NormalDistribution is the standard deviation, not the variance. To illustrate, the following computes the probability that an exponentially distributed random variable with parameter .5 takes value between 2 and 4.

```
Needs["Statistics`ContinuousDistributions`"];
f[x_] := PDF[ExponentialDistribution[.5], x];
NIntegrate[f[x], {x, 2, 4}]
```

0.232544

Cumulative Distribution Functions

Another way to characterize the probability law of a random variable is to measure how probability accumulates as we move through the state space.

DEFINITION 6.   The *cumulative distribution function* (c.d.f) of a random variable X is the function

$$F(x) = P[X \leq x]$$

In particular, in the case that X is discrete with p.m.f. p,

$$F(x) = \sum_{t \leq x} p(t) \tag{23}$$

and in the case that X is continuous with p.d.f. f,

$$F(x) = \int_{-\infty}^{x} f(t)\, d t \tag{24}$$

In the discrete case, the cumulative distribution function F is a step function whose jumps are located at the points of the state space of X. To see this, imagine moving to the right along a real axis on which the states of X are laid out in increasing order. Strictly between two successive states, no extra probability is accumulated, and so the c.d.f. remains constant until the next state is reached. But when that next state, say t, is reached, it contributes its probability $p(t)$ to the previous total, which implies that F jumps by an amount $p(t)$ at this point. This is illustrated in Figure A.4 for the discrete distribution with p.m.f. $p(t) = 1/3$ if $t = 1, 2, 3$. In the continuous case, $F(x)$ is the area under the graph of the density function f that lies to the left of x, which is shaded in Figure A.5.

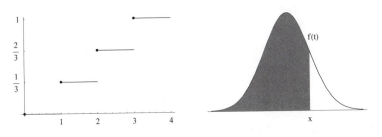

Figure A.4 – A discrete c.d.f.   **Figure A.5** – $F(x)$ is the area under the density up to x

Also, by the Fundamental Theorem of Calculus, for continuous distributions we have the important relation $F'(x) = f(x)$, where F is the c.d.f. of the distribution and f is the p.d.f.

EXAMPLE 2. (a) The c.d.f. of the discrete uniform distribution on $\{1, 2, ..., 10\}$ is

$$F(x) = \begin{cases} 0 & \text{if } x < 1 \\ k/10 & \text{if } k \le x < k+1, \text{ for } k = 1, ..., 9 \\ 1 & \text{if } x \ge 10 \end{cases}$$

(b) The c.d.f. of the exponential distribution with rate λ is

$$F(x) = P[X \le x] = \int_0^x \lambda e^{-\lambda t} \, dt = 1 - e^{-\lambda x}, \ x \ge 0$$

Often useful is the complementary result: $1 - F(x) = P[X > x] = e^{-\lambda x}$. ∎

Multivariate Distributions

The notion of the probability distribution of a random variable extends in a natural way to several random variables $X_1, X_2, ..., X_n$. Again we consider separately the discrete and continuous cases.

DEFINITION 7. (a) Discrete random variables $X_1, X_2, ..., X_n$ have the *joint probability mass function* $p(x_1, x_2, ..., x_n)$ if

$$P[X_1 = x_1, X_2 = x_2, ..., X_n = x_n] = p(x_1, x_2, ..., x_n)$$

(b) Continuous random variables $X_1, X_2, ..., X_n$ have the *joint probability density function* $f(x_1, x_2, ..., x_n)$ if, for all subsets $A_1, A_2, ..., A_n$ of the real line for which the integral is defined,

$$P[X_1 \in A_1, X_2 \in A_2, ..., X_n \in A_n] =$$
$$\int_{A_1} \int_{A_2} \cdots \int_{A_n} f(x_1, x_2, ..., x_n) \, dx_1 \, dx_2 \cdots dx_n$$

In the discrete case, notice that

$$P[X_1 \in A_1, X_2 \in A_2, ..., X_n \in A_n] =$$
$$\sum_{x_1 \in A_1} \sum_{x_2 \in A_2} \cdots \sum_{x_n \in A_n} p(x_1, x_2, ..., x_n) \tag{25}$$

which is similar in form to the corresponding probability in the continuous case.

We can define a joint c.d.f. by

$$F(x_1, x_2, \ldots, x_n) = P[X_1 \le x_1, X_2 \le x_2, \ldots, X_n \le x_n] \tag{26}$$

The one variable sum and integral in formulas (23) and (24) are simply replaced by multiple sums and integrals over the n variables.

Let $X_1, X_2, \ldots, X_n$ have joint discrete mass function p, and denote the state space of X_i by E_i. Then, by (25),

$$\begin{aligned} P[X_1 = x_1] &= P[X_1 = x_1, X_2 \in E_2, \ldots, X_n \in E_n] \\ &= \sum_{x_2 \in E_2} \cdots \sum_{x_n \in E_n} p(x_1, x_2, \ldots, x_n) \end{aligned}$$

Therefore, we can obtain the distribution of X_1 alone (called its *marginal distribution*) by adding the values of p over all other variables $x_2, x_3, \ldots, x_n$. Similarly, we can find the marginal of any other X_j by adding over all x_i for $i \ne j$. The joint marginal distribution of two of the random variables X_i and X_j can be found by adding over all x_k such that k is neither equal to i nor j, etc.

The idea of marginal distributions introduced in the last paragraph carries over directly to the continuous case, with sums replaced by integrals. For example, the marginal of X_1 is

$$f_1(x_1) = \int_{E_2} \cdots \int_{E_n} f(x_1, x_2, \ldots, x_n) \, dx_2 \cdots dx_n$$

In general, to get the joint density of a combination of the X_j's, integrate the joint density with respect to those variables x_i not included in the combination.

EXAMPLE 3. If X_1, X_2, X_3 have the joint density

$$f(x_1, x_2, x_3) = \begin{cases} 6 & \text{if } 0 < x_1 < x_2 < x_3 < 1 \\ 0 & \text{otherwise} \end{cases}$$

then the joint marginal density of X_1 and X_2 is

$$f(x_1, x_2) = \begin{cases} \int_{x_2}^1 6 \, dx_3 = 6(1 - x_2) & \text{if } 0 < x_1 < x_2 < 1 \\ 0 & \text{otherwise} \end{cases}$$

The marginal density of X_1 is

$$f_1(x_1) = \begin{cases} \int_{x_1}^1 6(1 - x_2) \, dx_2 = 3 x_1^2 - 6 x_1 + 3 & \text{if } 0 < x_1 < 1 \\ 0 & \text{otherwise} \end{cases} \quad \blacksquare$$

A.3 Conditional Probability and Independence

If it is known that an event A has occurred, then the sample space is essentially limited to A, and probabilities of other events change correspondingly.

By way of motivation, suppose there are ten equally likely outcomes in the sample space Ω depicted in Figure A.6. Five outcomes are in A, four are in B, and A and B share two outcomes. If A has occurred, then B now has two of the five equally likely outcomes in A, hence the new probability that B will occur is 2/5. Note that this is the same as

$$\frac{P[A \cap B]}{P[A]} = \frac{2/10}{5/10}$$

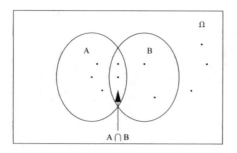

Figure A.6 – Conditional probability of event B given event A

DEFINITION 8. The *conditional probability* that B occurs given that A has occurred is defined by

$$P[B \mid A] = \frac{P[A \cap B]}{P[B]}$$

provided that $P[A] > 0$.

We obtain the following useful equation, called the *multiplication rule*, by simply rewriting the definition of conditional probability:

$$P[A \cap B] = P[A]\,P[B \mid A] \tag{27}$$

This property extends by induction to intersections of many events. Each conditional probability factor in the product conditions on all of the previous events. For example, in the case of three events,

$$P[A \cap B \cap C] = P[A]\,P[B \mid A]\,P[C \mid A \cap B] \tag{28}$$

EXAMPLE 4. Suppose that the time T until arrival of the next bus has the exponential distribution with rate λ. The probability that a bus does not arrive by time $t + s$, given that it has not arrived by time s is

$$P[T > t + s \mid T > s] = P[\{T > s\} \cap \{T > t + s\}] / P[T > s]$$
$$= P[T > t + s] / P[T > s]$$
$$= e^{-\lambda(t+s)} / e^{-\lambda s} = e^{-\lambda t} = P[T > t]$$

This property is called the *memoryless property* of the exponential distribution. Given that we have waited s time units for the bus, the probability that we wait for t more time units is the same as the probability that we wait for t time units had the last bus just left. Having waited for s time units does not help at all; the wait until the next bus is still an exponentially distributed random variable with rate λ. ∎

EXAMPLE 5. The multiplication rule is often useful for experiments that occur in stages. For instance, if a device can be either working or not working at each time period, and we have information about the probability that the device works at time 1, and the conditional probability that the device will be working at time 2 given that it works at time 1, then we can compute

P[device works at both times 1 and 2] =

   P[device works at time 1] · P[device works at time 2 | device works at time 1]

This kind of computation is fundamental to the study of Markov chains in Chapter 4. ∎

If the occurrence of an event A does not affect the probability of another event B, then we call these two events *independent*. We adopt the following definition, which does not require $P[A]$ and $P[B]$ to be non-zero.

DEFINITION 9. Events A and B are *independent* if

$P[A \cap B] = P[A] P[B]$

Notice that by the definition of conditional probability, if A and B are independent of each other and if $P[A] > 0$, then

$$P[B \mid A] = \frac{P[A \cap B]}{P[A]} = \frac{P[A] P[B]}{P[A]} = P[B] \tag{29}$$

This coincides with our intuition about the meaning of independence.

   The definition of independence extends to more than two sets in the following way.

DEFINITION 10. Events A_1, A_2, ..., A_n are called *mutually indepen-dent* if, for any $m = 2, 3, ..., n$ and any choice of m of these events,

$$P[A_{k_1} \cap A_{k_2} \cap \cdots \cap A_{k_m}] = P[A_{k_1}] \cdot P[A_{k_2}] \cdots P[A_{k_m}]$$

EXAMPLE 6. Suppose that a system consists of three components con-nected in series, as in Figure A.7. The system fails at the time T of first failure of any component. Let T_1, T_2, and T_3 be the failure times of the individual components. If the T_i's each have the $\Gamma(2, 1)$ distribution, and for all fixed t the events $\{T_i > t\}$ are mutually independent, find the density of T.

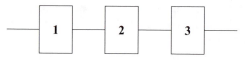

Figure A.7 – A series system

By (19), the p.d.f. of each random variable T_i is

$$f(t) = \begin{cases} t\,e^{-t} & \text{if } t > 0 \\ 0 & \text{otherwise} \end{cases}$$

A simple integration by parts yields that the c.d.f. associated to this density is

$$F[t] = \int_0^t x * E^{-x}\,dx$$

$1 - e^{-t}\,(1 + t)$

We will compute the c.d.f. of $T = \min\{T_1, T_2, T_3\}$, then differentiate it to obtain the probability density function of T. The key observation is that the system failure time T is greater than t if and only if all three component failure times are greater than t. The c.d.f. of T is

$$G(t) = P[T \leq t] = 1 - P[T > t]$$
$$= 1 - P[T_1 > t] \cdot P[T_2 > t] \cdot P[T_3 > t]$$
$$= 1 - (1 - P[T_1 \leq t])^3$$
$$= 1 - (e^{-t}(1 + t))^3$$

Therefore, the density of T is

```
Simplify[D[1 - (E^-t (1 + t))^3, t]]
```

$3 e^{-3t} t (1 + t)^2$

■

As illustrated by the last example, we are mostly concerned with independence in connection with random variables. The next definition follows along the lines of Definition 10.

DEFINITION 11. Random variables $X_1, X_2, ..., X_n$ are called *mutually independent* if for any subsets $B_1, B_2, ..., B_n$ of the state spaces of the random variables,

$$P[X_1 \in B_1, X_2 \in B_2, ..., X_n \in B_n] =$$
$$P[X_1 \in B_1] P[X_2 \in B_2] \cdots P[X_n \in B_n]$$

Independence of random variables can be shown to be equivalent to the factorization of the joint density (or mass function, in the discrete case) into the product of the marginal densities:

$$f(x_1, x_2, ..., x_n) = f_1(x_1) f_2(x_2) \cdots f_n(x_n)$$

Independence is also equivalent to the factorization of the joint c.d.f. into the product of the marginal c.d.f.'s. When a group of random variables is independent and each has the same probability distribution, we say that they are i.i.d (for *independent and identically distributed*). The statistical term for a group of n i.i.d. random variables is a *random sample* of size n.

In stochastic processes, we often have observations $X_1, X_2, ...$ made at times 1, 2, ... , respectively. These observations may or may not be independent. One might be interested in gaining information about X_{n+1} given one or more of the previous observations $X_1, X_2, ..., X_n$. For this reason, conditional distributions of random variables given other random variables play an important role in this subject.

DEFINITION 12. Let random variables X and Y have joint density (or mass function) $f(x, y)$, and let X have marginal density (or mass function) $f_1(x)$. The *conditional density* (or *conditional mass function*) of Y given $X = x$ is defined by

$$f(y \mid x) = \frac{f(x,y)}{f_1(x)}$$

for those x such that $f_1(x) > 0$, and it is left undefined otherwise.

The definition of conditional probability mass function for discrete random variables is consistent with the earlier definition of conditional probabilities of events. When the event $X = x$ has non-zero probability, we can write

$$P[Y = y \mid X = x] = \frac{P[X=x, Y=y]}{P[X=x]}$$

The ratio on the right is exactly the right side of the defining formula for conditional mass function. Thus, in the discrete case, $f(y \mid x)$ means $P[Y = y \mid X = x]$. The situation is not quite as simple in the continuous case, since the probability that a continuous random variable exactly equals a value x is zero. It is possible to take a conditional c.d.f. approach to justify the definition in the continuous case, but it is enough for us to work by analogy with the discrete case. You should simply understand that to calculate the conditional probability of the event $\{Y \in B\}$ given $X = x$, one integrates in the usual way, using the conditional density:

$$P[Y \in B \mid X = x] = \int_B f(y \mid x) \, dy \tag{30}$$

EXAMPLE 7. Consider the random variables X_1 and X_2 from Example 3. Suppose we observe the event that $X_1 = 1/4$. Let us find the conditional probability that $X_2 > 1/2$. We can use *Mathematica* to compute $f_1(1/4)$:

```
f₁[x_] := 3 x² - 6 x + 3;
f₁[1 / 4]
```

$$\frac{27}{16}$$

Then, since $f(x_2 \mid x_1 = 1/4) = f(1/4, x_2)/f_1(1/4)$, we have

```
(* P[X₂>1/2|X₁=1/4]  = *)

1/ (27/16) * ∫₁/₂¹  6 (1 - x2) dx2
```

$$\frac{4}{9}$$

■

Conditional distributions are defined similarly in the multiple variable case. For completeness, we give the definition next, together with a new idea called *conditional independence*.

DEFINITION 13. (a) The *conditional density* (or *conditional mass function*) of a set of random variables $Y_1, Y_2, \ldots, Y_m$ given $X_1 = x_1, X_2 = x_2, \ldots, X_n = x_n$ is

$$f(y_1, y_2, \ldots, y_m \mid x_1, x_2, \ldots, x_n) = \frac{f(x_1, x_2, \ldots, x_n, y_1, y_2, \ldots, y_m)}{f_n(x_1, x_2, \ldots, x_n)}$$

where f is the joint density of all of the X's and Y's, and f_n is the joint marginal of the X's.

(b) Two random variables Y and Z are *conditionally independent* given another random variable X if

$$P[Y \in A, Z \in B \mid X = x] = P[Y \in A \mid X = x] \cdot P[Z \in B \mid X = x]$$

for all subsets A and B of the state spaces of Y and Z, respectively.

The property of conditional independence is crucial in the study of Markov processes. In the discrete case, we can show easily that conditional independence of Y and Z implies the following equation about the conditional mass functions:

$$f(z \mid x, y) = f(z \mid x) \tag{31}$$

which essentially says that knowledge of both X and Y gives no more information than knowledge of X alone for the prediction of Z. To show (31), we can apply the definition of conditional independence to the singleton sets $A = \{y\}$ and $B = \{z\}$ to obtain

$$f(z \mid x, y) = f(x, y, z) / f_{12}(x, y)$$
$$= f_1(x) f(y \mid x) f(z \mid x) / f_{12}(x, y)$$
$$= f_{12}(x, y) f(z \mid x) / f_{12}(x, y) = f(z \mid x)$$

The same result is true in the continuous case, but a deeper study of the theory of integration is required to prove it.

EXAMPLE 8. A traveling salesman begins at a randomly selected city from among the four in Figure A.8. At time 1 he goes to some other city, and at time 2 to a third city such that the city he visits at time 2 is conditionally independent of the city at time 0, given the city at time 1. If the one-step conditional probabilities of visiting cities from other cities are the weights in the directed graph, find the probability that the salesman is in city 3 at time 2.

Figure A.8 – Space of states of a traveling salesman

Let X_0 be the initial position, let X_1 be the position at time 1, and let X_2 be the position at time 2. An inspection of the graph shows that there are only two ways to reach city 3 in two time steps, namely the paths 3, 4, 3 and 2, 4, 3. Thus, by the disjoint union property and the multiplication rule,

$$P[X_2 = 3]$$
$$= P[X_2 = 3, X_1 = 4, X_0 = 3] + P[X_2 = 3, X_1 = 4, X_0 = 2]$$
$$= P[X_2 = 3 \mid X_1 = 4, X_0 = 3] P[X_1 = 4 \mid X_0 = 3] P[X_0 = 3]$$
$$+ P[X_2 = 3 \mid X_1 = 4, X_0 = 2] P[X_1 = 4 \mid X_0 = 2] P[X_0 = 2]$$

But X_2 has been assumed to be conditionally independent of X_0 given X_1, so that as in the remark just before this example, the probability that $X_2 = 3$ depends only on the event $X_1 = 4$, not on the value taken on by X_0. Reading the one-step conditional probabilities from the graph, and noting that X_0 has probability 1/4 of being at each city, we see that

$$P[X_2 = 3] = 1\,(1/2)\,(1/4) + 1\,(1/2)\,(1/4) = 1/4 \quad \blacksquare$$

In the example above we have used one of the most important computational devices in probability theory, which we formalize in the following theorem.

THEOREM 1. (Law of Total Probability) Let $B_1, B_2, \ldots$ be pairwise disjoint events whose union is the entire sample space, and let A be another event. Then,

$$P[A] = \sum_n P[A \cap B_n] = \sum_n P[A \mid B_n]\, P[B_n] \tag{32}$$

provided the conditional probabilities in the last line exist. $\quad \blacksquare$

The first equation in (32) is apparent from Figure A.9 because the sets B_n break A into the union of disjoint pieces of the form $A \cap B_n$. The second equation in (32) is just the multiplication rule. This approach was used in the example, with A equal to the event $X_2 = 3$, and the sets B_1 and B_2 representing the two possible previous paths traveled by the salesman.

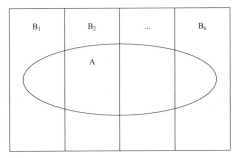

Figure A.9 – The Law of Total Probability

EXAMPLE 9. Let X have the marginal mass function $p_1(x)$, and suppose that the conditional mass function of Y given $X = x$ is $p(y \mid x) = P[Y = y \mid X = x]$. Then,

$$
\begin{aligned}
P[Y = y] &= \sum_x P[Y = y \mid X = x]\, P[X = x] \\
&= \sum_x p(y \mid x)\, p_1(x)
\end{aligned}
\tag{33}
$$

This method for computing probabilities involving Y is sometimes called "conditioning and un-conditioning on X." There is a corresponding continuous version. If $f(x, y)$ is the joint density of X and Y, $f(y \mid x)$ is the conditional density of Y given $X = x$, and $f_1(x)$ is the marginal density of X, then

$$P[Y \in B] = \int_{E_x} P[Y \in B \mid X = x] \, f_1(x) \, dx \quad \blacksquare \qquad (34)$$

A.4 Expectation

The next definition captures the notion of the average value of a function of a random variable.

DEFINITION 14. Let X be a random variable with p.d.f. $f(x)$ and state space E. The *expected value* of a function $g(X)$ is

$$E[g(X)] = \int_E g(x) \, f(x) \, dx$$

provided that the integral exists. If X is discrete with mass function $p(x)$, then we define

$$E[g(X)] = \sum_{x \in E} g(x) \, p(x)$$

Since $g(X)$ is a random variable that takes on possible values $g(x)$ for x in the state space E, we see that $E[g(X)]$ is a weighted average of the states of $g(X)$, where the weighting function is the density f, or the mass function p in the discrete case. As an elementary example, you can check that on the flip of two fair coins, the expected number of heads is one.

There are several expectations worth noting. We illustrate the continuous case; the discrete case merely replaces integrals by sums.

$$\mu = E[X] = \int x \, f(x) \, dx, \text{ called the } \textit{mean} \text{ of } X \qquad (35)$$

$$\sigma^2 = E[(X - \mu)^2] = \int (x - \mu)^2 \, f(x) \, dx, \text{ called the } \textit{variance} \text{ of } X \qquad (36)$$

By expanding the square, we easily obtain the computational formula

$$\sigma^2 = E[X^2] - \mu^2 \qquad (37)$$

We now define

$$E[X^n] = \int x^n \, f(x) \, dx, \text{ called the } n^{\text{th}} \textit{ moment} \text{ of } X \qquad (38)$$

$$M(t) = E[e^{tX}] = \int e^{tx} \, f(x) \, dx, \text{ called the } \textit{moment-generating} \atop \textit{function} \text{ of } X \qquad (39)$$

In the moment-generating function (m.g.f.), t is a real number belonging to the set of all numbers such that the integral exists. The m.g.f. is unique to the distribution, that is, no two distributions share the same m.g.f.

The mean measures the central tendency of the distribution, while the variance measures spread about the mean. The moment-generating function can be used to find the moments in the following way:

$$\frac{dM}{dt}\Big|_{t=0} = E[X\,e^{tX}]\big|_{t=0} = E[X]$$
$$\vdots \qquad\qquad\qquad\qquad\qquad (40)$$
$$\frac{d^n M}{dt^n}\Big|_{t=0} = E[X^n\,e^{tX}]\big|_{t=0} = E[X^n]$$

The following table summarizes the means, variances, and moment-generating functions of some of the most common distributions.

distribution	mean	variance	moment–generating function
Binomial(n, p)	np	$np(1-p)$	$((1-p)+pe^t)^n$
Geometric(p)	$1/p$	$(1-p)/p^2$	$pe^t(1-(1-p)e^t)^{-1}$, $t < -\ln(1-p)$
Poisson(μ)	μ	μ	$\exp(\mu(e^t-1))$
Uniform(a, b)	$(a+b)/2$	$(b-a)^2/12$	$(e^{tb}-e^{ta})/t(b-a)$, $t \neq 0$
Gamma(α, β)	$\alpha\beta$	$\alpha\beta^2$	$(1-\beta t)^{-\alpha}$, $t < 1/\beta$
Exponential(λ)	$1/\lambda$	$1/\lambda^2$	$(1-t/\lambda)^{-1}$, $t < \lambda$
Normal(μ, σ^2)	μ	σ^2	$\exp(\mu t + \frac{1}{2}\sigma^2 t^2)$

EXAMPLE 10.   (a)   As an example of the computation of means, let X have the exponential distribution with rate $\lambda = 3$. Then,

$$\boxed{(* \ \ \texttt{E[X]} \ = \ *) \ \int_0^\infty \texttt{x}\,(3\,\texttt{e}^{-3\,\texttt{x}})\,\texttt{dx}}$$

$$\frac{1}{3}$$

(b)   To illustrate the computation of moment-generating functions, let us verify the m.g.f. of the Poisson(μ) distribution. We have

$$M(t) = E[e^{tX}] \ = \ \sum_{k=0}^{\infty} e^{tk}\,\frac{e^{-\mu}\mu^k}{k!}$$
$$= \ e^{-\mu}\cdot\sum_{k=0}^{\infty}\frac{(e^t\mu)^k}{k!}$$
$$= \ e^{-\mu}\exp(\mu\cdot e^t) = \exp(\mu(e^t-1))$$

You may show easily that, upon differentiating M once and setting t to 0, we get $E[X] = \mu$.   Differentiating a second time and setting $t = 0$ gives

$E[X^2] = \mu + \mu^2$. From this, and (37), the variance of the Poisson(μ) distribution is $\sigma^2 = \mu + \mu^2 - \mu^2 = \mu$. ∎

We occasionally have situations where we must compute the expected value of a random variable that can only take on the values 0 and 1, called a *Bernoulli* or *indicator* random variable. Suppose that the probability that $X = 1$ is p, and consequently the probability that $X = 0$ is $q = 1 - p$. Then,

$$E[X] = 1 \cdot p + 0 \cdot (1 - p) = p \qquad (41)$$

Note that this is the special case of the expectation of a binomial random variable with $n = 1$.

When random variables X and Y are independent, then it can be shown that the expectation of their product is the product of their expectations. The next theorem takes this a little farther and establishes the basis for a powerful technique for finding the distribution of the sum of independent random variables.

THEOREM 2. Let X and Y be independent random variables, and let g and h be functions. Then, provided the expectations exist,

$$E[g(X) \cdot h(Y)] = E[g(X)] \cdot E[h(Y)]$$

In particular, the moment-generating function of the sum $X + Y$ is

$$M_{X+Y}(t) = E[e^{t(X+Y)}] = E[e^{t X}].E[e^{t Y}] = M_X(t) \cdot M_Y(t) \ \blacksquare$$

The proof is an easy consequence of the fact that if X and Y are independent, then their joint density function factors into the product of their marginal densities. The result extends easily by induction to n mutually independent random variables. The fact that the m.g.f. of the sum of independent random variables is the product of the individual m.g.f.'s of those variables often enables us to find the m.g.f. of the sum before we know its distribution. If this m.g.f. can be simplified and recognized, then one can recognize the distribution of the sum, since moment-generating functions are unique to the distribution.

EXAMPLE 11. Let X_1, X_2, and X_3 be independent Poisson random variables with parameters μ_1, μ_2, and μ_3, respectively. The moment-generating function of $X_1 + X_2 + X_3$ is

$$
\begin{aligned}
E[e^{t(X_1+X_2+X_3)}] &= E[e^{t X_1}] \cdot E[e^{t X_2}] \cdot E[e^{t X_3}] \\
&= \exp(\mu_1(e^t - 1)) \cdot \exp(\mu_2(e^t - 1)) \cdot \exp(\mu_3(e^t - 1)) \\
&= \exp((\mu_1 + \mu_2 + \mu_3)(e^t - 1))
\end{aligned}
$$

Since this is the moment-generating function of a Poisson distribution, we see that the sum $X_1 + X_2 + X_3$ is Poisson with parameter $\mu_1 + \mu_2 + \mu_3$. ∎

Conditional expectations are defined similarly to ordinary expectations. The only difference is that the weighting functions are conditional densities (or mass functions in the discrete case) instead of ordinary densities as in Definition 14.

DEFINITION 15. Let X and Y be jointly distributed continuous random variables, such that the conditional density of Y given $X = x$ is $f(y \mid x)$. Then the *conditional expectation* of a function $g(X, Y)$, given $X = x$, is

$$E[g(X, Y) \mid X = x] = \int g(x, y) f(y \mid x) \, d y$$

In the discrete case, if the conditional mass function of Y given $X = x$ is $p(y \mid x)$, then the conditional expectation is

$$E[g(X, Y) \mid X = x] = \sum g(x, y) p(y \mid x)$$

Notice that the definitions imply that

$$E[g(X, Y) \mid X = x] = E[g(x, Y) \mid X = x]$$

In addition, since the definition allows the function g to depend on X, we can obtain the following intuitive result:

$$
\begin{aligned}
E[g(X) \mid X = x] &= \int g(x) f(y \mid x) \, d y \\
&= g(x) \int f(y \mid x) \, d y \qquad (42) \\
&= g(x)
\end{aligned}
$$

That is, if the event $\{X = x\}$ is known to have occurred, then it is certain that $X = x$ and all references to X may be replaced by the constant x. In (42), $g(X)$ is essentially non-random, given $X = x$.

EXAMPLE 12. Referring again to the random variables X_1 and X_2 of Examples 3 and 7, the conditional expectation of X_2 given $X_1 = 1/2$ is

$$
\begin{aligned}
E[X_2 \mid X_1 = 1/2] &= \int_{1/2}^1 x_2 \, f(x_2 \mid 1/2) \, d x_2 \\
&= \int_{1/2}^1 x_2 \cdot \frac{f_{12}(1/2, x_2)}{f_1(1/2)} \, d x_2 \\
&= \tfrac{4}{3} \cdot \int_{1/2}^1 x_2 \cdot 6 (1 - x_2) \, d x_2 = 2/3
\end{aligned}
$$

after a simple integration. ■

One special property of conditional expectation comes up often. First, notice that

$$h(x) = E[g(X, Y) | X = x] \tag{43}$$

defines a function of x. Then $h(X)$ is a random variable, which we write as

$$h(X) = E[g(X, Y) | X] \tag{44}$$

We can take the expectation of this random variable. In the continuous case,

$$
\begin{aligned}
E[E[g(X, Y) | X]] = E[h(X)] &= \int h(x) f_1(x) \, d x \\
&= \int (\int g(x, y) f(y | x) \, d y) f_1(x) \, d x \\
&= \int \int g(x, y) f(x, y) \, d y \, d x \\
&= E[g(X, Y)]
\end{aligned}
\tag{45}
$$

In this derivation we have used the usual notations $f_1(x)$, $f(y | x)$, and $f(x, y)$ for the marginal density of X, the conditional density of Y given $X = x$, and the joint density, respectively. The discrete case is similar. Loosely stated, formula (45) says that the expected value of the conditional expectation is the ordinary expectation.

We have one final note in this very brief review of expectation. It is easy to show, using properties of sums and integrals, that expectation and conditional expectation are *linear operators*:

$$E[a X + b Y] = a E[X] + b E[Y] \tag{46}$$

Also, because the integral of a density function (or sum of a mass function) over the entire state space is one, it is a simple matter to prove that the expected value of a constant is that constant:

$$E[c] = c \tag{47}$$

A.5 Convergence Theorems

Let $A_1, A_2, \ldots$ be a sequence of events in a sample space Ω on which there is a probability measure P. Suppose that these sets are nested increasing, as shown in Figure A.10(a), i.e., for all $n \geq 1$, $A_n \subset A_{n+1}$. Then the "limiting set" approached by the sequence of sets is the union

$$\bigcup_{n=1}^{\infty} A_n$$

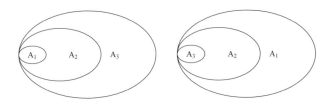

(a) A nested increasing sequence of sets   (b) A nested decreasing sequence of sets

Figure A.10 – Monotone continuity of probability

Suppose, on the other hand, that the sequence is nested decreasing, as in Figure A.10(b), that is, for all $n \geq 1$, $A_n \supset A_{n+1}$. Then the "limiting set" is the intersection

$$\bigcap_{n=1}^{\infty} A_n$$

How does the probability of the limiting event depend on the individual event probabilities? The answer is provided by the following theorem, usually called the *monotone continuity of probability*.

THEOREM 3. (a) If $A_1 \subset A_2 \subset A_3 \subset \cdots$, then

$$P[\bigcup_{n=1}^{\infty} A_n] = \lim_{n \to \infty} P[A_n]$$

(b) If $A_1 \supset A_2 \supset A_3 \supset \cdots$, then

$$P[\bigcap_{n=1}^{\infty} A_n] = \lim_{n \to \infty} P[A_n] \quad \blacksquare$$

EXAMPLE 13. Let X be a random variable. Clearly,

$$A_n = \{X \leq x + 1/n\} \supset \{X \leq x + 1/(n+1)\} = A_{n+1}$$

Also, the event $\{X \leq x\}$ occurs if and only if the event $\{X \leq x + 1/n\}$ occurs for all $n \geq 1$. Therefore, by part (b) of Theorem 3,

$$P[X \leq x] = P[\bigcap_{n=1}^{\infty} \{X \leq x + 1/n\}] = \lim_{n \to \infty} P[X \leq x + 1/n]$$

This shows that if F is the cumulative distribution function of X, then $F(x + 1/n) \longrightarrow F(x)$ as $n \longrightarrow \infty$. The same argument, applied to a general sequence of points converging from the right to x, shows that the c.d.f. of a random variable must be right-continuous. ■

Perhaps the most famous theorem in probability is the *Strong Law of Large Numbers*. To facilitate the statement of this result and some of its consequences to be encountered elsewhere in the book, we introduce the following notion. An event A is said to occur for *almost every outcome* ω in the sample space (alternatively, A occurs *almost everywhere*, abbreviated a.e.) if $P[A^c] = 0$. That is, A occurs almost everywhere if the set of outcomes for which A does not occur has probability zero.

THEOREM 4 (Strong Law of Large Numbers). Let X_1, X_2, X_3, ... be a sequence of independent, identically distributed random variables with finite mean $\mu = E[X_1]$. Then,

$$\lim_{n \to \infty} \sum_{k=1}^{n} X_k(\omega)/n = \mu$$

for almost every outcome ω. ■

EXAMPLE 14. Suppose that a coin with head probability p is flipped repeatedly. Define X_n to be 1 if the nth flip is a head, and 0 otherwise. If the flips are independent, then the sequence (X_n) satisfies the hypotheses of the Strong Law of Large Numbers. Also, $E[X_1] = p$, by (41). The sum of the first n X_k's is the total number of heads in the first n flips, hence the quantity

$$\overline{X} = \sum_{k=1}^{n} X_k / n$$

is the proportion of heads in the first n flips. The strong law implies that for all but some exceptional outcomes of probability zero, the proportion of heads in the first n flips converges to p as $n \longrightarrow \infty$. ■

There are occasions when expectation must be interchanged with limits. It is not always possible to do this, but the next two theorems give sufficient conditions under which the interchange can be done. The question is as follows. Let X_1, X_2, X_3, ... be a sequence of random variables that reach a limit X for almost every outcome. Does the sequence of expectations $E[X_1]$, $E[X_2]$, $E[X_3]$, ... approach $E[X]$?

THEOREM 5 (Monotone Convergence Theorem). If (X_n) is an increasing sequence of random variables whose expectations exist, then

$$E[\lim_{n \to \infty} X_n] = \lim_{n \to \infty} E[X_n] \quad ■$$

THEOREM 6 (Dominated Convergence Theorem).   Let (X_n) be a sequence of random variables. If there exists a random variable Y with finite expectation such that $|X_n| \le Y$ for all n, then

$$E[\lim_{n \to \infty} X_n] = \lim_{n \to \infty} E[X_n] \quad \blacksquare$$

The proofs are beyond the scope of this text, but can be found in advanced probability texts such as Chung [14] and Tucker [58]. The special case of the Dominated Convergence Theorem in which the dominating random variable Y is just a constant, that is, the sequence (X_n) is bounded, is called the *Bounded Convergence Theorem*.   The Monotone Convergence Theorem often arises in the context of infinite series of random variables, as in the following example.

EXAMPLE 15. Let Y_1, Y_2, Y_3, ... be an i.i.d. sequence of random variables with the probability mass function:

$$p(k) = P[Y = k] = \begin{cases} 1/3 & \text{if } k = 0,\ 1,\ \text{or } 2 \\ 0 & \text{otherwise} \end{cases}$$

Notice that the mean of the distribution is 1. Suppose Y_n represents a reward received at time n, and the value of that reward in present day dollars is $(.95)^n \cdot Y_n$. Find the expected present value of the total of all rewards.

The total discounted reward is a random variable given by

$$X = \sum_{k=1}^{\infty} (.95)^k\, Y_k$$

Then X is the limit of the sequence of partial sums $X_n = \sum_{k=1}^{n} (.95)^k\, Y_k$. Since the Y_k's are non-negative valued, the sequence $X_1, X_2, X_3, \ldots$ is increasing. By the Monotone Convergence Theorem,

$$
\begin{aligned}
E[X] &= E[\lim_{n \to \infty} X_n] \\
&= \lim_{n \to \infty} E[X_n] \\
&= \lim_{n \to \infty} \sum_{k=1}^{n} (.95)^k\, E[Y_k] \\
&= \lim_{n \to \infty} \sum_{k=1}^{n} (.95)^k \cdot 1 \\
&= 1/(1 - .95) - 1 = 19
\end{aligned}
$$

It is easy to check that the Dominated Convergence Theorem could also have been used to justify the interchange of limit and expectation in the second line of the computation. $\blacksquare$

Appendix B
Answers to Selected Exercises

Section 1.1

1. The adjacency matrix A and its third power are:

$$\begin{pmatrix} 1 & 0 & 0 & 0 \\ 1 & 0 & 1 & 0 \\ 0 & 1 & 0 & 1 \\ 0 & 0 & 0 & 1 \end{pmatrix} \quad \begin{pmatrix} 1 & 0 & 0 & 0 \\ 2 & 0 & 1 & 1 \\ 1 & 1 & 0 & 2 \\ 0 & 0 & 0 & 1 \end{pmatrix}$$

7. There is a mismatch in vertex degrees.

11. (a) The graph is strongly connected.
(b) The graph is also quasi-connected.

14. The connected components are $\{1, 2, 3, 4, 6, 14\}$, $\{5, 7, 8, 10, 13, 16\}$, and $\{9, 11, 12, 15\}$.

17. The components are: $\{1, 2, 10\}$, $\{3, 4, 6, 7\}$, and $\{5, 8, 9\}$.

19. (a) 1: 4, 5;   2: 4, 5, 6; 3: 5, 6;   4: 1, 2;   5: 1, 2, 3;   6: 2, 3
(b) 1: 2;   2: 3;   3: 2, 4, 5;   4: 5;   5: 1,4
(c) 1: 2, 4;   2: 1, 7;   3: 6;   4: 1, 5, 7;   5: 4, 8;   6: 3;   7: 2, 4, 8;   8: 5, 7

Section 1.2

1. A spanning tree has edges: $\{1, 2\}$, $\{2, 5\}$, $\{2, 3\}$, $\{4, 7\}$, $\{2, 6\}$, $\{1, 4\}$.

2. (a) The spanning tree has edges $\{5, 6\}$, $\{4, 5\}$, $\{3, 4\}$, $\{2, 6\}$, $\{1, 6\}$. The total cost of these edges is 15.
(b) The new spanning tree is $\{\{1, 2\}, \{2, 6\}, \{1, 4\}, \{5, 6\}, \{1, 3\}\}$. The total cost of these edges is 9.

5. The optimal tree uses edges
$\{1,2\}, \{1,3\}, \{1,6\}, \{1,7\}, \{1,8\}, \{1,9\}, \{1,10\}, \{3,4\}$, and $\{3,5\}$.

9. This list of degrees is not possible.

12. At most $\displaystyle\binom{n(n-1)/2}{n-1}$ trees can form.

13. One tree is $\{\{1, 2\}, \{1, 5\}, \{1, 6\}, \{2, 3\}, \{2, 4\}\}$. It is not unique.

15. The edge set is:
{(1, 2), (1, 3), (1, 4), (1, 5), (1, 6),   .
  (2, 7), (2, 8), (5, 9), (5, 12), (7, 10), (8, 11)}

16. The directed spanning tree contains edges
(1, 2), (1, 3), (1, 4), (2, 5), (4, 6), (4, 7), (5, 8), (6, 9).

Section 1.3

1. The Kruskal minimal spanning tree is
{{1, 4}, {6, 7}, {1, 2}, {1, 3}, {3, 6}, {7, 8}, {4, 5}}. Its total cost is 15.

2. The best set of edges is
{{3, 6}, {5, 8}, {1, 2}, {5, 7}, {2, 4}, {4, 7}, {5, 6}, {5, 9}}.
The total weight is 19.

4. The edges in the best tree are
{{2, 5}, {2, 7}, {3, 4}, {1, 5}, {10, 11}, {12, 14}, {8, 10},
   {2, 12}, {4, 5}, {6, 13}, {12, 15}, {2, 13}, {9, 15}, {1, 8}}.
The total weight of the spanning tree is 241.

8. The best edges are
{{1, 4}, {1, 2}, {8, 9}, {3, 4}, {6, 7}, {7, 9}, {9, 10}, {4, 6}, {5, 7}},
and the total weight is 25.

11. The spanning tree includes edges
  (1, 2), (1, 3), (1, 4), (2, 5), (4, 6), (5, 7), (6, 8). Edge {7, 8} may be substituted for edge {6, 8} with no loss.

12. The paths are 1,2; 1,3; 1,4; 1,5; 1,2,6; 1,3,7; 1,3,8; 1,5,9; 1,3,7,10.

13. The paths are 1,2 ; 1,3; 1,4; 1,5; 1,4,6.

17. The edges in the minimal directed spanning tree are
(1, 2), (1, 3), (1, 4), (2, 5), (4, 6), (4, 7), (6, 9), (7, 8).

Section 1.4

2. There is a maximal path 1, 2, 4, 8, 10, and another maximal path 1, 3, 7, 9, 10, both of total cost 14.

3. The unique critical path is 1, 3, 6, 10 of length 13.

4. The critical path is B, D, F, H, J, of cost 20.

6. The debugging takes 7 days, with critical tasks B, D, and F.

8. The project takes 13 days, and the critical, delay-causing tasks are B, D, F, H, and J.

9. The sequence of tasks A, B, C, E, H, I, J is critical, and the commercial takes 18 days in all.

11. Tasks A, D, F, H form a chain that must be done in succession, and therefore the project cannot be done in less than $2 + 5 + 4 + 3 = 14$ time units.

Section 1.5

1. (a) For $V_0 = \{1, 2\}$, the cut is $K = \{(1, 3), (1, 4), (2, 3), (2, 5)\}$ and its capacity is 8. For $V_0 = \{1, 2, 3\}$, the cut is $K = \{(1, 4), (2, 5), (3, 5), (4, 3)\}$ and   its   capacity   is   8.   For $V_0 = \{1, 4\}$,   the   cut   is $K = \{(1, 2), (1, 3), (4, 3), (4, 5)\}$ and its capacity is 10.
(b) Flow on $(1, 2)$: 4; flow on $(1, 3)$: 2; flow on $(1, 4)$: 2; flow on $(2, 3)$: 1; flow on $(2, 5)$: 3; flow on $(4, 3)$: 0; flow on $(3, 5)$: 3; flow on $(4, 5)$: 2.

4. The capacity of the cut is 19. It is not minimal.

8. Flow on $(1, 2)$: 3; flow on $(1, 3)$: 2; flow on $(2, 3)$: 0; flow on $(1, 4)$: 1; flow on $(2, 4)$: 3; flow on $(3, 4)$: 2.

9. The first augmenting path is 1, 2, 5 on which a flow of 3 may be added. Then we find path 1, 3, 5 on which a flow of 2 can be added. Then 1, 4, 5 receives a flow of 2. Finally 1, 2, 3, 5 is the last augmenting path, receiving a flow of 1 unit. Flow on $(1, 2)$: 4; flow on $(1, 3)$: 2; flow on $(2, 3)$: 1; flow on $(1, 4)$: 2; flow on $(2, 5)$: 3; flow on $(3, 5)$: 3; flow on $(4, 5)$: 2.

10. The maximal flow is 8.

11. The maximal flow is 11.

12. Flow on $(1, 2)$: 4; flow on $(1, 3)$: 3; flow on $(2, 4)$: 1; flow on $(3, 4)$: 3; flow on $(2, 5)$: 3; flow on $(4, 6)$: 4; flow on $(5, 6)$: 3.

Section 1.6

2. The matching $\{\{v_1, w_1\}, \{v_2, w_3\}, \{v_3, w_2\}\}$ is maximal.

3. One path is 5, 8, 2, 6, which leads to the matching $\{\{2, 6\}, \{3, 7\}, \{5, 8\}\}$.

4. The path 3, 7, 2, 8 is augmenting, producing the matching $\{\{1, 6\}, \{2, 8\}, \{3, 7\}, \{4, 9\}\}$.

8. (a) $L_1 = (5, 6, 4, 3, 4, 6, 0, 0, 0, 0, 0, 0)$. (b) The revised labeling is $(3, 6, 2, 3, 4, 6, 0, 2, 0, 0, 0, 0)$.

9. (a) $L_1 = \{4, 3, 5, 3, 6, 0, 0, 0, 0, 0\}$ (b) $L_2 = (4, 2, 4, 2, 6, 0, 1, 1, 0, 0)$.

12. There are two complete matchings, both of which are maximal, namely $\{\{1, 5\}, \{2, 6\}, \{3, 8\}, \{4, 7\}\}$, and $\{\{1, 6\}, \{2, 5\}, \{3, 8\}, \{4, 7\}\}$.

13. The maximal matching is $\{\{1, 6\}, \{2, 10\}, \{3, 8\}, \{4, 7\}, \{5, 9\}\}$.

14. One maximal matching is
$\{\{1, 8\}, \{2, 7\}, \{3, 9\}, \{4, 10\}, \{5, 11\}, \{6, 12\}\}$.

15. One maximal matching is $\{\{1, 6\}, \{2, 7\}, \{3, 8\}, \{4, 9\}, \{5, 10\}\}$.

16. An optimal matching is
$\{\{1, 9\}, \{2, 10\}, \{3, 11\}, \{4, 12\}, \{5, 14\}, \{6, 13\}, \{7, 16\}, \{8, 15\}\}$.
The optimal weight is 60.

Section 2.1

1. It is optimal to buy $x = 100$ Jeeps and $y = 0$ vans.

2. The minimum value of the objective is 3, taken on at $(3, 0)$.

3. The optimum is 12000 taken on at all points on the segment between $(1, 2)$ and $(2, 1)$.

4. The best arrangement is 16 mice and 8 rats.

5. This problem has no feasible solutions.

6. The minimum toxin dosage is 40, using none of substance 1 and 40 gms of substance 2.

7. The problem is unbounded.

9. (a) $(3/2, 1) = \frac{1}{2}(1, 2) + \frac{1}{2}(2, 0)$.
(b) $(1, 1) = \frac{2}{5}(0, 5/2) + \frac{1}{2}(2, 0) + \frac{1}{10}(0, 0)$ is one choice.

11. (a) One choice is $(3/2, 1) = \frac{1}{2}(1, 2) + \frac{1}{3}(3, 0) + \frac{1}{6}(0, 0)$.
(b) $(1, 1) = \frac{1}{2}(0, 1) + \frac{1}{4}(1, 2) + \frac{1}{4}(3, 0)$ is one of many choices.

14. The direction of most rapid increase is the vector $(3, 2)$. Moving in this direction, $x = 3/2$, $y = 1$ is the intersection point.

Section 2.2

2. (b) The answer is yes, with $t = 3/4$. (c) x_3 must be between 0 and 1/2, and $x_2 = 1/2 - x_3$.

8. One basic feasible solution is (0, 0, 0, 2, 4, 6); another is (2, 0, 0, 0, 2, 4).

Section 2.3

1. The point (0, 0, 5), corresponding to 5 miles of road, all repaired at level 3, is one optimal point. There is a second corner point solution (0, 10, 0), corresponding to 10 miles of road all at level 2. Any point on the line segment $t(0, 0, 5) + (1 - t)(0, 10, 0)$ is therefore also optimal.

3. One optimal solution is $x_1 = 0$, $x_2 = 0$, $x_3 = 0$, $x_4 = 5$. Another optimal solution is $x_1 = 3$, $x_2 = 0$, $x_3 = 0$, $x_4 = 7/2$. The set of all solutions is the set of all points on the line segment connecting these two.

4. The optimal combination of animals is $x_1 = 1000$ cattle, and $x_3 = 4000$ buffalo (no horses).

5. One optimal solution is $x_1 = 0$, $x_2 = 2$, $x_3 = 0$. Another is $x_1 = 0$, $x_2 = 0$, $x_3 = 2$.

7. No single family dwellings, 10 apartments, and a $500,000 profit.

13 (a) The optimal value is 10/3 taken on at (8/3, 2/3). (b) The conditions are $h_2 \geq \frac{1}{2} h_1 - 1$, and $8 + 2 h_1 \geq h_2$.

Section 2.4

1. Use 6 lbs. of feed 1.

2. The optimal solution is $y_1 = 8/5$, $y_2 = 8/5$, $g = 16/5$.

4. $y_1 = 4$ hours of calisthenics, $y_2 = 1$ hour of jogging, $y_3 = 1$ hour of biking, $y_4 = 0$ hours of rowing.

9. The optimal value of the objective is 120/7, taken on when $y_1 = 0$, $y_2 = 45/7$, $y_3 = 25/7$.

13. The maximum value for the dual problem of 8 is taken on at the point (4, 0). The minimum occurs at $y_1 = 0$, $y_2 = 0$, and $y_3 = 2$.

14 (a) The dual is

$$\text{maximize } f = 10\,x_1 + 7\,x_2$$

$$\text{subject to:} \quad \begin{array}{rcl} x_1 + 2\,x_2 & \leq & 2 \\ 4\,x_2 & = & -1 \\ 5\,x_1 & \leq & 1 \end{array}$$

$$x_1 \geq 0,\ x_2 \text{ unrestricted}$$

(b) The minimum value is 1/4.

Section 3.1

1. There are two optimal solutions, with $x_1 = 5$, $x_2 = 1$ and with $x_1 = 2$, $x_2 = 4$.

3. The unique optimal solution is $x_1 = 11$, $x_2 = 6$, $x_3 = 0$, $f = 17$.

4. One optimal solution is $x_1 = 2$, $x_2 = 0$, $x_3 = 2$, for the three pastry types, and another is $x_1 = 2$, $x_2 = 4$, $x_3 = 0$.

6. The minimum of the objective is -5, taken on at $x_1 = 0$, $x_2 = 5$.

7. The maximal solution is $x_1 = 13/3$, $x_2 = -8/3$, and the optimal value of f is 2/3.

9. The point $x_1 = 6$, $x_2 = 5$ is optimal for the original problem.

10. $5000 from the in-town bank and $5000 from the out-of-town bank, or $5000 from the in-town bank, $2500 from the savings and loan, and $2500 from the out-of-town bank. The minimal interest is $900.

12. (a) The optimal amounts are $x_1 = 60$, $x_2 = 20$, $x_3 = 20$ grams of A, B, and C, respectively.

13. The problem is:

$$\text{maximize } f = x_{12} + x_{13}$$
subject to:
$$x_{12} \leq 4,\ x_{13} \leq 5,\ x_{23} \leq 3,\ x_{24} \leq 2,$$
$$x_{34} \leq 6,\ x_{35} \leq 2,\ x_{45} \leq 3,\ x_{46} \leq 4,\ x_{56} \leq 4,$$
$$x_{12} = x_{23} + x_{24},\ x_{13} + x_{23} = x_{34} + x_{35},$$
$$x_{24} + x_{34} = x_{45} + x_{46},\ x_{45} + x_{35} = x_{56}$$
$$x_{ij} \geq 0 \text{ for all } i,\ j$$

Section 3.2

1. The optimal solution is
$x_{11} = 0$, $x_{12} = 0$, $x_{13} = 100$, $x_{21} = 150$,
$x_{22} = 50$, $x_{23} = 0$, $x_{31} = 0$, $x_{32} = 100$, $x_{33} = 0$.

2. The optimal solution is:
$x_{11} = 200$, $x_{12} = 0$, $x_{13} = 0$, $x_{14} = 800$,
$x_{21} = 600$, $x_{22} = 500$, $x_{23} = 400$, $x_{24} = 0$,
and $f = 49000$.

5. The optimal cost is 540, taken on at $x_{11} = 0$, $x_{12} = 60$, $x_{13} = 20$, $x_{14} = 0$,
$x_{21} = 50$, $x_{22} = 0$, $x_{23} = 0$, $x_{24} = 0$, $x_{31} = 0$, $x_{32} = 0$, $x_{33} = 60$, $x_{34} = 40$,
$x_{41} = 10$, $x_{42} = 40$, $x_{43} = 0$, $x_{44} = 0$.

12. The solutions are $x_{13} = 10$, $x_{22} = 4$, $x_{23} = 4$, $x_{34} = 7$, $x_{41} = 10$, $x_{52} = 4$,
$x_{51} = 2$, $x_{54} = 0$, where x_{ij} is the number of units of bread that truck i
delivers to supermarket j. The optimal cost is 126.

13. (c) The optimal cost is 11, taken on with $x_{13} = 1$, $x_{32} = 1$, and $x_{21} = 1$;
i.e., matching 1 with 3, 3 with 2, and 2 with 1.

Section 3.3

2. The key coefficients are

$$\mathbf{c}_b = (0, 4), \ \mathbf{c}_{nb} = (2, 0), \ B = \begin{pmatrix} 1 & 1 \\ 0 & 2 \end{pmatrix}, \ N = \begin{pmatrix} 1 & 0 \\ 1 & 1 \end{pmatrix}.$$

4. (a) The two optimality conditions are
$-1000 + \Delta_2 - \Delta_1 \le 0$, $-4400 - \Delta_1 \le 0$.
Individually, $\Delta_1 \ge -1000$ and $\Delta_2 \le 1000$.
(c) The new optimal solution is $x_1 = 0$, $x_2 = 12$, $f = 54000$.

7. (a) The inequalities are

$$\begin{cases} \frac{1}{3}(6200 + 3\Delta_1 - \Delta_2 - 2\Delta_3) \ge 0 \\ 1000 + \Delta_2 - \Delta_3 \ge 0 \\ 4000 + \Delta_3 \ge 0 \end{cases}$$

(b) $5(19000 + 3\Delta_2 + \Delta_3)$.

(c)
$$\Delta_1 \ge -6200/3$$

$\Delta_2 \leq 6200$ and $\Delta_2 \geq -1000$

$\Delta_3 \leq 6200/2 = 3100$ and $\Delta_3 \leq 1000 \Longleftrightarrow \Delta_3 \leq 1000$

8. (a) The inequalities are

$$\begin{cases} 5000 + \Delta_1 - 5\Delta_2 \geq 0 \\ 5 + \Delta_2/200 \geq 0 \end{cases}$$

(b) The new objective value is $20 + \frac{\Delta_2}{50}$.

(c)

$\Delta_1 \geq -5000$

$\Delta_2 \leq 1000$ and $\Delta_2 \geq -1000$

11. The current solution is still optimal iff $\frac{-\Delta_2}{8} - \frac{5\Delta_3}{4} \leq \frac{5}{4}$.

12. The old solution is still optimal under the perturbation iff $\Delta_1 \geq -3600$.

13. (a) $\Delta \geq -\frac{1}{3}$. (b) The problem becomes unbounded.

Section 4.1

1. $\begin{pmatrix} 0 & \frac{3}{5} & 0 & \frac{2}{5} \\ \frac{1}{3} & 0 & \frac{1}{3} & \frac{1}{3} \\ \frac{1}{3} & \frac{1}{3} & 0 & \frac{1}{3} \\ 0 & 1 & 0 & 0 \end{pmatrix}$

2. The row 1 column 4 element of T^2 is $1/5$.

4. $P[X_1 = F,\ X_2 = F,\ X_3 = E \mid X_0 = F] = \frac{27}{1000}$.

6(a) $A, A, B, C, A, B, A, B, C, D$.

(b) The transition matrix is $T = \begin{pmatrix} 1/2 & 1/2 & 0 & 0 \\ 1/2 & 0 & 1/2 & 0 \\ 1/2 & 0 & 0 & 1/2 \\ 0 & 0 & 0 & 1 \end{pmatrix}$

Section 4.2

1. (a) $\{3/4, 1/4, 0\}$. (b) The probabilities are $\left\{1 - \frac{3^{2-n}}{4}, \frac{3^{2-n}}{4}, 0\right\}$.

2 (a) $T^3(3, 4) = .121116$. (b) $\mathbf{p}^{(0)} \cdot T^5(3) = .186337$.

4. The limit as $n \to \infty$ of T^n is the matrix, both of whose rows are $\left\{\frac{q-1}{q+p-2}, \frac{p-1}{q+p-2}\right\}$.

6. $P[X_3 = 0 \mid X_0 = 5] = 109/120$ and $P[X_2 = 1, X_3 = 0 \mid X_0 = 5] = 13/60$.

7. $55/288$.

8. $n = 54$.

Section 4.3

1. $F_k(3, 1) = \frac{1}{3} \cdot (\frac{1}{2})^k \cdot (\frac{(3/2)^k - 1}{1/2} - 1)$.

2. $1/2, 0, 1/12, 1/18, 11/216$.

3. $F_k(1, 2) = 0$ for all k.

$F_k(3, 2) = \frac{3}{5} (\frac{1}{15})^{k-1}$, $k = 1, 2, 3, \ldots$

$F_k(2, 2) = \begin{cases} 1/4 & \text{if } k = 1 \\ (3/5)(1/15)^{k-2} \cdot (1/4) & \text{if } k \geq 2 \end{cases}$

5. $F_k(2, 4) = \begin{cases} 3/10 & \text{if } k = 1 \\ (3/10)^k + (3/10)(\frac{5}{2}(1/2)^{k-1}(1 - (3/5)^{k-1})) & \text{if } k \geq 2 \end{cases}$

6. $F_k(1, 4) = 2(\frac{1}{3})^{k-3} \cdot (1 - (3/4)^{k-2})$.

8. $E[T_4 \mid X_0 = 2] = 17/6$.

Section 4.4

1. The closed sets are $\{1, 2, 3, 4\}$, $\{2, 3, 4\}$, $\{3, 4\}$, and $\{4\}$.

2. Since states 1 and 2 form an irreducible set and a recurrence class. States $\{7, 8, 9\}$ comprise an irreducible set and a recurrence class. States 3, 4, 5, and 6 are transient.

4. States 1 and 2 are transient. State 3 has its own recurrence class C_1. The set $\{4, 5\}$ makes up a second recurrence class C_2.

5. The group of states $\{1, 5, 6\}$ is one recurrence class. State 3 is in a class by itself. Also, the group of states $\{2, 4, 7\}$ is a third recurrence class. State 8 is the only transient state.

6. States 1, 2, 6, and 7 are transient. State 8 is absorbing, and states 3, 4, and 5 form a recurrence class.

Section 4.5

1. The recurrence classes are $\{3, 4, 5\}$ and $\{8\}$, the latter of which has the trivial limiting distribution $\pi = (1)$. For class $\{3, 4, 5\}$, the limiting distribution is $\pi_3 = 1/9$, $\pi_4 = 2/3$, $\pi_5 = 2/9$.

2. The limiting probabilities for all states are $1/4$.

3. In the long run, $1/3$ of the vans occupy each district.

5. The long-run average cost is 647.22.

6. We should decide in favor of the second press.

12 (a) The long-run average salary per day for school system 1 is 34.29, and for system 2 is 30. On this basis, system 1 is the better choice. (b) For system 1, the long-run discounted reward vector $R^\alpha f_1$ is $\{761.38, 628.97\}$; and for system 2, it is $\{696.77, 551.61\}$. Again, for both possible starting states, the long-term reward for system 1 dominates that of system 2.

13. The maximum value occurs at the left endpoint $\epsilon = -.1$.

Section 4.6

1. $f_{21} = 1$, $f_{24} = 0$ $f_{31} = 6/7$, $f_{34} = 1/7$, where state 1 is E, state 2 is G, state 3 is F, and state 4 is P. If half are fair and half are good initially, the proportion reaching the excellent state is $\frac{13}{14}$.

2. Each of the two transient states has probability $1/3$ of being absorbed by class $\{1, 2, 3\}$ and $2/3$ of being absorbed by class $\{4, 5\}$.
This means that the limit of T^n is

$$\begin{pmatrix} 3/10 & 1/10 & 3/5 & 0 & 0 & 0 & 0 \\ 3/10 & 1/10 & 3/5 & 0 & 0 & 0 & 0 \\ 3/10 & 1/10 & 3/5 & 0 & 0 & 0 & 0 \\ 0 & 0 & 0 & 4/7 & 3/7 & 0 & 0 \\ 0 & 0 & 0 & 4/7 & 3/7 & 0 & 0 \\ 1/10 & 1/30 & 1/5 & 8/21 & 2/7 & 0 & 0 \\ 1/10 & 1/30 & 1/5 & 8/21 & 2/7 & 0 & 0 \end{pmatrix}$$

4. The smallest such value of p is about .542.

5. 100.

6. The limiting matrix is

$$\lim_{n\to\infty} T^n = \begin{array}{c} \\ 1 \\ 2 \\ 7 \\ 8 \\ 9 \\ 3 \\ 4 \\ 5 \\ 6 \end{array} \begin{array}{ccccccccc} 1 & 2 & 7 & 8 & 9 & 3 & 4 & 5 & 6 \\ \left(\begin{array}{ccccccccc} 2/3 & 1/3 & 0 & 0 & 0 & 0 & 0 & 0 & 0 \\ 2/3 & 1/3 & 0 & 0 & 0 & 0 & 0 & 0 & 0 \\ 0 & 0 & 2/5 & 1/5 & 2/5 & 0 & 0 & 0 & 0 \\ 0 & 0 & 2/5 & 1/5 & 2/5 & 0 & 0 & 0 & 0 \\ 0 & 0 & 2/5 & 1/5 & 2/5 & 0 & 0 & 0 & 0 \\ 2/9 & 1/9 & 4/15 & 2/15 & 4/15 & 0 & 0 & 0 & 0 \\ 0 & 0 & 2/5 & 1/5 & 2/5 & 0 & 0 & 0 & 0 \\ 0 & 0 & 2/5 & 1/5 & 2/5 & 0 & 0 & 0 & 0 \\ 0 & 0 & 2/5 & 1/5 & 2/5 & 0 & 0 & 0 & 0 \end{array}\right) \end{array}$$

7. About 22% of entering students graduate, 44% of second-year students, 73% of third-year students, 81% of fourth-year students, and 90% of fifth-year students.

Section 5.1

1. (a) .0243895; (b) .161668; (c) .00487791

3. The $\Gamma(k, 1/\lambda)$ distribution, mean k/λ, and variance k/λ^2.

4. (a) $\lambda s + N_t$; (b) $(\lambda t)(\lambda s) + \lambda t + (\lambda t)^2$.

5. $\binom{n}{k}(\frac{3}{8})^k (\frac{5}{8})^{n-k}$.

8. $\lambda \cdot \int_0^\infty f(u)\,du$.

9. λ approximately equal to .82 is the value at which the probability reaches .95.

11. 45.

12. 5.

14. 9 per minute.

Section 5.2

1. (c) The long-run proportion of time in state 1 is $\lambda/(\lambda + \mu)$.

4. 10 beds.

7. c is approximately .188598.

8. Under the stated conditions, $E[T] = 7\,(n/2)$. Also, $\mathrm{Var}(T) = 25\,(n/2)$.

9. $\{0.00155199, 0.0155199, 0.0775996, 0.258665, 0.646663\}$.

10. $p_0 = \left(1 + \frac{\lambda_0}{\mu_1} \cdot \frac{1}{1 - \lambda/\mu}\right)^{-1}$ and $p_j = \frac{\lambda_0 \, \lambda^{j-1}}{\mu_1 \, \mu^{j-1}} \left(1 + \frac{\lambda_0}{\mu_1} \cdot \frac{1}{1 - \lambda/\mu}\right)^{-1}$.

11. The limiting probabilities are

$$p_0 = \left(1 + \frac{3\lambda_0}{\lambda_L} + \frac{2\lambda_0}{\lambda_H}\right)^{-1}; \quad p_1 = \frac{3\lambda_0}{\lambda_L} \cdot p_0; \quad p_2 = \frac{2\lambda_0}{\lambda_H} \cdot p_0$$

Section 5.3

1. $F^{(n)}(t) = \int\limits_0^t \frac{\lambda^n}{(n-1)!} \, s^{n-1} \, e^{-\lambda s} \, d s.$

2. $G * F(t) = \begin{cases} 0 & \text{if} \quad t < 1 \\ \frac{1}{4} t - \frac{1}{4} & \text{if} \quad 1 \le t < 2 \\ \frac{1}{2} t - \frac{3}{4} & \text{if} \quad 2 \le t < 3 \\ \frac{1}{4} t & \text{if} \quad 3 \le t < 4 \\ 1 & \text{if} \quad t \ge 4 \end{cases}$

$F * G(t) = \begin{cases} 0 & \text{if} \quad t < 1 \\ \frac{1}{2} \frac{1}{2} (t - 1) & \text{if} \quad 1 \le t < 2 \\ \frac{1}{2} \left(\frac{1}{2} (t - 1) + \frac{1}{2} (t - 2)\right) & \text{if} \quad 2 \le t < 3 \\ \frac{1}{2} \left(1 + \frac{1}{2} (t - 2)\right) & \text{if} \quad 3 \le t < 4 \\ \frac{1}{2} (1 + 1) & \text{if} \quad t \ge 4 \end{cases}$

4. The long-run expected number of reports that can be finished per unit time is $1/(1/\lambda_1 + 1/\lambda_2 + 1/\lambda_3)$.

7. The expected number of renewals per unit time converges to $1/\mu = 1/(5/6 + c)$.

8. $m(t) = [\frac{2}{3} t]$.

13. The long-run cost per time is $\frac{c}{1/\lambda_1 + 1/\lambda_2}$.

15. Investment 1 is best.

17 (a) p; (c) p; (d) $3 t \cdot p$.

18. The final price itself is expected to be about $20 + \frac{(2 p - 1) t}{8}$.

Section 5.4

1. (a) $\lambda_i = \begin{cases} \lambda & \text{if } i = 0, 1, \ldots, 5 \\ 0 & \text{otherwise} \end{cases}$ $\mu_j = \begin{cases} j \cdot \mu & \text{if } j = 1, 2, 3, 4 \\ 4\mu & \text{if } j = 5, 6 \\ 0 & \text{otherwise} \end{cases}$

(b) $\{\frac{2}{15}, \frac{4}{15}, \frac{4}{15}, \frac{8}{45}, \frac{4}{45}, \frac{2}{45}, \frac{1}{45}\}$.

(c) Doubling the service rate changes the probability that the queue is full to $1/1045$.

2. The traffic intensity is $\frac{5\sqrt{\pi}}{2}$. The queue will have no limiting distribution.

3. The limiting distribution is Poisson with parameter $\lambda/\mu = \rho$.

5. $N = 6$ suffices.

6. $s = 6$ suffices.

7. The minimum cost occurs when $\rho = 1$.

8. The total probability of 3 or fewer is about $.692882$.

13. $p_n = = \begin{cases} \frac{(\lambda/\mu)^n}{n!} \cdot p_0 & \text{if } n \le 1 \\ \frac{(\lambda/\mu)^n}{1! \, 1^{n-1}} \cdot p_0 & \text{if } n > 1 \end{cases}$

14. $7/8$.

15. The limiting distribution is $\pi_j = (1 - \beta)\beta^j$, $j = 0, 1, 2, \ldots$, where $\beta = .122365$.

16. $p_0 = \left(1 + \left(\frac{\lambda/\mu}{1 - \lambda \, p/\mu}\right)\right)^{-1}$, $p_n = \frac{\lambda^n \, p^{n-1}}{\mu^n} \, p_0$, $n \ge 1$.

Section 5.5

1. (a) $.194712$; (b) $.841345$.

(c) $f(x_2, x_3, x_4) = \frac{1}{\sqrt{2\pi}} e^{\frac{-(x_4-x_3)^2}{2}} \cdot \frac{1}{\sqrt{2\pi}} e^{\frac{-(x_3-x_2)^2}{2}} \cdot \frac{1}{\sqrt{2\pi}} e^{\frac{-x_2^2}{2}}$

2. (a) $.357962$; (b) $.23906$.

4. $f(x) = \frac{e^{2\mu(M+N)} - e^{2\mu(M-x)}}{e^{2\mu(M+N)} - 1}$.

6. $\int_K^\infty (p_1 e^x - K) \; \frac{1}{\sqrt{2\pi\sigma^2 T}} \; e^{-(x-\mu T)^2/2\sigma^2 T} \, dx.$

10. After some experimentation we find the value to be about $a = 10.85$.

11. $g(y) = \frac{1}{y} \cdot \frac{1}{\sqrt{2\pi t}} \, e^{-(\log(y))^2/2t}$. The desired probability is .159983.

12. The density function is

$$g(y) = \frac{1}{\sqrt{2\pi t}} \, e^{-y^2/2t} + \frac{1}{\sqrt{2\pi t}} \, e^{-(2M-y)^2/2t}, \; y \leq M$$

13. The mean is 0 and the variance is $\sum_{i=0}^{n-1} x_i^2(t_{i+1} - t_i)$.

Section 6.1

1. For this policy, actions 0, 0, 1, and 0 are taken. The rewards are 1, 2, 2, 2.

3. A stationary policy can be created in 81 ways. In the second scenario, there are $81^5 = 3486784401$ policies.

4. The path 5, 8, 10 is the shortest path from 5 to 10, with cost 6. The shortest path from 6 to 10 is 6, 8, 10 with cost 4.

6. $V(A, \mathbf{u}) = 11.23$ and $V(B, u) = 8.96$.

7. There are 2^7 stationary policies.

8. There are $6!^3 = 373248000$ admissible policies.

9. The transition matrix under u is

$$T_u = \begin{pmatrix} 0 & 1 & 0 & 0 \\ 1/2 & 0 & 1/2 & 0 \\ 0 & 1 & 0 & 0 \\ 0 & 1 & 0 & 0 \end{pmatrix}$$

The expected total reward is 4.5.

10. The expected total reward is 11/2.

Section 6.2

1. It is optimal to make no rockets at any time.

2. (b) (i) When the terminal cost is 50 we manufacture as many rockets as possible at each time. (ii) Again at all times, 3 rockets should still be made. (iii) This time it is always optimal to make 0 rockets.

3. For each time n, $u_n(A) = 1$, $u_n(B) = 2$.

4. At all times, the optimizing actions are to turn the furnace on when the temperature is below 68, and turn it off otherwise.

5. The stationary policy that always takes action 0 is optimal.

6. $V_1 = (0, 15, 25, 35, 45, 50)$, with optimal action function
$u_1(i) = 0$, $i = 1, 2, 3, 4$; $u_1(5) = 1$.
   At time 0, $V_0 = (0, 75/4, 30, 40, 185/4, 205/4)$, with the same optimal action function $u_0(i) = 0$, $i = 1, 2, 3, 4$; $u_0(5) = 1$.

7. For $T = 5$, it is optimal to harvest a single unit of fish at time 0, when there are 5 units available.

8. At each time we see that the action function $u(1) = 1$, $u(2) = 2$ is optimal.

10. At each time, it is optimal for the investor to invest $2000 in Venture B.

11. The only stream of investments that results in a share of 15 at phase 3 is to save all the money for the last period: 0, 0, and $4 million and no better final share is possible.

12. At time 3, the optimal immigration policy is
$u_3(0) = 4$, $u_3(1) = 3$, $u_3(2) = 2$, $u_3(3) = 1$, $u_3(4) = 0$, $u_3(5) = 0$.
   Notice that immigration is permitted to bring the population up to 4 for states 0 through 4. At times 2, 1, and 0, the optimal actions are
$u_n(0) = 5$, $u_n(1) = 4$, $u_n(2) = 3$, $u_n(3) = 2$, $u_n(4) = 0$, $u_n(5) = 0$.
   Except for population 4, it is optimal to let immigration raise the population to 5.

Section 6.3

5. 58 steps are required for the initial function $w_0(1) = w_0(2) = 0$, yielding the vector [39.3275, 36.9745]; and for the initial function $w_0(1) = 40$, $w_0(2) = 30$, just 33 steps are required, resulting in the vector [39.3294, 36.9764].

6. The computation shows that n should be at least as large as 59 to guarantee convergence to within .1.

7. The stationary policy defined by $u(1) = 1$, $u(2) = 1$ is optimal.

8. With the new cost parameters, action 2 is now optimal at state 1.

9. (a) action 0 is optimal at state 1 and $W(1) = 3.14607$; (b) action 0 is still optimal at state 1 and $W(1) = 32.095$.

10. The policy **u** that uses action 0 at state 0, action 1 at state 1, and action 2 at states 2, 3, and 4 is optimal.

12. $w_1 = \{43, 143, 243, 343, 443, 543, 600\}$;
$w_2 = \{81.7, 181.7, 281.7, 381.7, 481.7, 562.35, 600\}$;
$w_3 = \{116.53, 216.53, 316.53, 416.53, 507.823, 571.058, 600\}$. Note that in all three cases, the optimizing actions suggest to advertise at all states other than state 6.

13. The policy $u(0) = 0, u(1) = 0, u(2) = 1, u(3) = 2$ is optimal. The optimal value function is $w = \{4.78571, 9.5, 10.5, 11.5\}$.

Section 6.4

1. The optimal policy is $u(1) = 1, u(2) = 2$.

2. (a) $u(i) = i$ is optimal. (b) The computation shows the optimality of the policy $u(i) = 0$ for $i < 2$ and $u(i) = i$ otherwise.

3. The policy $u(i) = 1, i = 0, ..., 5; u(6) = 0$ is optimal.

4. The policy that releases 0 when $i = 0, 1$; releases 1 when $i = 2$; and releases 2 when $i = 3$ is optimal.

6. The policy that replaces when the machine is in condition 3 or worse is optimal.

10. The policy that harvests 0 at levels 0, 1, 2, 3 and 4; and harvests 1 at level 5 is optimal.

13. The solution vector **V** to the DP equation has entries 34050/841, 35210/841, and 34850/841.

Section 6.5

1. The policy that stops at states $\{3, 5\}$ is intuitively optimal. The optimal values, that is $V(1), V(2), V(3), V(4)$, and $V(5)$, are 4, 4, 5, 3, and 3, respectively.

2. $V(i) = i; i = 1, 2, 3, 4$ and it is optimal to stop at every state.

3. The stopping set is $\{1, 2, 4\}$, and the value function is $V(1) = 6$, $V(2) = V(3) = 5$, $V(4) = V(5) = V(6) = 3$.

5. (c) It is optimal not to stop when his wealth is 1, 2, 3, 4, 5, but only when it is 0 or 6.

6. The only states for which $V(i) = f(i)$ are states 0 and 5, so we stop when the stock price reaches those values.

7. The optimal policy is to stop at rewards 0, 2000, or 3000.

8. (b) It is optimal to stop immediately at every state.

11. The contestant should answer the first question and then quit.

Section 6.6

1. The option is worth .12 at time 0.

2. For $u = .08$, the time 0 option value is .18; for $u = .09$, it is .24.

3. For $E = 22$, the time 0 option value is .41; for $E = 24$, it is .05.

4. For $\alpha = .95$, the time 0 option value is .10; for $\alpha = .9$, it is .08.

5. The put option value is .01 at time 0.

7. (a) The policy whose optimal actions are (1, 1, 1, 0, 1, 1, 0, 0, 1) at the states in their usual order is optimal when $r = 4$. (b) The same policy as in (a) is optimal when $r = 6$.

8. The parameter value $r = 3.1$ is the cutoff value to the nearest tenth.

9. The parameter value $r = 6.1$ is the cutoff value to the nearest tenth.

10. At time 2, the optimal action function is $u_2 = (1, 1, 1, 0, 1, 1, 0, 0, 1)$; the same action function is optimal at time 1; but at time 0, the optimal strategy changes to $u_0 = (2, 2, 2, 1, 2, 2, 0, 1, 2)$.

15. At all times it is optimal to buy as much stock as the checking account will afford.

16. The optimal action at each period is to sell all existing stock.

Appendix C
Glossary of *Mathematica* Commands

The commands below are grouped by the *Mathematica* package in which they reside. Also contained in the four packages are utility commands that are needed by the other commands, although the user does not interact directly with them. We list the usage messages that are identical to the on-line help messages that a user will see when querying for information about the commands.

KnoxOR`Graphs`

AddFlow
AddFlow[capacities, flows, augmentingpath, epsilon] takes the capacity and flow matrices for a maximal flow problem, and the augmenting path and amount of new flow epsilon to augment by, and it returns the new flow matrix.

AdjustComponents
AdjustComponents[u, v, components] is used by Kruskal and SpanningTree-OneStep. It accepts two vertices, u and v, and returns an updated components list obtained by setting the component number which is larger of that of u and v, and also all similar component numbers, to the smaller of that of u and v.

AugmentMatching
AugmentMatching[matching, augmentingpath] returns a revised matching that augments using a given augmenting path on the given previous matching M. It keeps all edges in the matching that were not on the augmenting path, and deletes edges that are, replacing them by edges in the augmenting path that were not in the original matching. Edges must be written with left-side vertices first and right-side vertices second.

ComputePathCosts
ComputePathCosts[theTree, theRoot] computes the costs of all paths to all vertices in the given tree from the given root.

ComputeSlacks
ComputeSlacks[theGraph, theTree, pathcosts] takes a directed graph, a directed spanning tree for that graph, and the list of path costs in the tree to each vertex and computes a matrix of slack values, that is, path cost to v minus the path cost to u plus the cost of omitted edge (u,v), for all omitted edges. Entries of the slack matrix are 0 for edges that are not omitted.

ConvertToAdjMatrix

ConvertToAdjMatrix[listofedges, numberofvertices, opts] takes a list of edges and converts it to an adjacency matrix. Options are GraphType → Undirected, which may be set to Directed; and Weighted → False, which can be set to True if the list of edges has a third component that gives the weight of the edge that is to be stored in the weighted adjacency matrix.

DirectedSpanningTree

DirectedSpanningTree[theGraph, initialTree, theRoot, opts] takes a given initial spanning tree of a given directed, quasi-connected graph, both in adjacency matrix form, and the vertex number of the root of the tree. It performs the full minimal directed spanning tree algorithm, displaying all intermediate graphs unless the option ShowTree is set to False, and returns the minimal spanning tree in adjacency matrix form. The display options of DisplayGraph may be passed in.

DirectedSpanningTreeFirstStep

DirectedSpanningTreeFirstStep[theGraph, initTree, theRoot, opts] displays the initial spanning tree supplied by the user, together with unused edges in the whole graph and slack values for the unused edges. The option Show-Tree is made True by default, and if so the first tree is displayed. All graphs are in adjacency matrix form. The display options of DisplayGraph may be passed in.

DirectedSpanningTreeOneStep

DirectedSpanningTreeOneStep[theGraph, currentTree, theRoot, newedge, opts] performs one step of the directed spanning tree algorithm, inserting the new edge into the current tree in the graph with the given root, and deleting the edge that had pointed to the same vertex as the new edge. The option ShowTree is made True by default, and if so the new tree is displayed. The value returned by the function is the new tree. All graphs are in adjacency matrix form. The display options of DisplayGraph may be passed in.

DisplayBipartiteGraph

DisplayBipartiteGraph[weightmatrix, opts] takes the weight matrix of a bipartite graph, in which the rows mean vertices on one side and the columns mean vertices on the other, and displays the graph. Its options are Show-Weights→True indicating the edge weights are to be shown; Labeling→ Automatic, which may be set to a list of vertex labels L(v) in the matching algorithm; Matching→None, which can be set to a list of edges in a current matching; and the options of DisplayGraph. EdgeLabels will be superceded though. DisplayBipartiteGraph computes the matrix of EdgeLabels to pass to DisplayGraph if ShowWeights is true. If Matching is set, then edges in the matching will be shown as solid, and the edges not used by the matching will be shown dashed. The VertexPositions and VertexLabelPositons options are preset to give the graph a satisfactory appearance, but may be changed by the user.

DisplayGraph
DisplayGraph[adjmatrix, opts] shows the graph associated with the given adjacency matrix. Options are GraphType → Undirected, which may be set to Directed to obtain a directed graph; VertexLabels → Automatic, which may be set to a list of labels for vertices; VertexPositions → Automatic, which may be set to a list of coordinates for the vertices; VertexLabelPositions → Automatic, which may be set to a list of values such as Above, Below, ToLeft, ToRight to indicate where the vertex labels should be positioned relative to the vertex points; EdgeLabels → Automatic, which can be set to a matrix whose elements are to be used as labels on the edges; EdgeLabelPositions → Automatic, which like VertexLabelPositions can be set to directional offsets from the midpoint of the edge; EdgeSeparation->.01, which controls the separation between double arrows in a directed graph; and EdgeStyle → Thickness[.005], which can apply a style to edges. It also accepts the options of SelfLoops, which are LoopPositions→Automatic, which can be set to a list of values such as Above, Below, ToLeft, and ToRight to indicate where the loops should be drawn relative to the points; and LoopSize→.05, which controls the size of loops by setting the percentage of the overall picture size that the loop radius will be.

EdgeLabelPositions
EdgeLabelPositions is an option for DisplayGraph, which may be set to a matrix of values such as Above, Below, ToLeft, ToRight to indicate where the edge labels should be positioned relative to the edge midpoints.

EdgeLabels
EdgeLabels is an option for DisplayGraph, which can be set to a matrix whose elements are to be used as labels on the edges.

Edge Separation
EdgeSeparation is an option for DisplayGraph, set to .01 by default, which controls the gap between the two arrows in a double edge

EdgeStyle
EdgeStyle is an option for DisplayGraph, set to Thickness[.005] by default, which can apply a style to edges.

EqualitySubgraph
EqualitySubgraph[weightmatrix, labeling] produces the weight matrix of the equality subgraph for a labeling in the maximal matching problem, given the weight matrix and the labeling.

FindAugmentingPath

FindAugmentingPath[capacities, flows, source, sink, opts] is a breadth-first search for the maximal flow problem. It takes the capacity matrix and flow matrix, and the source and sink vertices. The command returns the list{augmentingpath,epsilon}, or {{},0} if the augmenting path could not be found and the sink could not be labeled. It accepts one option, ShowLabels→ True, which displays a table of vertex labels found by BFS if it is kept at True, and suppresses the table if it is set to False.

FindChildren

FindChildren[A, parents] is a function used by ComputePathCosts to find all children of the vertices in the list parents. The argument A is the adjacency matrix of the graph.

FindDirectedRoot

FindDirectedRoot[digraph] returns the root of a directed graph if one exists, else Null and a message indicating that a root was not found.

FindNeighbors

FindNeighbors[capacities,vert] returns a list of children or parents of the given vertex in the graph indicated by the given capacity matrix.

GraphType

GraphType is an option for DisplayGraph. It is Undirected by default, and can be set to Directed for a directed graph.

Kruskal

Kruskal[adjlist, n, opts] performs Kruskal's algorithm to find a minimal undirected spanning tree, given a list of weighted edges, each element of the form {v1,v2,weight}, and the number of vertices. Options accepted are ShowTree → True to display the new tree, Weighted → True for a weighted graph, and the options of DisplayGraph.

Labeling

Labeling is an option for DisplayBipartiteGraph, set to Automatic by default, which can be set to a list of vertex labels in the maximal matching algorithm. The labels will be shown when the graph is displayed.

LoopPositions

LoopPositions is an option for DisplayGraph, which can be set to a list of values such as Above, Below, ToLeft, and ToRight to indicate where the loops should be drawn relative to the points. Its default value of Automatic positions all loops above the points.

LoopSize

LoopSize is an option for DisplayGraph, initialized to .05, which controls the size of loops by setting the percentage of the overall picture size that the loop radius will be.

Matching
Matching is an option for DisplayBipartiteGraph, set to None by default, which can be set to a list of edges in a matching. These edges will be shown as solid segments, and others will be shown as dashed.

MaximalDirectedSpanningTree
MaximalDirectedSpanningTree[theGraph,initialTree,theRoot,opts] takes a given initial spanning tree of a given directed, quasi-connected graph, both in adjacency matrix form, and the vertex number of the root of the tree. It performs the full maximal directed spanning tree algorithm, displaying all intermediate graphs unless the option ShowTree is set to False, and returns the maximal spanning tree in adjacency matrix form. The display options of DisplayGraph may be passed in.

MaximalFlow
MaximalFlow[capacities, source, sink, opts] takes the capacities, source, and sink in a maximal flow problem as its parameters. It returns the final flow matrix. The Option ShowSteps→True can be set to choose whether to display the intermediate steps or not. The command also accepts the ShowLabels→True option to display tables of breadth-first search labels, and it accepts the display options for DisplayGraph.

MaxSlack
MaxSlack[slackmatrix] returns a pair {maximum, {row,column}}, which are the maximum element and its position in the given slack matrix.

MinSlack
MinSlack[slackmatrix] returns a pair {minimum, {row,column}}, which are the minimum element and its position in the given slack matrix.

QuasiConnectedQ
QuasiConnectedQ[digraph] returns True or False respectively according to whether the given directed graph is quasi-connected or not.

ReviseLabeling
ReviseLabeling[weightmatrix, labeling, S, T] finds the Δ for a revised labeling in the maximal matching problem, given the weightmatrix of the bipartite graph and the sets of vertices S and T on the left and right sides of the graph that were found in the previous unsuccessful search for an augmenting path. It returns the result in the form of a list {Δ, newlabeling}.

ShowLabels

ShowLabels→True is a boolean option for FindAugmentingPath and MaximalFlow, which if True displays a table of vertex labels.

ShowSteps

ShowSteps→True is an option for MaximalFlow, which if set to True shows all the intermediate flow augmenting steps.

ShowTree

ShowTree is an option for SpanningTreeOneStep, True by default, which can be set to False to suppress the tree display.

ShowWeights

ShowWeights is an option for DisplayBipartiteGraph, initialized to False, which determines whether the edge weights should be displayed.

SortEdges

SortEdges[adjlist] takes a list of weighted edges, each element of the form {v1,v2,weight}, and sorts it into increasing order of weight. It is used by the Kruskal command to prepare the edges.

SpanningTreeOneStep

SpanningTreeOneStep[treelist, edgelist, edgenumber, componentlist, opts] is one step of Kruskal's algorithm for minimal undirected spanning trees. It takes a current tree, a list of edges of the whole graph, the number of the edge in that list to substitute, and the list of connected components of the vertices. If the new edge has vertices that belong to different components, it is added to the tree and the component numbers of all vertices like the one that has the larger component between the two that are incident on the new edge are adjusted down to the component number of the smaller. The command returns the revised tree and componentlist. Options accepted are ShowTree → True to display the new tree, Weighted → True for a weighted graph, and the options of DisplayGraph.

VertexLabelPositions

VertexLabelPositions is an option for DisplayGraph, which may be set to a list of values such as Above, Below, ToLeft, ToRight to indicate where the vertex labels should be positioned relative to the vertex points.

VertexLabels

VertexLabels is an option for DisplayGraph, which may be set to a list of labels for vertices.

VertexPositions
VertexPositions is an option for DisplayGraph, which may be set to a list of coordinates for the vertices.

Weighted
Weighted is an option for ConvertToAdjMatrix, which may be set to True to construct a weighted adjacency matrix. It is also an option for SpanningTree-OneStep, which may be set to False for an unweighted graph.

KnoxOR`LinearProgramming`

Dictionary
Dictionary[system, basiclist, nonbasiclist] takes the system of constraint equations with the objective equation adjoined, a list of basic variables including the objective variable, and a list of non-basic variables, and displays the equivalent dictionary system of equations. The basic variables are solved for and the system is well-aligned with variables in columns.

ObjectiveLines
ObjectiveLines is an option for PlotFeasibleRegion, which can be set to a list of constant values. Lines in which the objective function is set equal to each of the constants are displayed on the feasible region.

ObjectiveLineStyle
ObjectiveLineStyle is an option for PlotFeasibleRegion, which can be used to apply a plot style to the objective lines in the feasible region.

PlotFeasibleRegion
PlotFeasibleRegion[constrainteqns, xdomain, ydomain, corners, objective, opts] takes the list of constraint equations, both x and y domains for plotting, the list of corners to use to bound the polygon, and the name of the objective function. The option ShowTable → True shows a table of objective function values at the corners. The option ObjectiveLines → Automatic can be set to a list of constant values c that will result in the display of lines of constant objective value equal to c on the graph. The option ShadingStyle can be set to a style for the feasible region. The option ObjectiveLineStyle can be set to a plot style for these lines. Other options are those of ImplicitPlot.

ShadingStyle
ShadingStyle is an option for PlotFeasibleRegion, which applies a style to the feasible region.

ShowTable
ShowTable is an option for PlotFeasibleRegion, which, if set to True, shows a table of objective function values at the corner points.

SimplexOneStep

SimplexOneStep[tableau, varlist, enteringbasic, departingbasic, basicvaria-blelist] takes the current simplex tableau, in the usual *Mathematica* form of a list of lists, the list of all variable names, the names of the entering and departing basic variables, and the list of current basic variable names. It performs one simplex step and prints the new tableau, with both row headings for the basic variables and column headings for all variables. Then it returns a pair {newtableau, newbasicvariablelist} for use in the next step.

TransportationOneStep

TransportationOneStep[tableau, varlist, enteringbasic, pivotrow, basicvaria-blelist] takes the current transportation simplex tableau, the list of all variable names, the name of the entering basic variable, the row in which it is to be made basic, and the list of current basic variable names, some of which can be blank. It prints the new tableau, with both row headings for the basic variables and column headings for all variables. Then it returns a pair {newtableau, newbasicvariablelist} for use in the next step of the phase 1 transportation algorithm.

KnoxOR`StochasticProcesses`

AbsorptionProbability

AbsorptionProbability[transmatrix, transientstatelist, recurrenceclass] takes the transition matrix of a Markov chain, the list of transient states, and a list that is a recurrence class of the chain, and returns a list of probabilities of absorption into the recurrence class, with one entry for each intial transient state.

DotSize

DotSize is an option for PlotStepFunction. It controls the size of the dots on the graph.

FirstPassageTime

FirstPassageTime[transmatrix, j, time] accepts the transition matrix of a Markov chain, a target state number j, and a time, and returns a list with an element for each initial state, of probabilities that the time of first visit from the initial state to state j equals this time.

Histogram

Histogram[datalist, numrectangles] plots a histogram of a list of data, with a desired number of rectangles. It inherits some of the options of Generalized-BarChart and has four of its own. The option Type has any of the values Relative (default), Absolute, or Scaled, depending on whether you want bars to have heights that are relative frequencies, absolute frequencies, or relative frequencies divided by interval length. The option Endpoints may be set to a list {a,b} of real numbers with a<b to force the histogram to be plotted between these endpoints. Otherwise the command uses the min and max of the datalist as endpoints. The option NumDigits (initialized to 2) can be used to set the number of decimaldigits used in the tick marks on the x-axis. The option Distribution->Continuous may be reset to Discrete in order to force a histogram whose boxes are at the integer values between the lowest and highest integer data value. The user cannot override the PlotRange option, nor AxesOrigin, nor Ticks, nor BarOrientation, in the interest of having a well-formed graph.

LimitingProbs
LimitingProbs[transmatrix] takes the transition matrix of a regular Markov chain and returns the vector of limiting probabilities.

Nt
Nt[arrtimes, t] takes a list of arrival times of a Poisson process and a time t and returns the cumulative number of arrivals by time t.

PlotContsProb
PlotContsProb[density, domain, between] plots the area under the given function on the given domain between the points in the list between, which is assumed to consist of two points in increasing order. Options are the options that make sense for Show, and ShadingStyle->RGBColor[1,0,0], which can be used to give a style to the shaded area region.

PlotSimulateBrownianMotion
PlotSimulateBrownianMotion[x0, deltat, numpoints] takes an initial state x0, a timestep deltat, and a number of time points and simulates a standard Brownian motion. It produces a connected list plot of the path.

PlotStepFunction
PlotStepFunction[fn, domain, jumplist] plots a step function on the domain specified, with jumps at the points in jumplist, which is a list of sorted numbers. The step function is assumed to be right continuous, as a c.d.f. is. It accepts option DotSize→.017 to change the size of the dots, StepStyle→ RGBColor[0,0,0] to assign a style to the steps, and it inherits any options that make sense for Show.

ProportionOfTime
ProportionOfTime[processlist] takes the output of the SimBirthDeathProcess command in the form {jumptimelist,statelist} and finds the proportion of time that the process was in each of the states it visited, in the form of a list of pairs {state, proportion}.

ReachableSet
ReachableSet[transmatrix, state] returns the list of states reachable from the given state for the Markov chain with the given transition matrix.

SimBirthDeathProcess
SimBirthDeathProcess[x0, finaltime, birthrate, deathrate] takes an initial state x0, a finaltime to end simulation, a birthrate and a deathrate for a birth–death process, and returns a list of jumptimes and states passed through in the form {timelist, statelist}.

SimDiscreteDist
SimDiscreteDist[problist] is used by SimMarkovChain. It takes a list of numbers that forms a valid probability distribution and simulates a value having that distribution.

SimMarkovChain
SimMarkovChain[transmatrix, start, numsteps] returns a list of numsteps simulated states for a Markov chain with the given transition matrix and starting state.

SimulateNArrivals
SimulateNArrivals[lambda, n] returns a list of n simulated arrival times for a Poisson process with rate lambda.

StepSize
StepSize[deltat] is a function used by PlotSimulateBrownianMotion   to generate a random step size of either positive or negative deltat.

StepStyle
StepStyle is an option for PlotStepFunction, which gives a style to the steps. Its default is RGBColor[0,0,0], or black.

KnoxOR`DynamicProgramming`

DPEquation
DPEquation[TransMats, RewardMatrix, Val] takes the list of transition matrices, one for each action; the reward matrix as a function of state and action; and the current value function for the finite horizon stochastic dynamic programming problem, and returns a list, one for each state, of dynamic programming equation values for each action. The rowwise maxima or minima form the next value function.

DiscountedDPEquation
DiscountedDPEquation[TransMats, RewardMatrix, Val, alpha] takes the list of transition matrices, one for each action; the reward matrix as a function of state and action; the current value function for the infinite horizon stochastic

dynamic programming problem; and the discount factor alpha, and returns a list, one for each state, of dynamic programming equation values for each action. The rowwise maxima or minima form the next value function.

PolicyImprovementOneStep
PolicyImprovementOneStep[TransMats, RewardMatrix, alpha, policy] takes the list of transition matrices, one for each action; the reward matrix as a function of state and action; the discount factor alpha; and a current policy represented as a list whose i^{th} element is the number of the action taken when the state is i. It returns a list, one for each state, of dynamic programming equation values for each action. The actions at which the rowwise maxima or minima are taken on form the next policy.

References

(1) Aho, A., J. Hopcraft, and J. Ullman. *Data Structures and Algorithms*. Addison-Wesley, Reading, MA (1983).

(2) Barlow, R.E., and F. Proschan. *Mathematical Theory of Reliability*. Wiley, New York (1965).

(3) Barlow, R.E., and F. Proschan. *Statistical Theory of Reliability and Life Testing*. Holt, Reinhart and Winston, New York (1974).

(4) Baxter, M., and A. Rennie. *Financial Calculus: An Introduction to Derivative Pricing*. Cambridge University Press (1996).

(5) Bellman, R. *Dynamic Programming*. Princeton University Press, Princeton, NJ (1957).

(6) Bertsekas, D., and S. Shreve. *Stochastic Optimal Control*. Academic Press, New York (1978).

(7) Black, F., and M. Scholes. "The Pricing of Options and Corporate Liabilities," *Journal of Political Economy* 81, pp. 637–659 (1973).

(8) Blumenthal, R.M., and R.K. Getoor. *Markov Processes and Potential Theory*. Academic Press, New York (1968).

(9) Bradley, S., A. Hax, and T. Magnanti. *Applied Mathematical Programming*. Addison-Wesley, Reading, MA (1977).

(10) Breiman, L. *Probability and Stochastic Processes, 2nd ed.* Scientific Press, Palo Alto, CA (1986).

(11) Busacker, R., and T. Saaty. *Finite Graphs and Networks*. McGraw-Hill, New York (1965).

(12) Carter, M., and C Price. *Operations Research: A Practical Introduction*. CRC Press, Boca Raton, FL (2001).

(13) Chung, K.L. *Elementary Probability Theory with Stochastic Processes*. Springer-Verlag, New York (1979).

(14) Chung, K.L. *A Course in Probability Theory*. Academic Press, New York (1974).

(15) Cinlar, E. *An Introduction to Stochastic Processes*. Prentice Hall, Englewood Cliffs, NJ (1975).

(16) Dantzig, G. *Linear Programming and Extensions*. Princeton University Press, Princeton, NJ (1963).

(17) Derman, C. *Finite State Markovian Decision Processes*. Academic Press, New York (1970).

(18) Devore, J. *Probability and Statistics for Engineering and the Sciences, 3rd ed.* Brooks/Cole, Pacific Grove, CA (1991).

(19) Dierker, P., and W. Voxman, *Discrete Mathematics*. Harcourt, Brace, and Jovanovich, Orlando, FL (1986).

(20) Dossey, J., A. Otto, L. Spence, and C. VandenEynden, *Discrete Mathematics, 4th ed.* Addison Wesley, Boston (2002).

(21) Dreyfus, S., and A. Law. *The Art and Theory of Dynamic Programming*. Academic Press, New York (1977).

(22) Feller, W. *An Introduction to Probability Theory and Its Applications, Vol. 1 & 2*. John Wiley, New York (1968).

(23) Gale, D., H.W. Kuhn, and A.W. Tucker. "Linear Programming and the Theory of Games," in *Activity Analysis of Production and Allocation*, T.C. Koopmans, Ed. John Wiley, New York, pp. 317–329 (1951).

(24) Gass, S. *Linear Programming: Methods and Applications, 5th ed.* McGraw-Hill, New York (1985).

(25) Gaylord, R., S. Kamin, and P. Wellin. *An Introduction to Programming with Mathematica, 2nd ed.* Springer-Verlag, New York (1996).

(26) Gibbons, A. *Algorithmic Graph Theory*. Cambridge University Press (1985).

(27) Gribik, P., and K. Kortanek. *Extremal Methods of Operations Research*. Marcel Dekker, New York (1985).

(28) Gross, D., and C. Harris. *Fundamentals of Queueing Theory, 2nd ed.* Wiley, New York (1985).

(29) Hastings, K. *Probability and Statistics*. Addison-Wesley, Reading, MA (1997).

(30) Hastings, K. *Introduction to Probability with Mathematica*. Chapman & Hall/CRC Press, Boca Raton, FL (2001).

(31) Hillier, F., and G. Lieberman. *Introduction to Operations Research, 5th ed.* McGraw-Hill, New York (1990).

(32) Hoel, P., S. Port, and C. Stone. *Introduction to Probability Theory.* Houghton Mifflin, Boston (1971).

(33) Hoel, P., S. Port, and C. Stone. *Introduction to Statistical Theory.* Houghton Mifflin, Boston (1971).

(34) Hoel, P., S. Port, and C. Stone. *Introduction to Stochastic Processes.* Houghton Mifflin, Boston (1972).

(35) Hogg, R.V., and E. Tanis. *Probability and Statistical Inference, 3rd ed.* Macmillan, New York (1988).

(36) Hogg, R.V., and A. Craig. *Introduction to Mathematical Statistics, 4th ed.* Macmillan, New York (1978).

(37) Howard, R. *Dynamic Programming and Markov Processes.* MIT Press, Cambridge, MA (1960).

(38) Jeter, M. *Mathematical Programming.* Marcel Dekker, New York (1986).

(39) Johnson, L.A., and D.C. Montgomery. *O.R. in Production Planning, Scheduling, and Inventory Control.* Wiley, New York (1974).

(40) Karlin, S., and H. Taylor. *A First Course in Stochastic Process, 2nd ed.* Academic Press, New York (1975).

(41) Karlin, S., and H. Taylor. *A Second Course in Stochastic Processes.* Academic Press, New York (1981).

(42) Knuth, D.E. *The Art of Computer Programming, Vol. 1, 2nd ed.* Addison-Wesley, Reading, MA (1973).

(43) Merton, R. "Optimum Consumption and Portfolio Rules in a Continuous-Time Model," *Journal of Economic Theory* 3, pp. 373–413 (1971).

(44) Minieka, E. *Optimization Algorithms for Networks and Graphs.* Marcel Dekker, New York (1978).

(45) Mood, A., F. Graybill, and D. Boes. *Introduction to the Theory of Statistics*, 3rd ed. McGraw-Hill, New York (1974).

(46) Mott, J.L., A. Kandel, and T. Baker. *Discrete Mathematics for Computer Scientists & Mathematicians, 2nd ed.* Prentice Hall, Englewood Cliffs, NJ (1986).

(47) Papadimitriou, C.H., and K. Steiglitz. *Combinatorial Optimization.* Prentice Hall, Englewood Cliffs, NJ (1982).

(48) Parzen, E. *Modern Probability Theory and its Applications.* Holden-Day, San Francisco (1962).

(49) Rao, S.S. *Optimization Theory and Applications, 2nd ed.* Wiley Eastern, Ltd., New Delhi (1984).

(50) Rockafellar, R.T. *Convex Analysis.* Princeton University Press, Princeton, NJ (1972).

(51) Ross, S. *Simulation, 3rd ed.* Academic Press, San Diego (2002).

(52) Ross, S. *Stochastic Processes.* Wiley, New York (1983).

(53) Ross, S. *Introduction to Stochastic Dynamic Programming.* Academic Press, New York (1983).

(54) Ross, S. *Applied Probability Models with Optimization Applications.* Holden-Day, San Francisco (1970).

(55) Saaty, T. *Elements of Queueing Theory with Applications.* Dover, New York (1961).

(56) Strang, G. *Introduction to Applied Mathematics.* Wellesley-Cambridge Press, Wellesley, MA (1986).

(57) Swamy, M.N.S., and Thulasiraman, K. *Graphs, Networks, and Algorithms.* Wiley, New York (1981).

(58) Tucker, H. *A Graduate Course in Probability.* Academic Press, New York (1967).

(59) Walker, R. *Introduction to Mathematical Programming*, Prentice Hall, Upper Saddle River, NJ (1999).

(60) Wolfram, S. *The Mathematica Book, 3rd ed.* Cambridge University Press (1996).

(61) Winston, W.L. *Operations Research: Applications and Algorithms, 2nd ed.* PWS-Kent, Boston (1991).

Index